# HARNESSING AutoCAD® 2004 Exercise Manual

# HARNESSING AutoCAD® 2004 Exercise Manual

THOMAS A. STELLMAN
G.V. KRISHNAN

autodesk® press

Australia • Canada • Mexico • Singapore • Spain • United Kingdom • United States

THOMSON
DELMAR LEARNING

autodesk press

# Harnessing AutoCAD® 2004 Exercise Manual

Thomas A. Stellman / G.V. Krishnan

**Autodesk Press Staff**

**Vice President, Technology and Trades SBU:**
Alar Elken

**Editorial Director:**
Sandy Clark

**Senior Acquisitions Editor:**
James DeVoe

**Senior Development Editor:**
John Fisher

**Editorial Assistant:**
Mary Ellen Martino

**Executive Marketing Manager:**
Cynthia Eichelman

**Channel Manager:**
Fair Huntoon

**Marketing Coordinator:**
Sarena Douglass

**Production Director:**
Mary Ellen Black

**Production Manager:**
Andrew Crouth

**Production Editor:**
Tom Stover

**Art and Design Specialist:**
Mary Beth Vought



Printed in Canada
4 5 XX 06 05

For more information contact
Delmar Learning, Executive Woods,
5 Maxwell Drive, PO Box 8007,
Clifton Park, NY 12065-8007

Or find us on the World Wide Web
at www.delmarlearning.com or
www.autodeskpress.com



**Library of Congress Cataloging-in-Publication Data**
ISBN 1-4018-5080-4

# CONTENTS

# INTRODUCTION

## EXERCISE MANUAL FOR HARNESSING AUTOCAD 2004

This Exercise Manual is the companion book intended to be used with Harnessing AutoCAD 2004. Each chapter has exercises designed to allow you to practice and test your skills in the commands and concepts covered in the corresponding chapter of the main text. The combination of these books was written to be used both in the classroom as a textbook as well as in the industry by the professional CADD designer/drafter as a reference and learning tool. Whether you're new to AutoCAD or a seasoned user upgrading your skills, Harnessing AutoCAD and this Exercise Manual will help you rein in the power of AutoCAD to improve your professional skills and increase your productivity.

These chapters contain a project exercise and exercises for the following disciplines: mechanical, architectural, civil, electrical, and piping. Exercise icons and tabs mark the discipline sections.

Additional drawing exercises, such as those for the piping discipline, have been added to the exercise manual contained on the CD that accompanies the Harnessing AutoCAD 2004 book. Because of these additions, the numbering of the exercises on the CD may not correspond with the exercises found in the printed manual.

If you would like to contribute drawing exercises for possible inclusion in the next version of this exercise manual, please contact the CADD/Drafting team at Autodesk Press by visiting our website: *www.autodeskpress.com*.

CHAPTER 2

project EXERCISE

# Fundamentals I

## PROJECT EXERCISE

This project exercise provides point-by-point instructions for setting up the drawing, laying out the border, and then creating the objects shown in Figure P2–1. In this exercise you will apply the skills acquired in Chapters 1 and 2 to drawing basic architectural, mechanical, electronic, and piping objects.

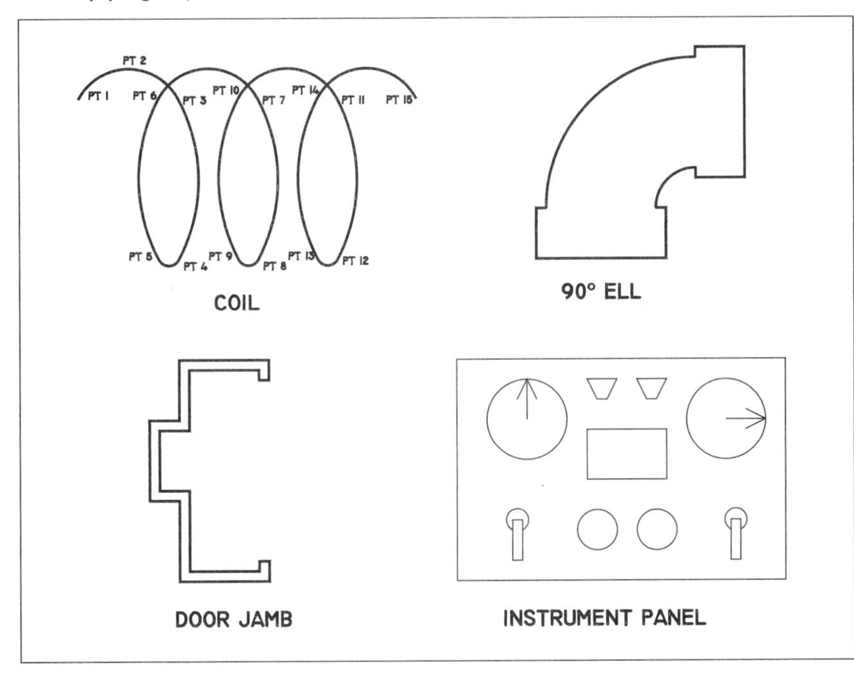


**Figure P2–1** *Completed project drawing*

In this project you will learn to do the following:

- Apply the "Use a Wizard" option of the Create a New Drawing dialog box.
- Set the Units used for the drawing.
- Set the Limits for the drawing Area
- Use the LINE, RECTANGLE, CIRCLE, and ARC commands.

**Note:** As you complete each step in the project procedure, place a check mark by the step to help you keep up with where you are in the project.

## SET UP THE DRAWING AND DRAW A BORDER

This procedure describes the steps required to set up the drawing and to draw the border.

**Step 1:** Invoke the AutoCAD program.

**Example:** Under Microsoft Windows, find the AutoCAD 2002 program in the Start > Programs menu and select it.

**Step 2:** Set the system variable STARTUP to 1, as shown below, that allows AutoCAD to create a new drawing using a wizard.

Command: startup (ENTER)
Enter a new value for STARTUP <default>: 1 (ENTER)

To create a new drawing, invoke the NEW command from the Standard toolbar, as shown in Figure P2–2, or select New from the File menu.

**Figure P2–2** *Invoking the* NEW *command from the Standard toolbar*

**Step 3:** AutoCAD displays the Create New Drawing dialog box, as shown in Figure P2–3. Select the Use a Wizard button and AutoCAD lists available wizards. Select Quick Setup wizard as shown in Figure P2-3 from the Select a Wizard list box and choose the OK button.

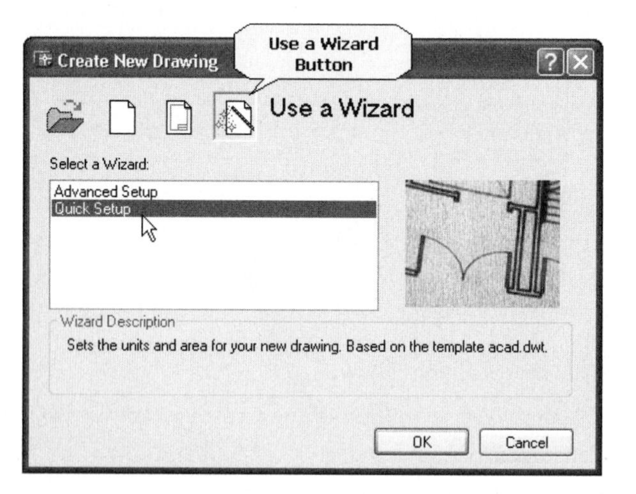


**Figure P2–3** *Create New Drawing dialog box*

**Step 4** AutoCAD displays the Quick Setup dialog box with the listing of options available for unit of measurement, as shown in Figure P2–4. Select the "Decimal" radio button, and choose the **Next >** button.

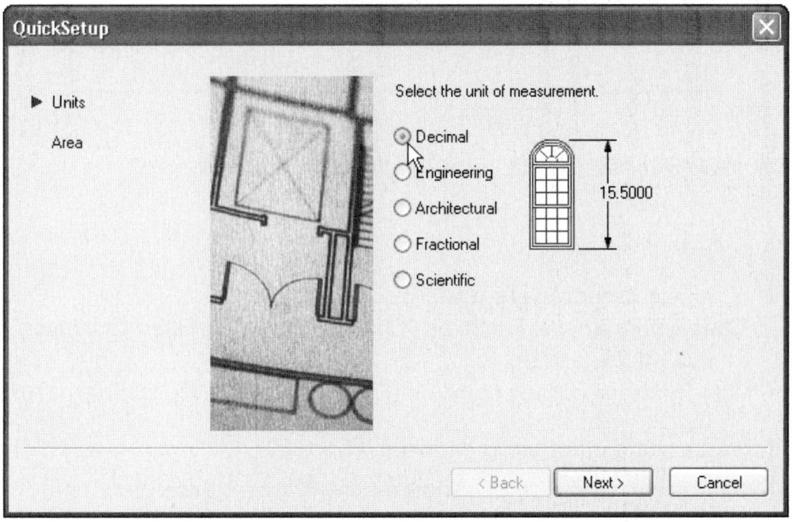

**Figure P2–4** *Quick Setup dialog box with the selection of the unit of measurement*

**Step 5** AutoCAD displays the Quick Setup dialog box with the display for Area setting, as shown in Figure P2–5. Under the **Width:** text box enter 12, and under the **Length:** text box enter 9.0. Choose the **Finish** button to close the Quick Step dialog box.

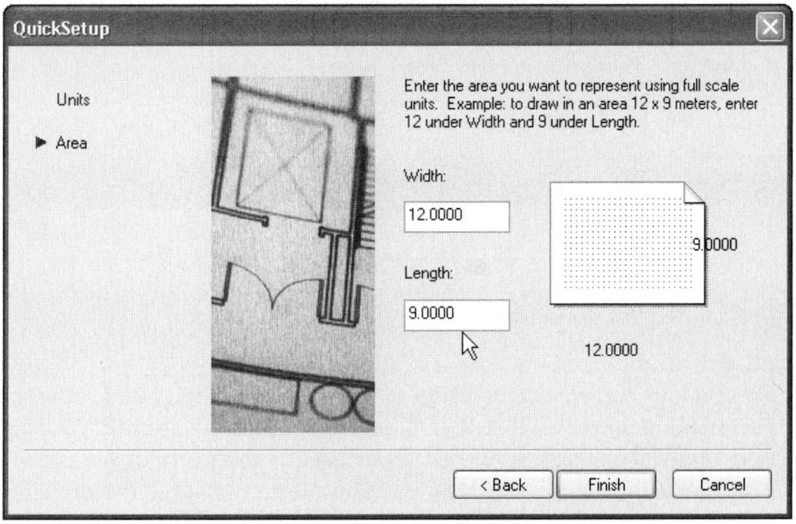

**Figure P2–5** *Quick Setup dialog box with the Area setting*

**Step 6** Invoke the ZOOM ALL command from the View menu. To draw the border, invoke the RECTANGLE command from the Draw toolbar, as shown in Figure P2–6, or enter **rectangle** at the "Command:" prompt and press ENTER.

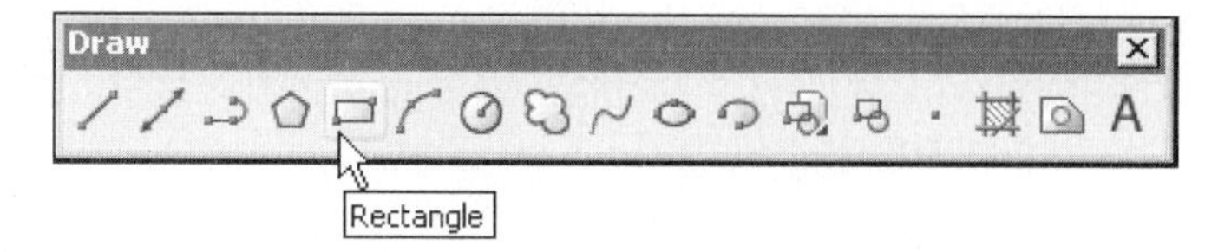


**Figure P2–6** *Invoking the* RECTANGLE *command from the Draw toolbar*

AutoCAD prompts:

Command: **rectangle** (ENTER)
Specify first corner point or [Chamfer/Elevation/Fillet/Thickness/Width]: **0.25,0.25** (ENTER)
Specify other corner point or [Dimensions]: **10.75,8.25** (ENTER)

AutoCAD draws a rectangle border as shown in Figure P2–7.

**Figure P2–7** *Completed border outline*

**Step 7** To save the current status of the drawing, invoke the SAVE command from the File menu or enter save at the "Command:" prompt. AutoCAD displays the Save Drawing As dialog box. Enter **PROJ2** as the name of the drawing in the **File name:** text box, and click the **Save** button to save the drawing and close the Save Drawing As dialog box. Make sure you are saving the drawing in the appropriate folder.

## DRAW THE DOOR JAMB

This procedure describes the steps required to draw the door jamb shown in Figure P2–8. The door jamb is drawn with the LINE command using rectangular coordinates and polar coordinates.

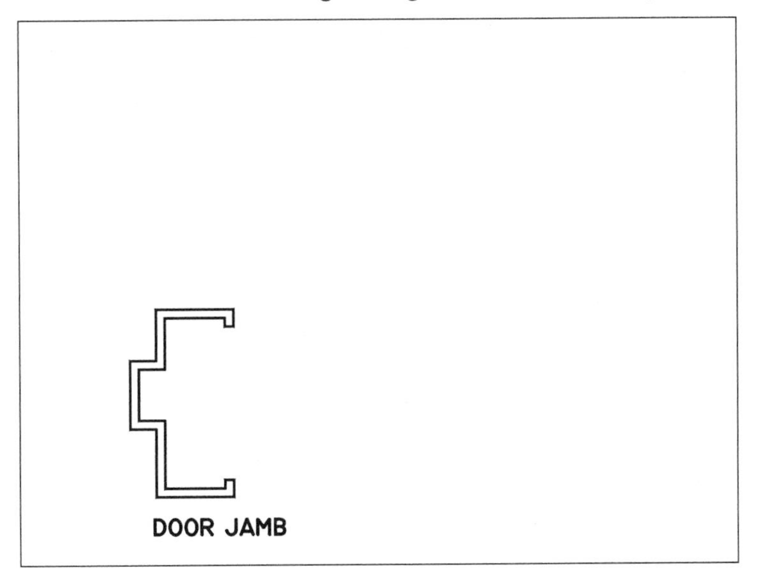


**Figure P2–8** *Door Jamb*

**Step 1** To draw the door jamb, invoke the LINE command from the Draw Toolbar, as shown in Figure P2–9, or enter **line** and press ENTER at the "Command:" prompt.

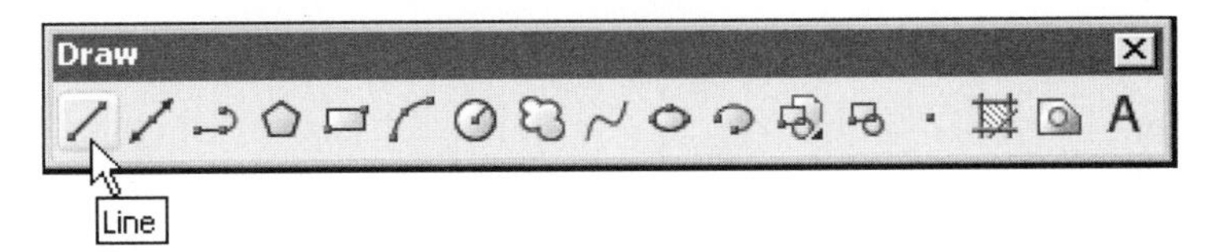


**Figure P2–9** *Invoking the LINE command from the Draw toolbar*

AutoCAD prompts:

Command: **line** (ENTER)
Specify first point: **2,1.5** (ENTER)
Specify next point or [Undo]: **2,2.5** (ENTER)
Specify next point or [Undo]: **1.625,2.5** (ENTER)
Specify next point or [Close/Undo]: **1.625,3.5** (ENTER)
Specify next point or [Close/Undo]: **2,3.5** (ENTER)
Specify next point or [Close/Undo]: **2,4.25** (ENTER)
Specify next point or [Close/Undo]: **3.125,4.25** (ENTER)
Specify next point or [Close/Undo]: **3.125,4** (ENTER)

Specify next point or [Close/Undo]: **3,4** (ENTER)
Specify next point or [Close/Undo]: **@.125<90** (ENTER)
Specify next point or [Close/Undo]: **@.875<180** (ENTER)
Specify next point or [Close/Undo]: **@.75<270** (ENTER)
Specify next point or [Close/Undo]: **@.375<180** (ENTER)
Specify next point or [Close/Undo]: **@0,–.75** (ENTER)
Specify next point or [Close/Undo]: **@.375,0** (ENTER)
Specify next point or [Close/Undo]: **@0,–1** (ENTER)
Specify next point or [Close/Undo]: **@.875,0** (ENTER)
Specify next point or [Close/Undo]: **@0,.125** (ENTER)
Specify next point or [Close/Undo]: **@.125,0** (ENTER)
Specify next point or [Close/Undo]: **@0,–.25** (ENTER)
Specify next point or [Close/Undo]: **@–1.125,0** (ENTER)
Specify next point or [Close/Undo]: *(press the* ENTER *to terminate the command sequence)*

**Step 2** To save the current status of the drawing, invoke the SAVE command from the File menu or enter **qsave** at the "Command:" prompt. AutoCAD saves the current status of the drawing.

#### DRAW THE INSTRUMENT PANEL

This procedure describes the steps required to draw the instrument panel shown in Figure P2–10. The instrument panel is drawn with the LINE, RECTANGLE, CIRCLE, and ARC commands.

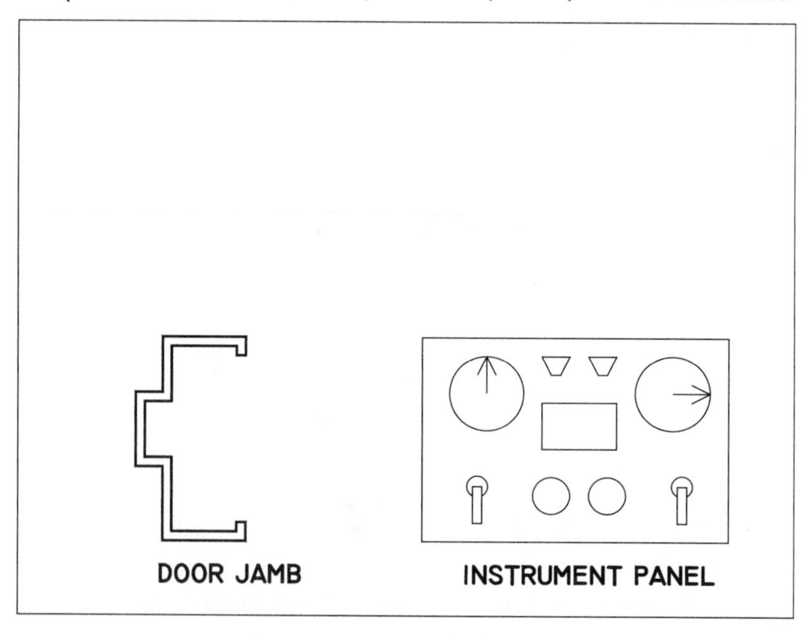


**Figure P2–10** *Door jamb and instrument panel*

**Step 1** To draw the main outline of the instrument panel and the central rectangle, invoke the RECTANGLE command from the Draw Toolbar or enter **rectangle** and press ENTER at the "Command:" prompt.

AutoCAD prompts:

Command: **rectangle** (ENTER)
Specify first corner point or [Chamfer/Elevation/Fillet/Thickness/Width]: **5.75,1.25** (ENTER)
Specify other corner point or [Dimensions]: **9.875,4** (ENTER)

Command: *(press* ENTER *to repeat the* RECTANGLE *command)*
Specify first corner point or [Chamfer/Elevation/Fillet/Thickness/Width]: **7.375,2.5** (ENTER)
Specify other corner point or [Dimensions]: **8.375,3.125** (ENTER)

**Step 2** To draw the trapezoidal shapes located in the top of the instrument panel, invoke the LINE command from the Draw Toolbar or enter **line** and press ENTER at the "Command:" prompt.

AutoCAD prompts:

Command: **line** (ENTER)
Specify first point: **7.375,3.75** (ENTER)
Specify next point or [Undo]: **@.375<0** (ENTER)
Specify next point or [Undo]: **7.625,3.5** (ENTER)
Specify next point or [Close/Undo]: **@.125<180** (ENTER)
Specify next point or [Close/Undo]: **c** (ENTER)

Command: *(press* ENTER *to repeat the* LINE *command)*
Specify first point: **8,3.75** (ENTER)
Specify next point or [Undo]: **@.375<0** (ENTER)
Specify next point or [Undo]: **8.25,3.5** (ENTER)
Specify next point or [Close/Undo]: **@.125<180** (ENTER)
Specify next point or [Close/Undo]: **c** (ENTER)

**Step 3** To draw the circular gauges with the arrow pointers located in the top of the instrument panel, first invoke the CIRCLE command from the Draw Toolbar as shown in Figure P2–11, or enter **circle** and press ENTER at the "Command:" prompt.

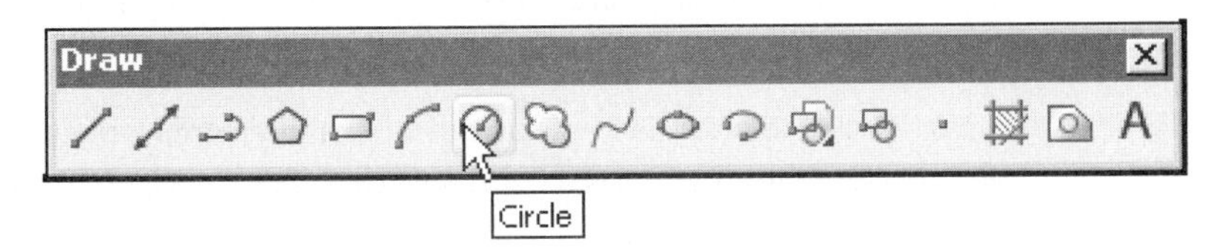


**Figure P2–11** *Invoking the* CIRCLE *command from the Draw toolbar*

AutoCAD prompts:

Command: **circle** (ENTER)
Specify center point for circle or [3P/2P/Ttr (tan tan radius)]: **6.625,3.25** (ENTER)
Specify radius of circle or [Diameter]: **@.5<0** (ENTER)

Command: *(press* ENTER *to repeat the* CIRCLE *command)*
Specify center point for circle or [3P/2P/Ttr (tan tan radius)]: **9.125,3.25** (ENTER)
Specify radius of circle or [Diameter]: **@.5<0** (ENTER)

To draw the arrow pointers, invoke the LINE command from the Draw Toolbar or enter **line** and press ENTER at the "Command:" prompt.

AutoCAD prompts:

Command: **line** (ENTER)
Specify first point: **6.5,3.5** (ENTER)
Specify next point or [Undo]: **6.625,3.75** (ENTER)
Specify next point or [Undo]: **6.75,3.5** (ENTER)
Specify next point or [Close/Undo]: *(press* ENTER *to terminate the command sequence)*

Command: *(press* ENTER *to repeat the* LINE *command)*
Specify first point: **6.625,3.25** (ENTER)
Specify next point or [Undo]: **6.625,3.75** (ENTER)
Specify next point or [Undo]: *(press* ENTER *to terminate the command sequence)*

Command: *(press* ENTER *to repeat the* LINE *command)*
Specify first point: **9.375,3.375** (ENTER)
Specify next point or [Undo]: **9.625,3.25** (ENTER)
Specify next point or [Undo]: **9.375,3.125** (ENTER)
Specify next point or [Close/Undo]: *(press* ENTER *to terminate the command sequence)*

Command: *(press* ENTER *to repeat the* LINE *command)*
Specify first point: **9.125,3.25** (ENTER)
Specify next point or [Undo]: **9.625,3.25** (ENTER)
Specify next point or [Undo]: *(press* ENTER *to terminate the command sequence)*

**Step 4** To draw the two circles located in the bottom of the instrument panel, first invoke the CIRCLE command from the Draw Toolbar or enter **circle** and press ENTER at the "Command:" prompt.

AutoCAD prompts:

Command: **circle** (ENTER)
Specify center point for circle or [3P/2P/Ttr (tan tan radius)]: **7.5,1.875** (ENTER)
Specify radius of circle or [Diameter]: **.25** (ENTER)

Command: *(press* ENTER *to repeat the* CIRCLE *command)*
Specify center point for circle or [3P/2P/Ttr (tan tan radius)]: **8.25,1.875** (ENTER)
Specify radius of circle or [Diameter]: **.25** (ENTER)

**Step 5** To draw the switches located in the bottom of the instrument panel, first invoke the RECTANGLE command from the Draw toolbar, or enter **rectangle** at the "Command:" prompt.

AutoCAD prompts:

Command: **rectangle** (ENTER)
Specify first corner point or [Chamfer/Elevation/Fillet/Thickness/Width]: **6.4375,1.5** (ENTER)
Specify other corner point or [Dimensions]: **6.5625,2** (ENTER)

Command: *(press* ENTER *to repeat the* RECTANGLE *command)*
Specify first corner point or [Chamfer/Elevation/Fillet/Thickness/Width]: **9.1875,1.5** (ENTER)
Specify other corner point or [Dimensions]: **9.3125,2** (ENTER)

Invoke the ARC command from the Draw Toolbar, as shown in Figure P2–12, or enter **arc** at the "Command:" prompt.

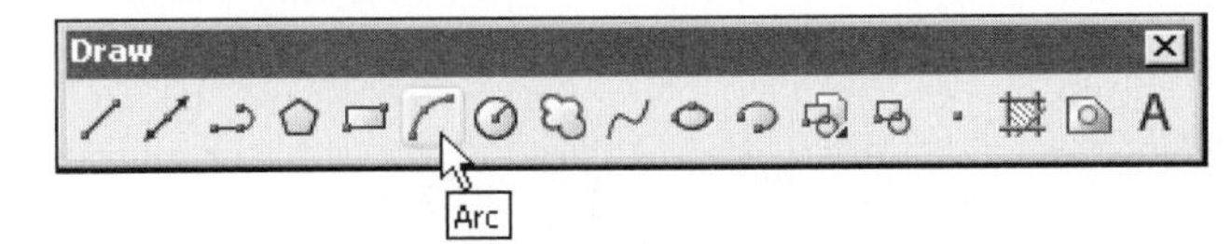


**Figure P2–12** *Invoking the* ARC *command from the Draw toolbar*

AutoCAD prompts:

Command: **arc** (ENTER)
Specify start point of arc or [CEnter]: **6.5625,1.875** (ENTER)
Specify second point of arc or [CEnter/ENd]: **c** (ENTER)
Specify center point of arc: **6.5,2** (ENTER)
Specify end point of arc or [Angle/chord Length]: **6.4375,1.875** (ENTER)

Command: *(press* ENTER *to repeat the* ARC *command)*
Specify start point of arc or [CEnter]: **9.3125,1.875** (ENTER)
Specify second point of arc or [CEnter/ENd]: **c** (ENTER)
Specify center point of arc: **9.25,2** (ENTER)
Specify end point of arc or [Angle/chord Length]: **9.1875,1.875** (ENTER)

**Step 6** To save the current status of the drawing, invoke the SAVE command from the File menu or enter **qsave** at the "Command:" prompt. AutoCAD saves the current status of the drawing.

### DRAW THE 90° ELL

This procedure describes the steps required to draw the 90-degree ell shown in Figure P2–13. The 90-degree ell is drawn with the LINE and ARC commands.

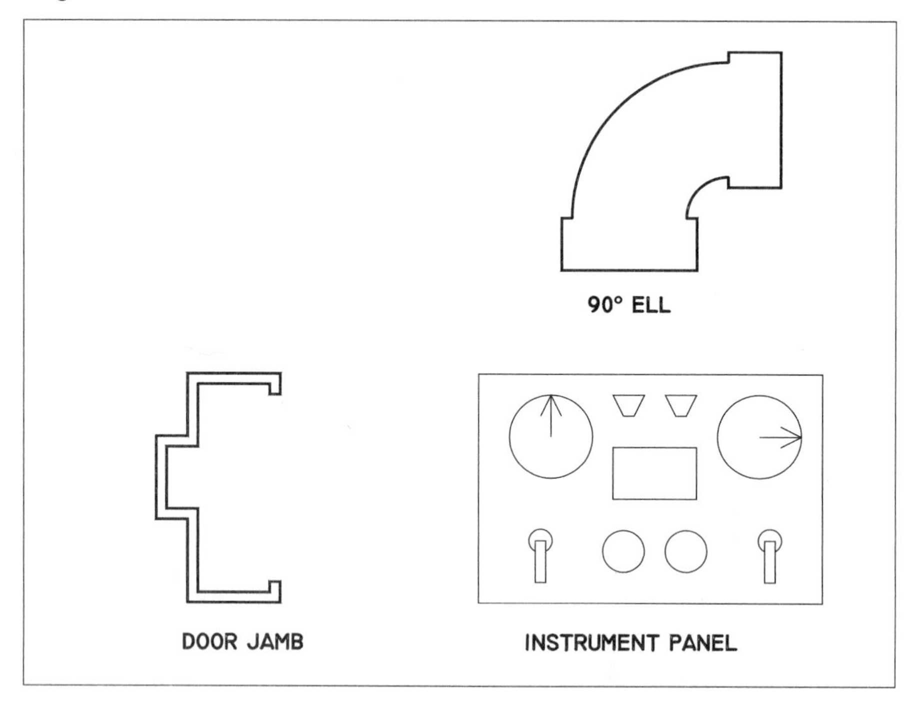


**Figure P2–13** *Door jamb, instrument panel, and 90-degree ell*

**Step 1** To draw the lower lines of the 90-degree ell, invoke the LINE command from the Draw toolbar or enter **line** and press ENTER at the "Command:" prompt.

AutoCAD prompts:

Command: **line** (ENTER)
Specify first point: **8.25,5.875** (ENTER)
Specify next point or [Undo]: **@.125<0** (ENTER)
Specify next point or [Undo]: **@.625<-90** (ENTER)
Specify next point or [Close/Undo]: **@1.625<180** (ENTER)
Specify next point or [Close/Undo]: **@.625<90** (ENTER)
Specify next point or [Close/Undo]: **@.125<0** (ENTER)
Specify next point or [Close/Undo]: *(press* ENTER *to terminate the command sequence)*

**Step 2** To draw the large outer arc, invoke the ARC command from the Draw toolbar or enter **arc** and press ENTER at the "Command:" prompt.

AutoCAD prompts:

Command: **arc** (ENTER)
Specify start point of arc or [CEnter]: **6.875,5.875** (ENTER)
Specify second point of arc or [CEnter/ENd]: **c** (ENTER)
Specify center point of arc: **@1.875<0** (ENTER)
Specify end point of arc or [Angle/chord Length]: **a** (ENTER)
Specify included angle: **–90** (ENTER)

**Step 3** To draw the upper lines of the 90-degree ell, invoke the LINE command from the Draw toolbar or enter **line** and press ENTER at the "Command:" prompt.

AutoCAD prompts:

Command: **line** (ENTER)
Specify first point: **8.75,7.75** (ENTER)
Specify next point or [Undo]: **@.125<90** (ENTER)
Specify next point or [Undo]: **@.625<0** (ENTER)
Specify next point or [Close/Undo]: **@1.625<-90** (ENTER)
Specify next point or [Close/Undo]: **@.625<180** (ENTER)
Specify next point or [Close/Undo]: **@.125<90** (ENTER)
Specify next point or [Close/Undo]: *(press* ENTER *to terminate the command sequence)*

**Step 4** To draw the smaller outer arc, invoke the ARC command from the Draw toolbar or enter **arc** and press ENTER at the "Command:" prompt.

AutoCAD prompts:

Command: **arc** (ENTER)
Specify start point of arc or [CEnter]: *(*enter @ *and press* enter *to select the last point as the start point)*
Specify second point of arc or [CEnter/ENd]: **c** (ENTER)
Specify center point of arc: **@.5<-90** (ENTER)
Specify end point of arc or [Angle/chord Length]: **@.5<180** (ENTER)

**Step 5** To save the current status of the drawing, invoke the SAVE command from the File menu or enter **qsave** at the "Command:" prompt. AutoCAD saves the current status of the drawing.

## DRAW THE COIL

This procedure describes the steps required to draw the coil shown in Figure P2–14. The coil is drawn with the ARC command.

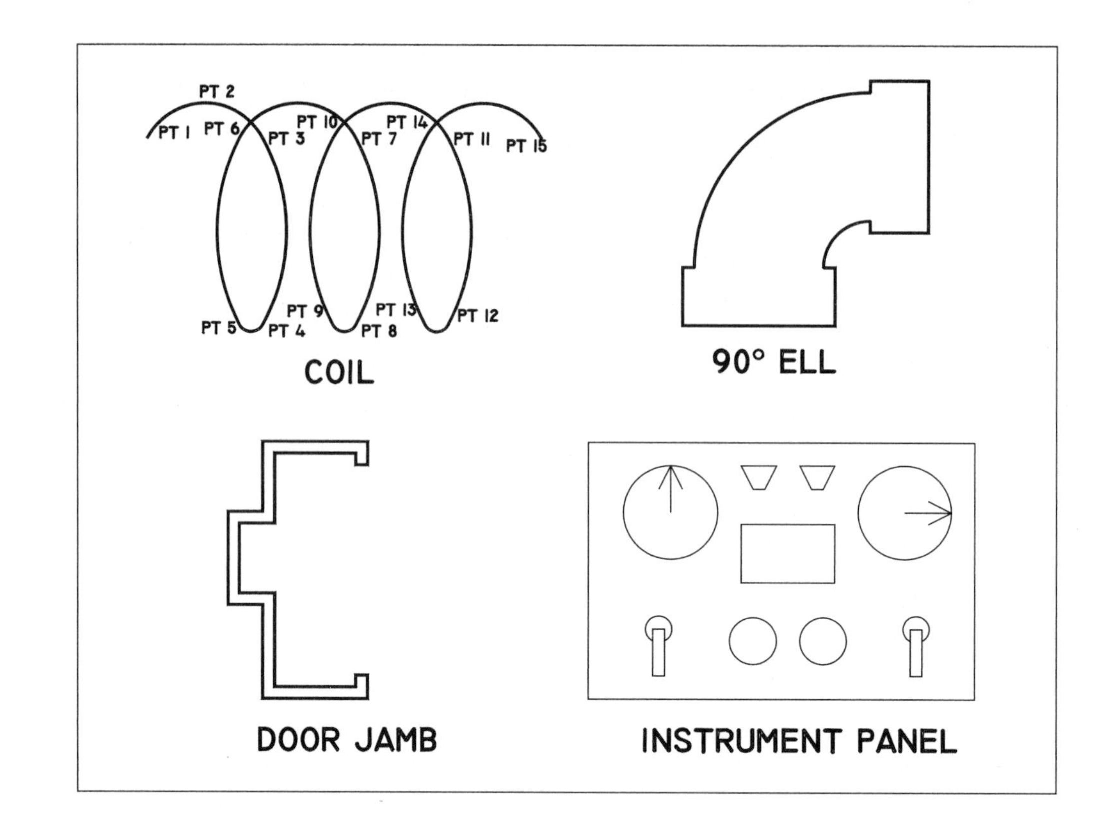


**Figure P2–14** *Door jamb, instrument panel, 90-degree ell, and coil*

**Step 1** To draw the coil, invoke the ARC command from the Draw toolbar or enter **arc** and press ENTER at the "Command:" prompt.

AutoCAD prompts:

Command: **arc** (ENTER)
Specify start point of arc or [CEnter]: **1,7.25** (ENTER)
Specify second point of arc or [CEnter/ENd]: **1.625,7.625** (ENTER)
Specify end point of arc: **2.25,7.25** (ENTER)

Command: (press ENTER to repeat the ARC command)
Specify start point of arc or [CEnter]: *(press ENTER to continue from the previous point)*
Specify end point of arc: **2.25,5.25** (ENTER)

Command: *(press ENTER to repeat the ARC command)*
Specify start point of arc or [CEnter]: *(press ENTER to continue from the previous point)*
Specify end point of arc: **2,5.25** (ENTER)

Command: *(press* ENTER *to repeat the* ARC *command)*
Specify start point of arc or [CEnter]: *(press* ENTER *to continue from the previous point)*
Specify end point of arc: **2,7.25** (ENTER)

Command: *(press* ENTER *to repeat the* ARC *command)*
Specify start point of arc or [CEnter]: *(press* ENTER *to continue from the previous point)*
Specify end point of arc: **3.25,7.25** (ENTER)

(Take a breath. This completes the first loop of the coil. Even though the Three-Point option was used for the leading arc of the first loop, the leading arc of the second loop, which is the trailing arc of the first loop, was drawn with the arc-arc continuation method.)

Command: *(press* ENTER *to repeat the* ARC *command)*
Specify start point of arc or [CEnter]: *(press* ENTER *to continue from the previous point)*
Specify end point of arc: **3.25,5.25** (ENTER)

Command: *(press* ENTER *to repeat the* ARC *command)*
Specify start point of arc or [CEnter]: *(press* ENTER *to continue from the previous point)*
Specify end point of arc: **3,5.25** (ENTER)

Command: *(press* ENTER *to repeat the* ARC *command)*
Specify start point of arc or [CEnter]: *(press* ENTER *to continue from the previous point)*
Specify end point of arc: **3,7.25** (ENTER)

Command: *(press* ENTER *to repeat the* ARC *command)*
Specify start point of arc or [CEnter]: *(press* ENTER *to continue from the previous point)*
Specify end point of arc: **4.25,7.25** (ENTER)

(Take another breath. This completes the second loop and starts the last loop of the coil.)

Command: *(press* ENTER *to repeat the* ARC *command)*
Specify start point of arc or [CEnter]: *(press* ENTER *to continue from the previous point)*
Specify end point of arc: **4.25,5.25** (ENTER)

Command: (press ENTER to repeat the ARC command)
Specify start point of arc or [CEnter]: *(press* ENTER *to continue from the previous point)*
Specify end point of arc: **4,5.25** (ENTER)

Command: *(press* ENTER *to repeat the* ARC *command)*
Specify start point of arc or [CEnter]: *(press* ENTER *to continue from the previous point)*
Specify end point of arc: **4,7.25** (ENTER)

Command: *(press* ENTER *to repeat the* ARC *command)*
Specify start point of arc or [CEnter]: *(press* ENTER *to continue from the previous point)*
Specify end point of arc: **5.25,7.25** (ENTER)

**Step 2** To save the current status of the drawing, invoke the SAVE command from the File menu or enter **qsave** at the command: prompt. AutoCAD saves the current status of the drawing.

Congratulations! You have just successfully applied several AutoCAD concepts in creating the drawing.

## EXERCISE 2–1

| Type of shape | Input Methods |
|---|---|
| Single Line Sequence | • Absolute rectangular coordinates<br>• Relative rectangular coordinates<br>• Relative polar coordinates |

Any one of the three listed methods can be used to enter the points that determine the shape in this exercise. This is not always the case, as you will see in the next few exercises. Also, all of the lines that determine the shape are sequentially connected; that is, the ending point of each line can be used as the starting point of another line. That means that this figure can be drawn without exiting the LINE command once it is invoked. Again, this is not the case with most shapes.

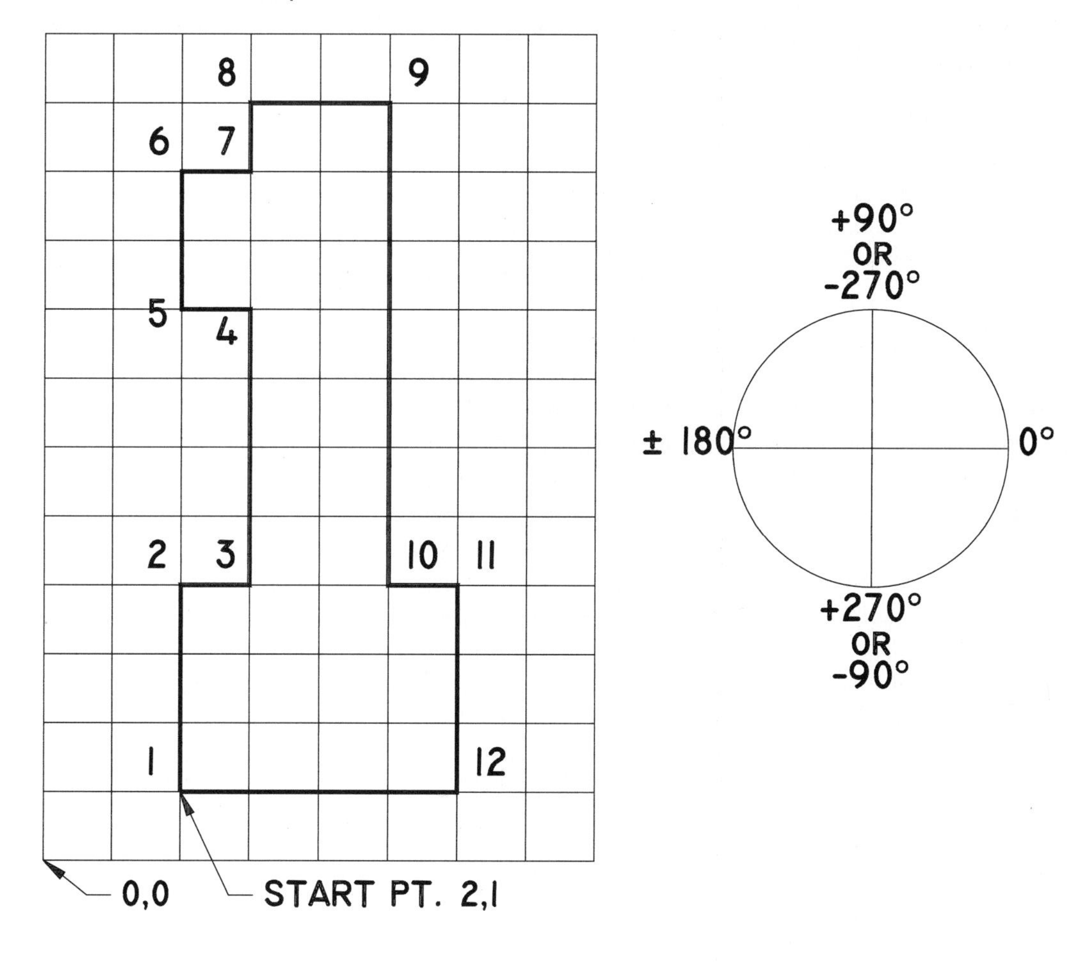

## Points of Interest

- With rectangular coordinates, the distances are entered as horizontal (*X*) and vertical (*Y*) displacements.
- Absolute rectangular coordinates are *X* and *Y* displacements from the origin (0,0).
- Relative rectangular coordinates are *X* and *Y* displacements from the last point entered.
- @ Represents the last point entered.
- Relative polar coordinates are a combination of a direction and a distance from the last point entered.
- Substituting rectangular coordinate input with polar coordinate input (or vice versa) is not always done easily without the application of trigonometric functions.

| **Drawing Setup** | Use Quick Setup wizard, set Units to Decimal, set Area to 9 x 13. Invoke the LINE command and then respond to the points as shown for each method. |
|---|---|

## Point by Point

| Absolute | Relative Rectangular | Relative Polar |
|---|---|---|
| 1. First point: **2,1** (ENTER) | 1. First point: **2,1** (ENTER) | 1. First point: **2,1** (ENTER) |
| 2. Next point: **2,4** (ENTER) | 2. Next point: **@0,3** (ENTER) | 2. Next point: **@3<90** (ENTER) |
| 3. Next point: **3,4** (ENTER) | 3. Next point: **@1,0** (ENTER) | 3. Next point: **@1<0** (ENTER) |
| 4. Next point: **3,8** (ENTER) | 4. Next point: **@0,4** (ENTER) | 4. Next point: **@4<90** (ENTER) |
| 5. Next point: **2,8** (ENTER) | 5. Next point: **@–1,0** (ENTER) | 5. Next point: **@1<180** (ENTER) |
| 6. Next point: **2,10** (ENTER) | 6. Next point: **@0,2** (ENTER) | 6. Next point: **@2<90** (ENTER) |
| 7. Next point: **3,10** (ENTER) | 7. Next point: **@1,0** (ENTER) | 7. Next point: **@1<0** (ENTER) |
| 8. Next point: **3,11** (ENTER) | 8. Next point: **@0,1** (ENTER) | 8. Next point: **@1<90** (ENTER) |
| 9. Next point: **5,11** (ENTER) | 9. Next point: **@2,0** (ENTER) | 9. Next point: **@2<0** (ENTER) |
| 10. Next point: **5,4** (ENTER) | 10. Next point: **@0,–7** (ENTER) | 10. Next point: **@7<270** (ENTER) |
| 11. Next point: **6,4** (ENTER) | 11. Next point: **@1,0** (ENTER) | 11. Next point: **@1<0** (ENTER) |
| 12. Next point: **6,1** (ENTER) | 12. Next point: **@0,–3** (ENTER) | 12. Next point: **@3<270** (ENTER) |
| 13. Next point: **2,1** (ENTER) | 13. Next point: **@–4,0** (ENTER) | 13. Next point: **@4<180** (ENTER) |
| 14. Command: (ENTER) | 14. Command: (ENTER) | 14. Command: (ENTER) |

## EXERCISE 2–2

| Type of shape | Input Methods |
|---|---|
| Multiple line sequences | • Absolute rectangular coordinates<br>• Relative rectangular coordinates<br>• Relative polar coordinates |

As in Exercise 2–1, the points that determine the shapes in this exercise can be entered by any one of the three listed methods here. But because these are three separate shapes, all lines are not sequentially connected. The LINE command must be invoked at least three times.

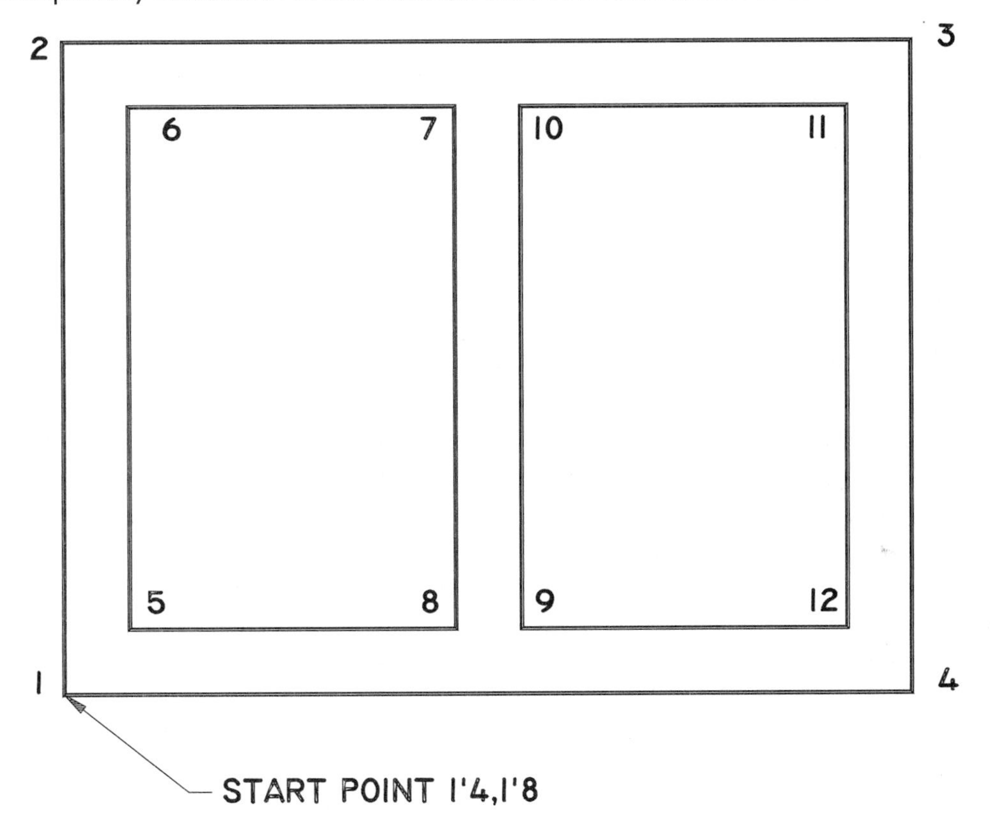


### Points of Interest

- Steps 5, 11, and 17 in the following tables illustrate how you can use the Close option of the LINE command to draw the last segment of a closed sequence of lines and exit the LINE command at the same time.

| Drawing Setup | Use Quick Setup Wizard, set Units to Architectural, set Area to 8' x 6'. |
|---|---|

Invoke the LINE command and then respond to the prompts as shown for each method. Steps 6 and 12 use automatic reinvoking of the previously used command by pressing (ENTER). In this case the LINE command is the previously used command.

| Absolute | Relative Rectangular | Relative Polar |
|---|---|---|
| 1. Start point: **1'4,1'8** (ENTER) | 1. Start point: **1'4,1'8** (ENTER) | 1. Start point: **1'4,1'8** (ENTER) |
| 2. Next point: **1'4,5'** (ENTER) | 2. Next point: **@0,3'4** (ENTER) | 2. Next point: **@3'4<90** (ENTER) |
| 3. Next point: **5'8,5'** (ENTER) | 3. Next point: **@4'4,0** (ENTER) | 3. Next point: **@4'4<0** (ENTER) |
| 4. Next point: **5'8,1'8** (ENTER) | 4. Next point: **@0,-3'4** (ENTER) | 4. Next point: **@3'4<-90** (ENTER) |
| 5. Next point: **c** (ENTER) | 5. Next point: **c** (ENTER) | 5. Next point:**c** (ENTER) |
| 6. Command: (ENTER) | 6. Command: (ENTER) | 6. Command: (ENTER) |
| 7. From point: **1'8,2'** (ENTER) | 7. From point: **1'8,2'** (ENTER) | 7. From point: **1'8,2'** (ENTER) |
| 8. Next point: **1'8,4'8** (ENTER) | 8. Next point: **@0,2'8** (ENTER) | 8. Next point: **@2'8<90** (ENTER) |
| 9. Next point: **3'4,4'8** (ENTER) | 9. Next point: **@1'8,0** (ENTER) | 9. Next point: **@1'8<0** (ENTER) |
| 10. Next point: **3'4,2'** (ENTER) | 10. Next point: **@0,–2'8** (ENTER) | 10. Next point: **@2'8<270** (ENTER) |
| 11. Next point: **c** (ENTER) | 11. Next point: **c** (ENTER) | 11. Next point: **c** (ENTER) |
| 12. Command: (ENTER) | 12. Command: (ENTER) | 12. Command: (ENTER) |
| 13. From point: **3'8,2'** (ENTER) | 13. From point: **3'8,2'** (ENTER) | 13. From point: **3'8,2'** (ENTER) |
| 14. Next point: **3'8,4'8** (ENTER) | 14. Next point: **@0,2'8** (ENTER) | 14. Next point: **@2'8<90** (ENTER) |
| 15. Next point: **5'4,4'8** (ENTER) | 15. Next point: **@1'8,0** (ENTER) | 15. Next point: **@1'8<0** (ENTER) |
| 16. Next point: **5'4,2'** (ENTER) | 16. Next point: **@0,–2'8** (ENTER) | 16. Next point: **@2'8<270** (ENTER) |
| 17. Next point: **c** (ENTER) | 17. Next point: **c** (ENTER) | 17. Next point: **c** (ENTER) |

## EXERCISE 2–3

| Type of shape | Input Methods |
|---|---|
| Multiple line sequences | • Absolute rectangular coordinates for the first line segment of the first sequence<br>• Relative polar coordinates for the remaining segments of the first sequence<br>• Relative polar coordinates for entering the starting point of the second and third separate single line sections of the shape |

Unlike the cases of Exercises 2–1 and 2–2, all of the points that determine the figure in this exercise cannot be entered with complete accuracy by all three of the methods listed (see Important Point below). Except for points A1, A2, A5, and A8, the points in the shape in this exercise can be accurately specified only by polar coordinates. For example, the coordinates of point A3 (2 units at 240° from point A2) are 5,1.267949192431. This is derived from the *X* coordinate of $6 - 2 \cos 240°$ and the *Y* coordinate of $5 - 2 \sin 240°$. Even though the *X* coordinate happens to be a round number in this case, it is an exception rather than the rule. Keying in numbers like 1.267949192431 takes time and is subject to error. Any fewer digits might cause problems when endpoints of sequences are expected to end exactly at some other point. You cannot see the difference on a plot, but some of the computer functions that depend on this type of accuracy might not perform as expected. So not only is it easier to enter the polar coordinates @2<240, it is also quicker, more accurate, and less subject to error.

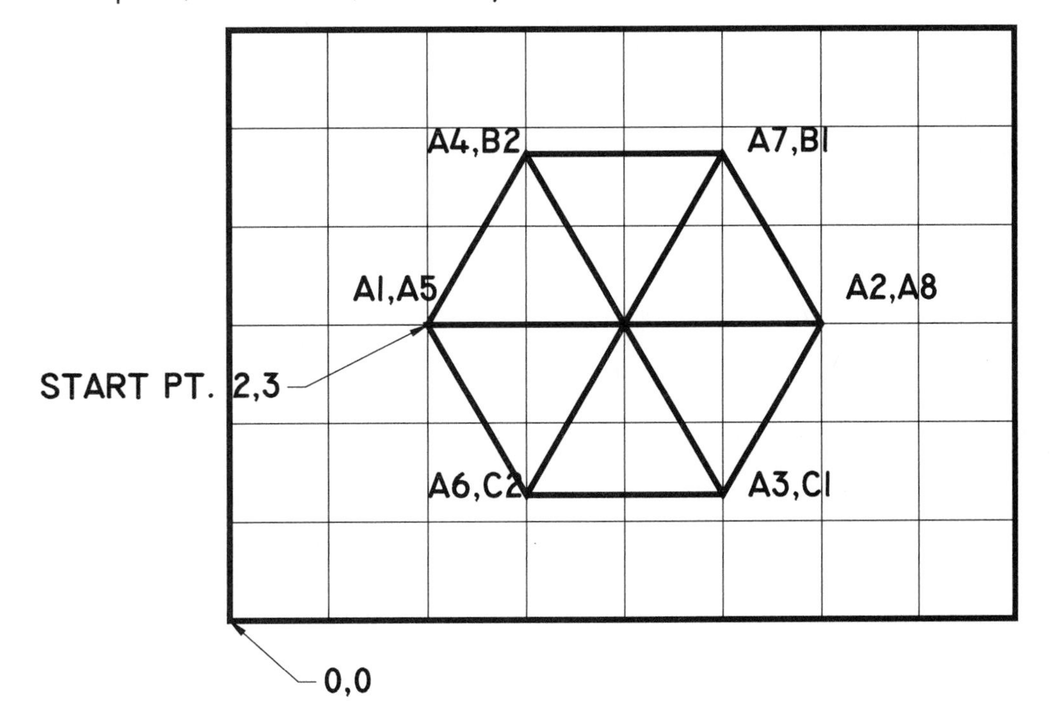


general EXERCISE

The LINE command must be invoked at least three times. And in order to enter the starting point for the second separate part of the shape (the line from B1 to B2), the last segment of the first sequence needs to end where shown, at point A8.

## Important Point

AutoCAD has more advanced commands and features to make drawing these early exercises more easily than used here.

## Points of Interest

- Steps 11 and 15 in the following Point by Point instructions show how you can specify starting points of lines (in this case points B1 and C1) by using the @ symbol and polar coordinate "distance and direction" data. Remember, the @ symbol means "last point."
- This method is possible because we ended the previous lines in predetermined points (A8 and B2) whose direction and distance from B1 and C1 we knew. If we had wanted to start segment B1 at B2, it would have been difficult to establish it from A8 by means of polar coordinates.
- Only the second point can be entered easily and accurately by means of any of the three methods (absolute, relative rectangular, or relative polar coordinates).

| **Drawing Setup** | Use Quick Setup wizard, set Units to Decimal, set Area to 10 X 7. |
|---|---|

Invoke the LINE command and then respond to the points prompt as shown. Steps 10 and 14 reinvoke the LINE command for separate sequences.

## Point by Point

1. Specify first point: **2,3** (ENTER) *(starting point of line A1–A2)*
2. Specify next point or [Undo]: **@4<0** (ENTER) *(endpoint of line A1–A2)*
3. Specify next point or [Undo]: **@2<240** (ENTER) *(endpoint of line A2–A3)*
4. Specify next point or [Close/Undo]: **@4<120** (ENTER) *(endpoint of line A3–A4)*
5. Specify next point or [Close/Undo]: **@2<240** (ENTER) *(endpoint of line A4–A5)*
6. Specify next point or [Close/Undo]: **@2<300** (ENTER) *(endpoint of line A5–A6)*
7. Specify next point or [Close/Undo]: **@4<60** (ENTER) *(endpoint of line A6–A7)*
8. Specify next point or [Close/Undo]: **@2<300** (ENTER) *(endpoint of line A7–A8)*
9. Specify next point or [Close/Undo]: (ENTER) *(exits the LINE command)*
10. Command: (ENTER) *(automatic recall of LINE command for B1 to B2)*
11. Specify first point: **@2<120** (ENTER) *(starting point of line B1-B2)*
12. Specify next point or [Undo]: **@2<180** (ENTER) *(endpoint of line B1–B2)*
13. Specify next point or [Undo]: (ENTER) *(exits the LINE command)*
14. Command: (ENTER) *(automatic recall of LINE command for C1 to C2)*
15. Specify first point: **@4<300** (ENTER) *(starting point of line C1 to C2)*
16. Specify next point or [Undo]: **@2<180** (ENTER) *(endpoint of line C1 to C2)*
17. Specify next point or [Undo]: (ENTER) *(exits the LINE command)*

## EXERCISE 2–4

| Type of shape | Input Methods |
|---|---|
| Single line sequence | • Absolute rectangular coordinates for the first line segment of the sequence<br>• Relative polar coordinates for the remaining segments Important Point |

AutoCAD has more advanced commands and features to make drawing these early exercises easier than we've shown here. These methods are pabulum. Solid food will come later.

### Points of Interest

Step 9 in the following Point by Point instructions can be substituted by simply selecting the Close option.

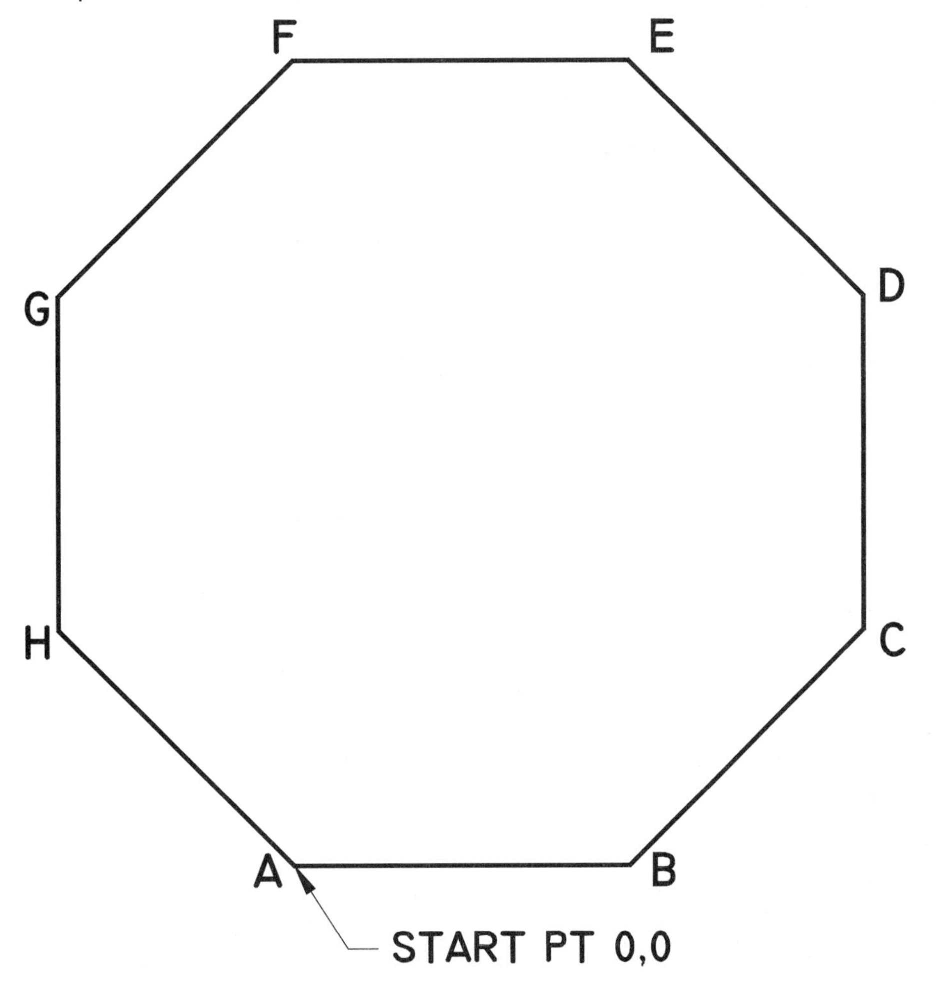

| Drawing Setup | Use Start from Scratch option, then set Units to Architectural and Limits to -4'4",-1'8" for lower left corner and 10'0",11'4" for upper right corner. |
|---|---|
| **Useful Point** | Don't forget to ZOOM ALL after setting the limits. |

Invoke the LINE command and respond to the "Specify first point:" as shown below.

## Point by Point

1. Specify first point: **0,0** (ENTER) *(starting point of line A–B)*
2. Specify next point or [Undo]: **@4'<0** (ENTER) *(endpoint of line A–B)*
3. Specify next point or [Undo]: **@4'<45** (ENTER) *(endpoint of line B–C)*
4. Specify next point or [Close/Undo]: **@4'<90** (ENTER) *(endpoint of line C–D)*
5. Specify next point or [Close/Undo]: **@4'<135** (ENTER) *(endpoint of line D–E)*
6. Specify next point or [Close/Undo]: **@4'<180** (ENTER) *(endpoint of line E–F)*
7. Specify next point or [Close/Undo]: **@4'<225** (ENTER) *(endpoint of line F–G)*
8. Specify next point or [Close/Undo]: **@4'<270** (ENTER) *(endpoint of line G–H)*
9. Specify next point or [Close/Undo]: **@4'<315** (ENTER) *(endpoint of line H–A)*
10. Specify next point or [Close/Undo]: (ENTER) *(exits the LINE command)*

## EXERCISE 2–5

| Type of shapes | Input Methods |
|---|---|
| Multiple line/arc sequences | • Absolute rectangular coordinates<br>• Relative rectangular coordinates<br>• Relative polar coordinates |

Some—but not all—of the points that determine the shape in this exercise can be entered by any one of the three listed methods. This exercise is primarily to emphasize when the line-arc continuation and the arc-line continuation options can be applied and when they cannot.

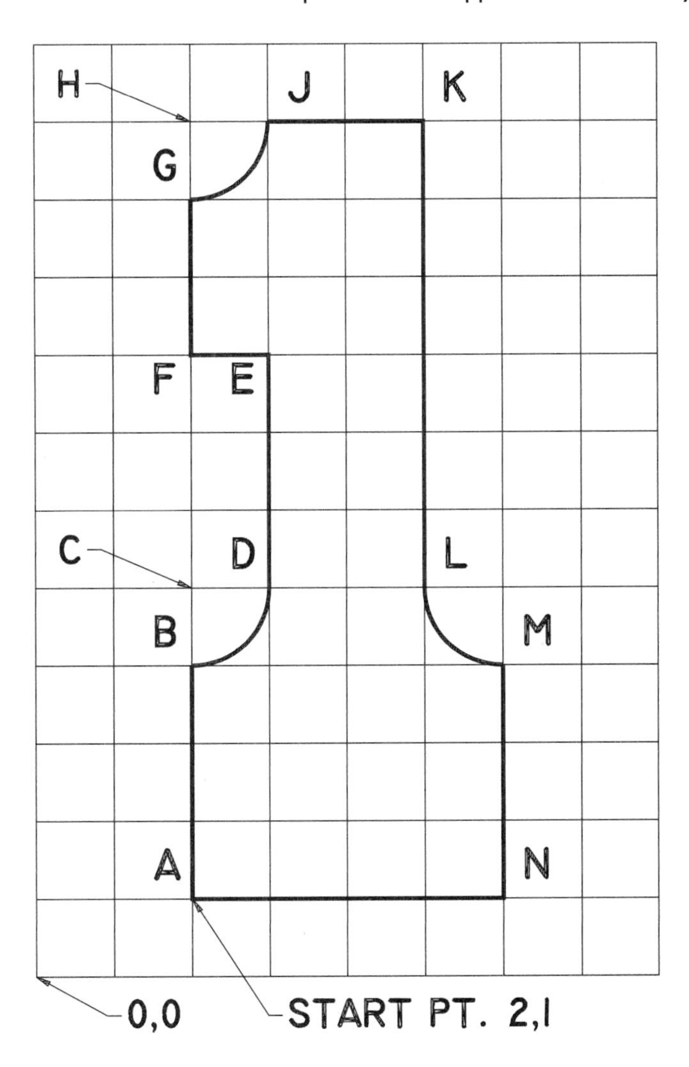

## Points of Interest

- The default line-line continuation of the LINE command does not require that lines drawn subsequent to the previous line segment start in the same direction that the previous line ends.
- In line-line continuation you may proceed in any desired direction with the continuing line.
- In line-arc and arc-line continuations, the connecting line/arc must start in the same direction that the previous arc/line ends.

| **Drawing Setup** | Use Quick Setup Wizard, set Units to Decimal, set Area to 9 x 13. |
|---|---|

## Point by Point

Invoke the LINE command, and AutoCAD prompts:

Command: **line** (ENTER)
Specify first point: **2,1** (ENTER) *(starting point of line A–B)*
Specify next point or [Undo]: **2,4** (ENTER) *(end point of the line A–B)*
Specify next point or [Close]: (ENTER) *(exits the LINE command)*

Invoke the ARC command, and AutoCAD prompts:

Command: **arc** (ENTER)
Specify start point of arc or [CEnter]: **2,4** (ENTER) *(starting point of the arc B–D)*
Specify second point of arc or [CEnter/ENd]: **c** (ENTER)
Specify center point of arc: **2,5** (ENTER) *(arc center for the arc B–D)*
Specify end point of arc or [Angle/chord Length]: **3,5** (ENTER) *(end point of the arc B–D)*

Invoke the LINE command, and AutoCAD prompts:

Command: **line** (ENTER)
Specify first point: (ENTER) *(invoke the continuation option)*
Length of line: **3** (ENTER) (length of the line D–E)
Specify next point or [Undo]: **2,8** (ENTER) (end point of the line E–F)
Specify next point or [Undo]: **2,10** (ENTER) (end point of the line F–-G)
Specify next point or [Close/Undo]: (ENTER) (exits the LINE command)

Invoke the ARC command, and AutoCAD prompts:

Command: **arc** (ENTER)
Specify start point of arc or [CEnter]: **2,10** (ENTER) *(starting point of the arc G–J)*
Specify second point of arc or [CEnter/ENd]: **c** (ENTER)
Specify center point of arc: **2,11** (ENTER) *(arc center for the arc G–J)*

Specify end point of arc or [Angle/chord Length]: **3,11** (ENTER) *(end point of the arc G–J)*

Invoke the LINE command, and AutoCAD prompts:

Command: **line** (ENTER)
Specify first point: **3,11** (ENTER) *(start point of the line J–K)*
Specify next point or [Undo]: **5,11** (ENTER) *(end point of the line J–K)*
Specify next point or [Undo]: **5,5** (ENTER) (end point of the line K–L)
Specify next point or [Close/Undo]: (ENTER) (exits the LINE command)

Invoke the ARC command, and AutoCAD prompts:

Command: **arc** (ENTER)
Specify start point of arc or [CEnter]: (ENTER) *(invoke the continuation option to start the arc at L)*
Specify end point of arc: **6,4** (enter) *(endpoint of the arc L–M)*

Invoke the LINE command, and AutoCAD prompts:

Command: **line** (ENTER)
Specify first point: **6,4** (ENTER) *(start point of the line M–N)*
Specify next point or [Undo]: **6,1** (ENTER) *(end point of the line M–N)*
Specify next point or [Undo]: **2,1** (ENTER) *(end point of the line N–A)*
Specify next point or [Close/Undo]: (enter) *(exits the* line *command)*

## EXERCISE 2–6

| Type of shapes | Input Methods |
|---|---|
| Single line sequence<br>Multiple arcs | • Absolute rectangular coordinates<br>• Three-point option for arcs |

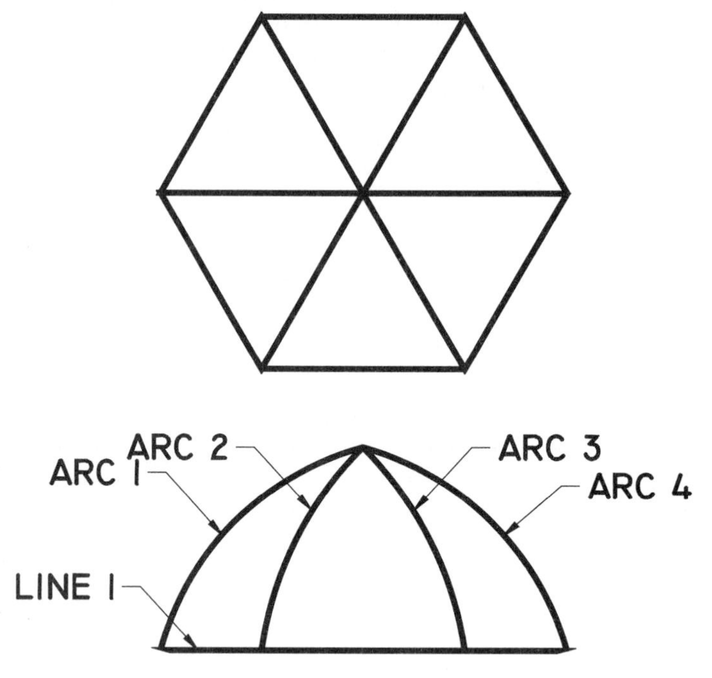


### Points of Interest

- The top view of this tent-shaped object gives some clues to its true shape, but not enough. A side view is required.
- In the front view the two outside arcs are represented in their true 2D appearance. However, the two inner arcs represent arcs that are rotated 60°. Their true representation would be ellipses. Since we have not covered ellipses, we will use arcs to give close approximations.

### Important Points

- Advanced features like the ELLIPSE command and 3D drawing may used to make the drawing exact if necessary.
- Note that arcs can be drawn both clockwise and counterclockwise with the three point option. This is not the case with all options of the ARC command.

| Drawing Setup | Open the drawing created in Exercise 2–3 |
|---|---|

## Point by Point

Invoke the LINE command, and AutoCAD prompts:

> Command: **line** (ENTER)
> Specify first point: **4,0.5** (ENTER) *(starting point of line 1)*
> Specify next point or [Undo]: **8,0.5** (ENTER) *(end point of the line 1)*
> Specify next point or [Undo]: (ENTER) (exits the LINE command)

Invoke the ARC command, and AutoCAD prompts:

> Command: **arc** (ENTER)
> Specify start point of arc or [CEnter]: **4,0.5** (ENTER) *(starting point of the arc 1)*
> Specify second point of arc or [CEnter/ENd]: **5,2** (ENTER) *(second point of the arc 1)*
> Specify end point of arc or [Angle/chord Length]: **6,2.5** (enter) *(end point of the arc 1)*
>
> Command: *(press* ENTER *to invoke the* arc *command again)*
> Specify start point of arc or [CEnter]: **5,0.5** (ENTER) *(starting point of the arc 2)*
> Specify second point of arc or [CEnter/ENd]: **5.5,1.875** (ENTER) *(second point of the arc 2)*
> Specify end point of arc or [Angle/chord Length]: **6,2.5** (enter) *(end point of the arc 2)*
>
> Command: (press ENTER to invoke the arc command again)
> Specify start point of arc or [CEnter]: **7,0.5** (ENTER) *(starting point of the arc 3)*
> Specify second point of arc or [CEnter/ENd]: **6.5,1.875** (ENTER) *(second point of the arc 3)*
> Specify end point of arc or [Angle/chord Length]: **6,2.5** (enter) *(end point of the arc 3)*
>
> Command: *(press* (ENTER) *to invoke the* arc *command again)*
> Specify start point of arc or [CEnter]: **8,0.5** (ENTER) *(starting point of the arc 4)*
> Specify second point of arc or [CEnter/ENd]: **7,2** (ENTER) *(second point of the arc 4)*
> Specify end point of arc or [Angle/chord Length]: **6,2.5** (ENTER) *(end point of the arc 4)*

# EXERCISE 2–7

| Type of shape | Input Methods |
| --- | --- |
| Multiple circles | • Absolute rectangular coordinates<br>• Center-Radius option for circles |

## Points of Interest

- The main octagon in this figure was drawn precisely by using the relative polar coordinate point entry method. However, only the coordinates of the bottom line are nice, whole units. The other endpoints around the polygon cannot easily be entered exactly from the keyboard.
- The most significant point of this figure is probably its center. The relative distance/direction from the center to the endpoints of the bottom line are calculated by trigonometry. Therefore, at this level we will simply enter "very close" coordinates for the center. We will do the same for the "bolt circle" (circle of circles).

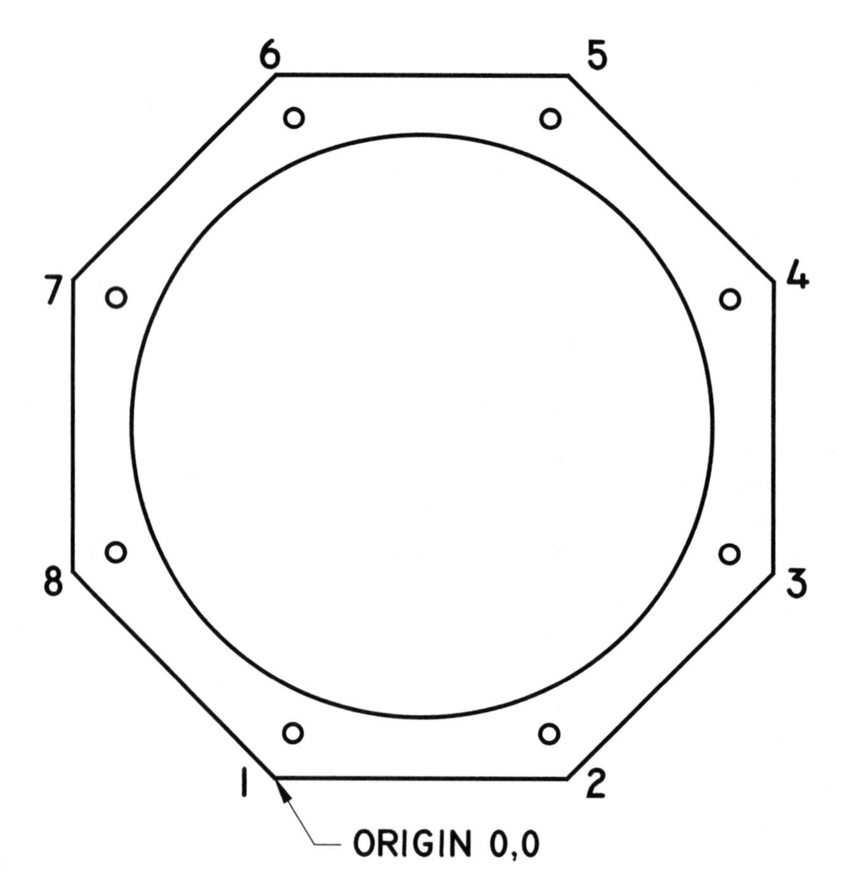

## Important Points

- Advanced features like the ARRAY command may be used to make the drawing exact if necessary.
- Note that arcs can be drawn both clockwise and counterclockwise with the three-point option. This is not the case with all options of the ARC command.

| **Drawing Setup** | Open the drawing made in Exercise 2–4. |
|---|---|
| **Power Point** | Each center of the smaller circles can be established relative to the previous one drawn. But in order to start the first small circle, its center must be established by its absolute coordinates or by its relation to another point. We know that it is 2" distance at 68.5° from the left end of the bottom line (the origin 0,0). So, in order to make the origin the "last point," we will just start a line there and then not complete it. This will establish it as the point used by the "@" symbol. The next chapter will eliminate this trickery with the Tracking feature. |

## Point by Point

Invoke the CIRCLE command, and AutoCAD prompts:

Command: **circle** (ENTER)
Specify center point for circle or [3P/2P/Ttr (tan tan radius)]: **2'0,4'10** (ENTER) *(center point of the large circle)*
Specify radius of circle or [Diameter]: **48** (ENTER) *(radius of the large circle)*

Invoke the LINE command, and AutoCAD prompts:

Command: **line** (ENTER)
Specify start point: **0,0** (ENTER) *(establish the last point at 0,0)*
Specify next point or [Undo]: *(press ENTER terminate the command sequence)*

Invoke the CIRCLE command, and AutoCAD prompts:

Command: **circle**
Specify center point for circle or [3P/2P/Ttr (tan tan radius)]: **@8<67.5** (ENTER) *(center point of the first small circle)*
Specify radius of circle or [Diameter]: **1.5** (ENTER) *(radius of the first small circle)*

Command: *(press ENTER to invoke the CIRCLE command again)*
Specify center point for circle or [3P/2P/Ttr (tan tan radius)]: **@42<0** (ENTER) *(center point of the second small circle)*
Specify radius of circle or [Diameter]: **1.5** (ENTER) *(radius of the second small circle)*

Command: *(press* ENTER *to invoke the* CIRCLE *command again)*
Specify center point for circle or [3P/2P/Ttr (tan tan radius)]: **@42<45** (ENTER)
*(center point of the third small circle)*
Specify radius of circle or [Diameter]: **1.5** (ENTER) *(radius of the second third circle)*

Command: *(press* ENTER *to invoke the* CIRCLE *command again)*
Specify center point for circle or [3P/2P/Ttr (tan tan radius)]: **@42<90** (ENTER)
*(center point of the fourth small circle)*
Specify radius of circle or [Diameter]: **1.5** (ENTER) *(radius of the fourth small circle)*

(Draw the fifth through eighth small circles by continuing with the method explained above in the previous circles but adding 45 degrees to the angle each time)

## EXERCISES 2–8 THROUGH 2–14

Create the drawings (orthographic projections) according to the settings given in the following table:

| Settings | Value |
|---|---|
| 1. Units<br>2. Limits | Decimal<br>Lower left corner: 0,0<br>Upper right corner: 12,9 |

**Note:** Grid lines in the drawings are spaced 0.25 units apart.

### EXERCISE 2–8

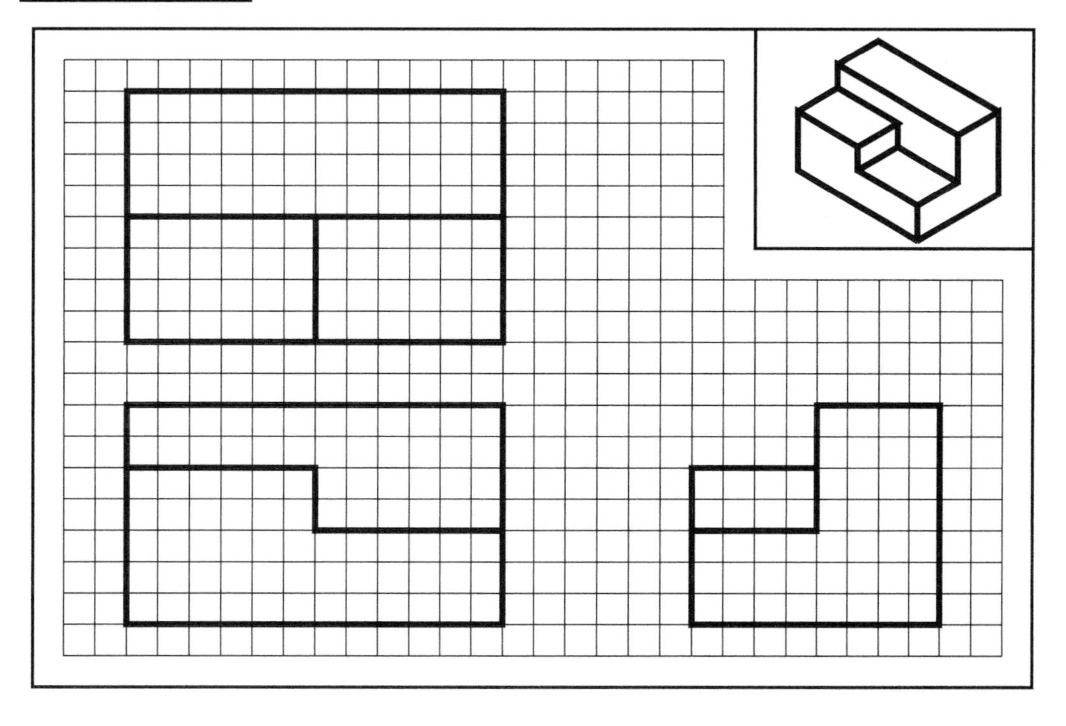

## EXERCISE 2–9

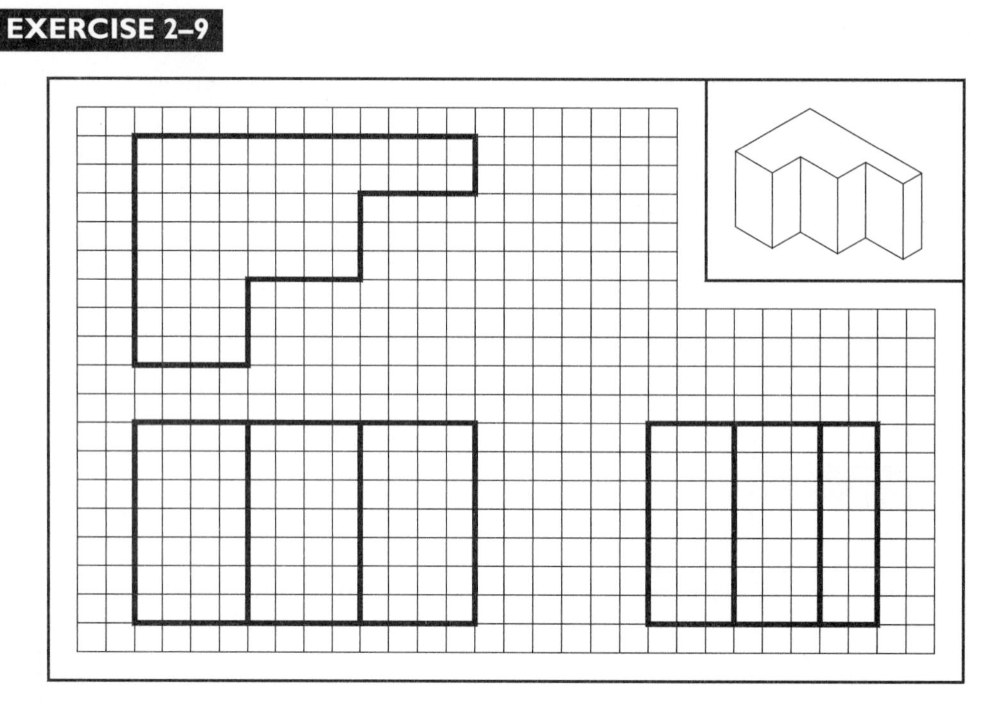

## EXERCISE 2–10

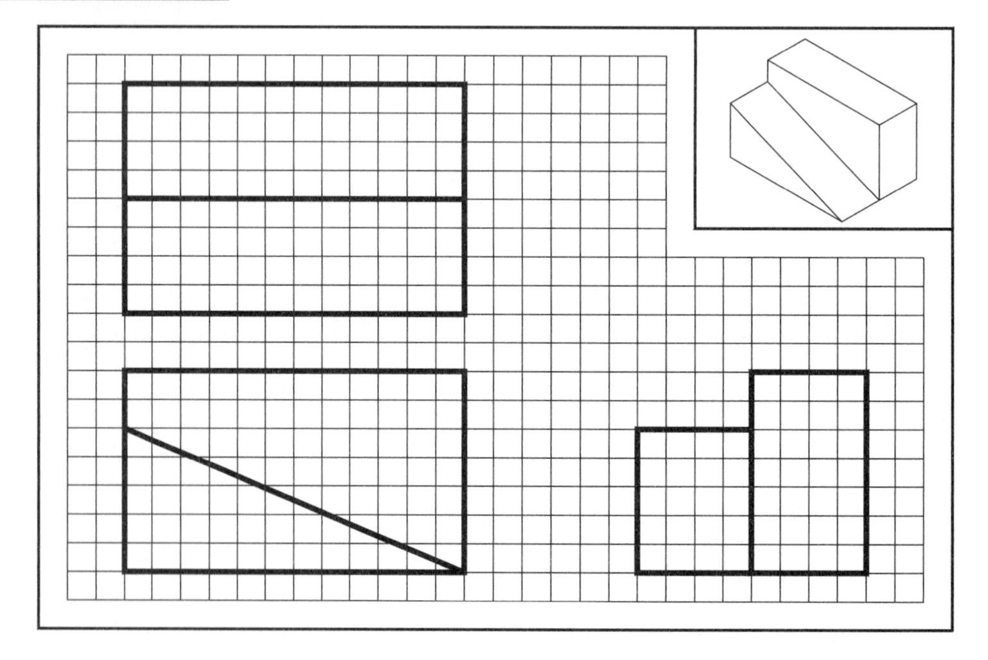

## EXERCISE 2–11

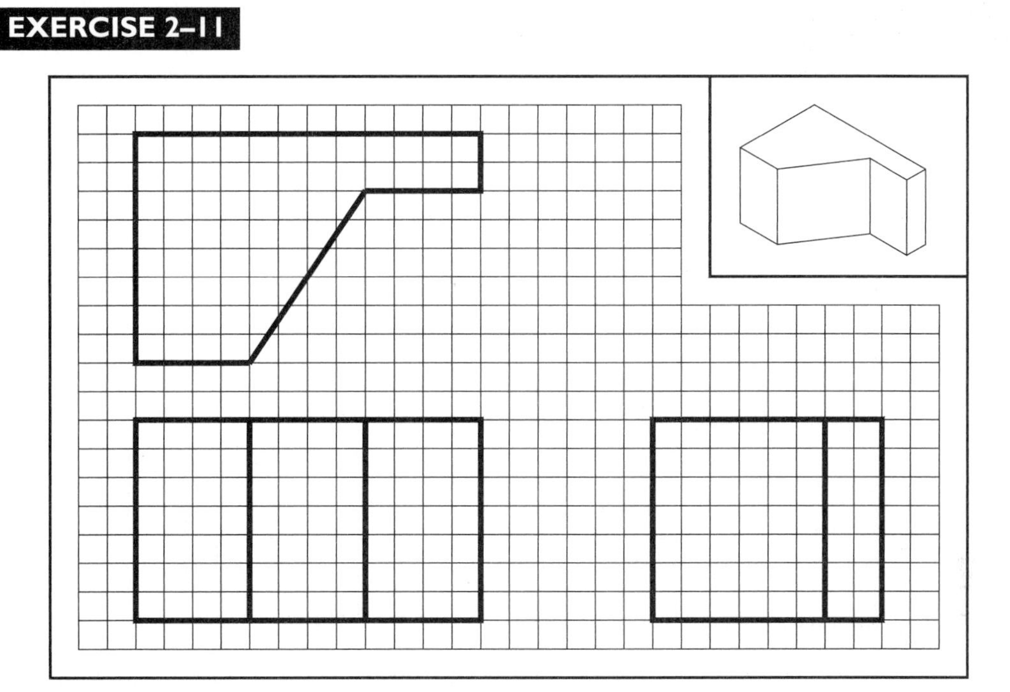

## EXERCISE 2–12

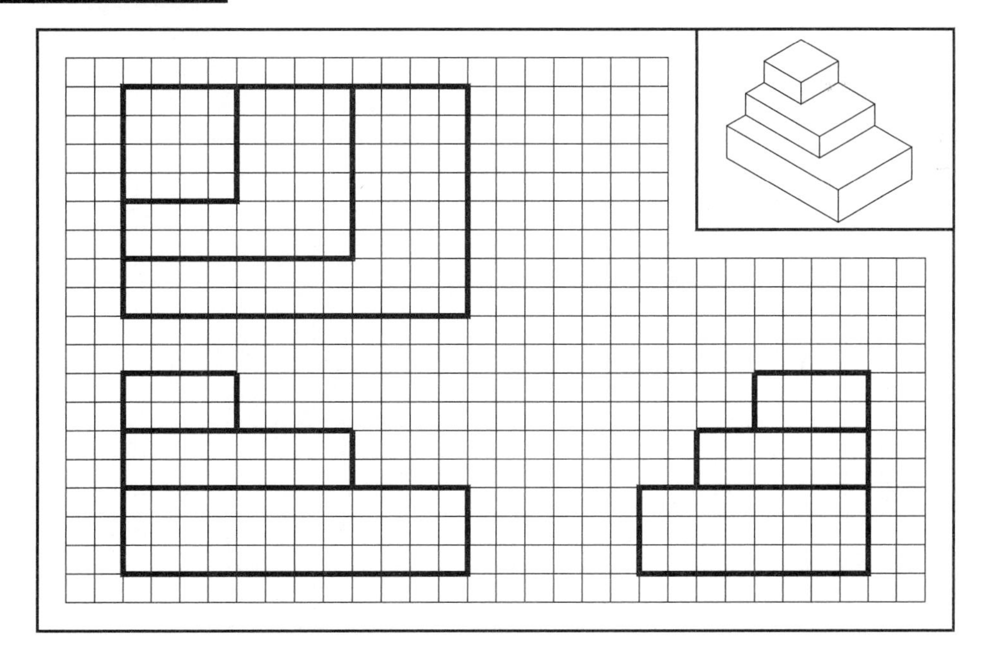

## EXERCISE 2–13

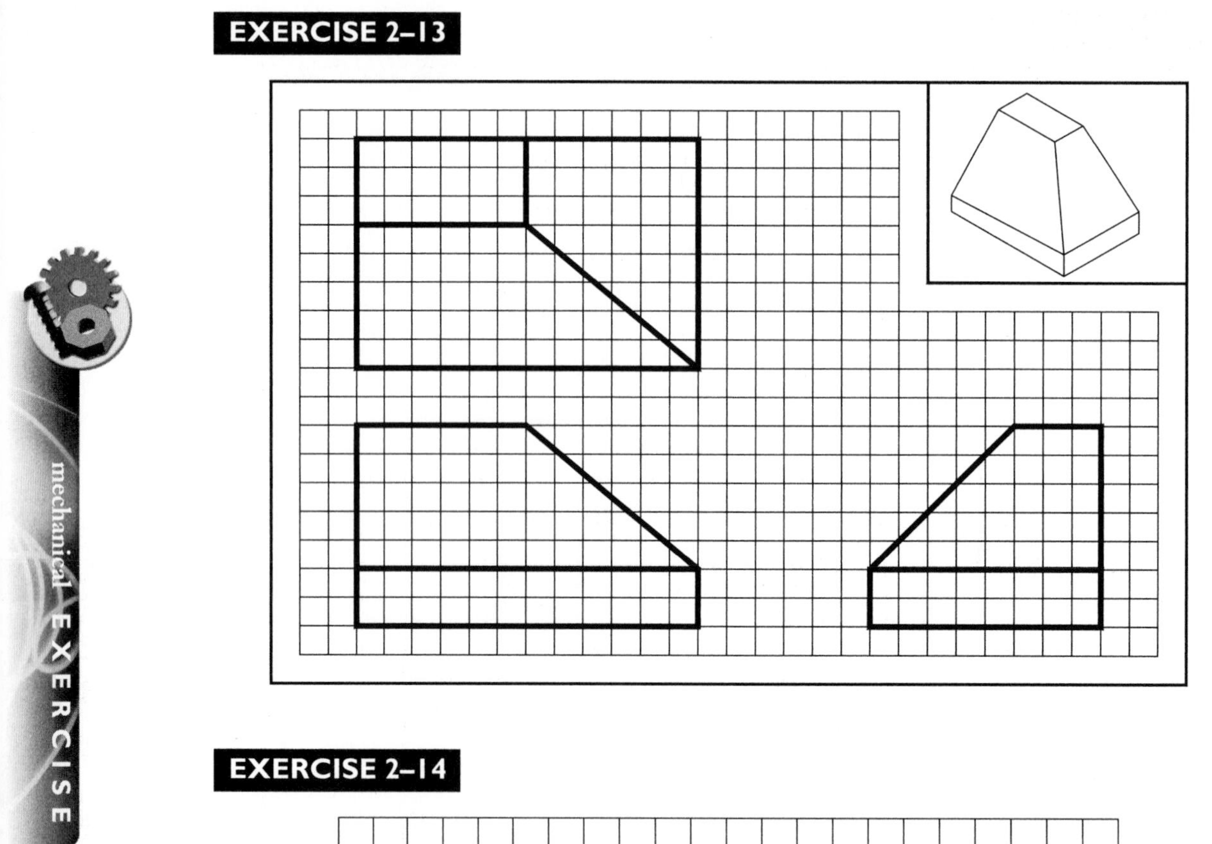

## EXERCISE 2–14

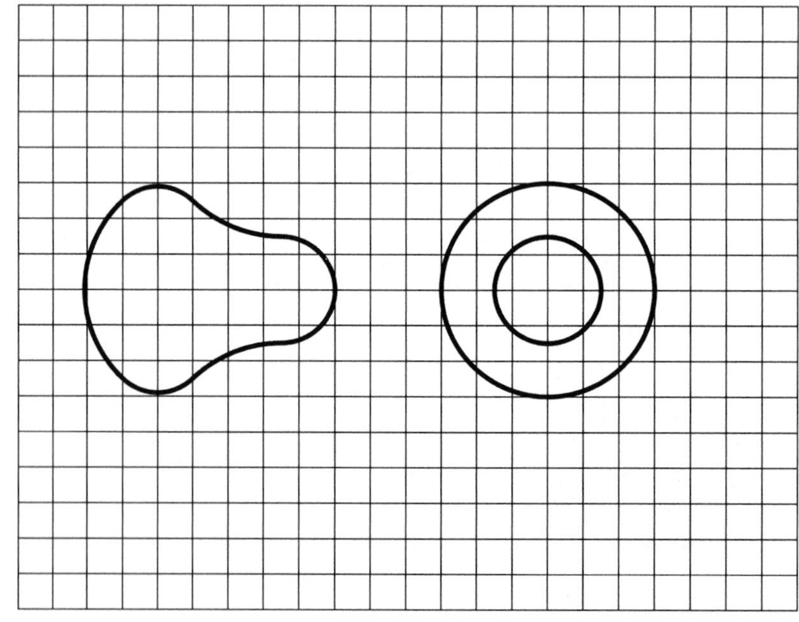

## EXERCISE 2–15

| Type of shape | Input methods |
|---|---|
| Multiple line/arc sequences | • Absolute rectangular coordinates<br>• Relative rectangular coordinates<br>• Relative polar coordinates |

- This exercise is to be done without line-arc or arc-line continuation. Later, you will learn to use the command that draws Polylines; its options facilitate polyline-polyarc-polyline continuation.
- Some architectural illusionism is introduced.

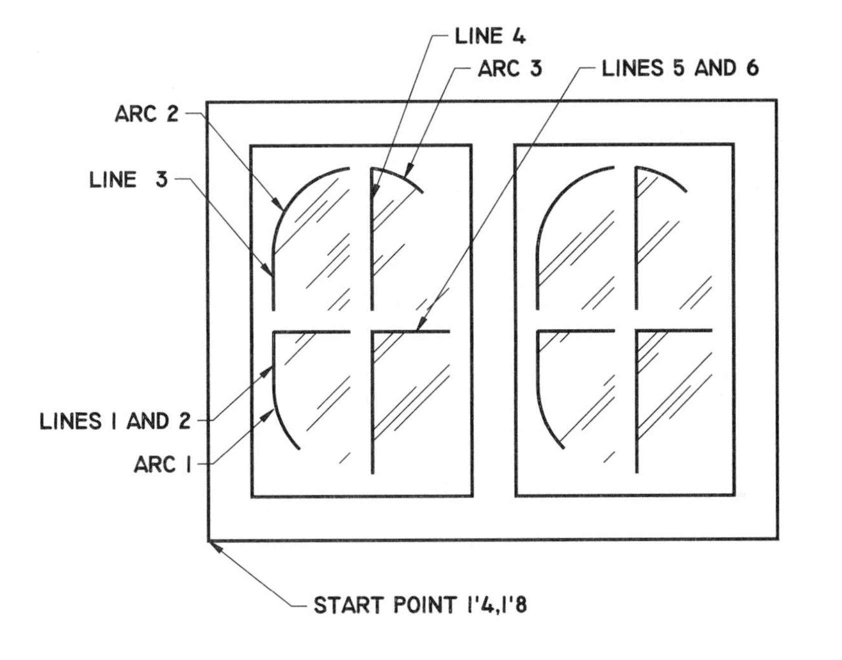


| Drawing Setup | Open the drawing created in Exercise 2–2. |
|---|---|

### Point by Point

Invoke the LINE command, and AutoCAD prompts:

Command: **line** (ENTER)
Specify first point: **2'5,3'3** (ENTER) *(starting point of line 1)*
Specify next point or [Undo]: **1'10,3'3** (ENTER) *(end point of the line 1)*
Specify next point or [Undo]: **1'10,2'10** (ENTER) (end point of the line 2)
Specify next point or [Close/Undo]: (ENTER) (exits the LINE command)

Invoke the ARC command, and AutoCAD prompts:

Command: **arc** (ENTER)

Specify start point of arc or [CEnter]: **1'10,2'10** (ENTER) *(starting point of the arc 1)*
Specify second point of arc or [CEnter/ENd]: **c** (ENTER)
Specify Center point of arc: **2'6,2'10** (ENTER) *(arc center for the arc 1)*
Specify end point of arc or [Angle/chord Length]: **a** (ENTER)
Specify Included angle: **45** (ENTER) *(arc included angle for the arc 1)*

Invoke the LINE command, and AutoCAD prompts:

Command: **line** (ENTER)
Specify first point: **1'10,3'5** (ENTER) *(starting point of line 3)*
Specify next point or [Undo]: **1'10,3'10** (ENTER) *(end point of the line 3)*
Specify next point or [Close/Undo]: (ENTER) *(exits the LINE command)*

Invoke the ARC command, and AutoCAD prompts:

Command: **arc** (ENTER)
Specify start point of arc or [CEnter]: **1'10,3'10** (ENTER) *(starting point of the arc 2)*
Specify second point of arc or [CEnter/ENd]: **c** (ENTER)
Specify Center point of arc: **2'6,3'10** (ENTER) *(arc center for the arc 2)*
Specify end point of arc or [Angle/chord Length]: **a** (ENTER)
Specify Included angle: **–82.82** (ENTER) *(arc included angle for the arc 2)*

Invoke the LINE command, and AutoCAD prompts:

Command: **line** (ENTER)
Specify first point: **2'7,3'5** (ENTER) *(starting point of line 4)*
Specify next point or [Undo]: **2'7,4'6** (ENTER) *(end point of the line 4)*
Specify next point or [Undo]: (enter) *(exits the* line *command)*

Invoke the ARC command, and AutoCAD prompts:

Command: **arc** (ENTER)
Specify start point of arc or [CEnter]: **2'7,4'6** (ENTER) *(starting point of the arc 3)*
Specify second point of arc or [CEnter/ENd]: **c** (ENTER)
Specify Center point of arc: **2'6,3'10** (ENTER) *(arc center for the arc 21)*
Specify end point of arc or [Angle/chord Length]: **a** (ENTER)
Specify Included angle: **–45** (ENTER) *(arc included angle for the arc 2)*

Invoke the LINE command, and AutoCAD prompts:

Command: **line** (ENTER)
Specify first point: **2'7,2'2** (ENTER) *(starting point of line 5)*
Specify next point or [Undo]: **2'7,3'3** (ENTER) *(end point of the line 5)*
Specify next point or [Undo]: **3'2,3'3** (ENTER) *(end point of the line 6)*
Specify next point or [Close/Undo]: (ENTER) *(exits the LINE command)*

(Second window: repeat the preceding points, adding 2'–0" to each X coordinate; slashed (glass) lines can be drawn freehand with the LINE command.)

## EXERCISE 2–16

Create the accompanying drawing of the reflected ceiling plan with light fixtures according to the settings given in the following table (Do not add dimension):

| Settings | Value |
|---|---|
| 1. Units<br>2. Limits | Architectural<br>Lower left corner: 0,0<br>Upper right corner: 36',24' |

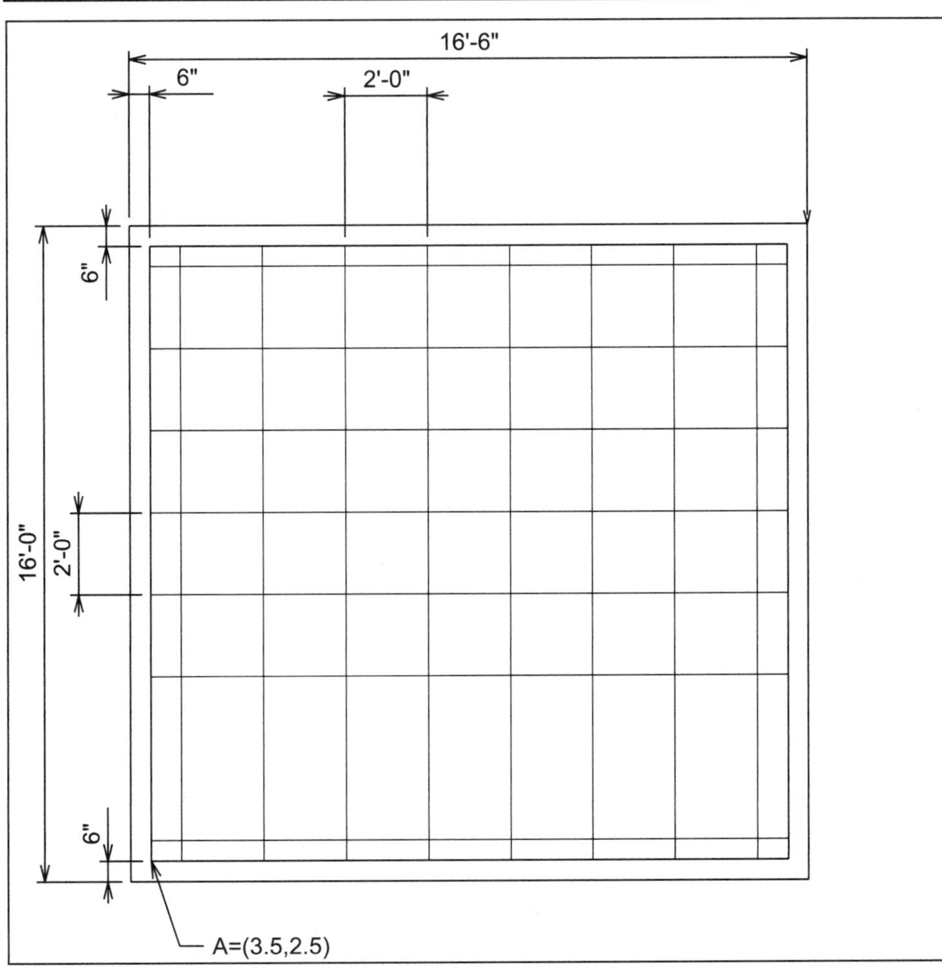


**Figure Ex2–16** *Reflected ceiling plan with light fixtures*

## EXERCISE 2–17

Create the accompanying drawing of the stair riser tread profile according to the settings given in the following table (Do not add dimension):

| Settings | Value |
|---|---|
| 1. Units<br>2. Limits | Decimal<br>Lower left corner: 0,0<br>Upper right corner: 48,36 |

**Hint:** Coordinates for a few points are given in the figure. Coordinates values for missing points can be calculated from the given dimensions.

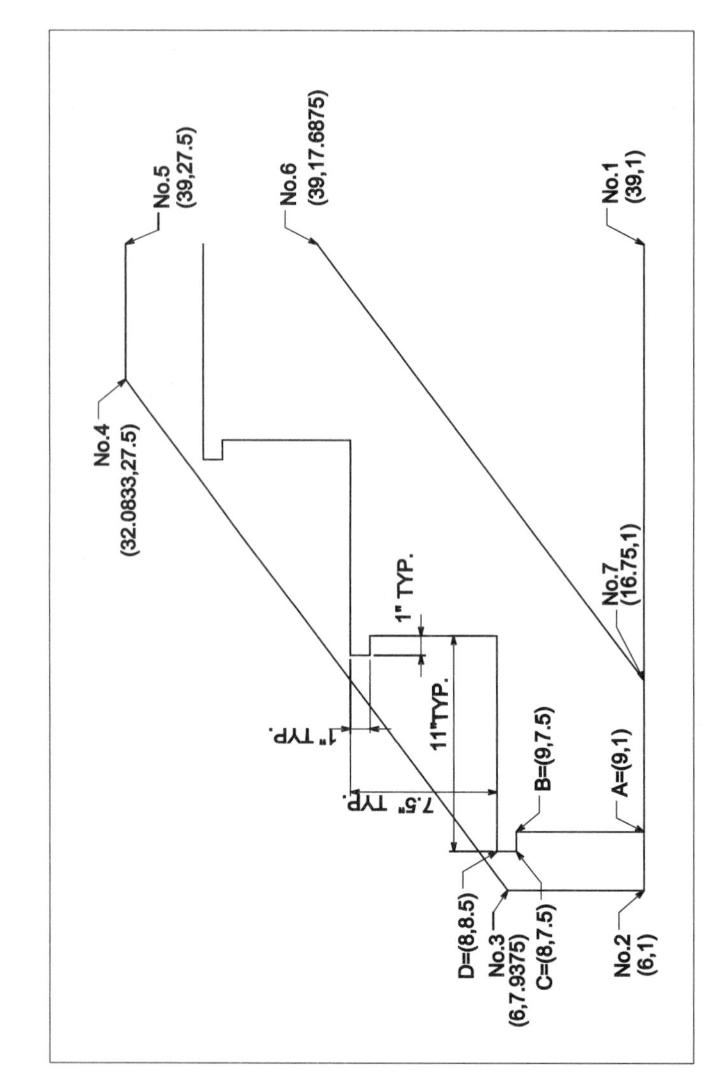


**Figure Ex2–17** *Stair riser tread profile*

## EXERCISE 2–18

Create the accompanying drawing of a stair in plan view according to the settings given in the following table (Do not add dimension):

| Settings | Value |
|---|---|
| 1. Units<br>2. Limits | Architectural<br>Lower left corner: 0,0<br>Upper right corner: 22',17' |

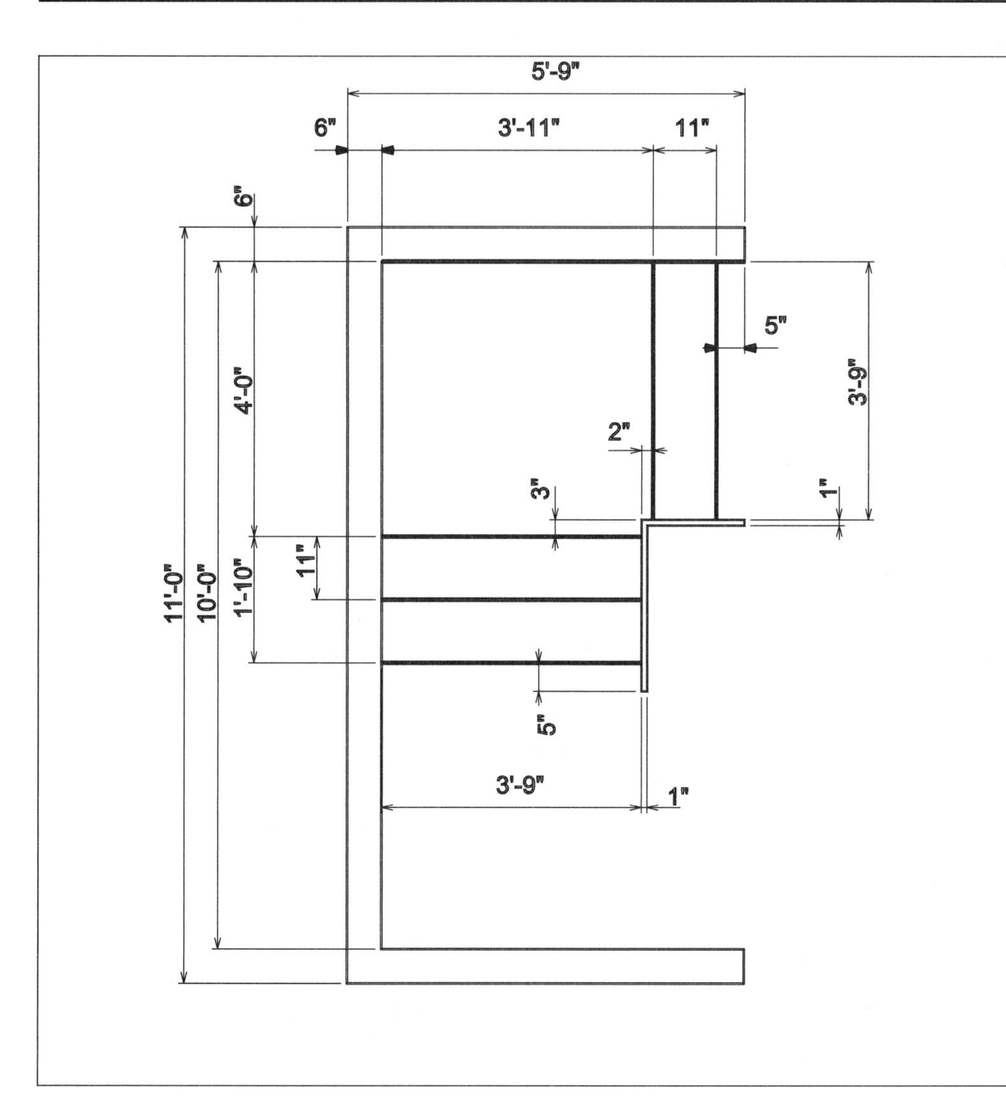


**Figure Ex2–18** *Stair in plan view*

## EXERCISE 2–19

Create the accompanying drawing of a masonry wall and footing section view according to the settings given in the following table (Do not add dimension):

| Settings | Value |
|---|---|
| 1. Units<br>2. Limits | Architectural<br>Lower left corner: 0,0<br>Upper right corner: 12',9' |

**Hint:** Coordinates for a few points are given in the figure. Use the given coordinates and dimensions in developing the values for the remaining points.

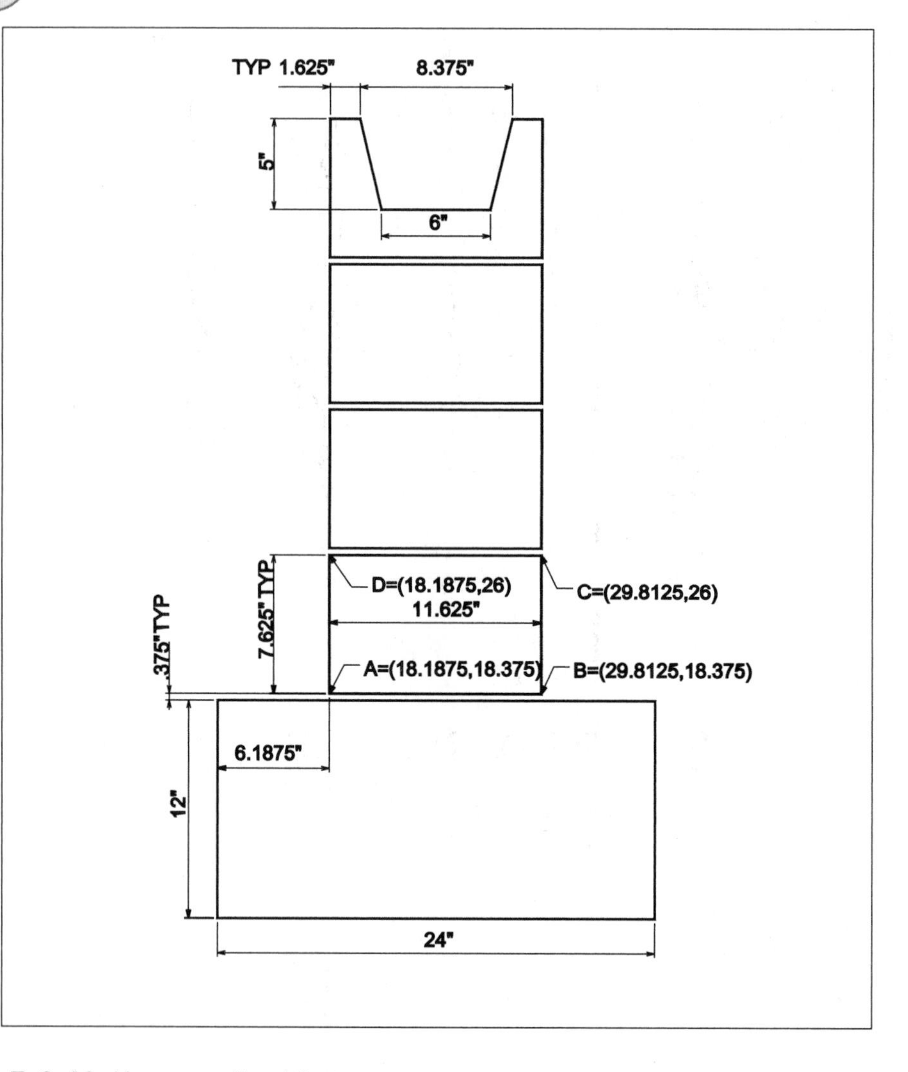

**Figure Ex2–19** *Masonry wall and footing section view*

## EXERCISE 2–20

Create the accompanying drawing of a Cul-de-Sac according to the settings given in the following table (Do not add dimension):

| Settings | Value |
|---|---|
| 1. Units<br>2. Limits | Architectural<br>Lower left corner: 0,0<br>Upper right corner: 200', 220' |

**Hint:** Refer to the table below for distances and angles for the various points referred in the figure.

| Variable | Distances | Angles |
|---|---|---|
| A | 22.873' | - |
| B | 83.036' | - |
| C | - | 35° |
| D | - | 35° |
| RE | 100.000' | - |
| RF | 50.000' | - |

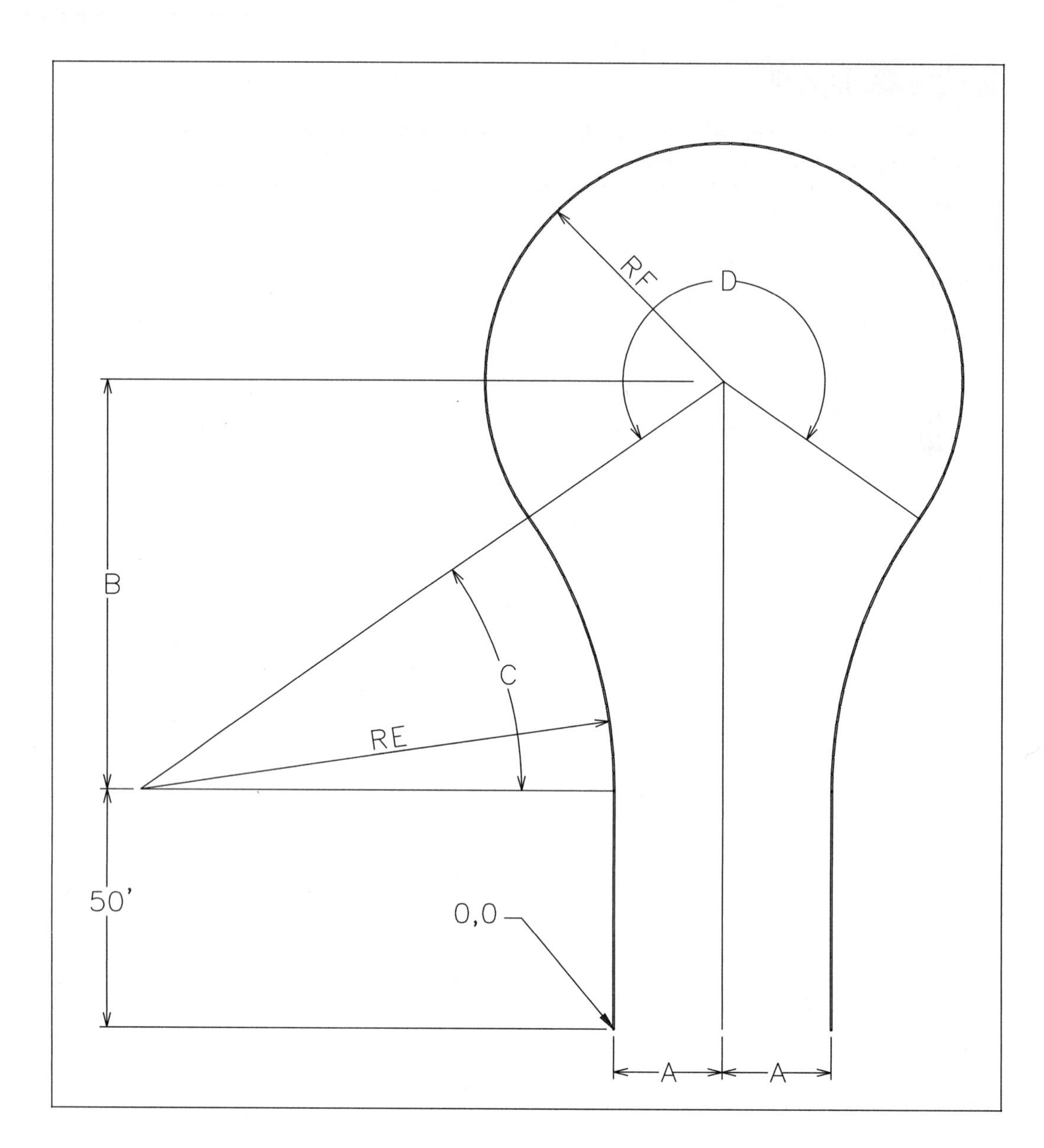


**Figure Ex2–20** *Cul-de-Sac*

## EXERCISE 2–21

Create the accompanying drawing of a concrete slopped retaining wall (section view) according to the settings given in the following table (Do not add dimension):

| Settings | Value |
|---|---|
| 1. Units<br>2. Limits | Architectural<br>Lower left corner: 0,0<br>Upper right corner: 12',9' |

**Hint:** Refer to the table below for coordinates for the various points referred in the figure. Calculate the missing coordinates from the given coordinates and dimensions.

| Letter | Coordinate |
|---|---|
| A | (6.0', .5') |
| B | (8.0', .5') |
| H | (2.0124',5.6667') |
| I | (5.5291',2.1167') |

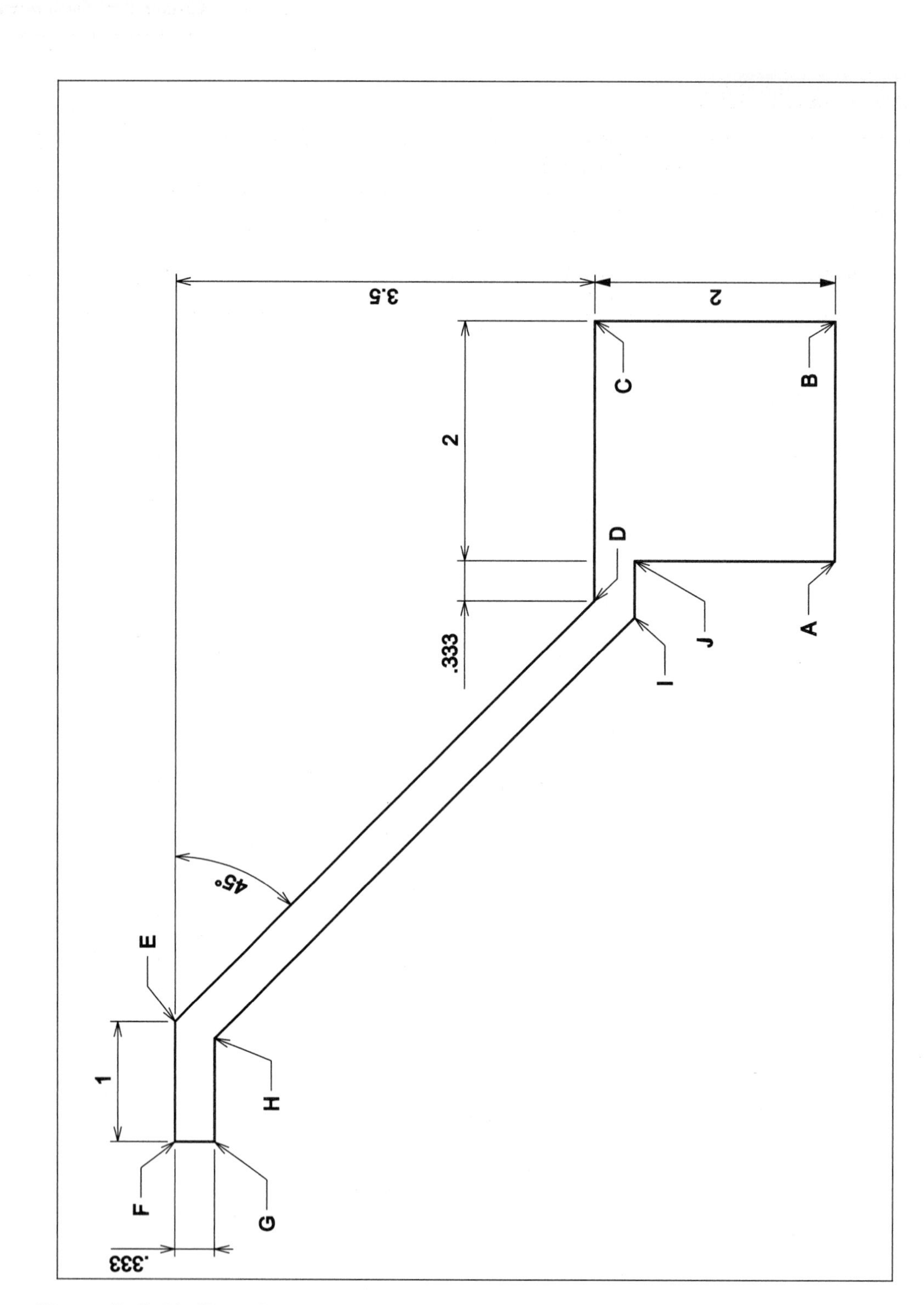


**Figure Ex2–21** *Slopped retaining wall*

## EXERCISE 2–22

Create the accompanying drawing of a section through trench drain according to the settings given in the following table (Do not add dimension):

| Settings | Value |
|---|---|
| 1. Units<br>2. Limits | Architectural<br>Lower left corner: 0,0<br>Upper right corner: 12',9' |

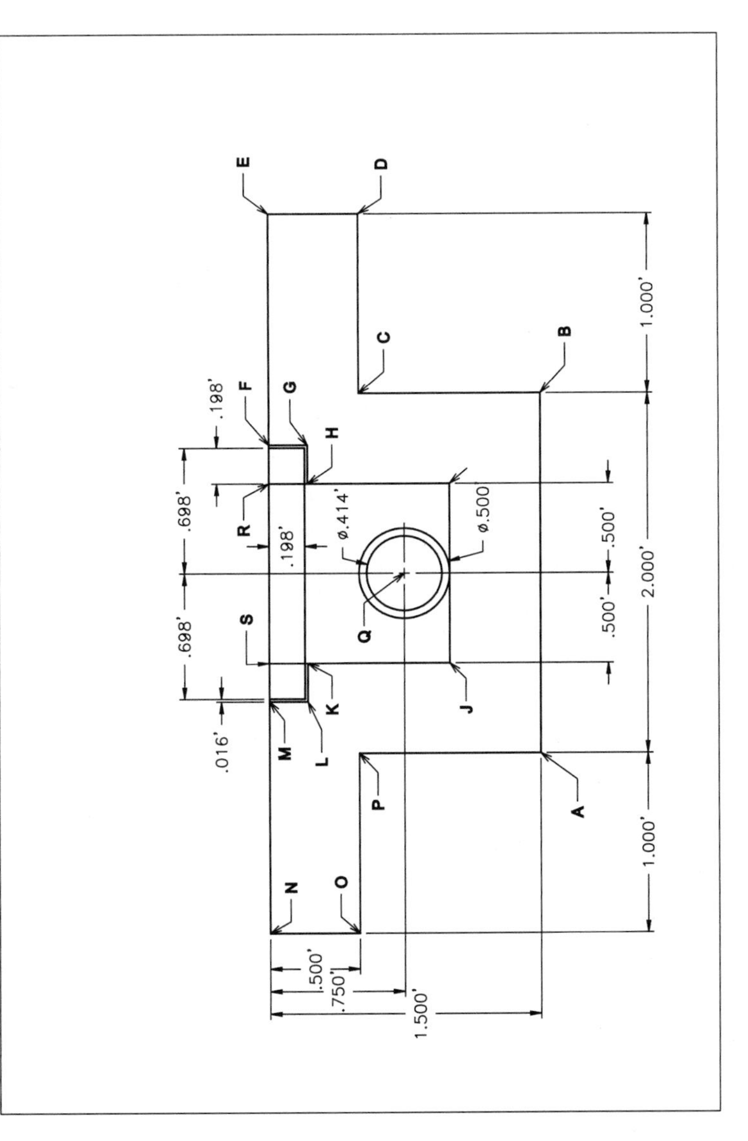


**Figure Ex2–22** *Trench drain*

## EXERCISE 2–23

Create the accompanying drawing of a concrete encased column according to the settings given in the following table (Do not add dimension):

| Settings | Value |
|---|---|
| 1. Units<br>2. Limits | Architectural<br>Lower left corner: 0,0<br>Upper right corner: 3',3' |

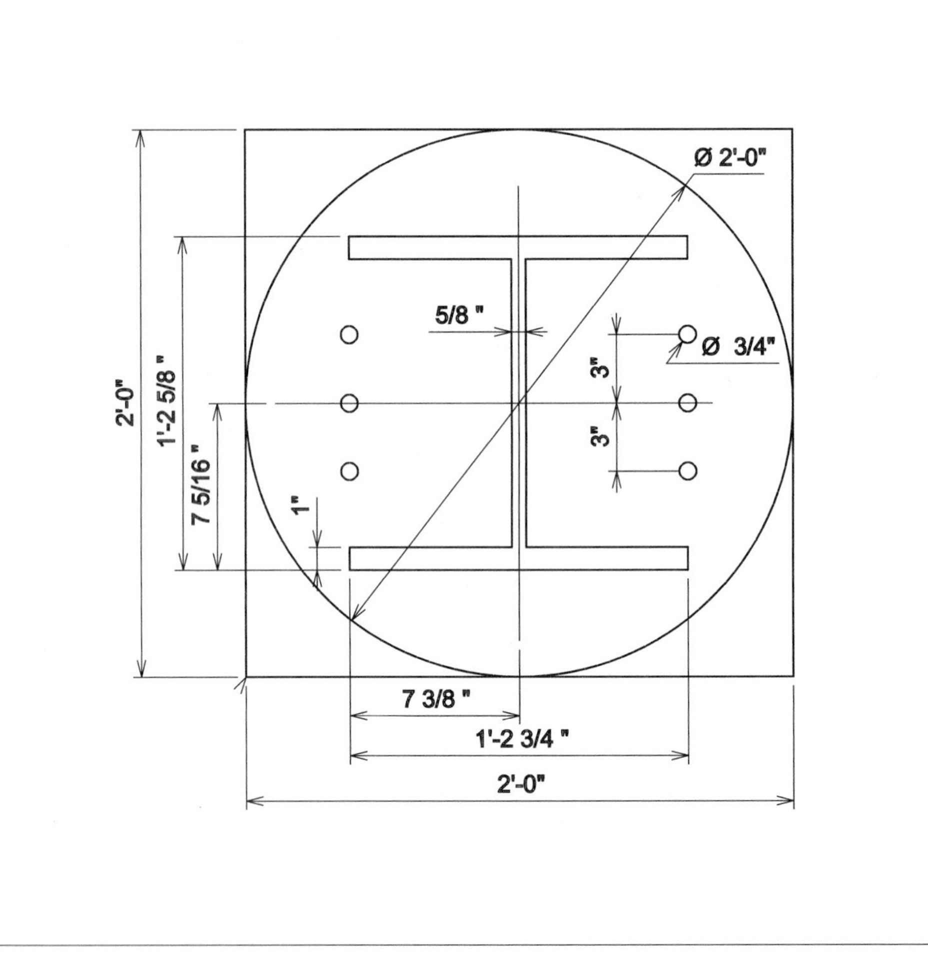


**Figure Ex2–23** *Concrete Encased Column*

## EXERCISE 2–24

Create the accompanying drawing of an incandescent light fixture according to the settings given in the following table (Do not add dimension):

| Settings | Value |
|---|---|
| 1. Units<br>2. Limits | Decimals<br>Lower left corner: -.25,-.25<br>Upper right corner: 1.75,1.75 |

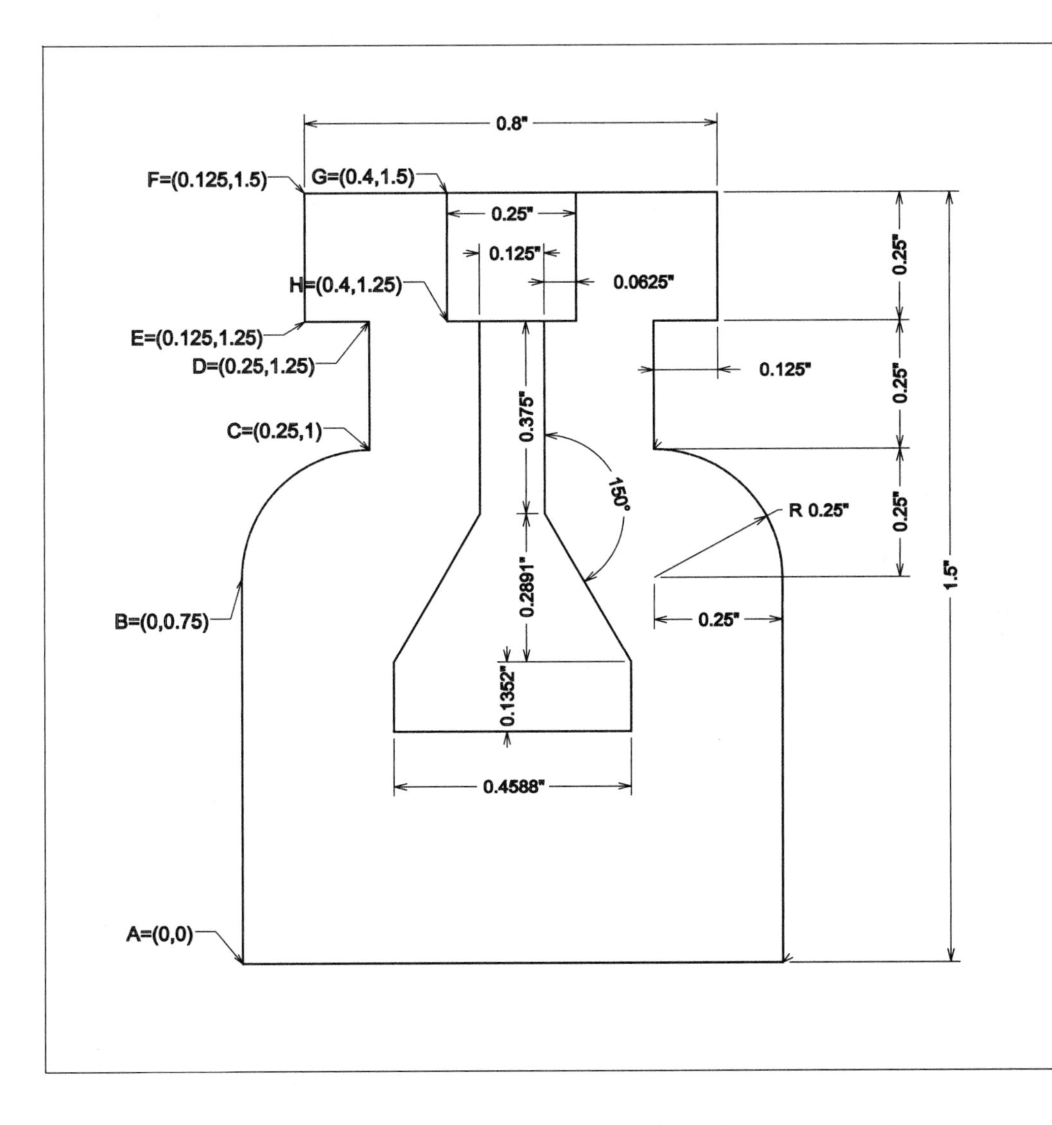


**Figure Ex2–24** *Incandescent light fixture*

## EXERCISE 2–25

Create the accompanying fluorescent light fixture drawing according to the settings given in the following table (Do not add dimension):

| Settings | Value |
|---|---|
| 1. Units | Decimal |
| 2. Limits | Lower left corner: 0,0<br>Upper right corner: 32,18 |

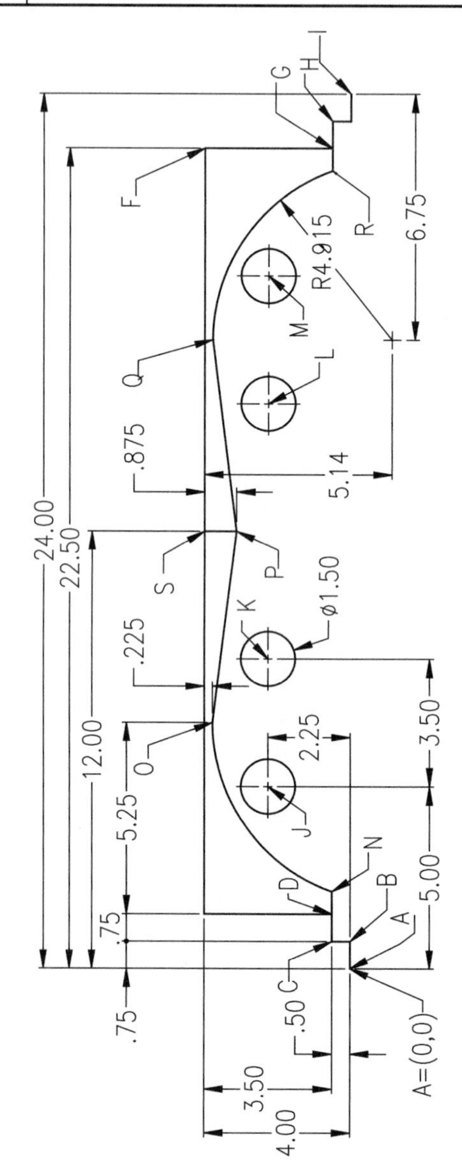

**Figure Ex2–25** *Fluorescent light fixture*

## EXERCISE 2–26

Create the accompanying drawing of a circuit breaker box (elevation view) according to the settings given in the following table (Do not add dimension):

| Settings | Value |
|---|---|
| 1. Units<br>2. Limits | Architectural<br>Lower left corner: 0,0<br>Upper right corner: 2.5', 3' |

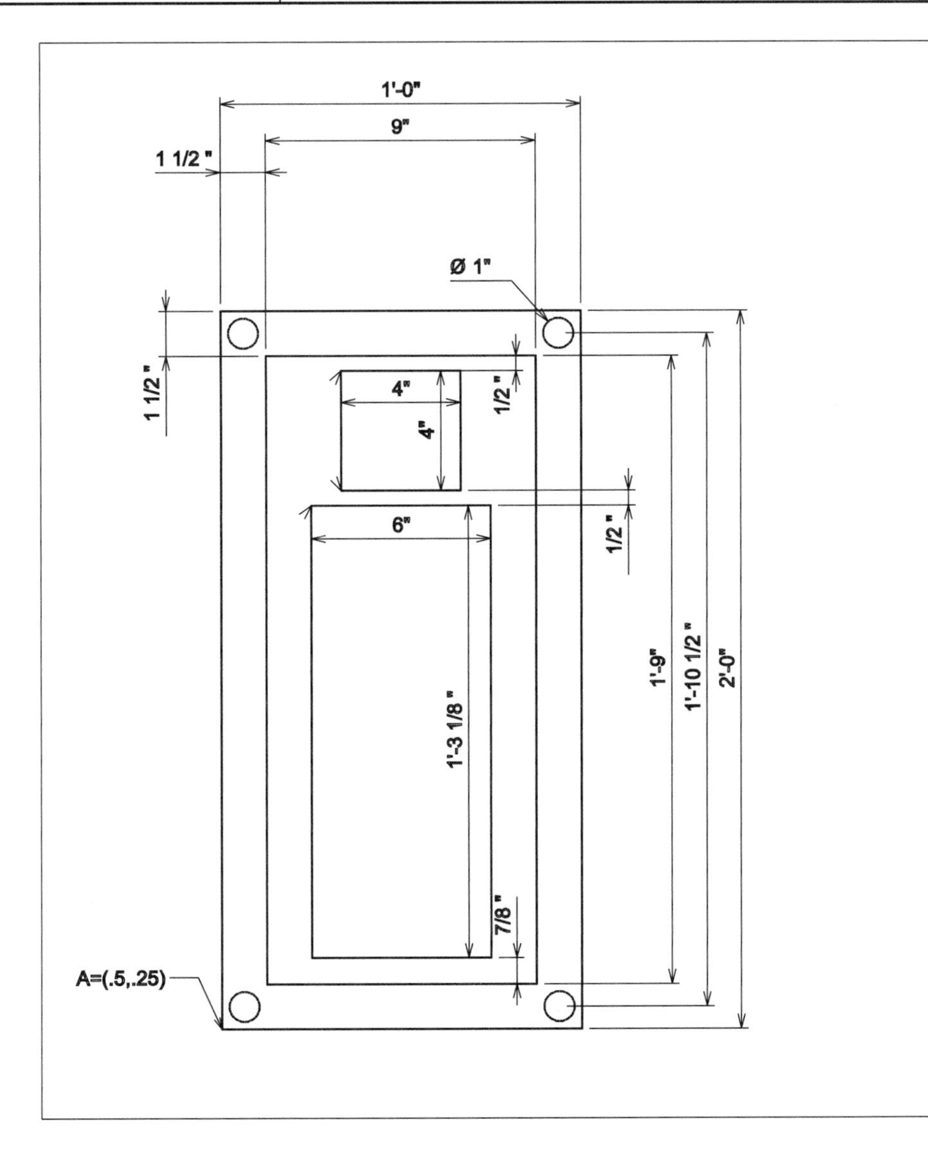


**Figure Ex2–26** *Circuit breaker box*

## EXERCISE 2–27

Create the accompanying drawing of a switch-wiring diagram according to the settings given in the following table:

| Settings | Value |
|---|---|
| 1. Units<br>2. Limits | Decimals<br>Lower left corner: 0,0<br>Upper right corner: 5,2.5 |

**Hint:** The diameter of the small circles is 0.0625, the diameter of the large circle is 0.3162.

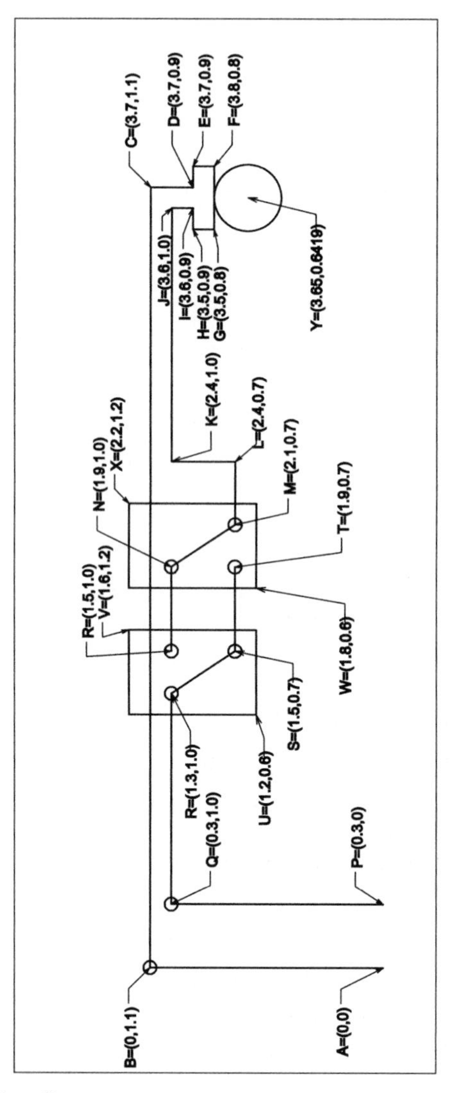

**Figure Ex2–27** *Switch-wiring diagram*

CHAPTER 3

# Fundamentals II



## PROJECT EXERCISE

This project exercise provides point-by-point instructions for setting up the drawing with layers and then creating the objects shown in Figure P3–1. In this exercise you will apply the skills acquired in Chapters 1, 2, and 3.

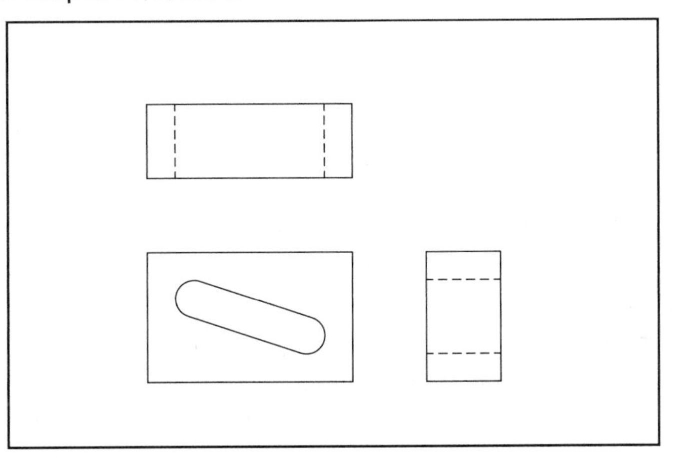

**Figure P3–1** *Completed project drawing*

In this project you will learn to do the following:

- Set up the drawing, including setting up the layers.
- Use the LINE and RECTANGLE commands with Object Snap modes

### SET UP THE DRAWING AND DRAW A BORDER

**Step 1:** Start the AutoCAD program.

**Step 2:** Set the system variable STARTUP to 1, as shown below, to allow AutoCAD to create a new drawing using a wizard.

Command: **startup** (ENTER)
Enter a new value for STARTUP <default>: 1 (ENTER)

To create a new drawing, invoke the NEW command from the Standard toolbar or select New from the File menu.

AutoCAD displays the Create New Drawing dialog box. Select the **Use a Wizard** button and AutoCAD lists available wizards. Select Quick Setup wizard from the Select a Wizard list box and choose the **OK** button.

AutoCAD displays the Quick Setup dialog box with the listing of the options available for unit of measurement. Select the "Decimal" radio button, and choose the **Next>** button.

AutoCAD displays the Quick Setup dialog box with the display for Area setting. Under the **Width:** text box enter **18**, and under the **Length:** text box enter **12**. Choose the **Finish** button to close the Quick Step dialog box. Invoke the ZOOM ALL command from the View menu.

**Step 3:** Invoke the LAYER command from the Layers toolbar, or choose Layer from the Format menu. AutoCAD displays the Layer Properties Manager dialog box.

Create four layers, rename them as shown in the table, and assign appropriate colors and linetypes.

| Layer Name | Color | Linetype | Lineweight |
|---|---|---|---|
| Border | Cyan | Continuous | Default |
| Object | Green | Continuous | Default |
| Hidden | Blue | Hidden | Default |
| Const | Red | Continuous | Default |

Set Border as the current layer, and close the Layer Properties Manager dialog box.

**Step 4:** To draw the border, invoke the RECTANGLE command from the Draw toolbar or enter **rectangle** at the Command: prompt and press (ENTER). AutoCAD prompts:

Command: **rectangle** (ENTER)
Specify first corner point or [Chamfer/Elevation/Fillet/Thickness/Width]: **0.25,0.25** (ENTER)
Specify other corner point or [Dimensions]: **17.75,11.75** (ENTER)

AutoCAD draws a rectangle border.

## DRAW THE OBJECTS

**Step 1:** From the Object Properties toolbar, choose the Layer Control down-arrow icon to display the layer list box. Select the Object layer as the current layer.

**Step 2:** Invoke the RECTANGLE command from the Draw toolbar to draw a rectangle.

Command: **rectangle**
Specify first corner point or [Chamfer/Elevation/Fillet/Thickness/Width]: **2,7.5**
Specify other corner point or [Dimensions]: **7.5,9.5**

Invoke the RECTANGLE command again from the Draw toolbar to draw another rectangle.

Command: **rectangle**
Specify first corner point or [Chamfer/Elevation/Fillet/Thickness/Width]:
**9.5,2**
Specify other corner point or [Dimensions]: **11.5,5.5**

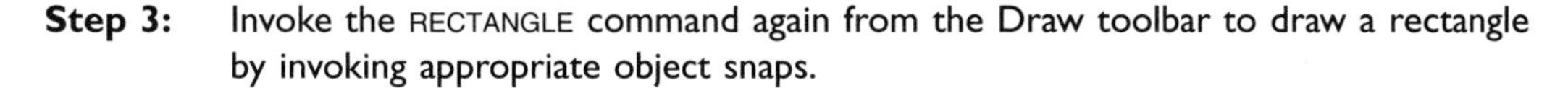

**Step 3:** Invoke the RECTANGLE command again from the Draw toolbar to draw a rectangle by invoking appropriate object snaps.

Command: **rectangle**
Specify first corner point or [Chamfer/Elevation/Fillet/Thickness/Width]:
*(invoke the APPint object snap, and select line 1 and line 4, as shown in Figure P3–2)*
Specify other corner point or [Dimensions]: *(invoke the App Int Object Snap mode, and select line 2 and line 3, as shown in Figure P3–2)*

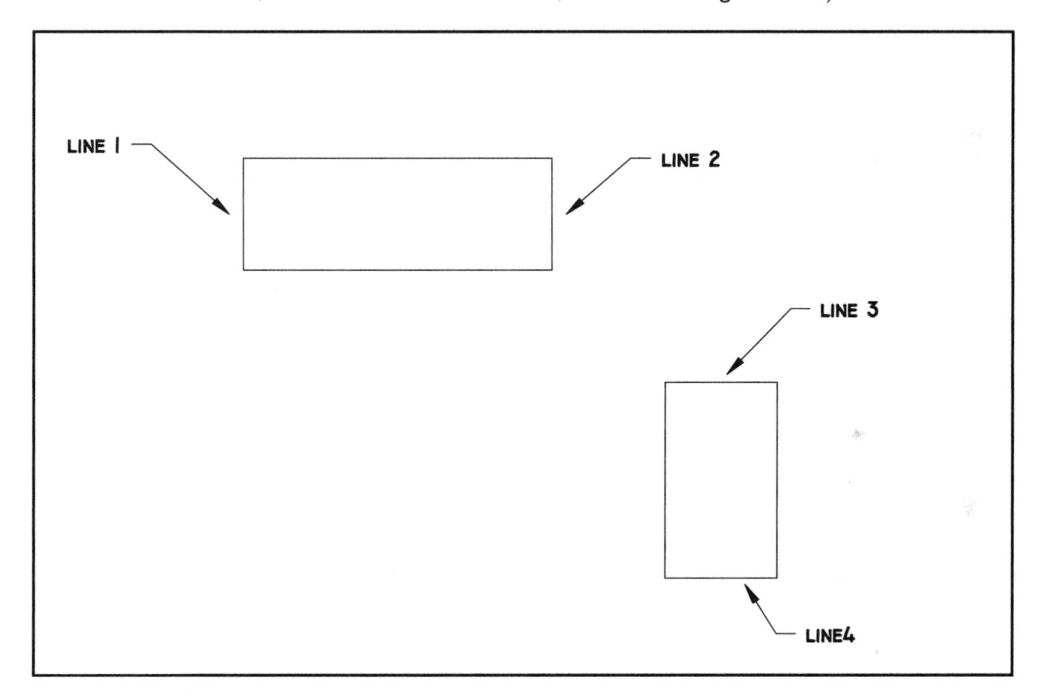


**Figure P3–2** *Identifying the lines to draw a rectangle*

After drawing the rectangle, the drawing should look like Figure P3–3.

**Step 4:** From the Layers toolbar, choose the Layer Control down-arrow icon to display the layer list box. Select the Const layer as the current layer.

**Step 5:** Invoke the CIRCLE command from the Draw toolbar and draw two circles.

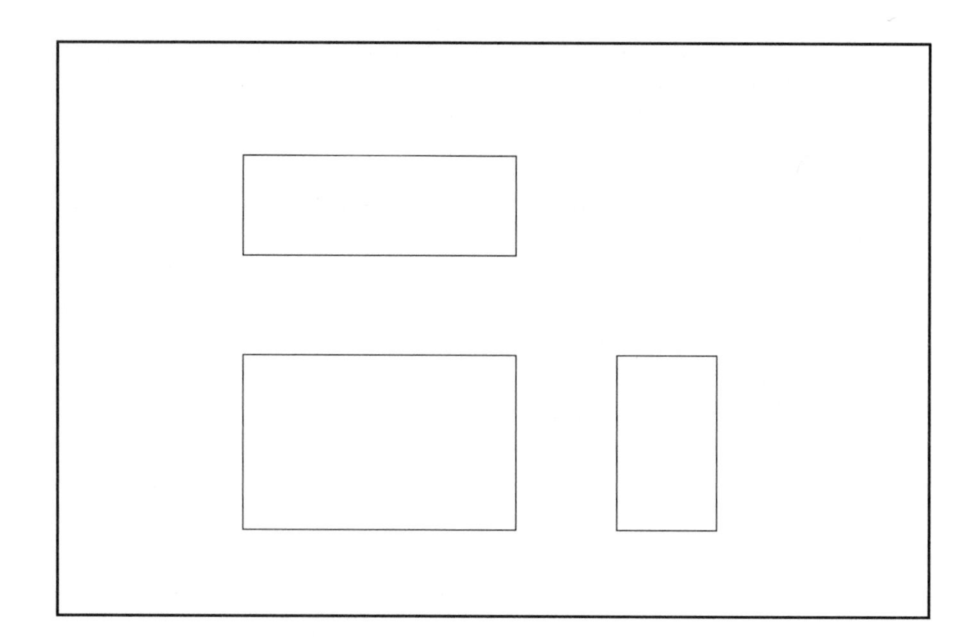

**Figure P3–3** *Completed drawing of rectangles*

Respond to the prompts in the command area as follows:

Command: **circle**
Specify center point for circle or [3P/2P/Ttr (tan tan radius)]: **3.25,4.25** (ENTER)
Specify radius of circle or [Diameter]: **.5** (ENTER)

Command: (ENTER)
Specify center point for circle or [3P/2P/Ttr (tan tan radius)]: **6.25,3.25** (ENTER)
Specify radius of circle or [Diameter]: **.5** (ENTER)

**Step 6:** Invoke the LINE command from the Draw toolbar and draw series of lines by invoking appropriate object snaps.

Command: **line**
Specify first point: *(invoke the QUAdrant object snap, and select a point QUA 1, as shown in Figure P3–4)*
Specify next point or [Undo]: *(invoke the PERpendicular object snap, and select line 5, as shown in Figure P3–4)*
Specify next point or [Undo]: (ENTER)

Command: (ENTER)
Specify first point: *(invoke the QUAdrant object snap, and select a point QUA 2, as shown in Figure P3–4)*

Specify next point or [Undo]: *(invoke the PERpendicular object snap, and select line 5, as shown in Figure P3–4)*
Specify next point or [Undo]:(ENTER)

Command: (ENTER)
Specify first point: *(invoke the QUAdrant object snap, and select a point QUA 3, as shown in Figure P3–4)*
Specify next point or [Undo]: *(invoke the PERpendicular object snap, and select line 6, as shown in Figure P3–4)*
Specify next point or [Undo]: (ENTER)

Command: (ENTER)
Specify first point: *(invoke the QUAdrant object snap, and select a point QUA 4, as shown in Figure P3–4)*
Specify next point or [Undo]: *(invoke the PERpendicular object snap, and select line 6, as shown in Figure P3–4)*
Specify next point or [Undo]: (ENTER)

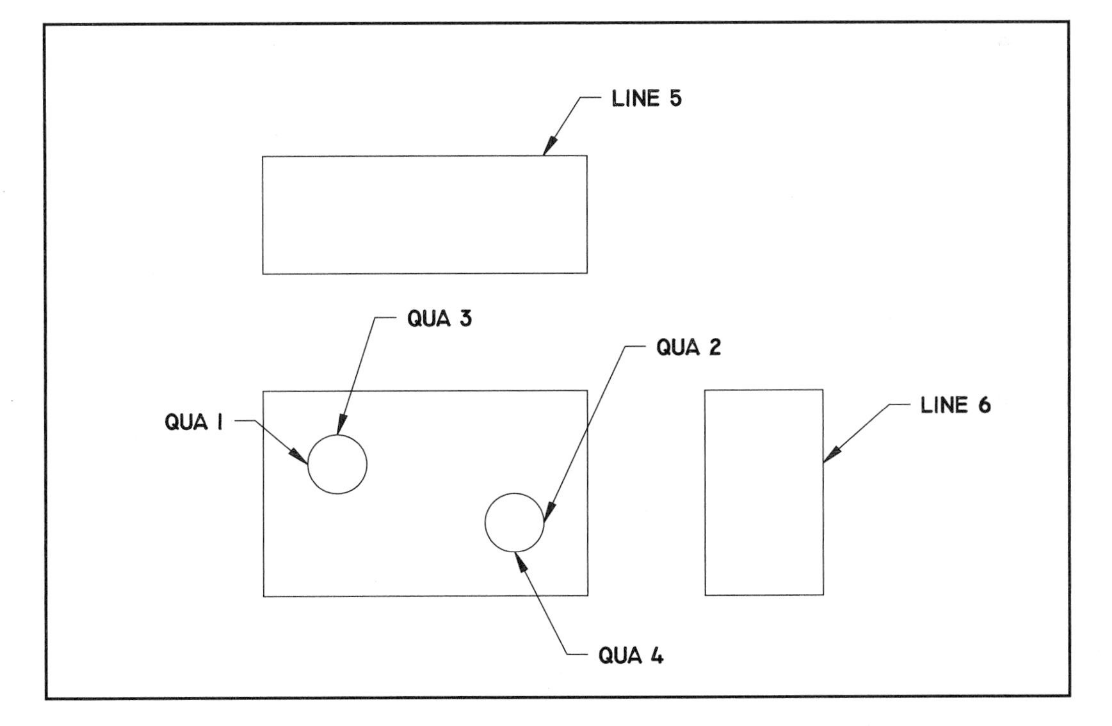


**Figure P3–4** *Identifying the points to draw additional lines*

The drawing should look like Figure P3–5.

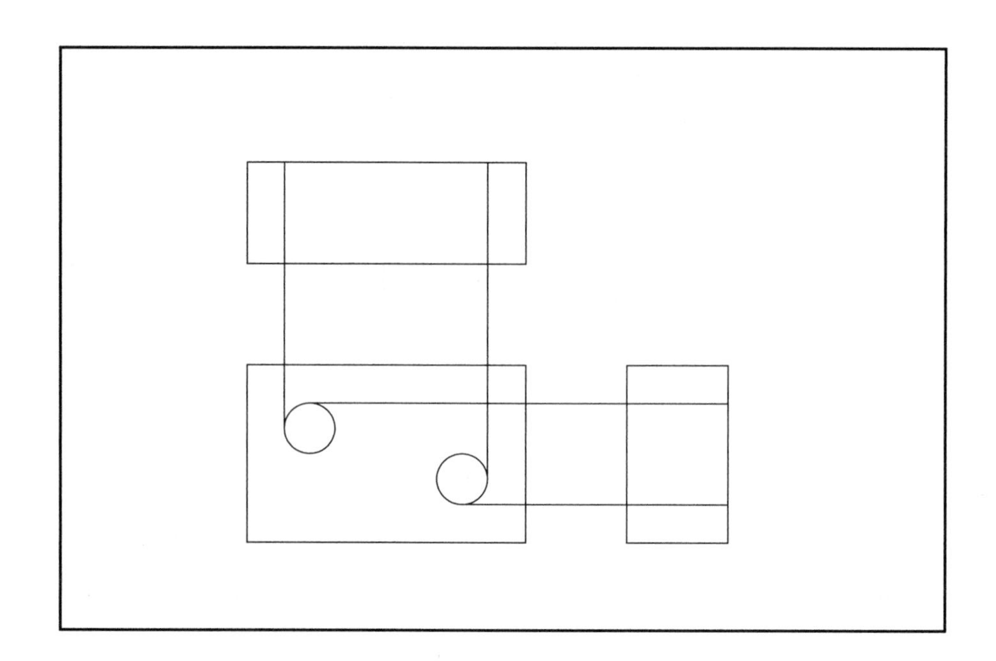

**Figure P3–5** *Completed drawing*

**Step 7:** From the Layers toolbar, choose the Layer Control down-arrow icon to display the layer list box. Select the Hidden layer to make it the current layer.

**Step 8:** Invoke the LINE command from the Draw toolbar. Draw a series of lines by invoking appropriate object snaps.

> Command: **line**
> Specify first point: *(invoke the INTersection object snap, and select point 1, as shown in Figure P3–6)*
> Specify next point or [Undo]: *(invoke the INTersection object snap, and select point 2, as shown in Figure P3–6)*
> Specify next point or [Undo]: (ENTER)

Draw three additional lines:

| From: | To: |
|---|---|
| point 3 | point 4 |
| point 5 | point 6 |
| point 7 | point 8 |

**Step 9:** Invoke the ERASE command from the Modify toolbar, and erase line 7, line 8, line 9, and line 10, as shown in Figure P3–7.

The completed drawing should look like Figure P3–8.

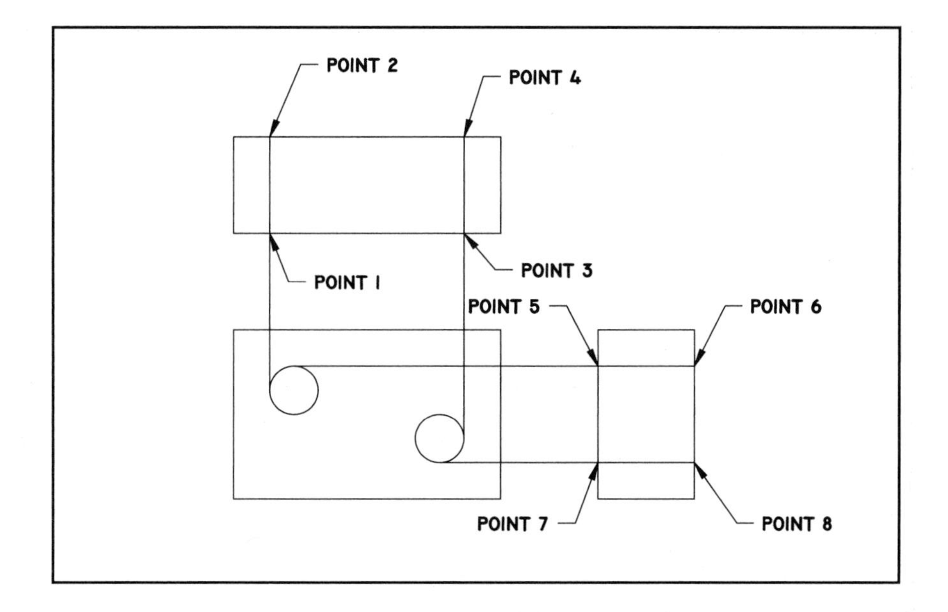


**Figure P3–6** *Identifying the points to draw additional lines*

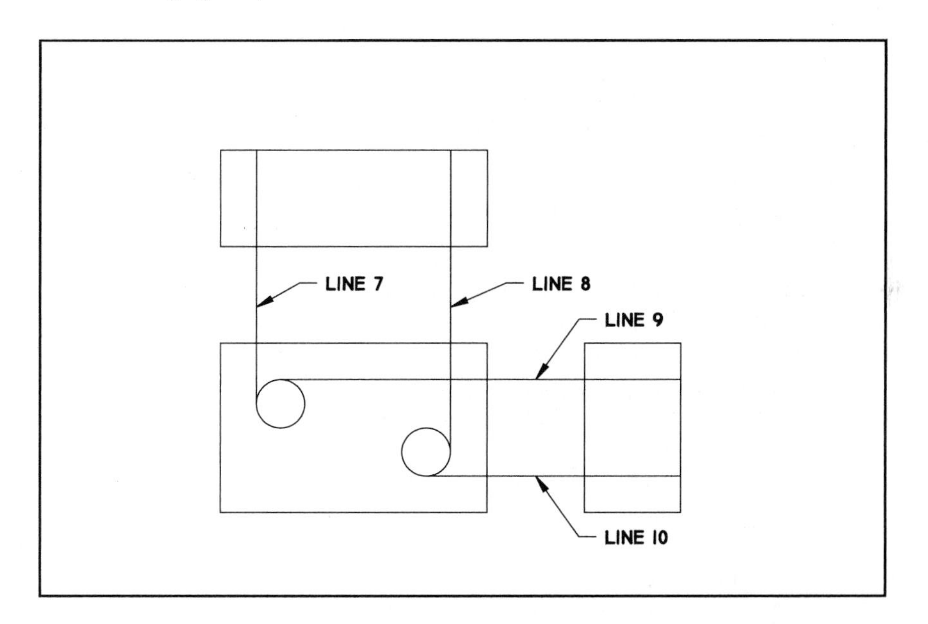


**Figure P3–7** *Identifying the lines to erase*

**Step 10:** From the Layers toolbar, choose the Layer Control down-arrow icon to display the layer list box. Select the Object layer to make it the current layer.

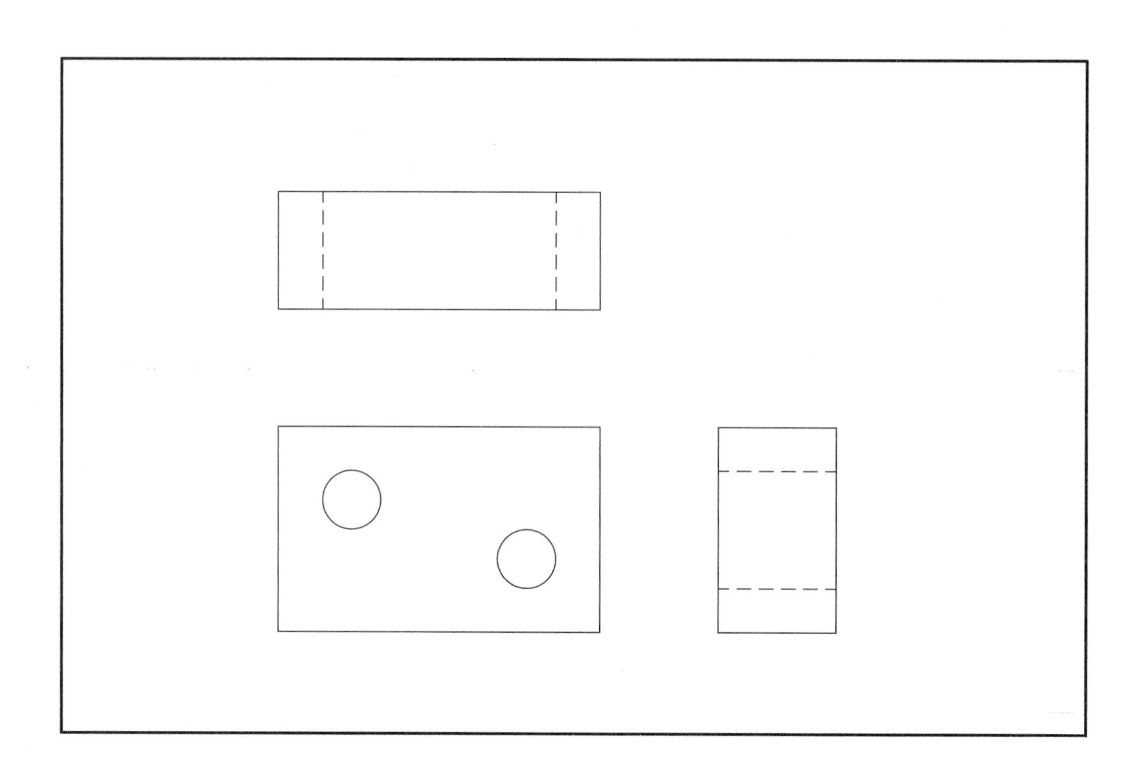

**Figure P3–8** *Completed drawing*

**Step 11:** Invoke the LINE command from the Draw toolbar and draw a series of lines.

Command: **line**
Specify first point: *(invoke the TANgent object snap, and select a point in the upper half of circle 1, as shown in Figure P3–9)*
Specify next point or [Undo]: *(invoke the TANgent object snap, and select a point in the upper half of circle 2, as shown in Figure P3–9)*
Specify next point or [Undo]: (ENTER)

Command: (ENTER)
Specify first point: *(invoke the TANgent object snap, and select a point in the lower half of circle 1, as shown in Figure P3–9)*
Specify next point or [Undo]: *(invoke the TANgent object snap, and select a point in the lower half of circle 2, as shown in Figure P3–9)*
Specify next point or [Undo]: (ENTER)

**Step 12:** Invoke the ARC 3-Points command from the Draw toolbar.

Command: **arc**
Specify start point of arc or [CEnter]: *(invoke the ENDpoint object snap and select endpoint 1. as shown in Figure P3–10)*

Specify second point of arc or [CEnter/ENd]: *(invoke the NEArest object snap and select a point on circle 1, where indicated in Figure P3–10)*
Specify end point of arc: *(invoke the ENDpoint object snap and select endpoint 2, as shown in Figure P3–10)*

Command: (ENTER)
Specify start point of arc or [CEnter]: *(invoke the ENDpoint object snap and select endpoint 3, as shown in Figure P3–10)*
Specify second point of arc or [CEnter/ENd]: *(invoke the NEArest object snap and select a point on circle 2, where indicated in Figure P3–10)*
Specify end point of arc: *(invoke the ENDpoint object snap and select endpoint 4, as shown in Figure P3–10)*

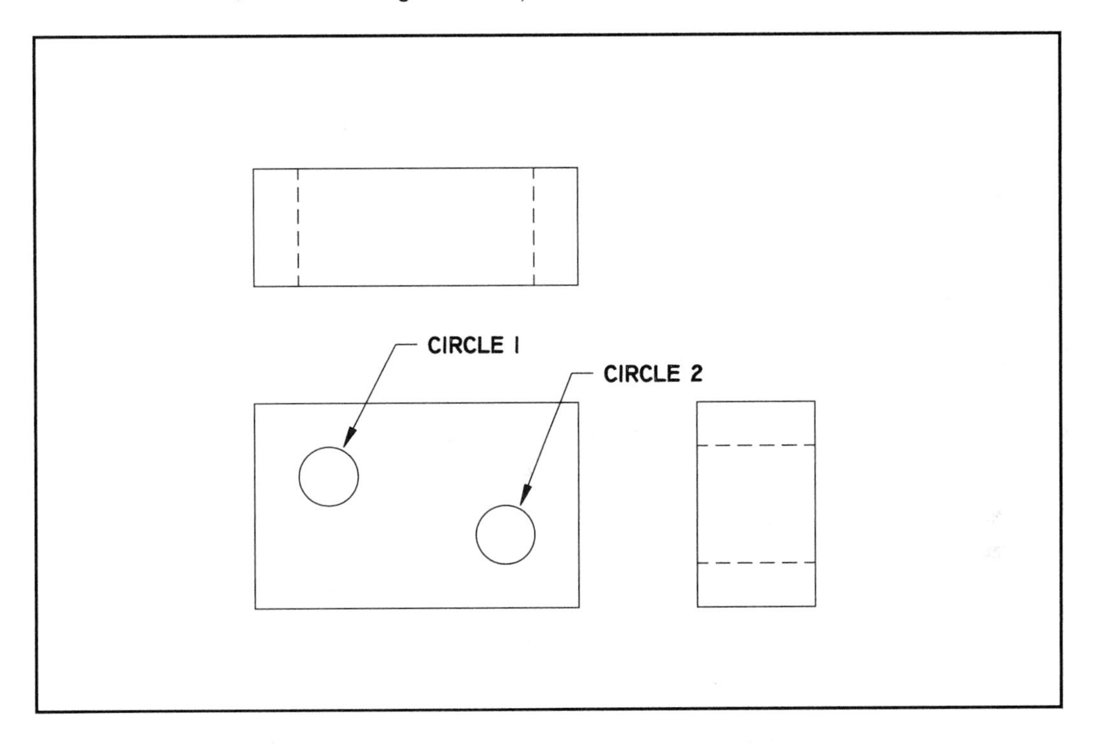


**Figure P3–9** *Identifying the object snap points to draw additional lines*

**Step 13:** From the Layers toolbar, choose the Layer Control down-arrow icon to display the layer list box. Set the Const layer to OFF.

The completed drawing should look like Figure P3–11.

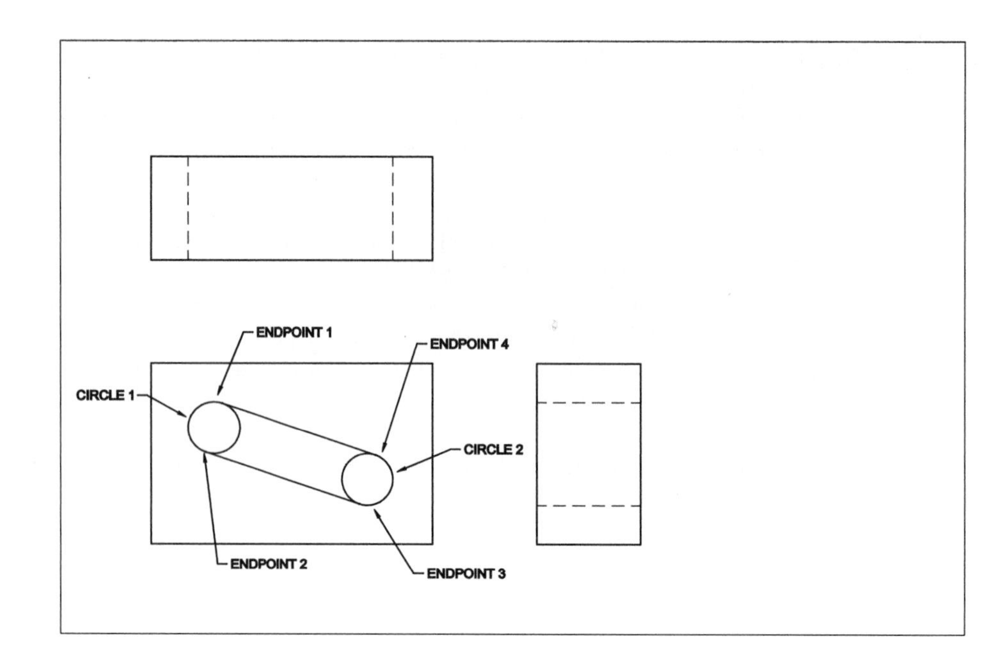


**Figure P3–10** *Identifying the Object Snap points to draw arcs*

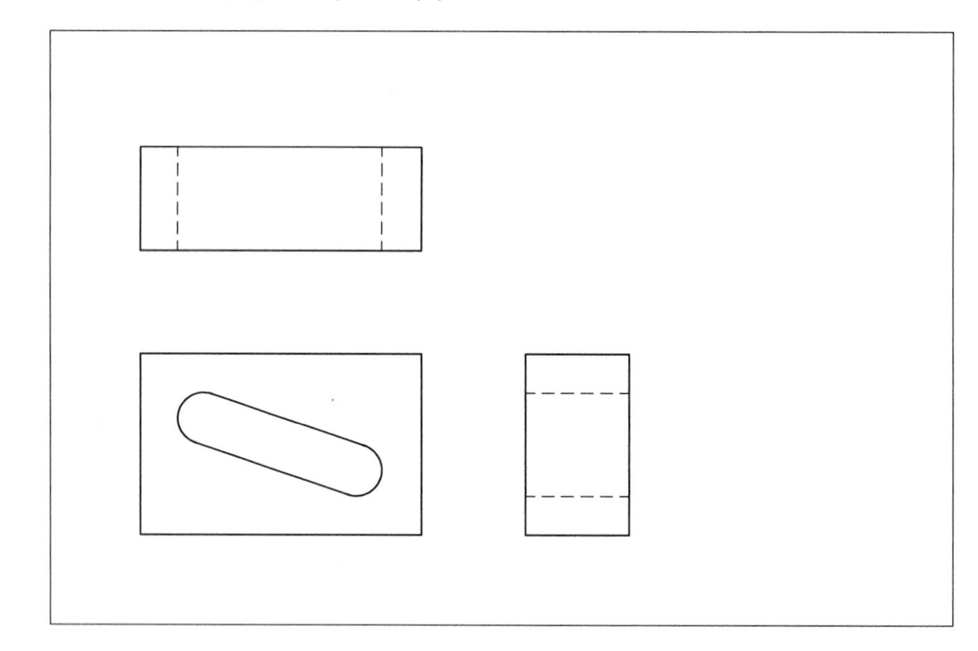

**Figure P3–11** *Completed drawing*

**Step 14:** Invoke the SAVE command from the File menu to save the current drawing as PROJ3.DWG.

## EXERCISE 3–1

Create the drawing shown in Figure Ex3–1 according to the settings given in the following table:

| Settings | Value | | |
|---|---|---|---|
| 1. Units | Decimal with two decimal places | | |
| 2. Limits | Lower left corner: 0,0 | | |
| | Upper right corner: 18,12 | | |
| 3. Grid | 0.25 | | |
| 4. Snap | 0.25 | | |
| 5. Layers | *NAME* | *COLOR* | *LINETYPE* |
| | Solid | Red | Continuous |
| | Hidden | Blue | Hidden |

**Hint:** To draw the border, use the RECTANGLE command with diagonally opposite corners at coordinates 0.25,0.25 and 16.75,11.75

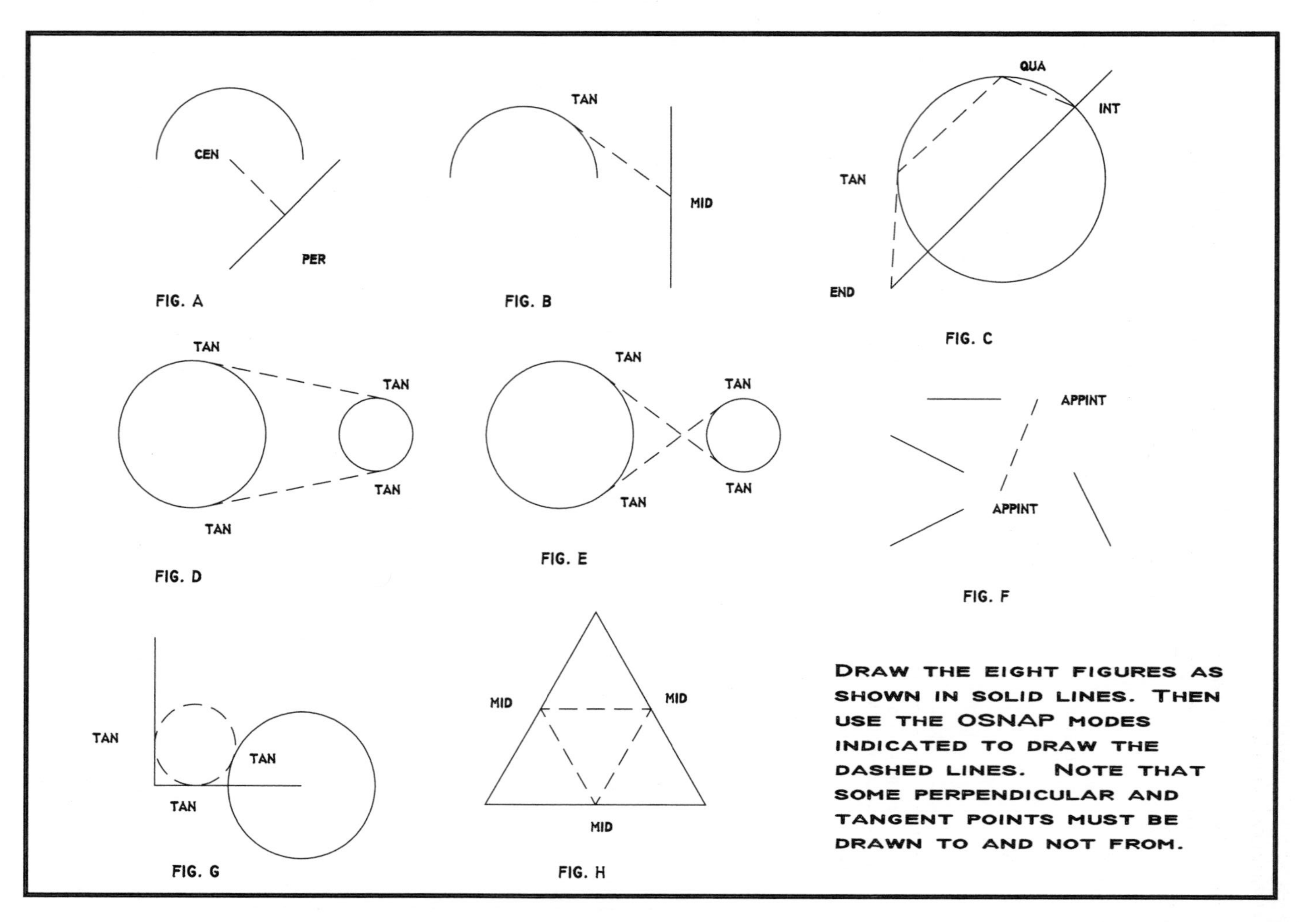

CEN
PER
FIG. A
TAN
MID
FIG. B
QUA
INT
TAN
END
FIG. C
TAN
TAN
TAN
TAN
FIG. D
TAN
TAN
TAN
TAN
FIG. E
APPINT
APPINT
FIG. F
TAN
TAN
TAN
FIG. G
MID
MID
MID
FIG. H
DRAW THE EIGHT FIGURES AS SHOWN IN SOLID LINES. THEN USE THE OSNAP MODES INDICATED TO DRAW THE DASHED LINES. NOTE THAT SOME PERPENDICULAR AND TANGENT POINTS MUST BE DRAWN TO AND NOT FROM.


**Figure Ex3–1**

## EXERCISE 3–2

Create the drawing shown in Figure Ex3–2 according to the settings given in the following table:

| Settings | Value | | |
|---|---|---|---|
| 1. Units | Decimal with two decimal places | | |
| 2. Limits | Lower left corner: 0,0<br>Upper right corner: 17,11 | | |
| 3. Grid | 0.50 | | |
| 4. Snap | 0.25 | | |
| 5. Layers | *NAME* | *COLOR* | *LINETYPE* |
| | Border | Red | Continuous |
| | Object | Green | Continuous |
| | Center | Blue | Center |

**Hints:** Use the LINE and CIRCLE commands to complete the drawing. Make sure to draw the objects in the appropriate layers. Do not dimension the drawing.

A good place to start is the center of the three circles. With the radius of 4 units, the center could be at coordinates 5,5.

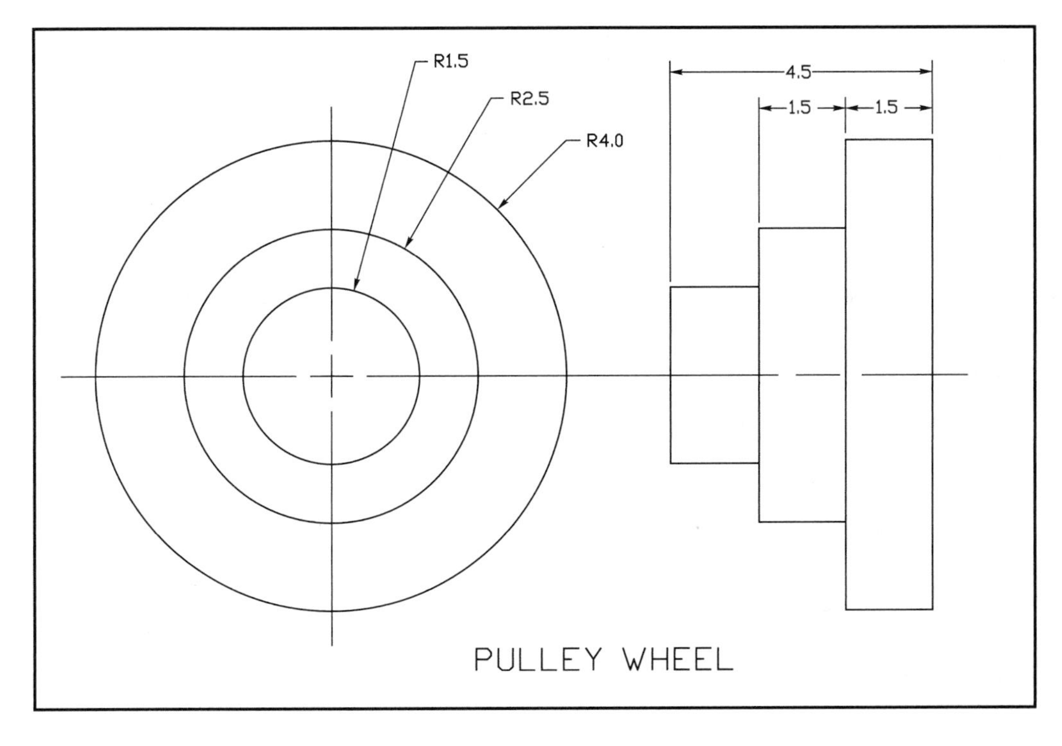

**Figure Ex3–2**

## EXERCISE 3–3

Create the drawing shown in Figure Ex3–3 according to the settings given in the following table:

| Settings | Value | | |
|---|---|---|---|
| 1. Units | Decimal with two decimal places | | |
| 2. Limits | Lower left corner: 0,0<br>Upper right corner: 22,17 | | |
| 3. Grid | 0.50 | | |
| 4. Snap | 0.25 | | |
| 5. Layers | *NAME* | *COLOR* | *LINETYPE* |
| | Border | Red | Continuous |
| | Object | Green | Continuous |
| | Center | Blue | Center |

**Hints:** Use the LINE and CIRCLE commands to complete the drawing. Make sure to draw the objects in the appropriate layers. Do not dimension the drawing.

A good place to start is the center of the three circles. With the radius of 5.5 units, the center could be at coordinates 7,7.

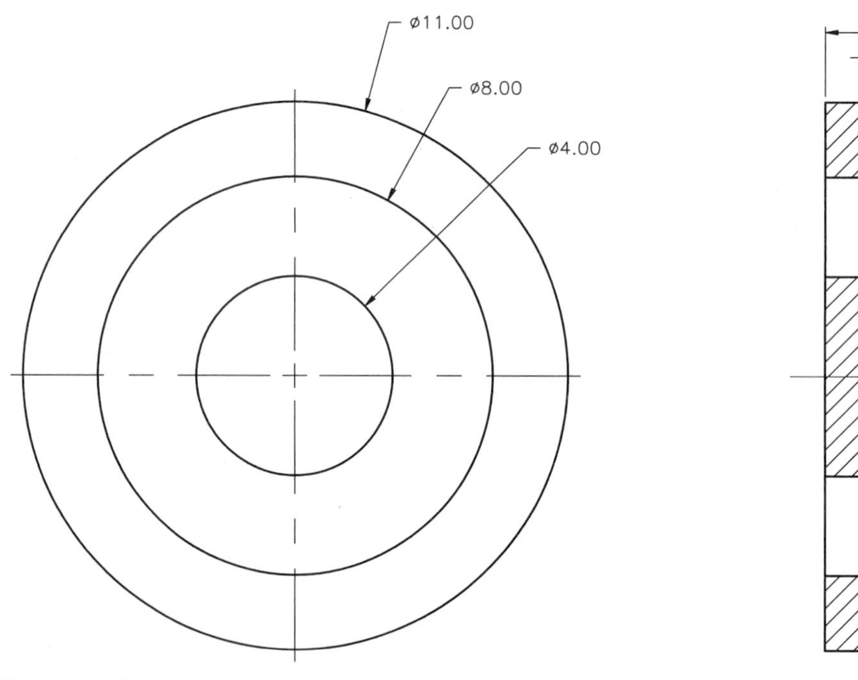

**Figure Ex3–3**

## EXERCISE 3–4

Create the drawing shown in Figure Ex3–4 according to the settings given in the following table:

| Settings | Value | | |
|---|---|---|---|
| 1. Units | Decimal with two decimal places | | |
| 2. Limits | Lower left corner: 0,0 | | |
| | Upper right corner: 24,18 | | |
| 3. Grid | 1.00 | | |
| 4. Snap | 0.50 | | |
| 5. Layers | *NAME* | *COLOR* | *LINETYPE* |
| | Border | Red | Continuous |
| | Object | Green | Continuous |
| | Center | Blue | Center |

**Hints:** Use the LINE and CIRCLE and ARC commands to complete the drawing. Make sure to draw the objects in the appropriate layers. Do not dimension the drawing.

A good place to start is the center of the large circle. The large circle is 4.5 units from the bottom of the plate and 8.5 units from the left side, so its center could be at coordinates 10,6.

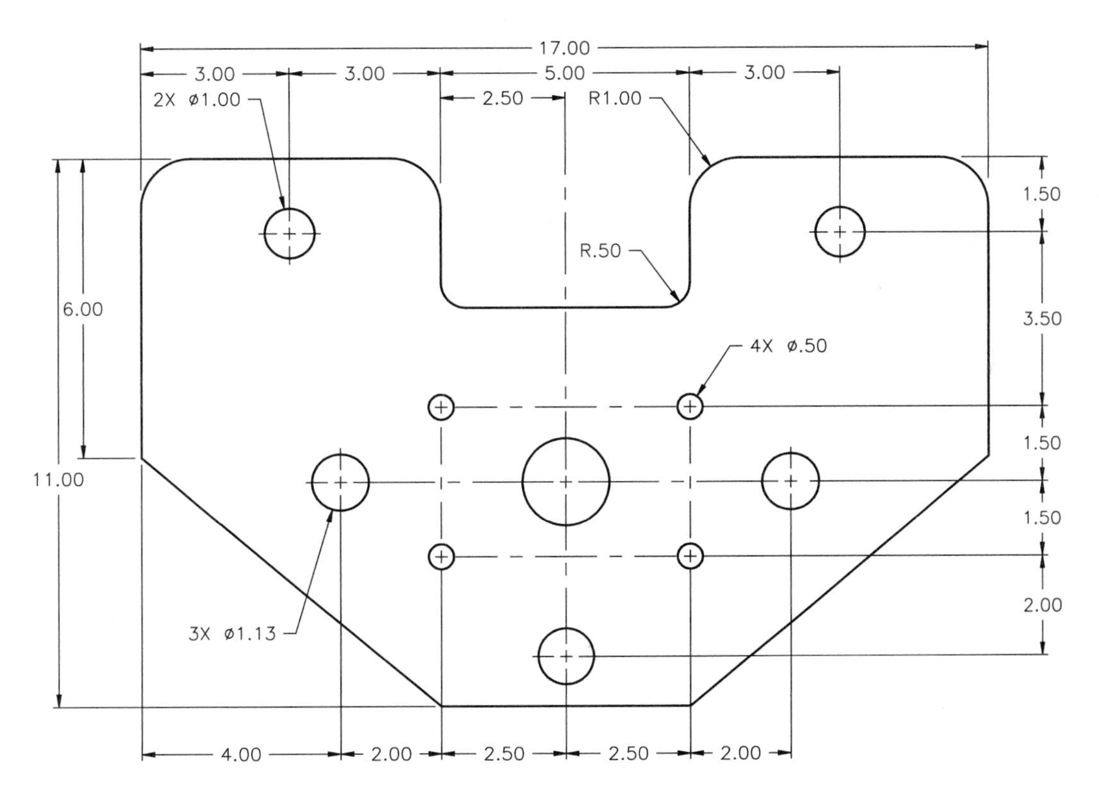

**Figure Ex3–4**

## EXERCISE 3–5

Create the drawing shown in Figure Ex3–5 according to the settings given in the following table:

| Settings | Value |
|---|---|
| 1. Units | Decimal |
| 2. Limits | Lower left corner: –2,–2<br>Upper right corner: 10,7 |
| 3. Grid | 1.0 |
| 4. Snap | 1.0 |

**Hint:** Draw a sequence of lines from A (7,1) through B, C, and D, to E (3,4) with the Snap set to 1 (or 0.50, 0.25, or 0.125) and set to ON. Using the SNAP Rotate option, select point E as the new origin and point A to designate the rotation angle. Using the Continue option of the LINE command, draw the lines from E through F and G and back to point A. Points F and G can be specified only with the Snap rotated, as explained, and the Snap set to ON.

### POINT OF INTEREST

- This exercise seems simple enough. However, it is not too often that two systems of coordinates share the same common points like A and E in this exercise. This is because E-A is a 5-unit hypotenuse of a right triangle whose base and altitude are 4 and 3 units.

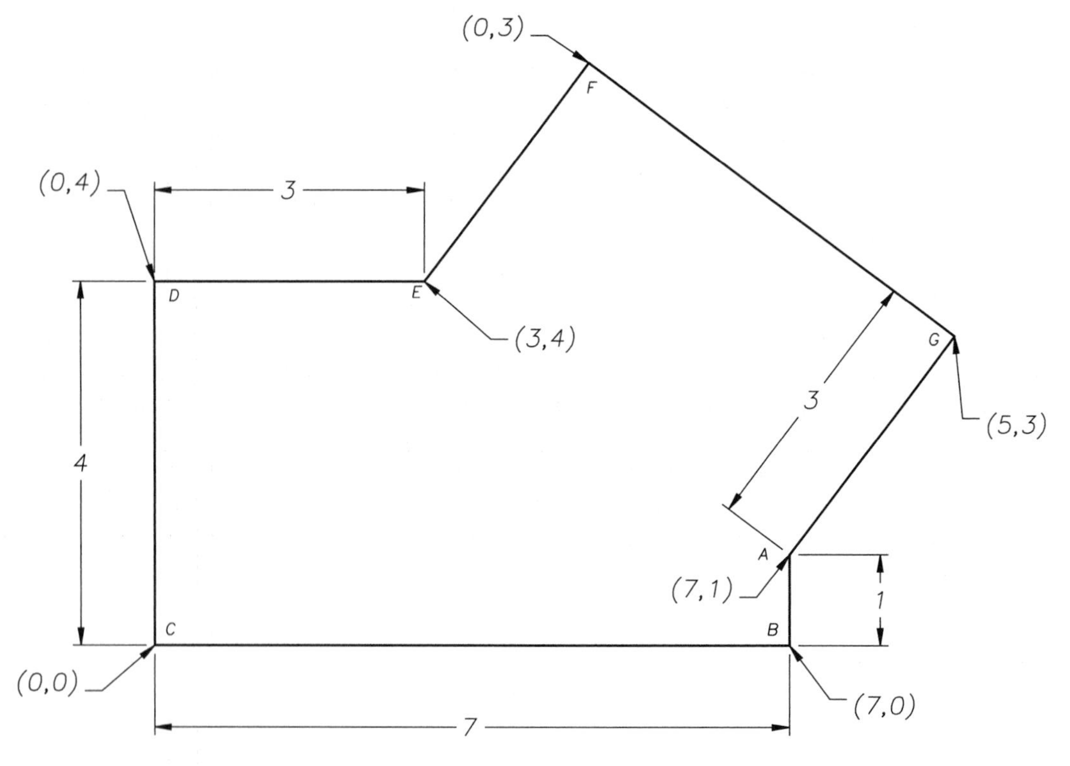

**Figure Ex3–5**

## EXERCISE 3–6

In the following exercise the outline of the house shown in Figure Ex3–6a will be drawn. Create the drawing according to the settings given in the following table:

| Settings | Value |
|---|---|
| 1. Units | Linear UNITS set to Architectural<br>Angular UNITS set to Surveyor's |
| 2. Limits | Lower left corner: –20',–10'<br>Upper right corner: 124',86 |
| 3. Grid | 5' |
| 4. Snap | Set to Rotated snap at an angle of 4d45'08". |

**Hint:** To draw the property lines, you can use the steps in the example following the text section on the Direct Distance Option regarding the Rotate option of the SNAP command. Do not draw the text; provided for reference only. After invoking the LINE command, begin tracking at TK1 (0,0) (refer to Figure Ex3–5b). Turn Ortho ON and track through point TK2 to TK3/A by forcing the cursor in the right direction and entering the proper distance (35.3' with the cursor to the East for TK2, and 7.1' with the cursor to the North for TK3/A). Draw the first line, A-B, by forcing the cursor to the north and entering the distance of 16.4'. Continue the lines around the outline of the house without exiting the LINE command. Each line's length can be entered when the cursor is forced in the right direction, thus using the DIRECT DISTANCE option for the endpoint of each line.

### POINT OF INTEREST

- It is important to realize that using the Rotate option of the SNAP command to establish a new snap origin and snap angle does not change the *X* and *Y* coordinates of the points on the screen. Nor does it change input angles. These can be done with the more complicated UCS command. This means that if you use Absolute Coordinate entry, AutoCAD will use the coordinates based on the origin and angle of rotation of the coordinate system in effect before changing the snap origin and/or angle of rotation. Also, Relative Rectangular or Polar Coordinate input is based on the unrotated system of coordinates. For example, if the snap origin and angle were changed to 4,4 (*X*,*Y* coordinates) and 45 degrees, respectively, and you were specifying a point relative to a point at the new snap origin as @4,0 or @4<0, the specified point would be at the *X*,*Y* coordinates of 8,4. They would not be 4 units distance at 45 degrees from 4,4 as you might expect. A rotated snap grid with a relocated origin is for use primarily in conjunction with the Snap and/or Ortho set to ON. Otherwise, SNAP Rotate is not that functional, as you will learn with experience.

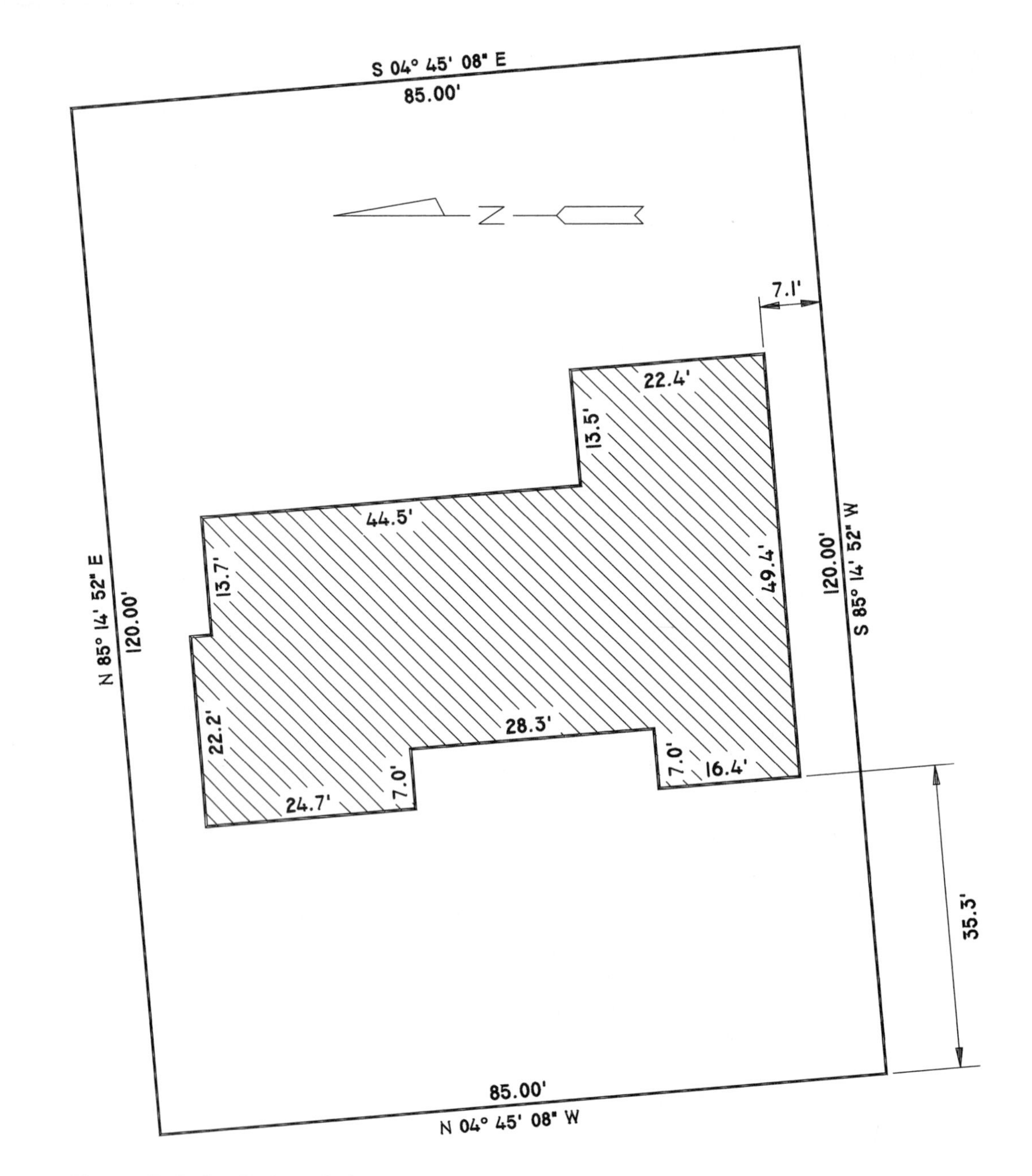


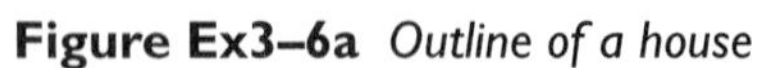

**Figure Ex3–6a** *Outline of a house*

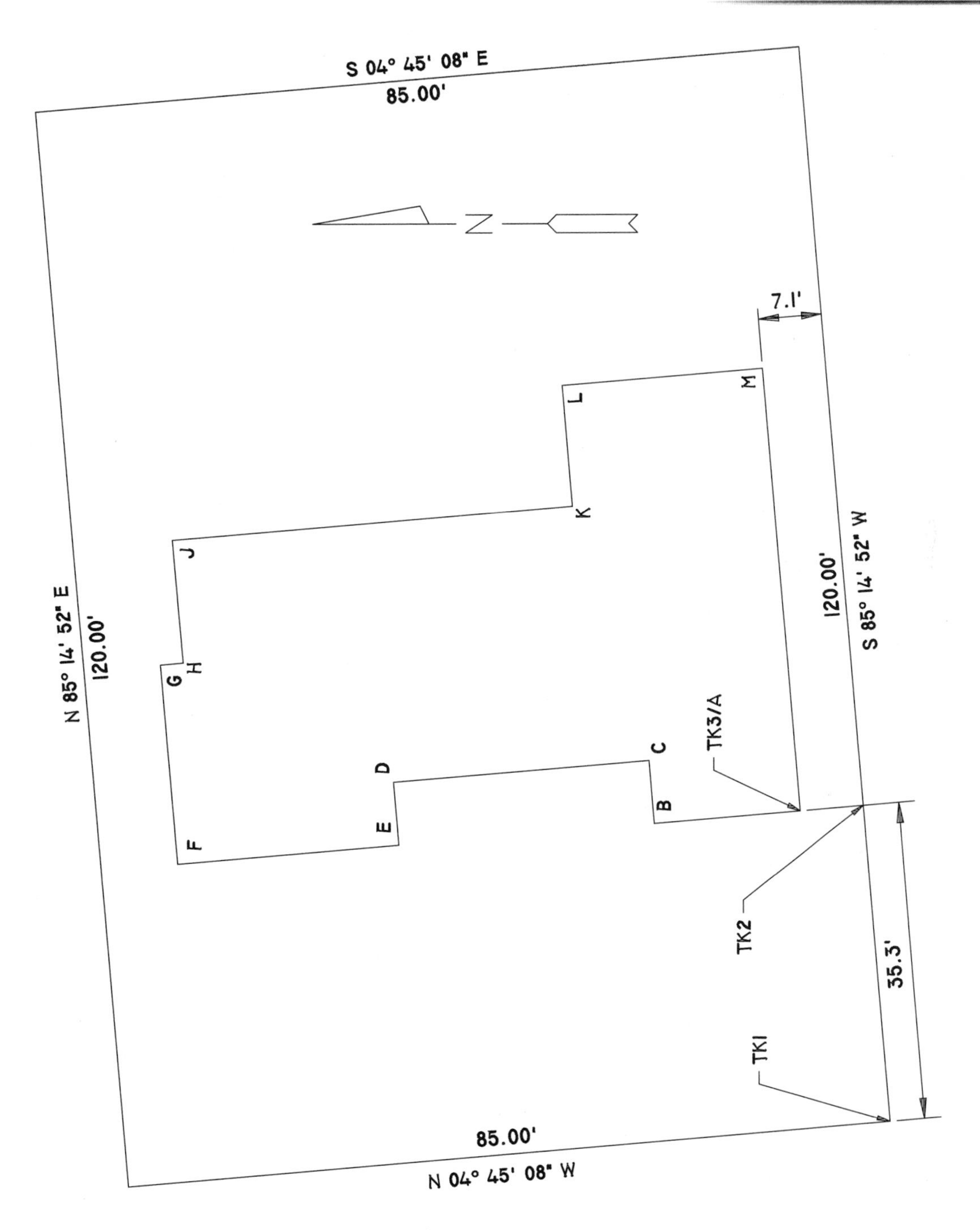


**Figure Ex3–6b** *Outline of a house with dimensions*

## EXERCISE 3–7

Create the stair plan drawing shown in Figure Ex3–7, according to the settings given in the following table:

| Settings | Value | | |
|---|---|---|---|
| 1. Units | Architectural | | |
| 2. Limits | Lower left corner: 0,0 | | |
| | Upper right corner: 22',17' | | |
| 3. Grid | 6" | | |
| 4. Snap | 3" | | |
| 5. Layers | *NAME* | *COLOR* | *LINETYPE* |
| | Border | Green | Continuous |
| | Stair riser | Blue | Hidden |
| | tread | White | Continuous |
| | text | Blue | Continuous |
| | wall | White | Continuous |
| | door | Red | Continuous |

**Hint:** Use the LINE, RECTANGLE, CIRCLE, and ARC commands to complete the drawing. Make sure to draw the objects in the appropriate layers. Do not dimension the drawing.

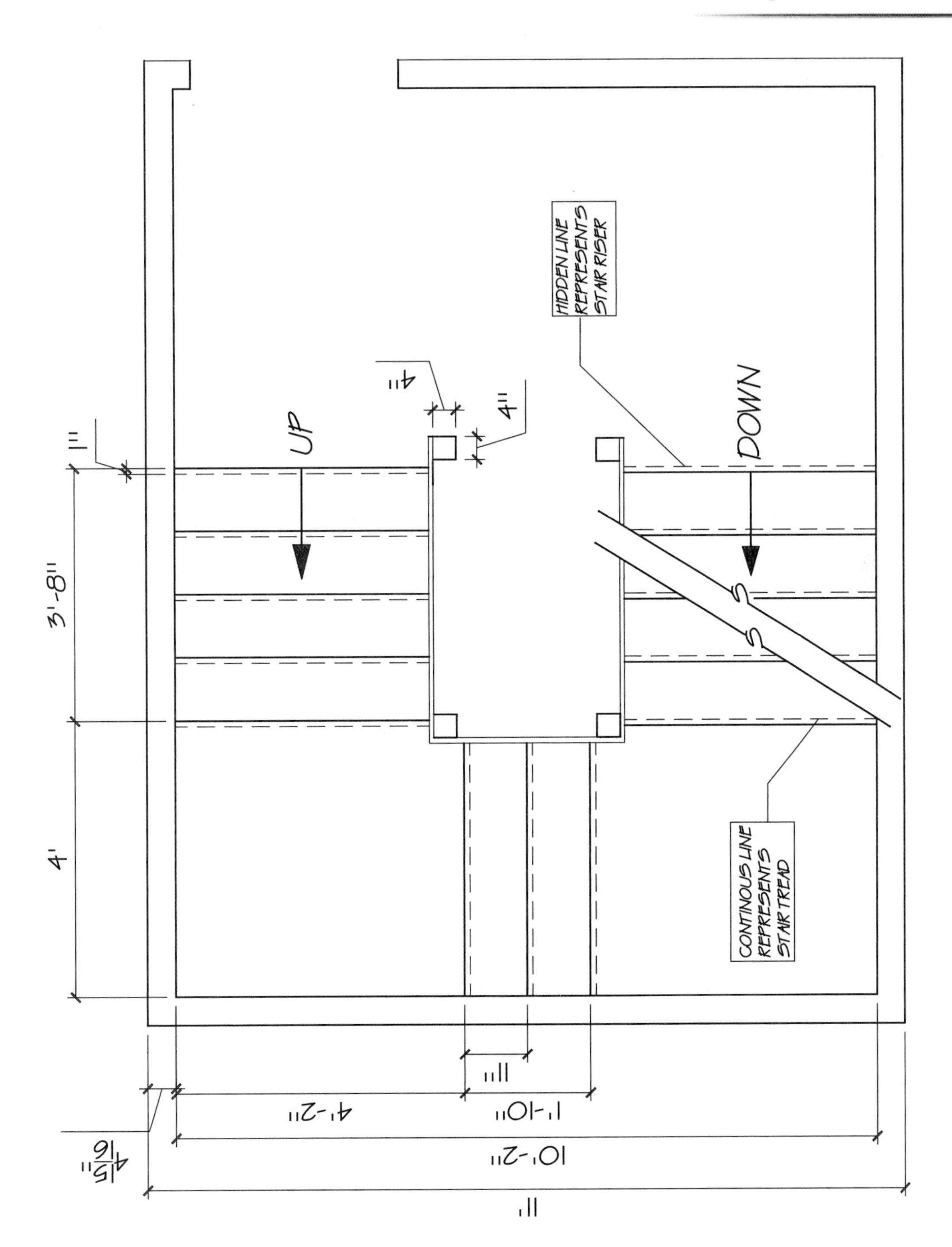


**Figure Ex3–7** *Stair Plan*

# EXERCISE 3–8

Create the spiral stair design plan drawing shown in Figure Ex3–8, according to the settings/design dimensions given in the following table:

| Settings | Value | | |
|---|---|---|---|
| 1. Units | Architectural | | |
| 2. Limits | Lower left corner: 0,0 | | |
| | Upper right corner: 12',9' | | |
| 3. Grid | 2 | | |
| 4. Snap | 1 | | |
| 5. Layers | *NAME* | *COLOR* | *LINETYPE* |
| | Border | Green | Continuous |
| | handrail | White | Continuous |
| | tread | White | Continuous |
| | text | Blue | Continuous |
| | landing | White | Continuous |

**Hint:** Use THE LINE, CIRCLE, and ARC commands to complete the drawing. Make sure to draw the objects in the appropriate layers. Do not dimension the drawing.

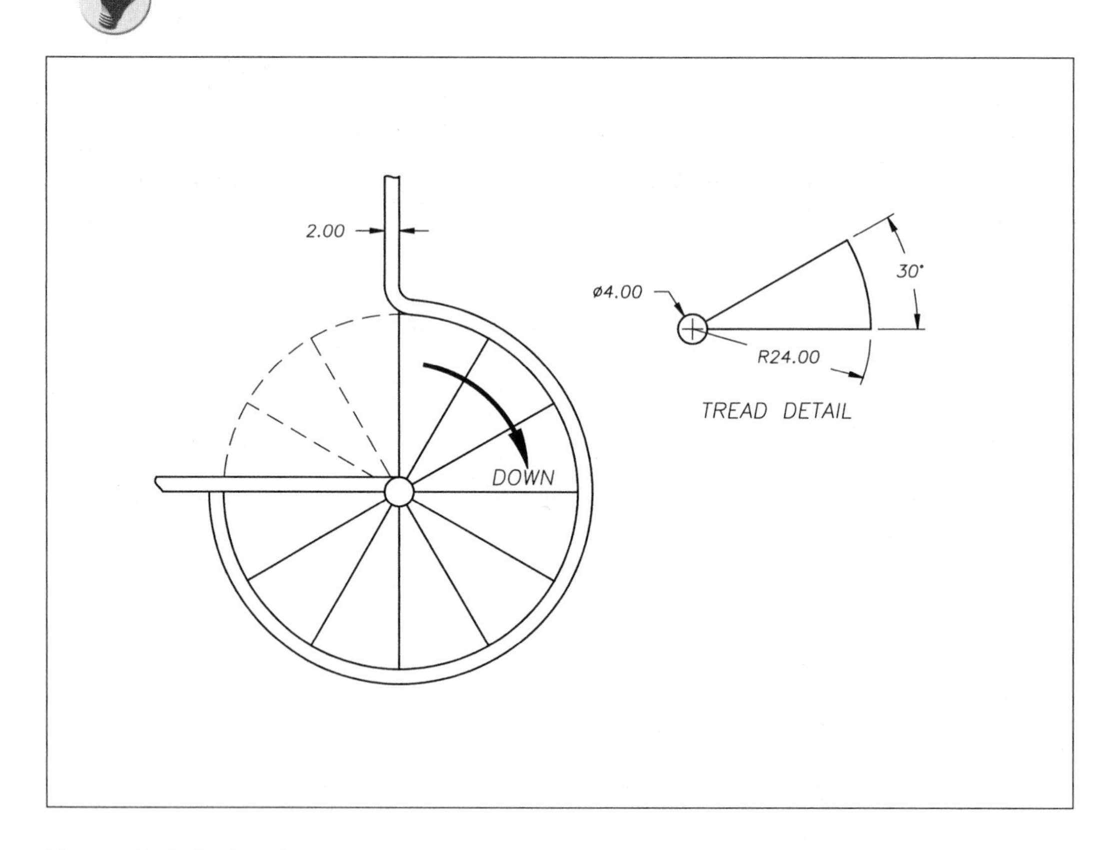

**Figure Ex3–8** *Sprial stair Design plan*

## EXERCISE 3–9

Create the pre-fabricated walk-in refrigerator design shown in Figure Ex3–9, according to the settings/design dimensions given in the following table:

| Settings | Value | | | |
|---|---|---|---|---|
| 1. Units | Architectural | | | |
| 2. Limits | Lower left corner: 0,0 | | | |
| | Upper right corner: 220,180 | | | |
| 3. Grid | 10 | | | |
| 4. Snap | 5 | | | |
| 5. Layers | *NAME* | *COLOR* | *LINETYPE* | |
| | border | Green | Continuous | |
| | panels | White | Continuous | |
| | door | Red | Continuous | |
| | shelf | Blue | Continuous | |
| 6. Walk in Refrigerator Design Dimensions | A | OUTSIDE WALL PANEL LENGTH | 23" | |
| | B | INSIDE WALL SHELF | 15" | |
| | C | INSIDE WALL SHELF | 18" | |
| | D | DOOR | 46" | |
| | E | INSIDE WALL SHELF | 20" | |
| | F | INSIDE WALL THICKNESS 4" | | |
| | G | OUTSIDE WALL PANEL LENGTH | 11.5" | |
| | H | OUTSIDE WALL THICKNESS | 4" | |
| | I | OUTSIDE WALL THICKNESS | 4" | |

**Hint:** Use the LINE, RECTANGLE, and ARC commands to complete the drawing. Make sure to draw the objects in the appropriate layers. Do not dimension the drawing.

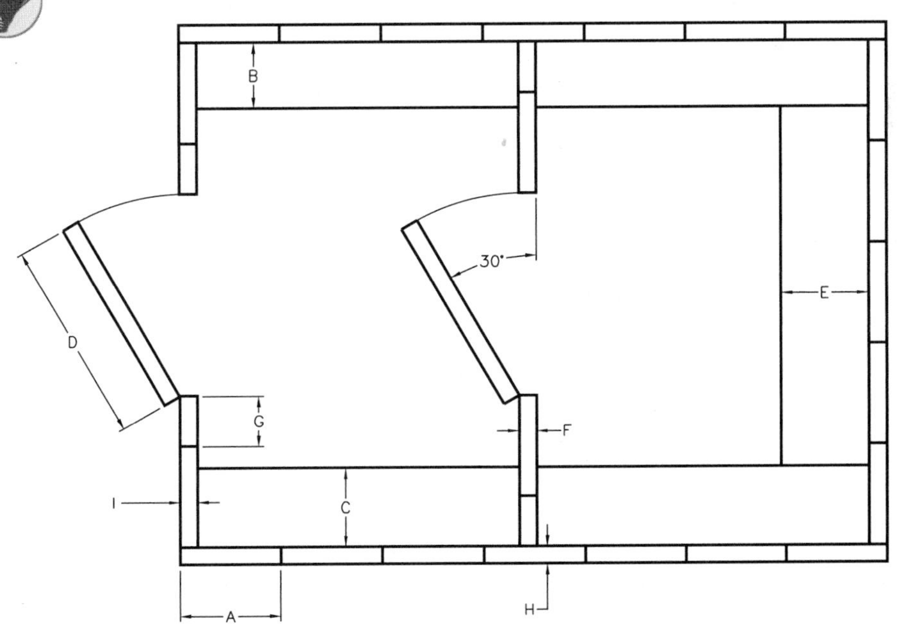


**Figure Ex3–9** *Walk-in Refrigerator Design*

architectural EXERCISE

## EXERCISE 3–10

Create the bell pier foundation shown in Figure Ex3–10 according to the settings/specifications given in the following table:

| Settings | Value | | |
|---|---|---|---|
| 1. Units | Architectural | | |
| 2. Limits | Lower left corner: 0',0' | | |
| | Upper right corner: 22',17' | | |
| 3. Grid | 6 | | |
| 4. Snap | 3 | | |
| 5. Layers | *NAME* | *COLOR* | *LINETYPE* |
| | Object | Green | Continuous |
| | Dim | Blue | Hidden |

**Hint:** Use the LINE and RECTANGLE commands to complete the drawing. Make sure to draw the objects in the appropriate layers. Do not dimension the drawing.

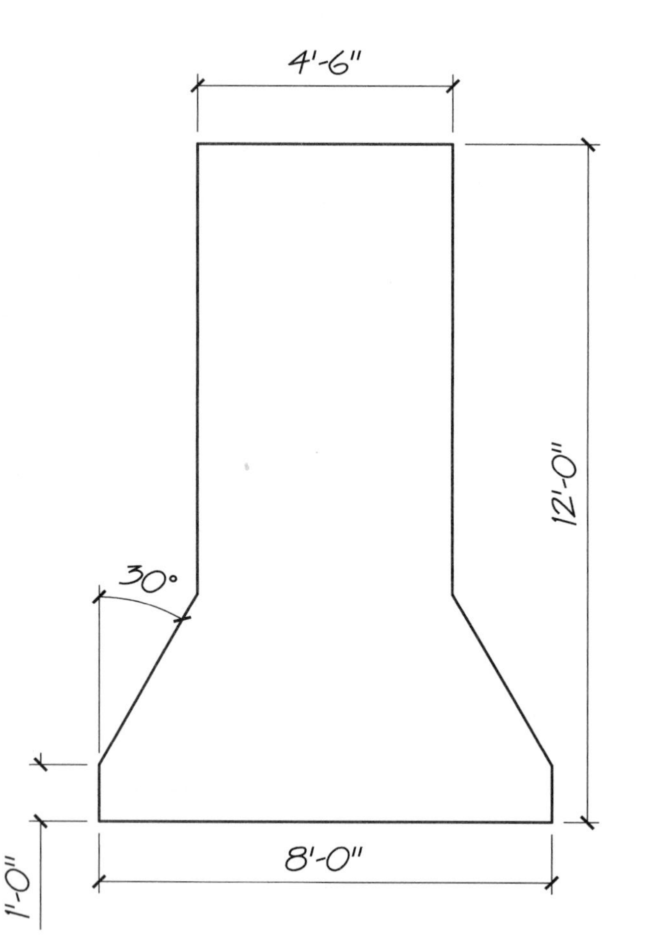


**Figure Ex3–10** *Bell Pier Foundation*

## EXERCISE 3–11

Create the typical concrete beam shown in Figure Ex3–11 according to the settings/specifications given in the following table:

| Settings | Value | | |
|---|---|---|---|
| 1. Units | Architectural | | |
| 2. Limits | Lower left corner: 0',0'<br>Upper right corner: 12',9' | | |
| 3. Grid | | | |
| 4. Snap | | | |
| 5. Layers | *NAME* | *COLOR* | *LINETYPE* |
| | Border | Red | Continuous |
| | object | White | Continuous |
| | reinf | White | Continuous |
| 6. Specifications | A | 39" | |
| | B | 24" | |
| | C | 12" | |
| | D | 3" | |
| | E | .75" | |

**Hint:** Use the LINE, RECTANGLE, CIRCLE, and ARC commands to complete the drawing. Make sure to draw the objects in the appropriate layers. Do not dimension the drawing.

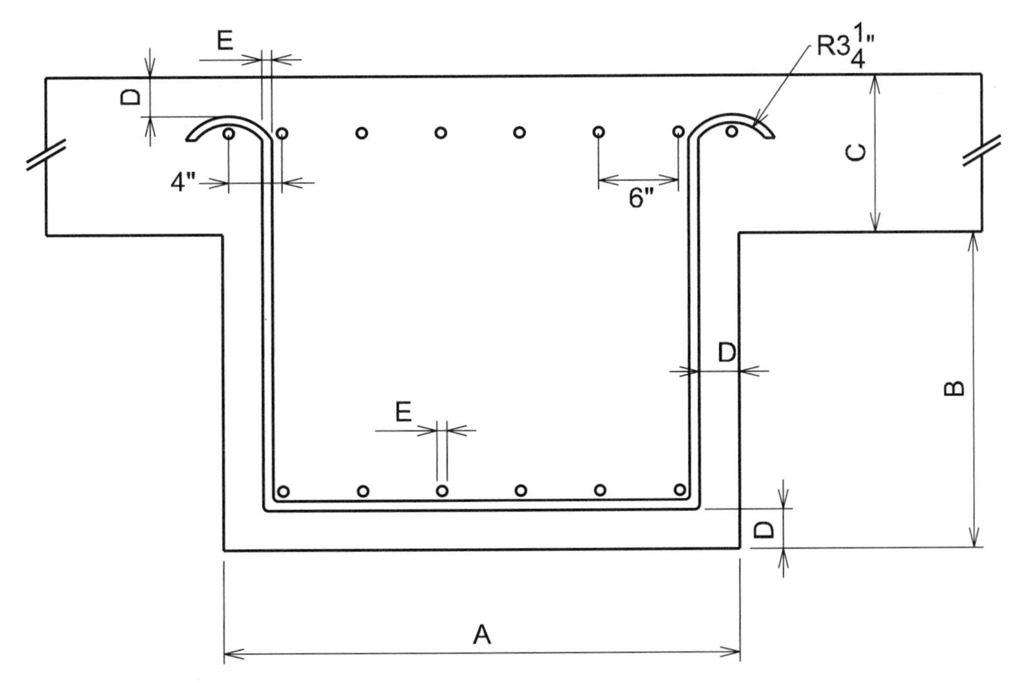


**Figure Ex3–11** *Typical concrete beam with reinforcing*

# EXERCISE 3–12

Create the residential driveway as shown in figure according to the settings/specifications given in the following table:

| Settings | Value | | | |
|---|---|---|---|---|
| 1. Units | Architectural | | | |
| 2. Limits | Lower left corner: 0',0' | | | |
| | Upper right corner: 70',50' | | | |
| 3. Grid | 2' | | | |
| 4. Snap | 1' | | | |
| 5. Layers | *NAME* | *COLOR* | *LINETYPE* | |
| | Border | white | Continuous | |
| | cline | White | Center | |
| | road | Red | Continuous | |
| | bldgwall | Green | Continuous | |
| | retwall | Blue | Continuous | |
| 6. Specifications | A | DRIVEWAY WIDTH | | 14'-0" |
| | B | STREET WIDTH | | 11'-0" |
| | C | GARAGE DOOR WIDTH | | 12'-0" |
| | D | WALL LENGTH | | 15'-0" |
| | E | WALL LENGTH | | 10'-0" |
| | G | WALL DISTANCE FROM DRIVEWAY | | 2'-0" |
| | R | RADIUS | | 10'-0" |

**Hint:** Use the LINE, RECTANGLE, and ARC commands to complete the drawing. Make sure to draw the objects in the appropriate layers. Do not dimension the drawing.

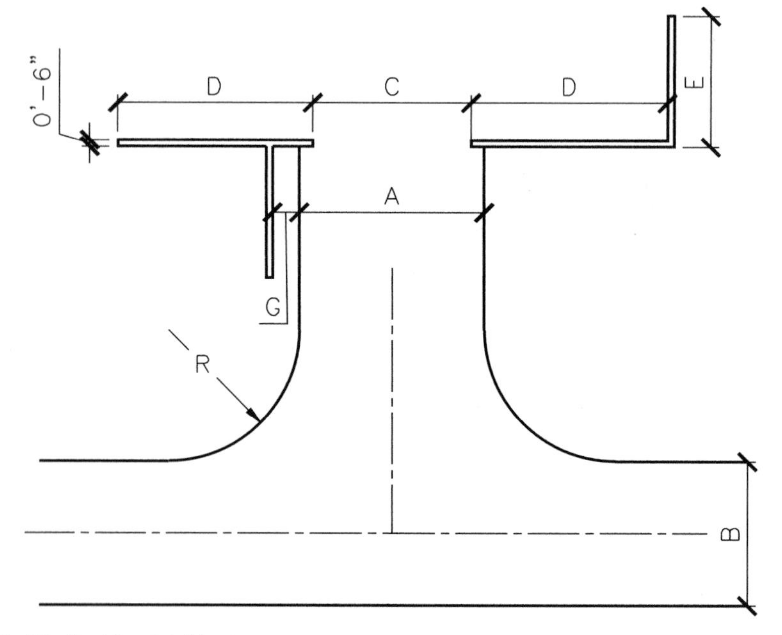

**Figure Ex3–12** *Residential driveway*

## EXERCISE 3–13

Create the Surface Drainage diagrammatic layout shown in Figure Ex3–13 according to the settings/specifications given in the following table:

| Settings | Value | | |
|---|---|---|---|
| 1. Units | Architectural | | |
| 2. Limits | Lower left corner: 0',0'<br>Upper right corner: 200',150' | | |
| 3. Grid | 10' | | |
| 4. Snap | 5' | | |
| 5. Layers | *NAME* | *COLOR* | *LINETYPE* |
| | Border | Green | Continuous |
| | cline | White | Center |
| | road | Red | Continuous |
| | cbasins | White | Continuous |
| | ssewer | Blue | Phantom |
| | mssewer | Whit | Hidden |
| | junction box | Green | Continuous |
| 6. Specifications | A | Primary Road | 24'-0" |
| | B | Secondary Road | 14'-0" |
| | R1 | Radius | 10'-0" |
| | R2 | Radius | 12'-0" |

**Hint:** Use the LINE, RECTANGLE, and ARC commands to complete the drawing. Make sure to draw the objects in the appropriate layers. Do not dimension the drawing.

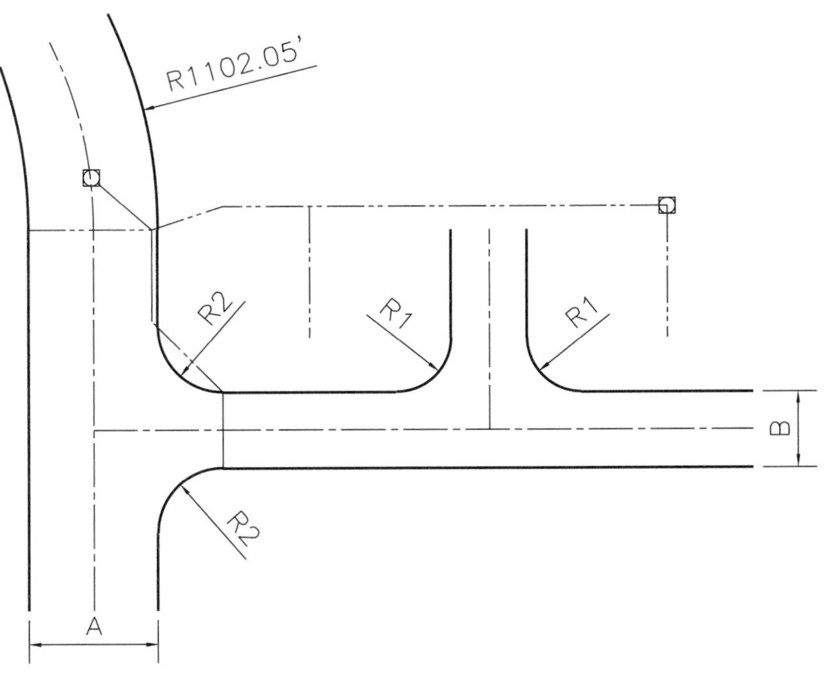


**Figure Ex3–13** *Surface Drainage Diagrammatic layout*

## EXERCISE 3–14

Create the uninterruptible power supply schematic diagram shown in Figure Ex3–14 according to the settings/specifications given in the following table:

| Settings | Value | | |
|---|---|---|---|
| 1. Units | Decimal | | |
| 2. Limits | Lower left corner: 0,0 | | |
| | Upper right corner: 22,17 | | |
| 3. Grid | 0.5 | | |
| 4. Snap | 0.25 | | |
| 5. Layers | *NAME* | *COLOR* | *LINETYPE* |
| | Border | White | Continuous |
| | Switches | White | Continuous |
| | Wiring | Red | Continuous |

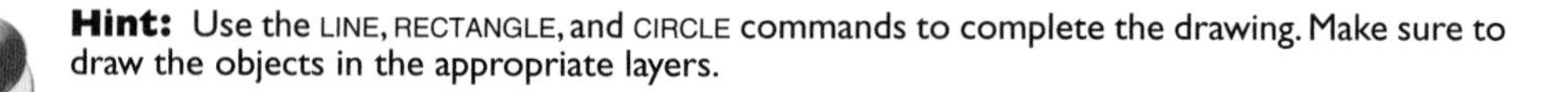

**Hint:** Use the LINE, RECTANGLE, and CIRCLE commands to complete the drawing. Make sure to draw the objects in the appropriate layers.

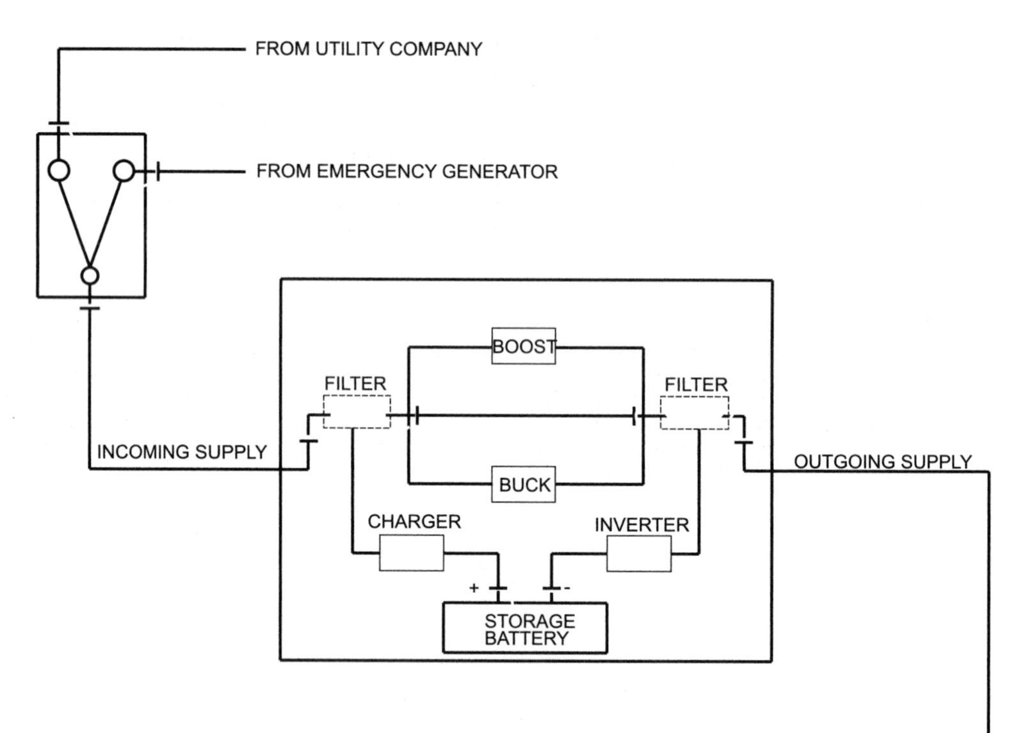


**Figure Ex3–14** *An uninterruptible power supply diagram*

## EXERCISE 3–15

Create the SCSI ID Post drawing shown in Figure Ex3–15 according to the settings given in the following table:

| Settings | Value | | |
|---|---|---|---|
| 1. Units | Decimal | | |
| 2. Limits | Lower left corner: 0,0 | | |
| | Upper right corner: 12,9 | | |
| 3. Grid | .5 | | |
| 4. Snap | .25 | | |
| 5. Layers | *NAME* | *COLOR* | *LINETYPE* |
| | Border | Green | Continuous |
| | ID0 | Red | Continuous |
| | ID1 | Red | Continuous |
| | ID2 | Red | Continuous |
| | ID3 | Red | Continuous |
| | ID4 | Red | Continuous |
| | ID5 | Red | Continuous |
| | ID6 | Red | Continuous |
| | ID7 | Red | Continuous |
| | ID8 | Red | Continuous |
| | ID9 | Red | Continuous |
| | ID10 | Red | Continuous |
| | ID11 | Red | Continuous |
| | ID12 | Red | Continuous |
| | ID13 | Red | Continuous |
| | ID14 | Red | Continuous |
| | ID15 | Red | Continuous |

**Hint:** ID0 represents the posts with no jumper pins in use. The requirements of this drawing are to create the posts and then the connector jumper pins on the appropriate layers. Each SCSI ID number should correspond to a single layer. You will be required to use more than one jumper pin for the different ID Numbers. The final result of this drawing will be evident by turning on and off the appropriate layers to create the corresponding SCSI ID numbers showing their jumper assignments. The diagram on the left shows the ID numbers and their corresponding post assignments.

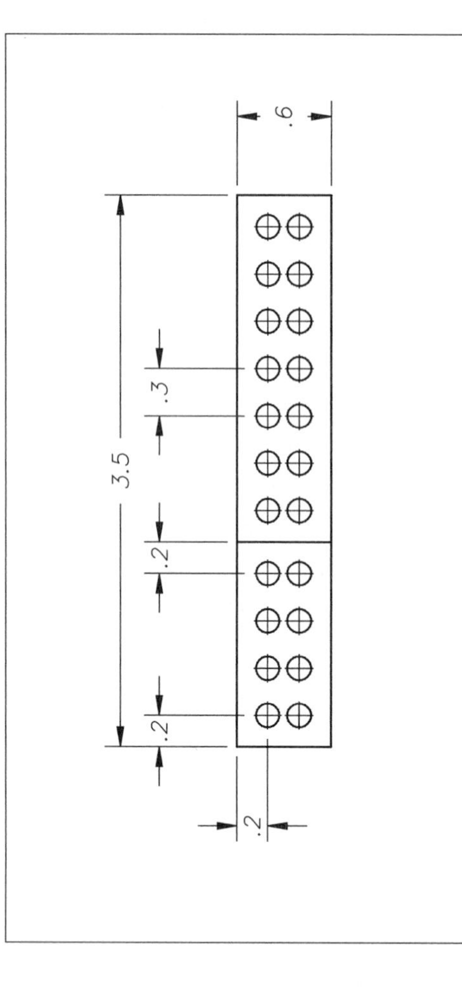


ID 1 ID 2 ID 3 ID 4 ID 5 ID 6 ID 7 ID 8 ID 9 ID 10 ID 11 ID 12 ID 13 ID 14 ID 15

**Figure Ex3–15** *SCSI ID Post drawing*

## EXERCISE 3–16

Create the wire diagram showing the pin assignments of a Ethernet cross over cable and a null-modem cable as shown in Figure Ex3–16 according to the settings/specifications given in the following table:

| Settings | Value | | |
|---|---|---|---|
| 1. Units | Decimal | | |
| 2. Limits | Lower left corner: 0,0 | | |
| | Upper right corner: 42,36 | | |
| 3. Grid | 1 | | |
| 4. Snap | .5 | | |
| 5. Layers | *NAME* | *COLOR* | *LINETYPE* |
| | Border | White | Continuous |
| | Ethernet Cable | Red | Continuous |
| | Ethernet Cable Text | Red | Continuous |
| | Null-modem Cable | Blue | Continuous |
| | Text | Blue | Continuous |

**Hint:** Use the LINE, RECTANGLE, and ARC commands to complete the drawing. Make sure to draw the objects in the appropriate layers.

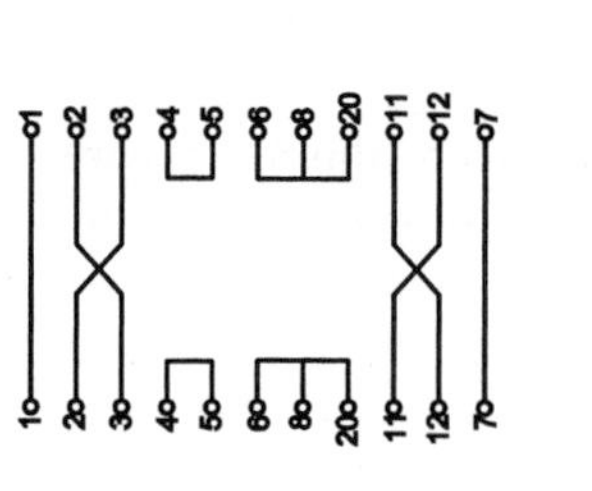


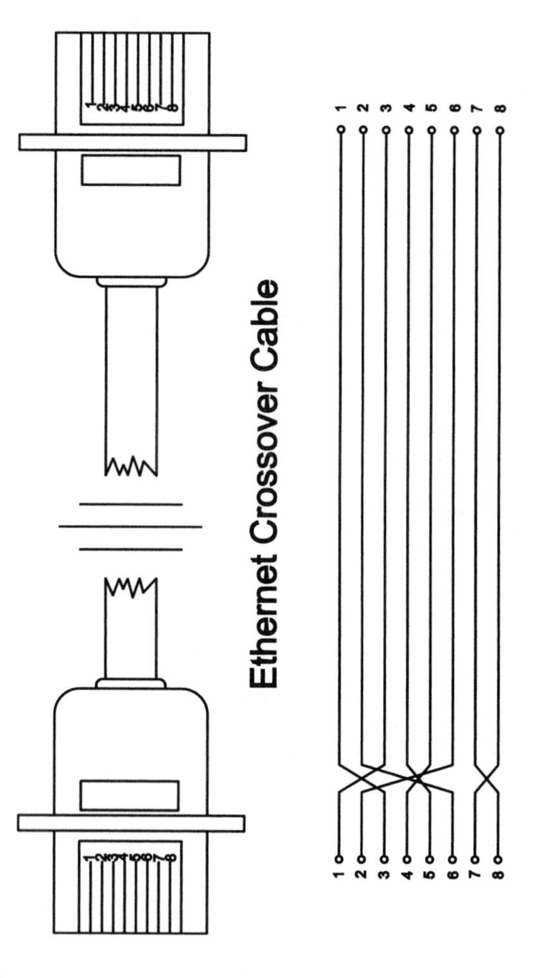


**Figure Ex3–16** *Wire diagram showing the pin assignments of a Ethernet cross over cable and a null-modem cable*

## EXERCISE 3–17

Create the circuit board of a 25 pin gender changer as shown in Figure Ex3–17 according to the settings given in the following table:

| Settings | Value | | |
|---|---|---|---|
| 1. Units | Decimal | | |
| 2. Limits | Lower left corner: 0,0<br>Upper right corner: 42,36 | | |
| 3. Grid | 1 | | |
| 4. Snap | .5 | | |
| 5. Layers | *NAME* | *COLOR* | *LINETYPE* |
| | Border | White | Continuous |
| | Board | White | Continuous |
| | Posts | Red | Continuous |
| | Etch wires | Blue | Continuous |

**Hint:** Use the LINE, RECTANGLE, and CIRCLE commands to complete the drawing. Make sure to draw the objects in the appropriate layers.

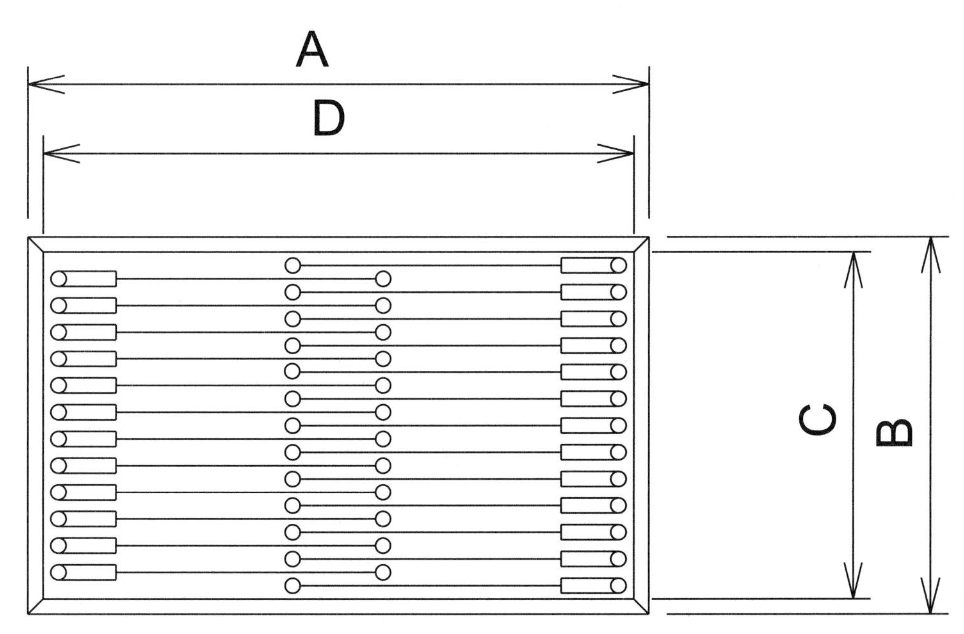


**Figure Ex3–17** *Circuit board of a 25 pin gender changer*

electrical EXERCISE

CHAPTER 4

# Fundamentals III



## PROJECT EXERCISE

This project exercise provides point-by-point instructions for setting up the drawing with layers and then creating the objects shown in Figure P4–1.

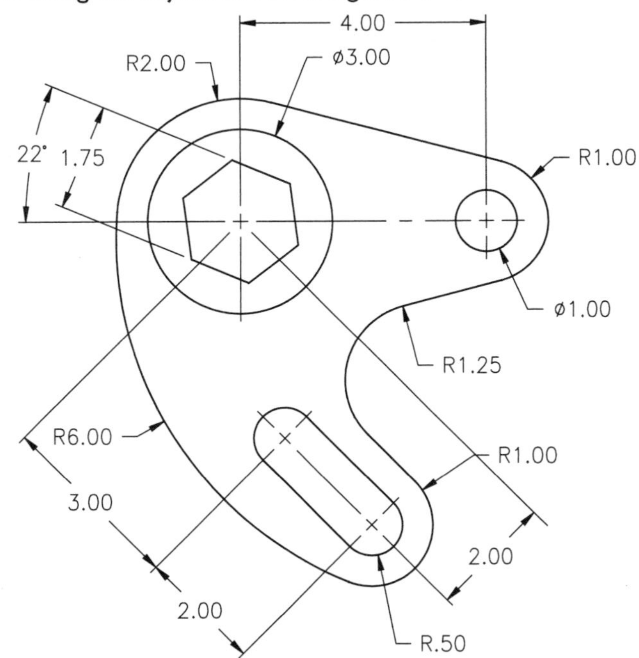


**Figure P4–1** *Completed project drawing*

In this project you will learn how to do the following:

- Set up the drawing, including Units, Limits and Layers
- Use the LINE, RECTANGLE, POLYGON, and CIRCLE commands to create objects
- Use the Osnap modes of int, tan, ttr, and per
- Use the TRIM, OFFSET, BREAK, MIRROR, and ERASE commands to modify objects

## SET UP THE DRAWING AND DRAW A BORDER

**Step 1:** Make sure system variable STARTUP is set to 1, that allows AutoCAD to create a new drawing using a wizard. To create a new drawing, invoke the NEW command from the Standard toolbar or select New from the File menu.

**Step 2:** AutoCAD displays the Create New Drawing dialog box. Select the **Use a Wizard** button and AutoCAD lists available wizards. Select Quick Setup wizard from the Select a Wizard list box and choose the **OK** button.

AutoCAD displays the Quick Setup dialog box with the display of the Units section. Select the "Decimal" radio button and choose the **Next>** button.

AutoCAD displays the Quick Setup dialog box with the display of the Area section. Under the **Width:** text box enter **18**, and under the **Length** text box enter **12**. Choose the **Finish** button to close the Quick Setup dialog box. Invoke the ZOOM ALL command from the View menu.

**Step 3:** Invoke the LAYER command from the Layerss toolbar, or select Layer... from the Format menu. AutoCAD displays the Layer Properties Manager dialog box.

Create three layers, and rename them as shown in the following table, assigning appropriate color and linetype.

| Layer Name | Color | Linetype |
|---|---|---|
| Border | Red | Continuous |
| Centerline | Green | Center |
| Object | Blue | Continuous |

Set Border as the current layer, and close the Layer Properties Manager dialog box.

**Step 4:** Open the Drafting Settings dialog box from the Tools menu, and set Grid to **0.5** and Snap to **0.5**, and then set the grid to ON.

**Step 5:** Invoke the RECTANGLE command to draw the border (17" by 11"), as shown in Figure P4–2.

Command: **rectangle** (ENTER)
Specify first corner point or [Chamfer/Elevation/Fillet/Thickness/Width]: **.5,.5** (ENTER)
Specify other corner point: **@17,11** (ENTER)

**Step 6:** Begin the layout of the drawing by drawing the centerlines as shown in Figure P4–3. Set the CENTERLINE layer as the current layer. Invoke the LINE command from the Draw toolbar and draw lines 1, 2, 3, and 4.

Command: **line** (ENTER)
Specify first point: **4,8** (ENTER)
Specify next point or [Undo]: **@8.5<0** (ENTER)
Specify next point or [Undo]: (ENTER)
Command: (ENTER)

Specify first point: **7,5.5** (ENTER)

Specify next point or [Undo]: **@5<90** (ENTER)
Specify next point or [Undo]: (ENTER)
Command: (ENTER)

Specify first point: **7,8** (ENTER)
Specify next point or [Undo]: **@3.5<225** (ENTER)
Specify next point or [Undo]: (ENTER)

Command: (ENTER)
Specify first point: **7,8** (ENTER)
Specify next point or [Undo]: **@6.5<315** (ENTER)
Specify next point or [Undo]: (ENTER)

**Figure P4–2** *Border for a mechanical drawing*

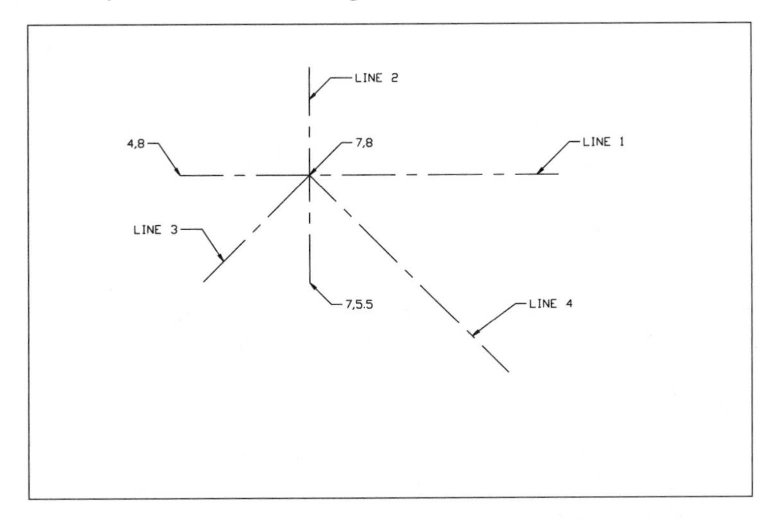


**Figure P4–3** *Placement of centerlines*

**Step 7:** Use the OFFSET command to construct the additional centerlines necessary for completion of the project. Invoke the OFFSET command from the Modify toolbar, and construct centerlines 2A, 3A, 3B, and 4A, as shown in Figure P4–4.

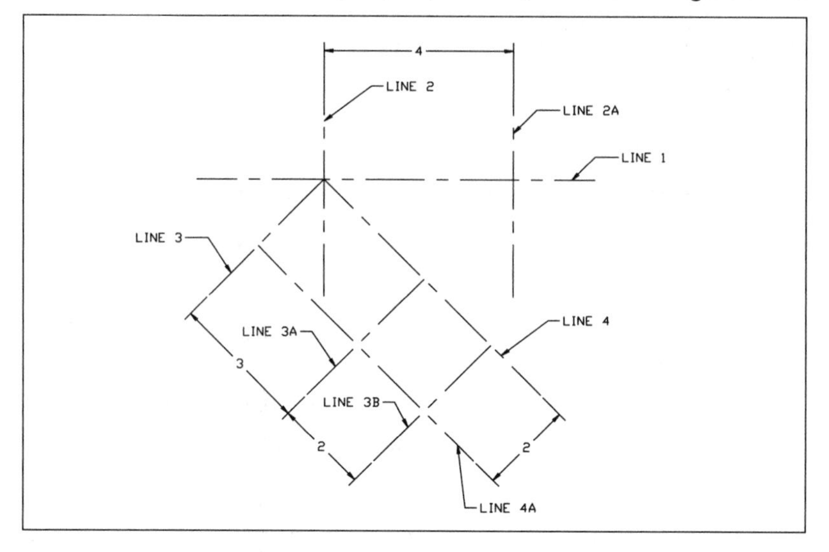


**Figure P4–4** *Placement of additional centerlines*

Command: **offset** (ENTER)
Specify offset distance or [Through] <1.0000>: **4**
Select object to offset or <exit>: *(select LINE 2)*
Specify point on side to offset: *(specify to the right of LINE 2 to construct LINE 2A, as shown in Figure P4–4)*
Select object to offset or <exit>: (ENTER)

Command: (ENTER)
OFFSET
Specify offset distance or [Through] <4.0000>: **3** (ENTER)
Select object to offset or <exit>: *(select LINE 3)*
Specify point on side to offset: *(specify a point to the right of line 3 to construct LINE 3A, as shown in Figure P4–4)*
Select object to offset or <exit>: (ENTER)

Command: (ENTER)
OFFSET
Specify offset distance or [Through] <3.0000>: **2** (ENTER)
Select object to offset or <exit>: *(select line 3A)*
Specify point on side to offset: *(specify a point to the right of line 3A to construct LINE 3B, as shown in Figure P4–4)*
Select object to offset or <exit>: *(select LINE 4)*
Specify point on side to offset: *(specify a point below LINE 4 to construct line 4A, as shown in Figure P4–4)*
Select object to offset or <exit>: (ENTER)

**Step 8:** Erase lines 3 and 4, as shown in Figure P4–4, by invoking the ERASE command from the Modify toolbar.

Command: **erase** (ENTER)
Select objects: *(select lines 3 and 4)*

Your drawing will appear as shown in Figure P4–5.

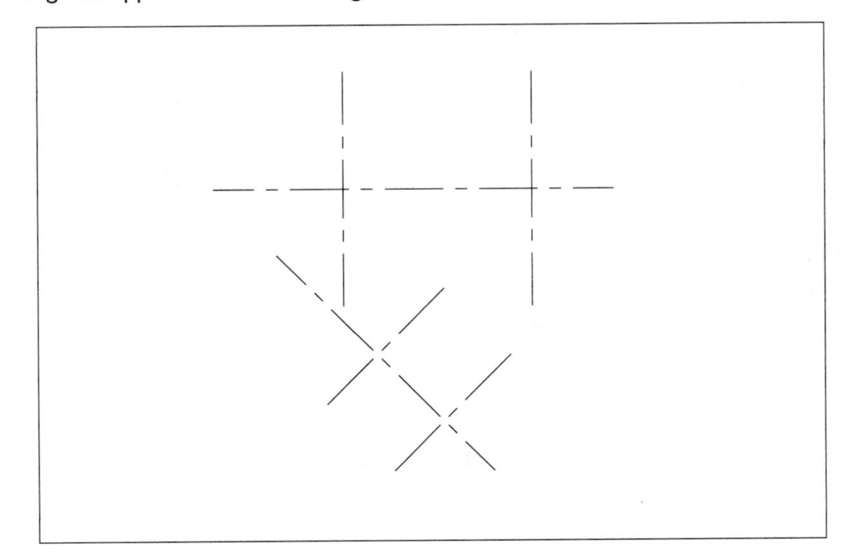

**Figure P4–5** *Drawing after lines 3 and 4 have been erased*

**Step 9:** The intersection of the centerlines you laid out in the previous steps will provide the center points for the circles. Set the OBJECT layer as the current layer. Invoke the CIRCLE command from the Draw toolbar and draw the three circles shown in Figure P4–6.

Command: **circle** (ENTER)
CIRCLE Specify center point for circle or [3P/2P/Ttr (tan tan radius)]: **7,8** (ENTER)
Specify radius of circle or [Diameter]: **2** (ENTER)

Command: (ENTER)
CIRCLE Specify center point for circle or [3P/2P/Ttr (tan tan radius)]: **11,8** (ENTER)
Specify radius of circle or [Diameter] <2.0000>: **1** (ENTER)

Command: (ENTER)
CIRCLE Specify center point for circle or [3P/2P/Ttr (tan tan radius)]: *(use the Object snap tool "intersection" to snap to the intersection of lines 3B and 4A to identify the center of circle 3)*
Specify radius of circle or [Diameter] <1.0000>: (ENTER)

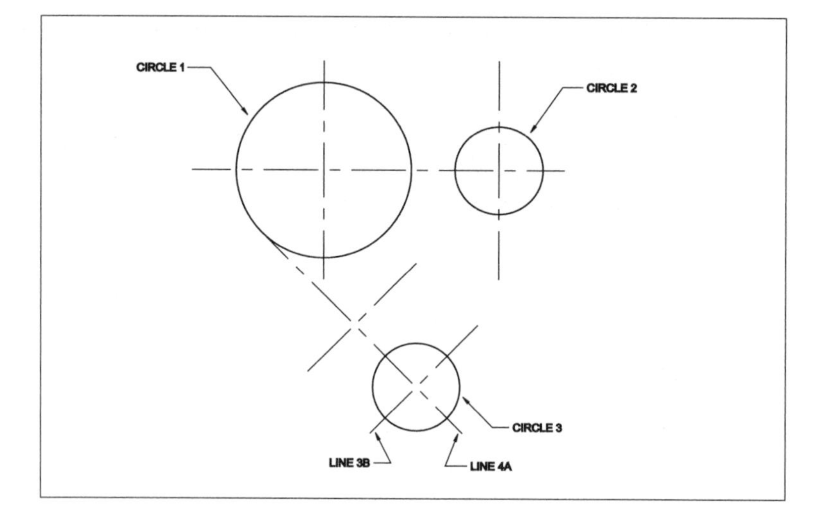

**Figure P4–6** *Placement of circles*

**Step 10:** Invoke the LINE command from the Draw toolbar, and draw LINE 5 tangent to two circles (use the Osnap tool "tangent"), as shown in Figure P4–7.

Command: **line** (ENTER)
Specify first point: *(invoke the tan object snap and select the upper part of CIRCLE 1)*
Specify next point or [Undo]: *(invoke the tan object snap and select the upper part of CIRCLE 2)*
Specify next point or [Undo]: *(press* ENTER *to terminate the command sequence)*

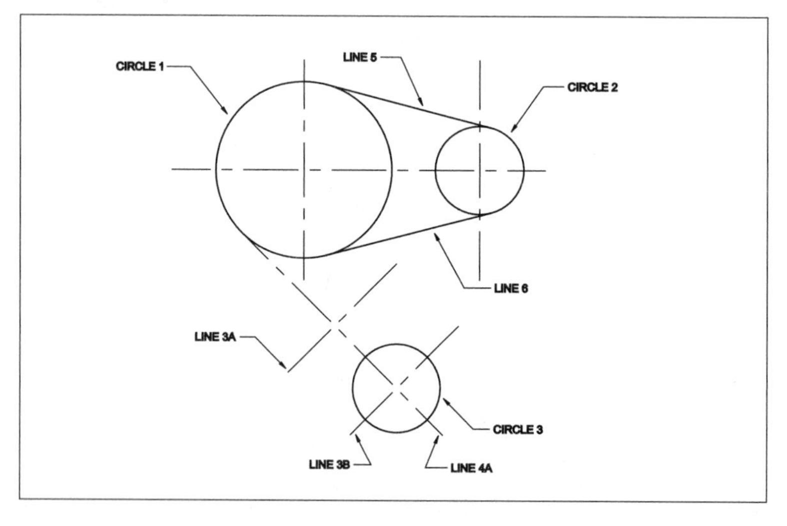


**Figure P4–7** *Lines 5 and 6 drawn tangent to two circles*

**Step 11:** Invoke the MIRROR command to create line 6 from line 5 as shown in Figure P4–7.

Command: **mirror** (ENTER)
Select objects: *(select LINE 5)*
Select objects: *(press* ENTER *to terminate the selection of objects)*
Specify first point of mirror line: **7,8** (ENTER)
Specify second point of mirror line: **11,8** (ENTER)
Delete source objects? [Yes/No] <N>: (ENTER)

**Step 12:** Invoke the LINE command and draw LINE 7 using the Object snap modes of intersection and perpendicular as shown in Figure P4–8.

Command: **line** (ENTER)
Specify first point: *(invoke the intersection object snap, and select the intersection of LINE 3B and CIRCLE 3 as indicated in Figure P4–8)*
Specify next point or [Undo]: *(invoke the perpendicular object snap, and select LINE 3A as indicated in Figure P4–8)*
Specify next point or [Undo]: *(press* (ENTER) *to terminate the command sequence)*

Your drawing will appear as shown in Figure P4–8.

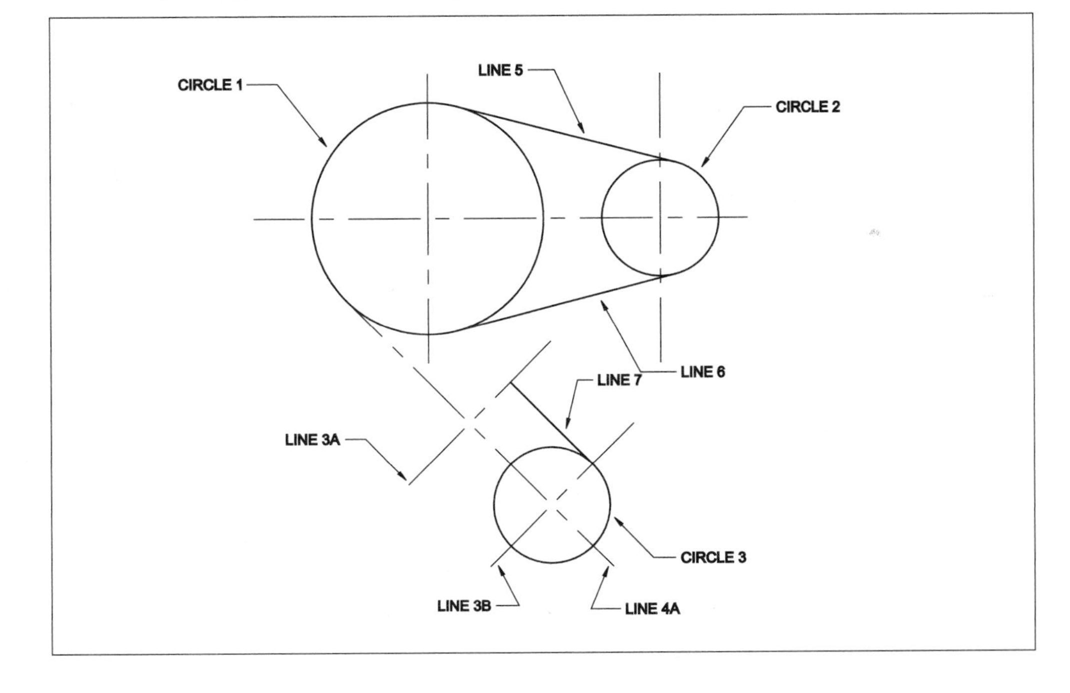


**Figure P4–8** *Drawing after drawing LINE 7*

**Step 13:** Draw two circles, CIRCLE 4 and CIRCLE 5, as shown in Figure P4–9, by invoking the CIRCLE command's Tan Tan Radius option.

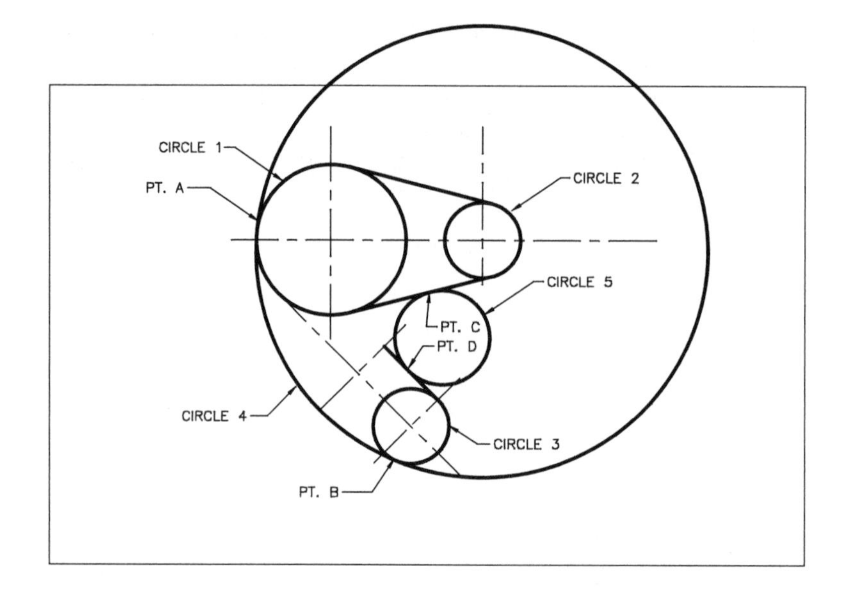

**Figure P4–9** *Placement of circle 4 and circle 5 tangent to two objects*

Command: **circle** (ENTER)
Specify center point for circle or [3P/2P/Ttr (tan tan radius)]: **t** (ENTER)
Specify point on object for first tangent of circle: *(select CIRCLE 1 at point A, as shown in Figure P4–9)*
Specify point on object for second tangent of circle: *(select CIRCLE 3 at point B, as shown in Figure P4–9)*
Specify radius of circle: **6** (ENTER)

Invoke the CIRCLE command again:

Command: **circle** (ENTER)
Specify center point for circle or [3P/2P/Ttr (tan tan radius)]: **t** (ENTER)
Specify point on object for first tangent of circle: *(select the line at point C, as shown in Figure P4–9)*
Specify point on object for second tangent of circle: *(select the line at point D, as shown in Figure P4–9)*
Specify radius of circle <6.0000>: **1.25** (ENTER)

Your drawing will appear as shown in Figure P4–9.

**Step 14:** Invoke the TRIM command from the Modify toolbar to modify the two circles previously drawn in Step 13.

Command: **trim** (ENTER)
Current settings: Projection = UCS Edge = None
Select cutting edges ...
Select objects: *(select CIRCLE 1, CIRCLE 3, LINE 6, and LINE 7, as in Figure P4–10, as the cutting edges, and press* ENTER*)*

Select object to trim or shift-select to extend or [Project/Edge/Undo]:
*(select CIRCLE 5 at point A, as shown in Figure P4–10)*
Select object to trim or shift-select to extend or [Project/Edge/Undo]:
*(select CIRCLE 4 at point B, as shown in Figure P4–10)*
Select object to trim or shift-select to extend or [Project/Edge/Undo]: *(press* ENTER *to complete the command sequence)*

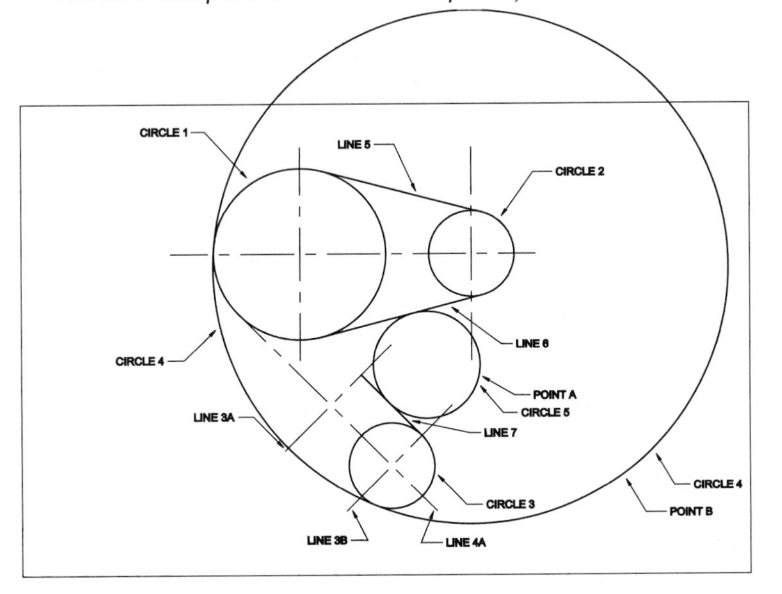


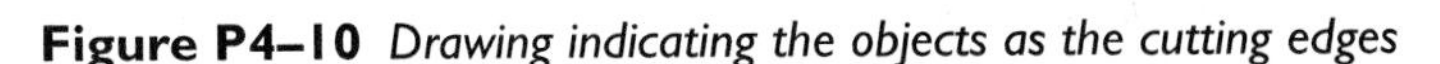
**Figure P4–10** *Drawing indicating the objects as the cutting edges*

Your drawing will appear as shown in Figure P4–11.

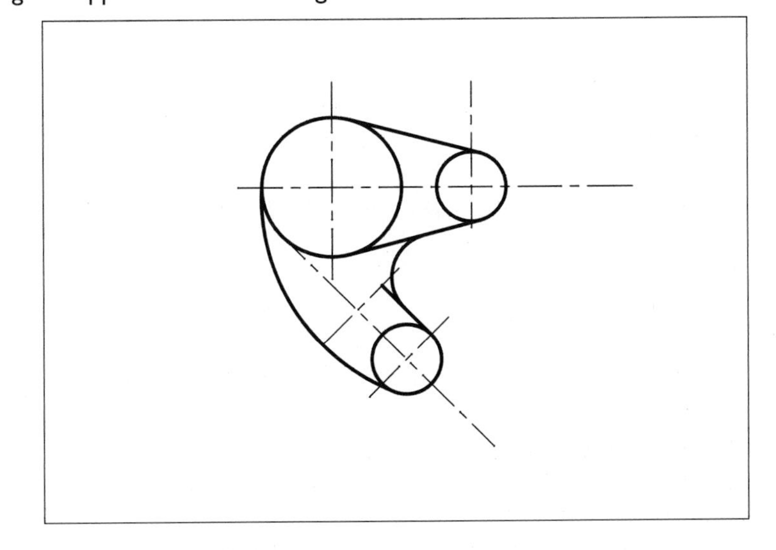

**Figure P4–11** *Circles modified with the* TRIM *command*

**Step 15:** Use the TRIM command to modify the circles and lines to achieve the layout shown in Figure P4–12. Invoke the TRIM command and select lines 5, 6, 7, and arc 1 as cutting edges, as shown in Figure P4–13. Select circles 1, 2, and 3, as shown in Figure P4–13, as objects to trim.

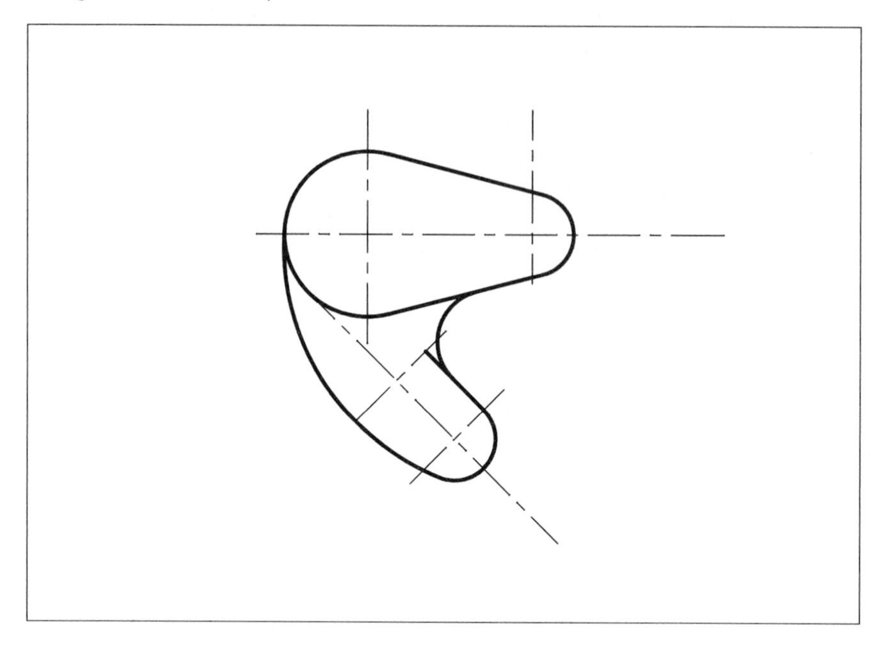

**Figure P4–12** *Drawing as it will look after modifying the circles and lines*

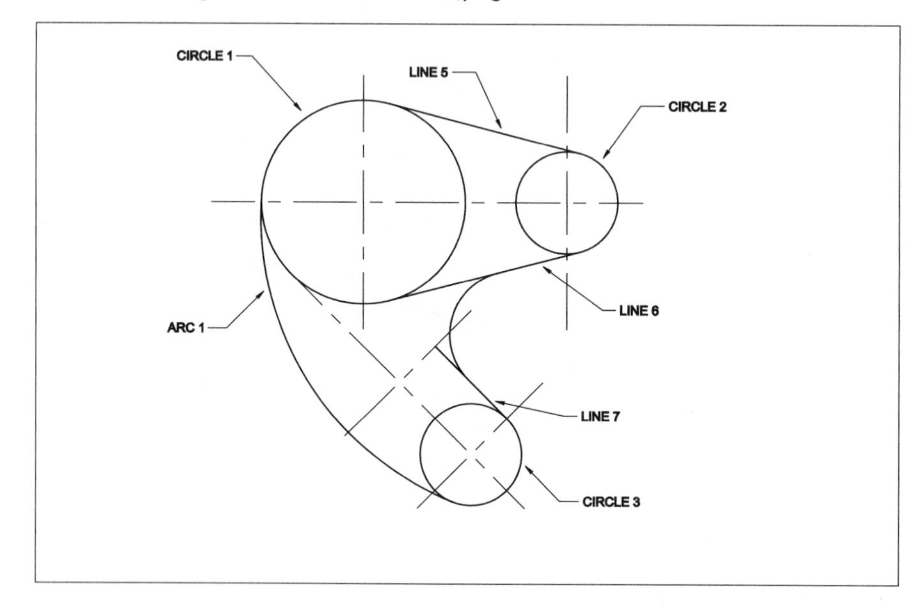

**Figure P4–13** *Drawing showing the selection of the objects as cutting and trim edges*

**Step 16:** Continue to use the TRIM command to modify the figure to achieve the layout shown in Figure P4–14. Invoke the TRIM command again and select arcs 1 and 2 as cutting edges, as shown in Figure P4–15. Select lines 6, 7, and arc 3 to trim, as shown in Figure P4–15. Your drawing will appear as shown in Figure P4–14.

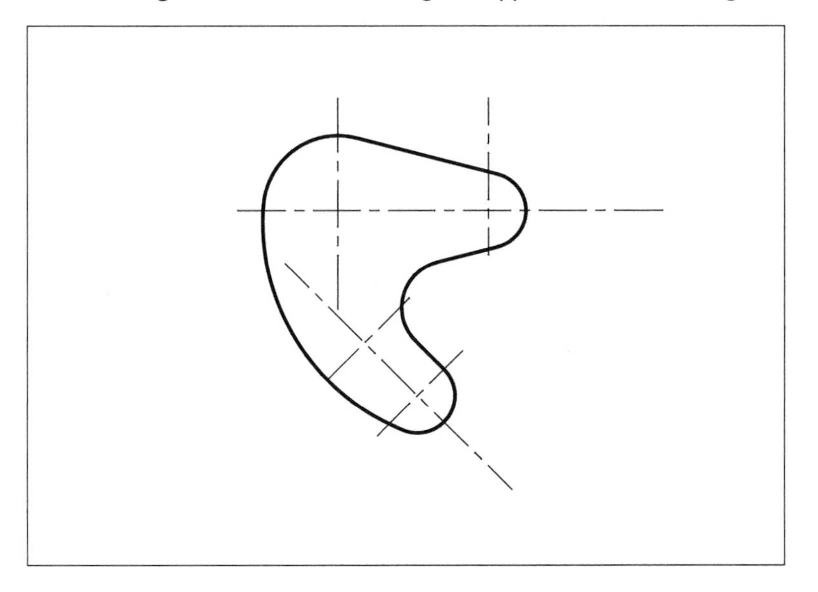

**Figure P4–14** *Drawing as it will look after modifying the arcs and lines*

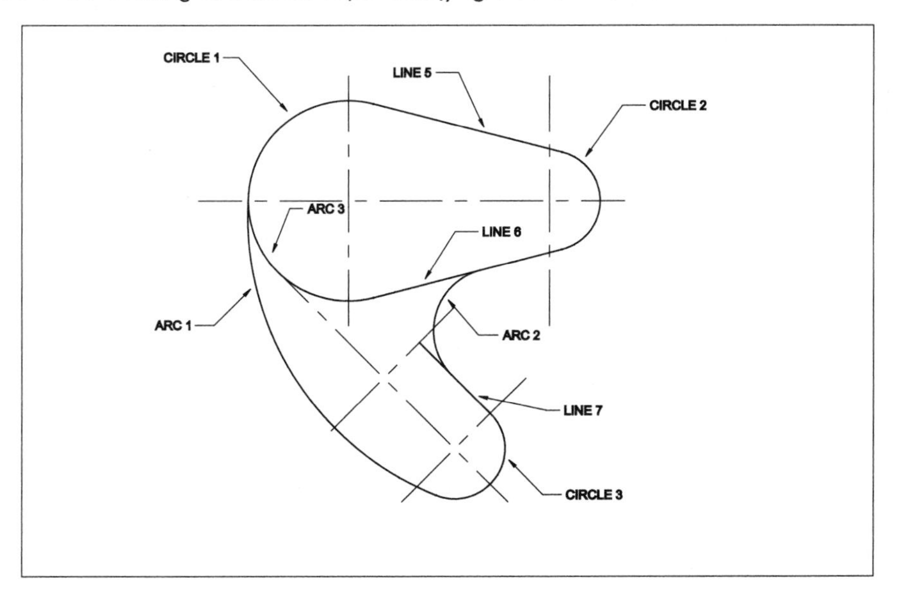

**Figure P4–15** *Drawing showing the selection of the objects as cutting and trim edges*

**Step 17:** Draw the circles necessary to complete the design. Invoke the CIRCLE command and draw circles with points 1, 2, and 3 as center points and with a radius of 0.5. Draw another circle with center point at point 4 and with a radius of 1.5, as shown in Figure P4–16.

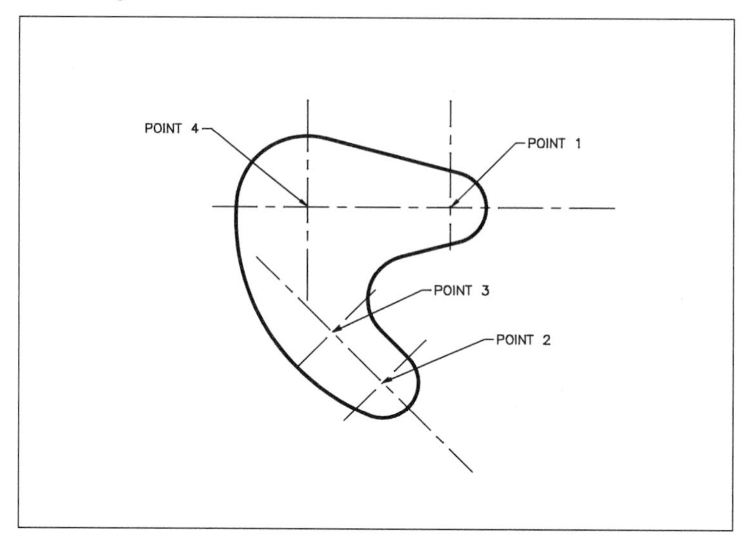


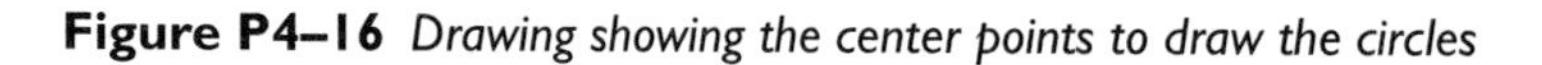

**Figure P4–16** *Drawing showing the center points to draw the circles*

**Note:** Make sure to use the object snap tool intersection when selecting the center points.

After drawing the circles, your drawing will appear as shown in Figure P4–17.

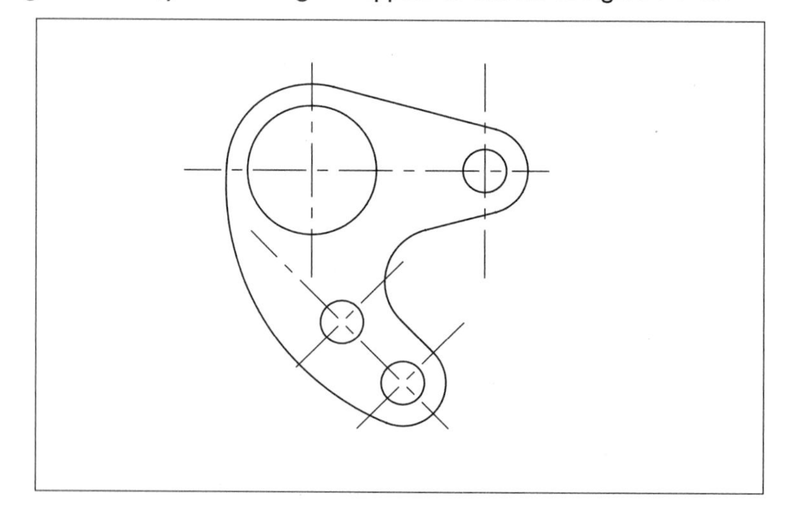

**Figure P4–17** *Design with the addition of circles*

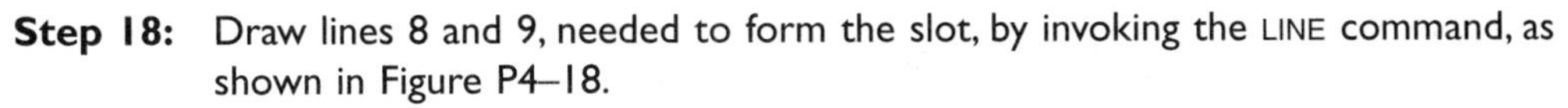

**Step 18:** Draw lines 8 and 9, needed to form the slot, by invoking the LINE command, as shown in Figure P4–18.

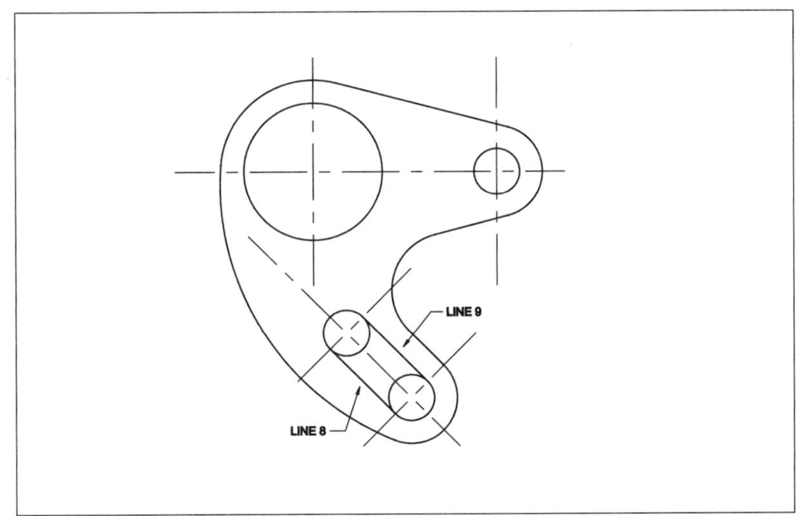


**Figure P4–18** *Addition of lines drawn to form the slot*

**Note:** Use the object snap tool intersection to snap to the intersection of the centerlines and the circle, as shown in Figure P4–18.

**Step 19:** Invoke the TRIM command and select lines 8 and 9 as cutting edges. Next, select the small circles as objects to trim at points 5 and 6, as shown in Figure P4–19.

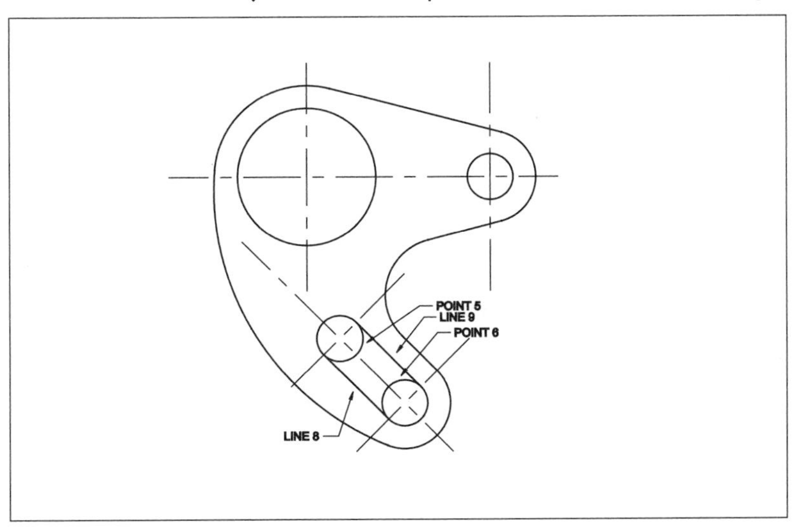


**Figure P4–19** *Trim edges and circles to trim*

Your drawing will appear as shown in Figure P4–20.

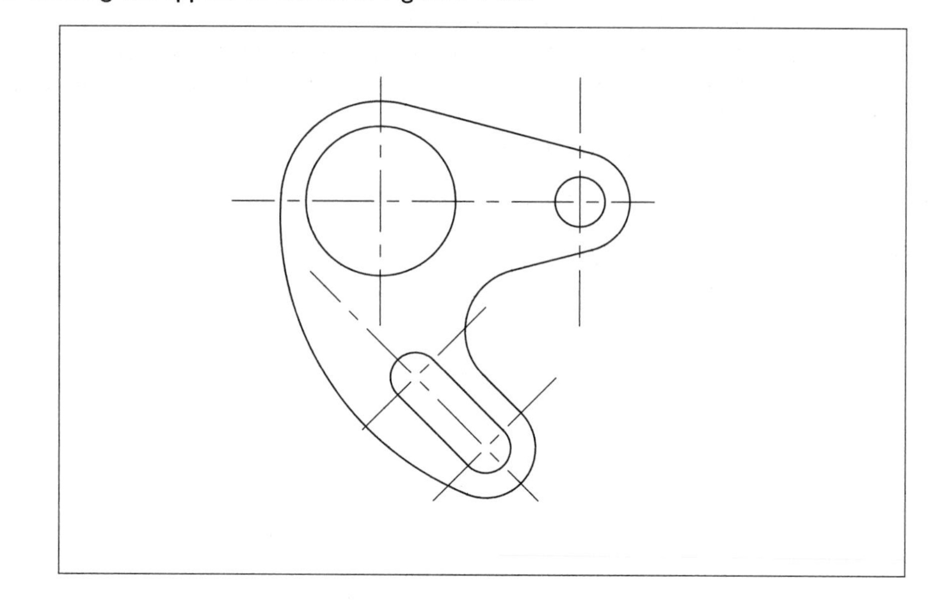

**Figure P4–20** *Drawing after the circles have been trimmed*

**Step 20:** Invoke the POLYGON command from the Draw toolbar to draw the polygon shown in Figure P4–21.

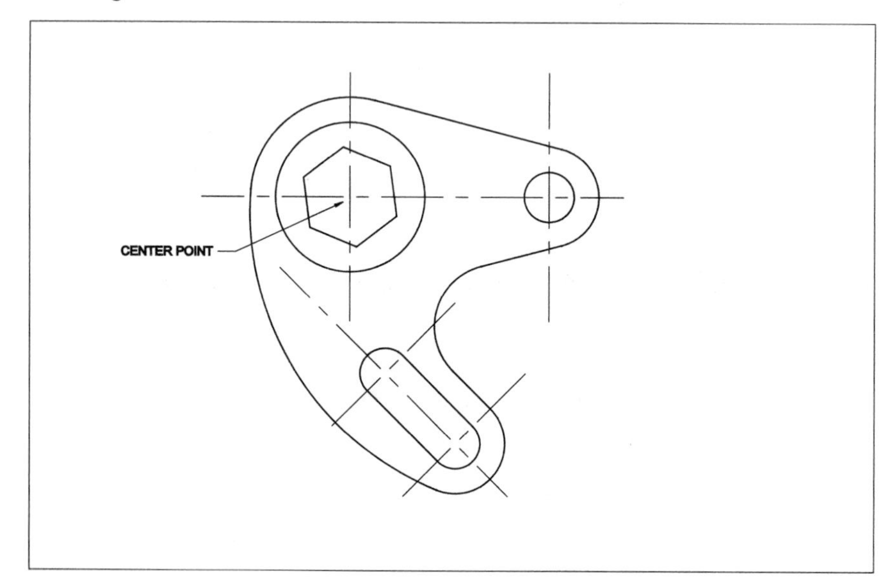


**Figure P4–21** *Design with a six-sided polygon*

Command: **polygon** (ENTER)
Enter number of sides <4>: **6** (ENTER)
Specify center of polygon or [Edge]: *(select center point as in Figure P4–21)*
Enter an option [Inscribed in circle/Circumscribed about circle] <I>: **c** (ENTER)
Specify radius of circle: **@0.875<67.5** (ENTER)

Your drawing will appear as shown in Figure P4–21.

**Step 21:** Invoke the BREAK command from the Modify toolbar and select points 7 and 8, as shown in Figure P4–22, to break the centerline.

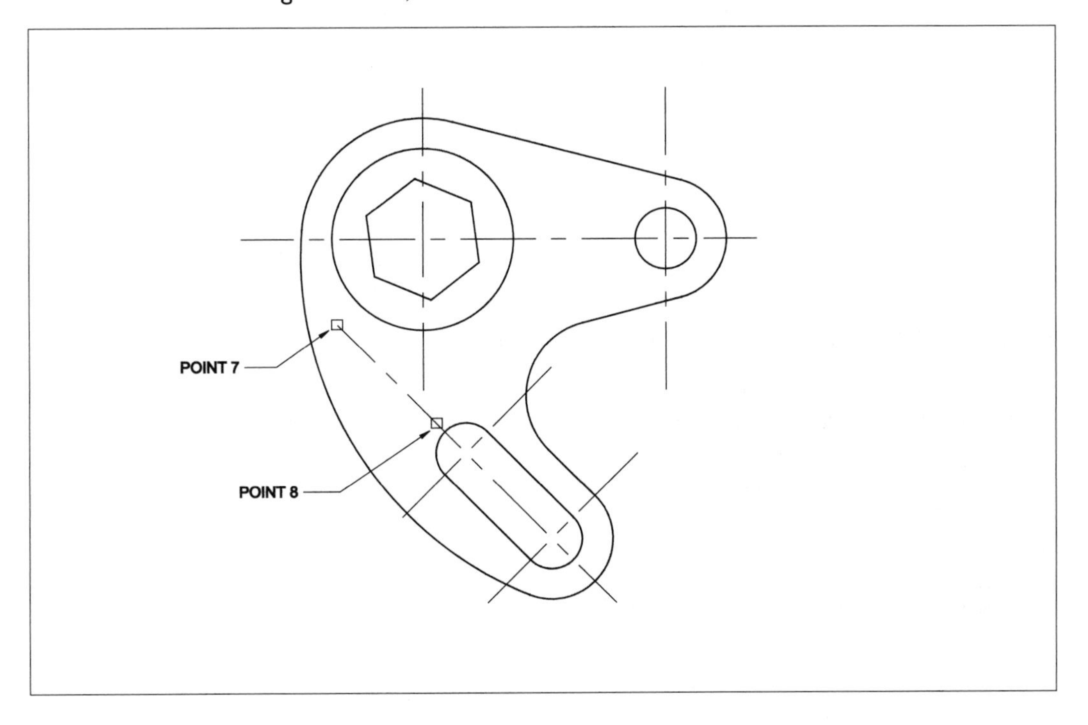


**Figure P4–22** *Showing the points where the BREAK command will remove part of the centerline*

Your drawing will appear as shown in Figure P4–23 after removing the specified portion of the centerline.

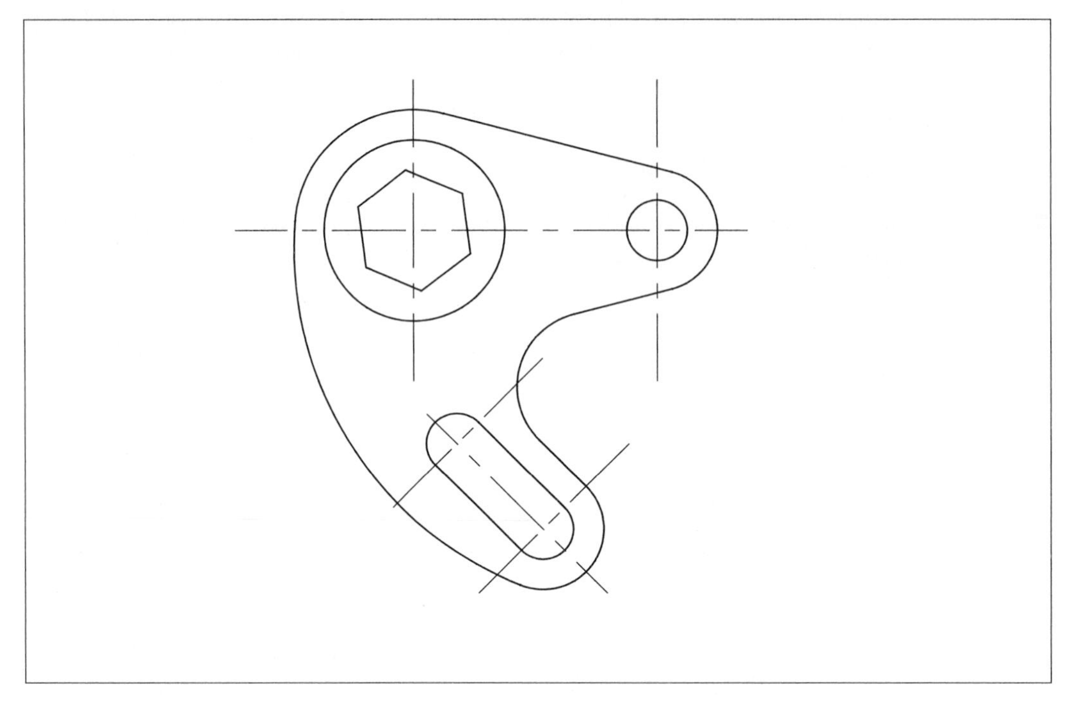

**Figure P4–23** *Drawing with the specified portion of the centerline removed*

**Step 22:** Save the drawing as PROJ4.DWG.

Congratulations! You have just successfully applied several AutoCAD concepts in creating a rather complex mechanical drawing.

## EXERCISE 4–1

Create the drawing of the sleeve block shown in Figure Ex4–1, according to the settings given in the following tables.

| Settings | Value |
|---|---|
| 1. Units | Decimal |
| 2. Limits | Lower left corner: 0,0<br>Upper right corner: 12,9 |
| 3. Grid | 0.50 |
| 4. Snap | 0.25 |

| Settings | Value | | |
|---|---|---|---|
| 5. Layers | *NAME* | *COLOR* | *LINETYPE* |
| | Object | Green | Continuous |
| | Center | Red | Center |
| | Hidden | Magenta | Hidden |
| | Text | Blue | Continuous |

**Hints:** The object lines for the sleeve block can be drawn by using the LINE, POLYGON, and CIRCLE commands, and the text objects can be drawn with the TEXT or MTEXT commands. The arcs can be created using the FILLET command. Do not dimension this drawing. Use the TEXT command to label the views only.

Use two circles and two lines as the basis of the slot. Then use the TRIM command to remove the inner halves of the circles, leaving the outer arcs that, with the two lines, comprise the slot.

Other arcs (.5 and .125 radii) can be created by using the FILLET command with the correctly specified radius.

You will probably have to object snap to the "intersections" of the apexes of the hexagon in the top view and draw lines perpendicular to the base of the front view in order to locate them correctly. Then use the TRIM command to remove the unwanted portions of the lines.

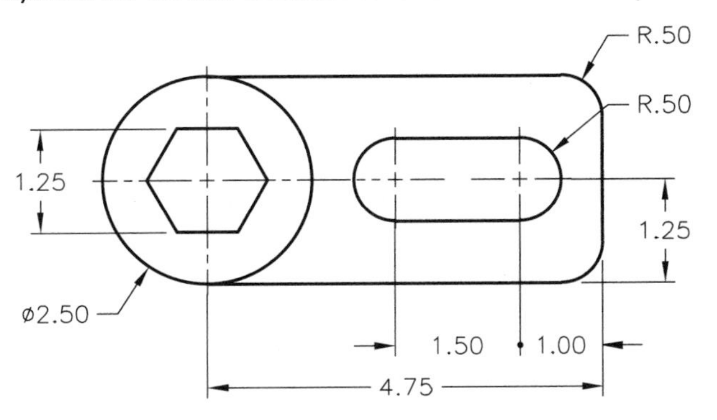

R.125
R.125
3.00
1.00

FRONT VIEW

**Figure Ex4–1** *Sleeve block*

## EXERCISES 4–2 & 4–3

Create drawings of the brackets and clamps shown in Figures Ex4–2 and Ex4–3 according to the settings given in the following table.

| Settings | Value | | |
|---|---|---|---|
| 1. Units | Decimal | | |
| 2. Limits | Lower left corner: 0,0 | | |
| | Upper right corner: 144,108 | | |
| 3. Grid | 8 | | |
| 4. Snap | 4 | | |
| 5. Layers | *NAME* | *COLOR* | *LINETYPE* |
| | Object | Green | Continuous |
| | Center | Red | Center |
| | Hidden | Magenta | Hidden |
| | Text | Blue | Continuous |

**Hint:** To lay out the views, you need to estimate the dimensions of the spaces between the objects. Then add up the total of the objects and estimated distances between to determine the spaces left over around the whole grouping. Do not dimension.

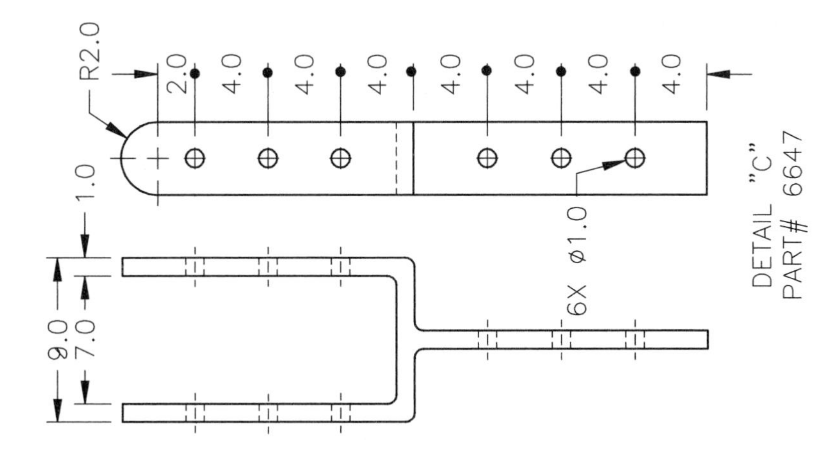


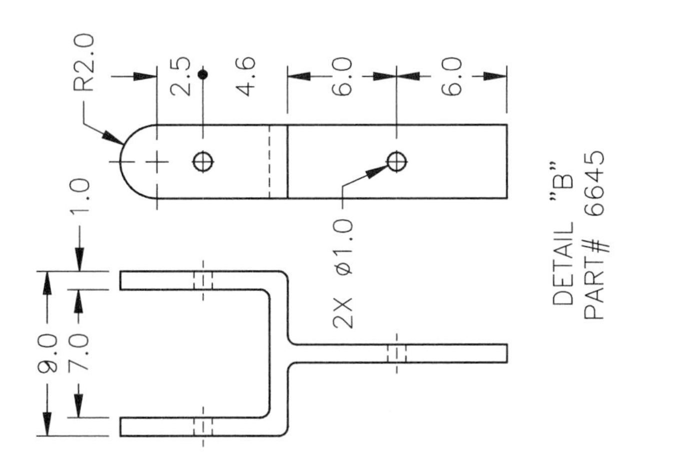


ALL ROUNDS AND FILLETS R.5

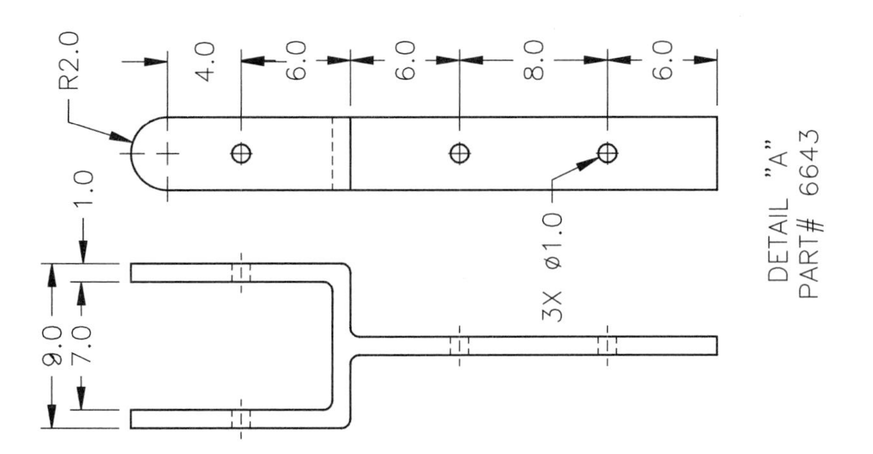


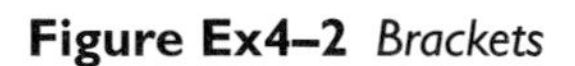
**Figure Ex4–2** *Brackets*

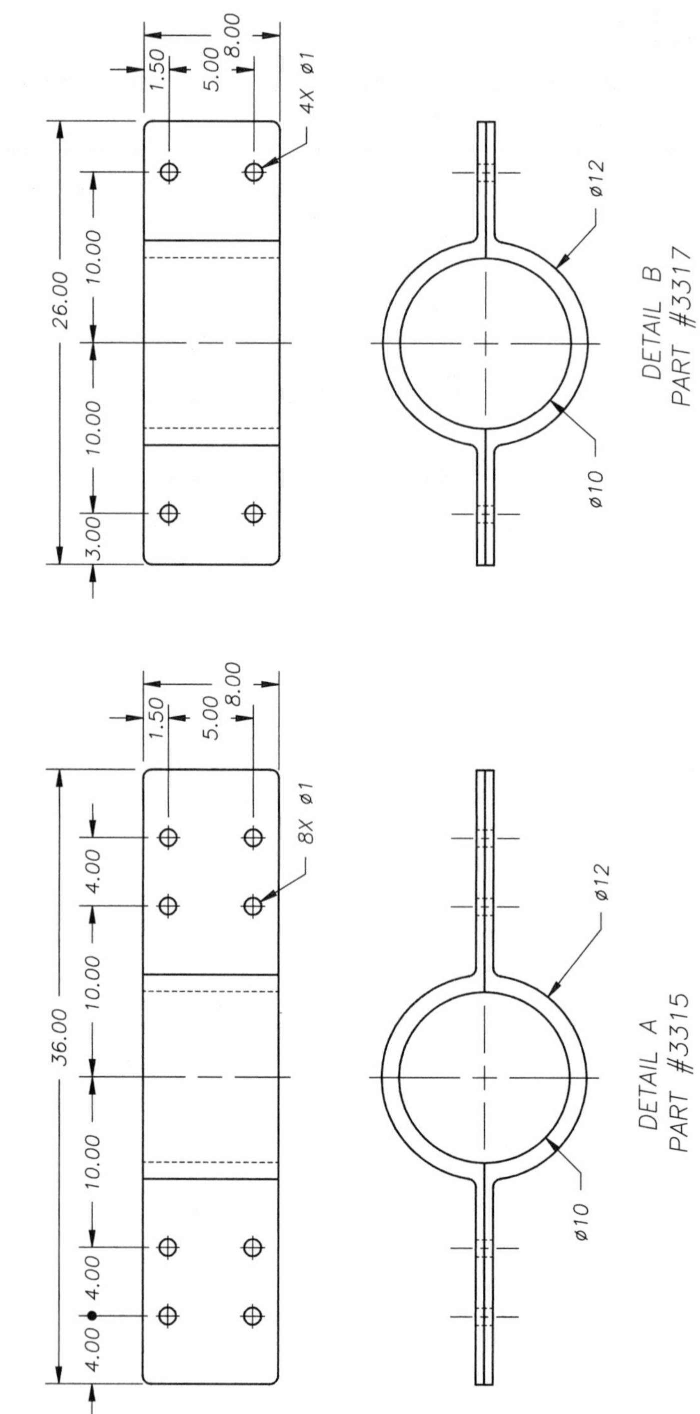

**Figure Ex4–3** *Clamps*

## EXERCISE 4–4

Create a drawing of the support bracket (top, front, and section view) shown in Figure Ex4–4 according to the settings given in the following table:

| Settings | Value | | |
|---|---|---|---|
| 1. Units | Decimal with 2 decimal places | | |
| 2. Limits | Lower left corner: 0,0 | | |
| | Upper right corner: 30,20 | | |
| 3. Grid | .25" | | |
| 4. Snap | .125" | | |
| 5. Layers | *NAME* | *COLOR* | *LINETYPE* |
| | Border | Cyan | Continuous |
| | Object | Red | Continuous |
| | Center | Green | Continuous |
| | Openings | Magenta | Continuous |
| | Hidden | Blue | Hidden |
| | Notes | Magenta | Continuous |

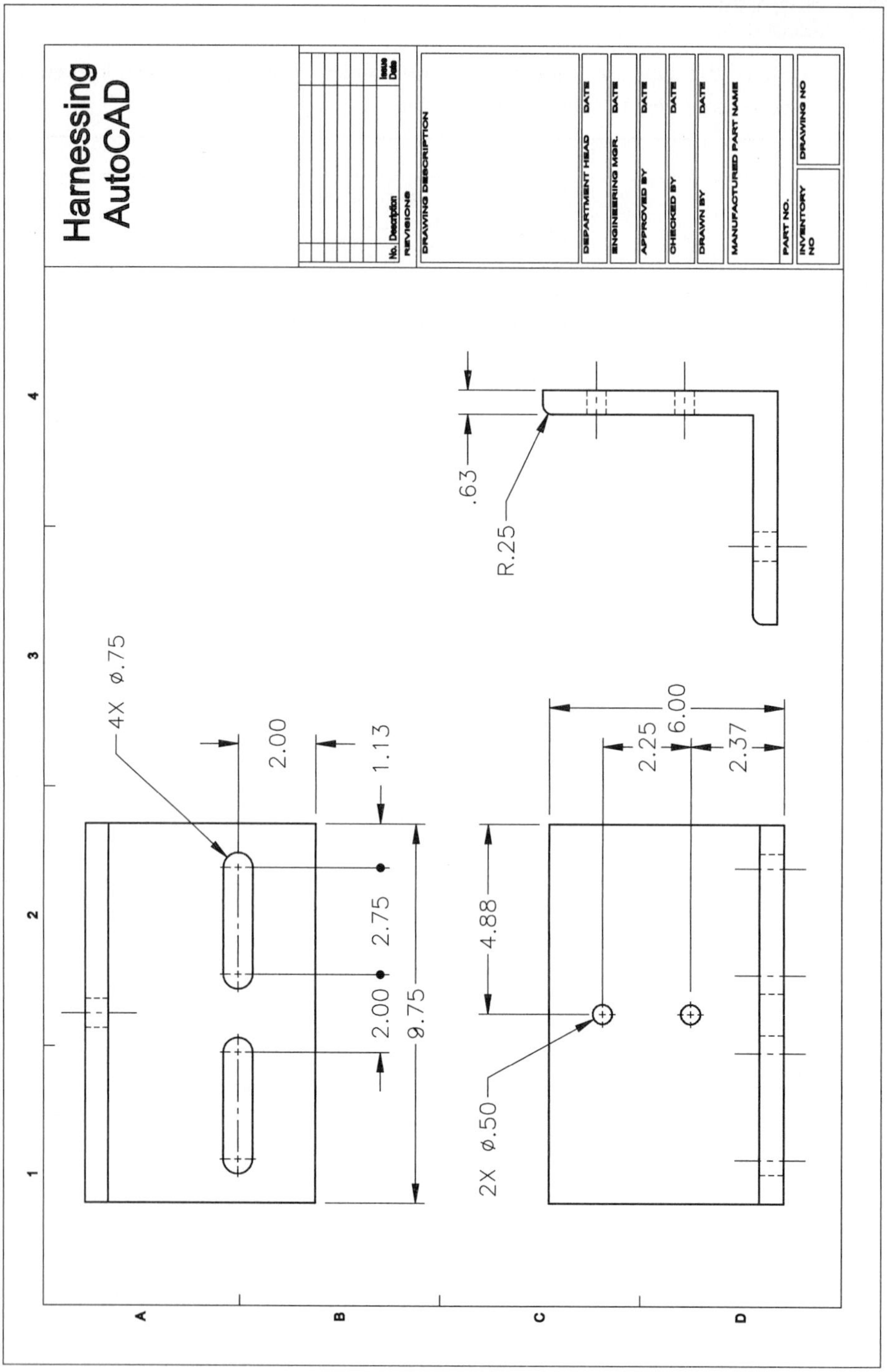

Harnessing AutoCAD

No. Description
REVISIONS
Issue Date
DRAWING DESCRIPTION
DEPARTMENT HEAD DATE
ENGINEERING MGR. DATE
APPROVED BY DATE
CHECKED BY DATE
DRAWN BY DATE
MANUFACTURED PART NAME
PART NO.
INVENTORY NO
DRAWING NO

1 2 3 4
A B C D

4X ⌀.75
2.00
1.13
2.00 2.75
9.75
2X ⌀.50
4.88
2.25
6.00
2.37
.63
R.25

**Figure Ex4–4** *Support Bracket*

## EXERCISE 4–5

Create a drawing of the clevis (two views) shown in Figure Ex4–5 according to the settings given in the following table:

| Settings | Value | | |
|---|---|---|---|
| 1. Units | Decimal with 3 decimal places | | |
| 2. Limits | Lower left corner: 0,0 | | |
| | Upper right corner: 34,26 | | |
| 3. Grid | .25" | | |
| 4. Snap | .125" | | |
| 5. Layers | *NAME* | *COLOR* | *LINETYPE* |
| | Border | Red | Continuous |
| | Object | Green | Continuous |
| | Center | Magenta | Continuous |
| | Hole | Blue | Continuous |
| | Hidden | Blue | Hidden |

**Hint:** Use the table below for the dimensions.

Dimensions in this table are given in inches.

| Clevis Number | Max. D | Max. p | B | N | A | Grip | W | T |
|---|---|---|---|---|---|---|---|---|
| 8 | 3 | 4 | 8 | 4 | 10 | Plate + ¼" * | 4 | 1 ½"(+ 1/8"-0) |

*The thickness of the Plate = 3 ¾".

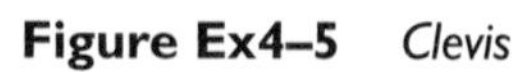


**Figure Ex4–5** *Clevis*

## EXERCISE 4–6

Create a structural steel framing plan, as shown in Figure Ex4–6, according to the settings given in the following table.

| Settings | Value | | |
|---|---|---|---|
| 1. Units | Architectural | | |
| 2. Limits | Lower left corner: –10'–0",–10'–0"<br>Upper right corner: 50'–0",35'–0" | | |
| 3. Grid | 12" | | |
| 4. Snap | 6" | | |
| 5. Layers | *NAME* | *COLOR* | *LINETYPE* |
| | Construction | Cyan | Continuous |
| | Border | Red | Continuous |
| | Columns | Green | Continuous |
| | Beams | Magenta | Continuous |
| | Text | Blue | Continuous |

**Hints:** Plan ahead before drawing the layout. The columns and beams in the structural steel framing plan can be drawn by using the PLINE command, and the text objects can be drawn with the TEXT or MTEXT commands. Construction lines can be drawn by using the XLINE and RAY commands combined with the ARRAY command.

Draw the border from coordinates –9'–6",–9'–6" to 49'–6",34'–6" using the RECTANGLE command.

Draw a vertical construction line by means of the XLINE command through a point whose X coordinate is 0'–0".

Copy the vertical construction line using the Multiple option of the COPY command.

Draw two horizontal construction lines by means of the RAY command from the point whose coordinates are 24'–0",0'–0". One of the lines is drawn to the left and the other to the right. Use the ARRAY command with the Rectangular option set to four rows for the left construction ray, at a spacing of 8'–0", and five rows of the right construction ray, at a spacing of 6'–0". The drawing should look like Figure Ex4–6a.

Using the PLINE command, draw the columns with a width of 2" and the beams with a width of 1". Draw the beams from intersection to intersection of the construction lines, and then, using the BREAK command, remove the sections of the polylines either 6" or 12" (make sure Snap is set to ON) from the intersections, depending on which direction the column is oriented.

After you have finished drawing, either the Construction layer can be turned OFF or the construction lines can be erased. Your completed drawing should look like the finished framing plan. Do *not* dimension the drawing.

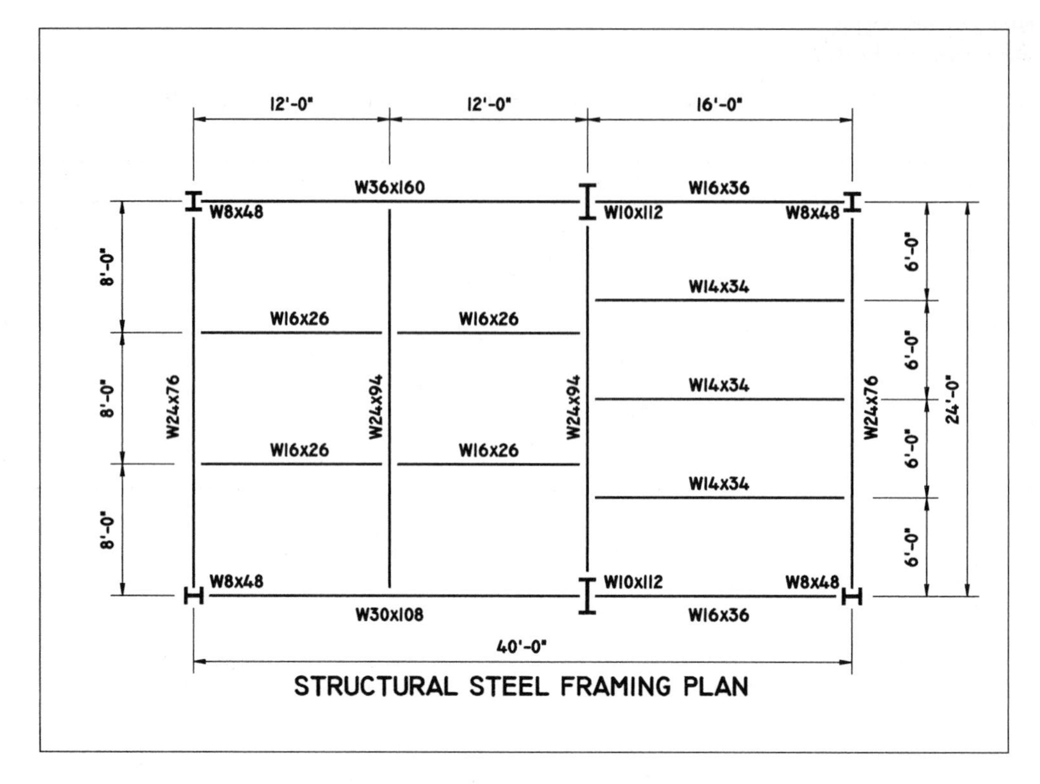


**Figure Ex4–6** *The completed drawing*

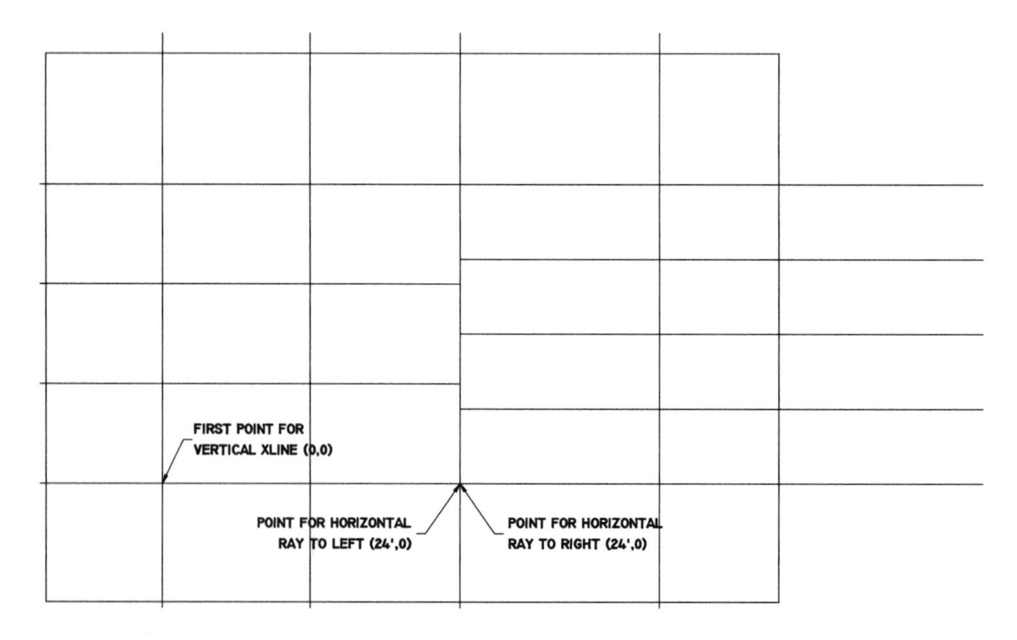


**Figure Ex4–6b** *Construction line layout*

## EXERCISE 4–7

Create a drawing of the ornamental fence shown in Figure Ex4–7 according to the settings given in the following table:

| Settings | Value | | |
|---|---|---|---|
| 1. Units | Architectural | | |
| 2. Limits | Lower left corner: 0,0 | | |
| | Upper right corner: 18',14' | | |
| 3. Grid | 1" | | |
| 4. Snap | .5" | | |
| 5. Layers | *NAME* | *COLOR* | *LINETYPE* |
| | Border | Red | Continuous |
| | Minorvertical | Green | Continuous |
| | Majorvertical | Magenta | Continuous |
| | Circles | Blue | Continuous |

**Hint:** Some of the commands that can be used to create this drawing are XLINE, RAY, POLYGONS, COPY, ARRAY, MIRROR, TRIM, BREAK, and EXTEND commands. Do not dimension.

architectural EXERCISE

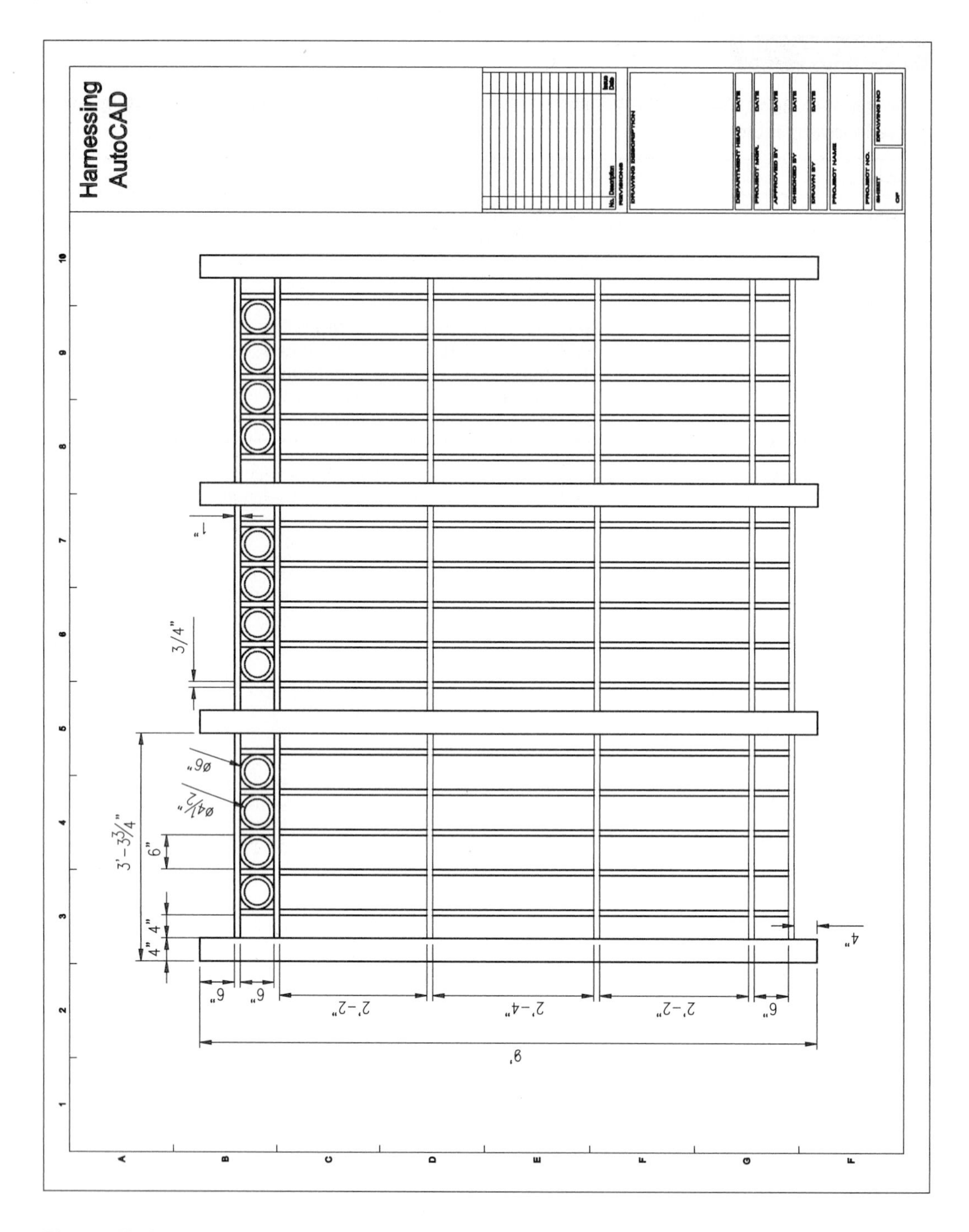


**Figure Ex4–7** *Ornamental fence*

## EXERCISE 4–8

Create a drawing of the stair handrails shown in Figure Ex4–8 according to the settings given in the following table:

| Settings | Value | | |
|---|---|---|---|
| 1. Units | Architectural | | |
| 2. Limits | Lower left corner: 0,0 | | |
| | Upper right corner: 12',9' | | |
| 3. Layers | *NAME* | *COLOR* | *LINETYPE* |
| | Border | Red | Continuous |
| | Stringer | Green | Continuous |
| | Handrail | Magenta | Continuous |
| | Notes | Blue | Continuous |
| | RiserandTread | White | Hidden |

**Hint:** Some of the commands that can be used to create this drawing are XLINE, RAY, COPY, TRIM, BREAK, and EXTEND commands. Do not dimension.

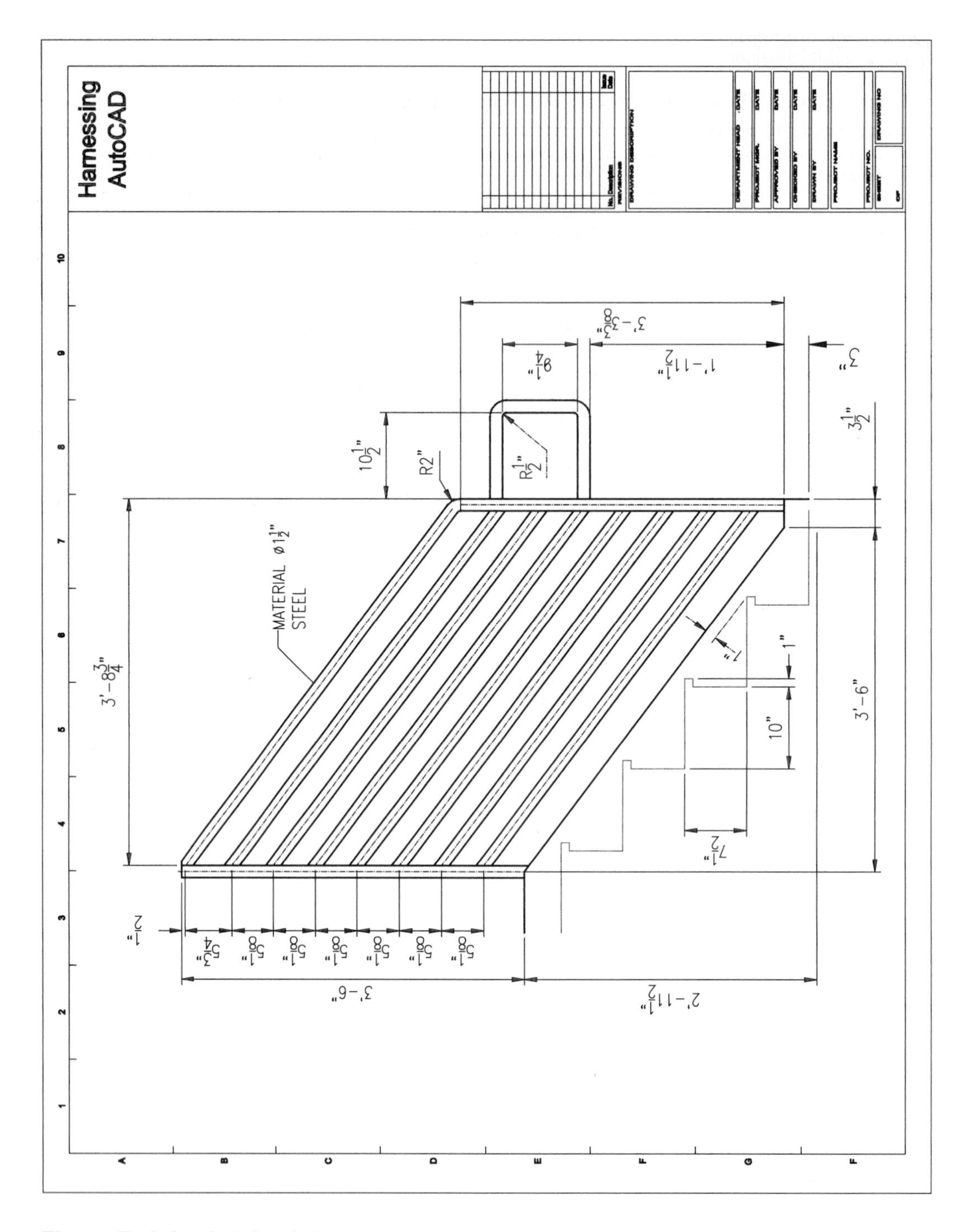


**Figure Ex4–8** *Stair handrail*

## EXERCISE 4–9

Create a drawing of the recessed store front (plan and elevation views) shown in Figure Ex4–9, according to the settings given in the following table:

| Settings | Value | | |
|---|---|---|---|
| 1. Units | Architectural | | |
| 2. Limits | Lower left corner: 0,0 | | |
| | Upper right corner: 38',30' | | |
| 3. Grid | 2" | | |
| 4. Snap | 1" | | |
| 5. Layers | *NAME* | *COLOR* | *LINETYPE* |
| | Border | Red | Continuous |
| | glass | Green | Continuous |
| | doors | Magenta | Continuous |
| | walls | Blue | Continuous |
| | mullions | White | Continuous |

**Hint:** Some of the commands that can be used to create this drawing are XLINE, RAY, POLYGONS, COPY, ARRAY, MIRROR, TRIM, BREAK, and EXTEND commands. Do not dimension.

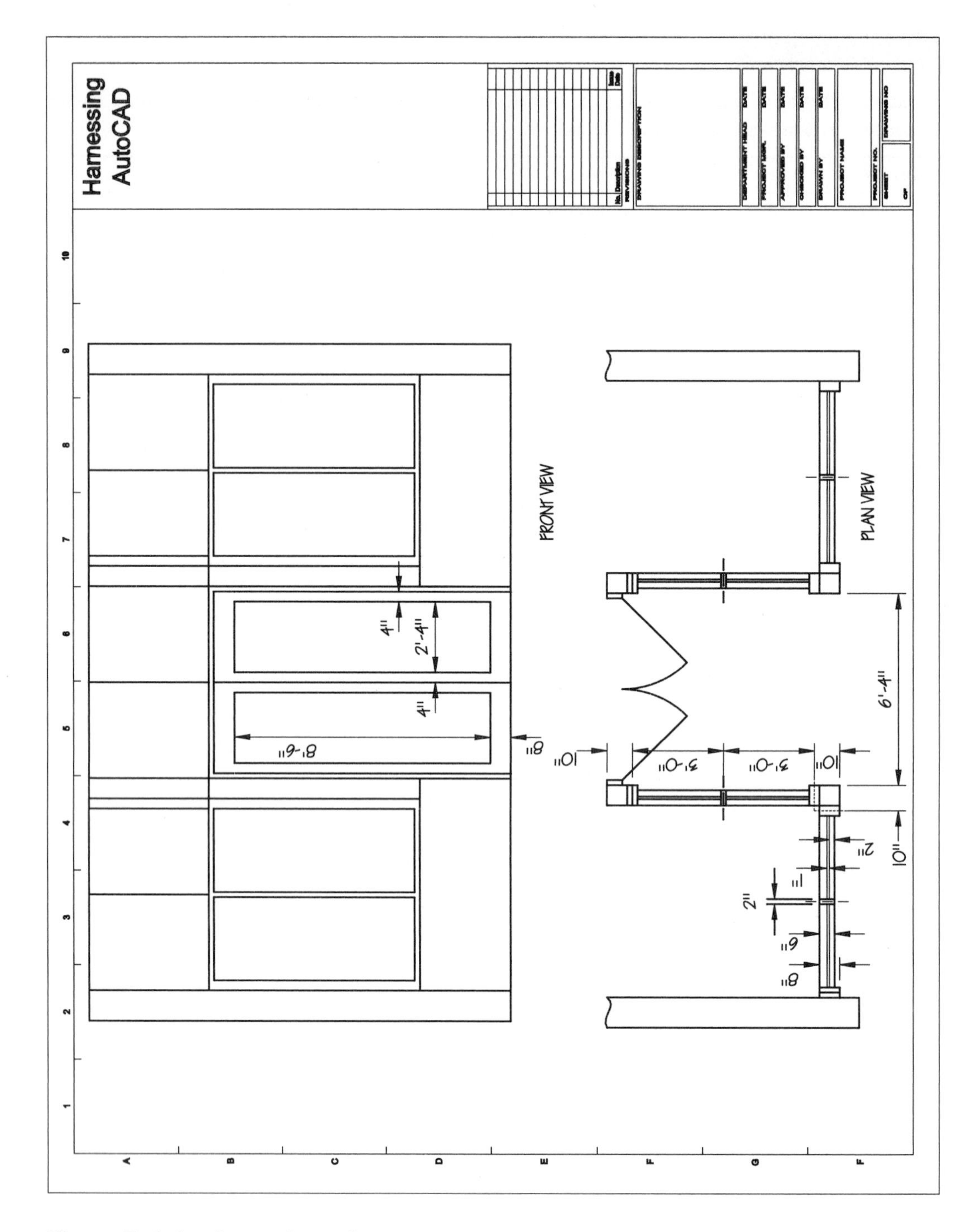


**Figure Ex4–9** *Recessed store front*

## EXERCISE 4–10

Create the drawing of the vending machine room (plan view) shown in Figure Ex4–10, according to the settings given in the following table:

| Settings | Value | | |
|---|---|---|---|
| 1. Units | Architectural | | |
| 2. Limits | Lower left corner: 0,0 | | |
| | Upper right corner: 30',22' | | |
| 3. Grid | 1" | | |
| 4. Snap | .5" | | |
| 5. Layers | *NAME* | *COLOR* | *LINETYPE* |
| | Border | Red | Continuous |
| | equipment | Green | Continuous |
| | walls | Magenta | Continuous |
| | shelves | Blue | Continuous |
| | doors | White | Continuous |

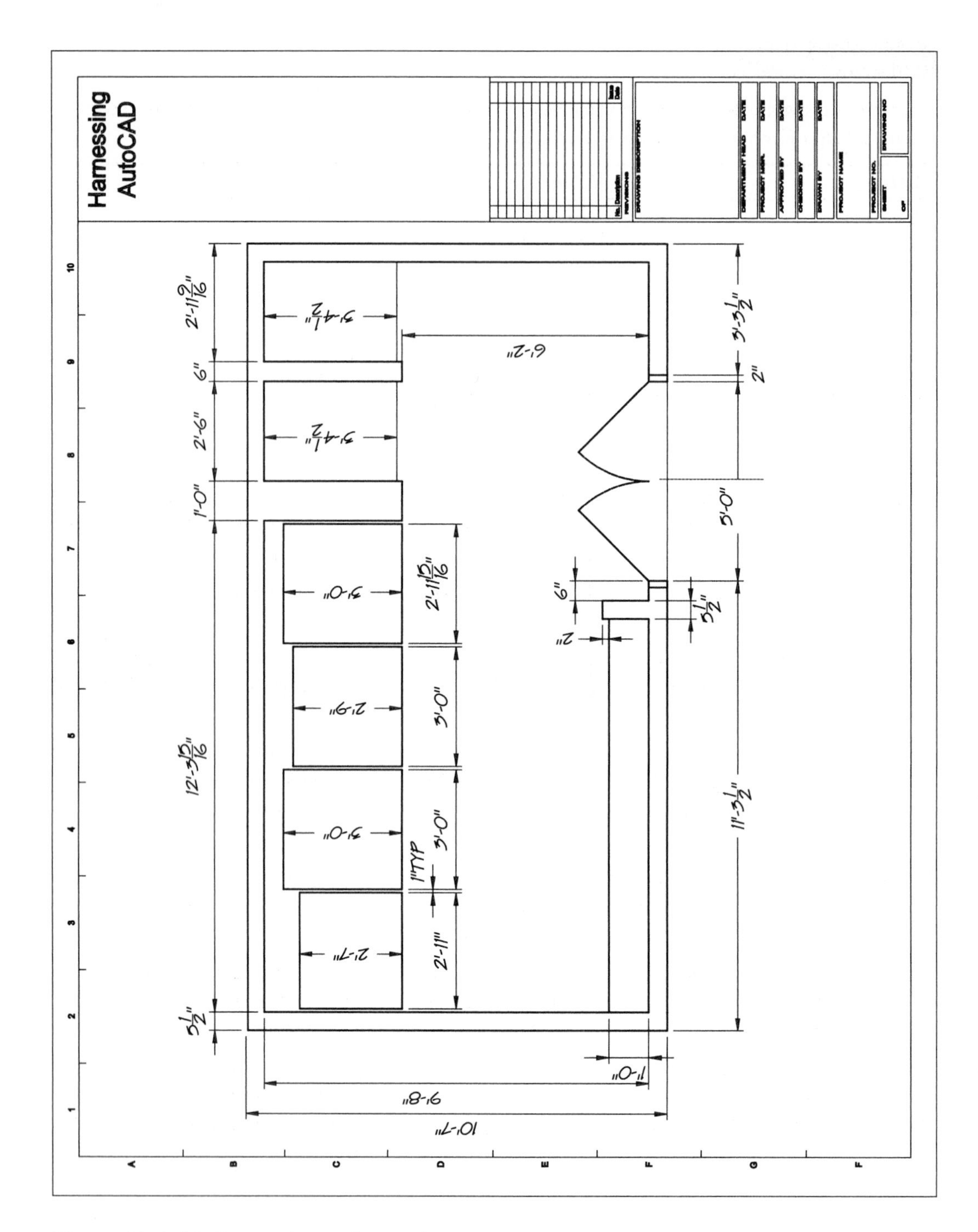


**Figure Ex4–10** *Recessed store front*

## EXERCISE 4–11

Create the survey plot plan drawing shown in Figure Ex4–11 with the settings given in the following table. (refer to Exhibit A for description)

| Settings | Value | | |
|---|---|---|---|
| 1. Units/Angle | Decimal/Surveyor | | |
| 2. Limits | Lower left corner: 0,0 | | |
| | Upper right corner: 680,440 | | |
| 3. Dimscale | 40 | | |
| 4. Ltscale | 20 | | |
| 5. Layers | *NAME* | *COLOR* | *LINETYPE* |
| | Property | Green | Continuous |
| | Text | Red | Center |

### EXHIBIT A: METES & BOUNDS

### Tracts 1

That certain 0.3443 acres of land, said tract being in the Ruben White Survey Abstract 24, same being out Lot 5, Blk. 5 Section 2 of DreamLand Place Subdivision, Harris County, Texas, Vol. Pg. Harris County Map Records, and being more particularly described by metes and bounds as follows;

*Commencing* at the intersection of F.M. 1942 (Gulf Pump Road) variable right of way, ie 60' at Lot 5, and the southwest corner of Blanchard Drive, a 60' right of way.

*THENCE* South 89 degrees 30' minutes 00" seconds West along the North line of F.M. 1942, a distance of 120.00' to a point for a corner, a SET #4 rebar and the *PLACE OF BEGINNING THE TRACT HEREIN DESCRIBED;*

THENCE continuing South 89 degrees 30' minutes 00" seconds West along the North line of F.M. 1942, and a South line of lot 5, a distance of 75.00' to a found #4 rebar for a corner in the West line of this Tract 1 and the East line of Tract 2, same being the Westerly portion of Lot 5, a point for a corner;

*THENCE* North 00 degrees 05' minutes 00" seconds West along the east line of Tract 2 and the West line of this tract a distance of 200.01 to a set #4 rebar in the center line of a drainage easement and the North line of Lot 5 a set #4 rebar for a corner;

*THENCE* North 89 degrees 30' minutes 00" seconds East along the North line of this tract and the South line of lot 7, a distance 75.00 to a set #4 rebar in the West line of Lots 3&4, on Blanchard Street, a point for a corner;

*THENCE* South 00 degrees 05' minutes 00" seconds East along the West line of Lots 3, 2 &1, on Blanchard Street, a distance of 200.00' to a set #4 rebar in the North right of way line of F.M. 1942 for a corner and *THE PLACE OF BEGINNING AND CONTAINING 0.3443 acres OF LAND OR 15,000.00 square feet of land.*

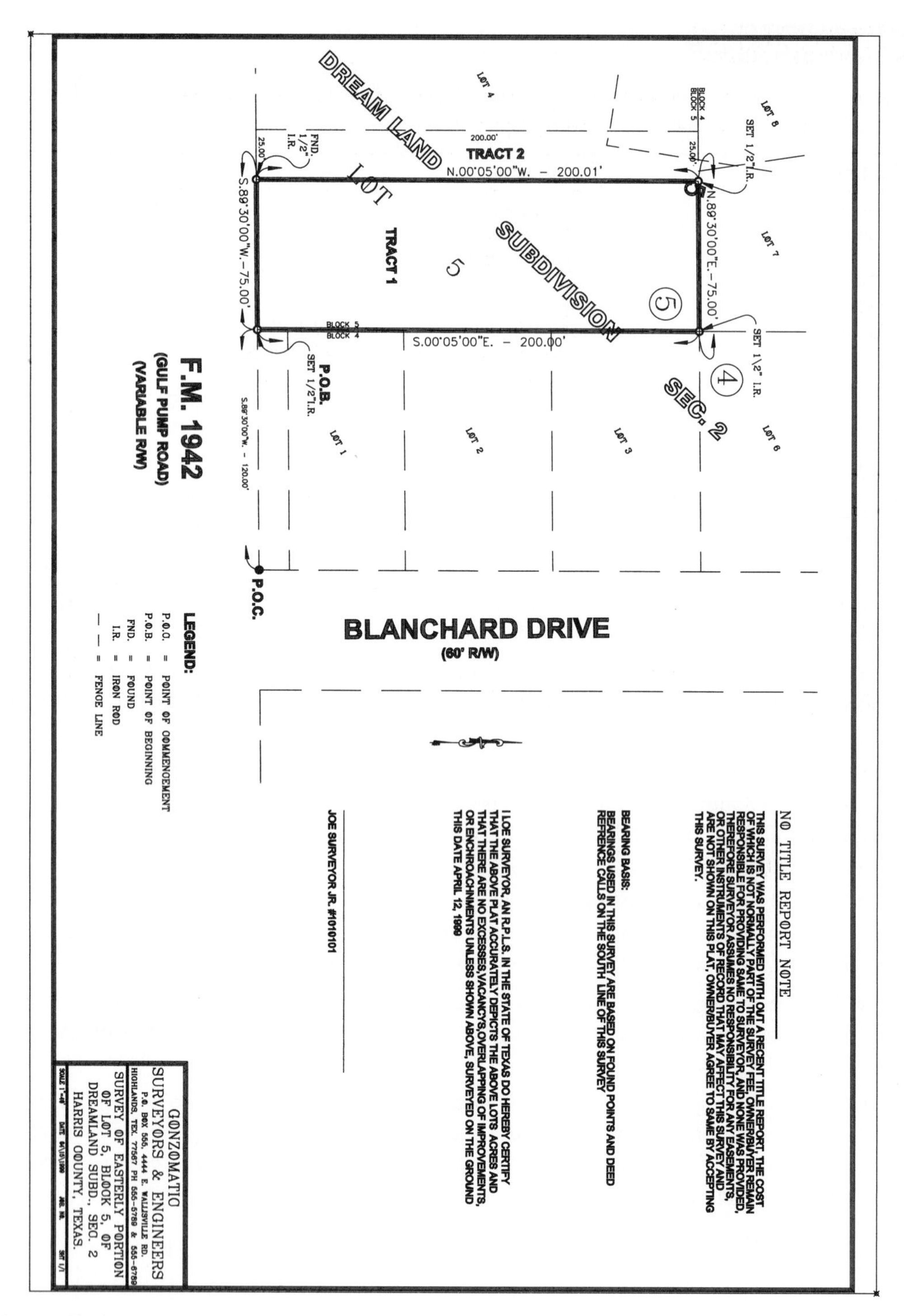


**Figure Ex4–11** *Survey Plot Plan*

## EXERCISE 4–12

Create a drawing of the section through a security retaining wall shown in Figure Ex4–12 according to the settings given in the following table:

| Settings | Value | | |
|---|---|---|---|
| 1. Units | Architectural | | |
| 2. Limits | Lower left corner: 0,0<br>Upper right corner: 20',15' | | |
| 3. Grid | 6" | | |
| 4. Snap | 3" | | |
| 5. Layers | *NAME* | *COLOR* | *LINETYPE* |
| | Border | Red | Continuous |
| | Center | Green | Continuous |
| | wallfooting | Magenta | Continuous |
| | earth | Blue | Continuous |
| | reinforcing | White | Continuous |
| | road | White | Continuous |

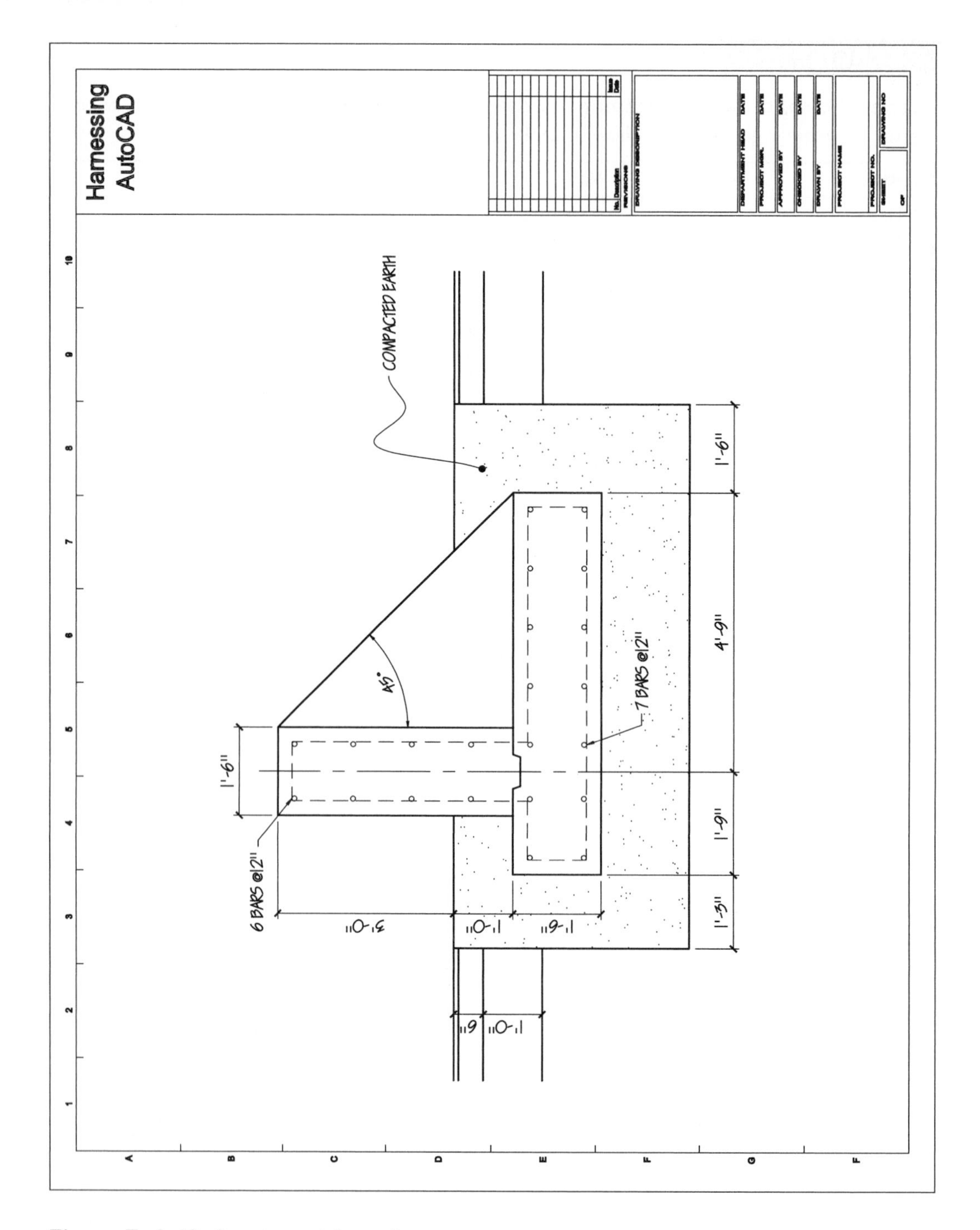


**Figure Ex4–12** *Security retaining wall*

## EXERCISE 4–13

Create a drawing of the parking lot security card access entrance shown in Figure Ex4–13 according to the settings given in the following table:

| Settings | Value | | |
|---|---|---|---|
| 1. Units | Architectural | | |
| 2. Limits | Lower left corner: 0,0 | | |
| | Upper right corner: 38',30' | | |
| 3. Grid | 4" | | |
| 4. Snap | 2" | | |
| 5. Layers | *NAME* | *COLOR* | *LINETYPE* |
| | Border | Red | Continuous |
| | Curbs | Green | Continuous |
| | bollard | Magenta | Continuous |
| | gate | Blue | Continuous |
| | cardreader | White | Continuous |
| | pressurepad | White | Continuous |
| | sign | White | Continuous |

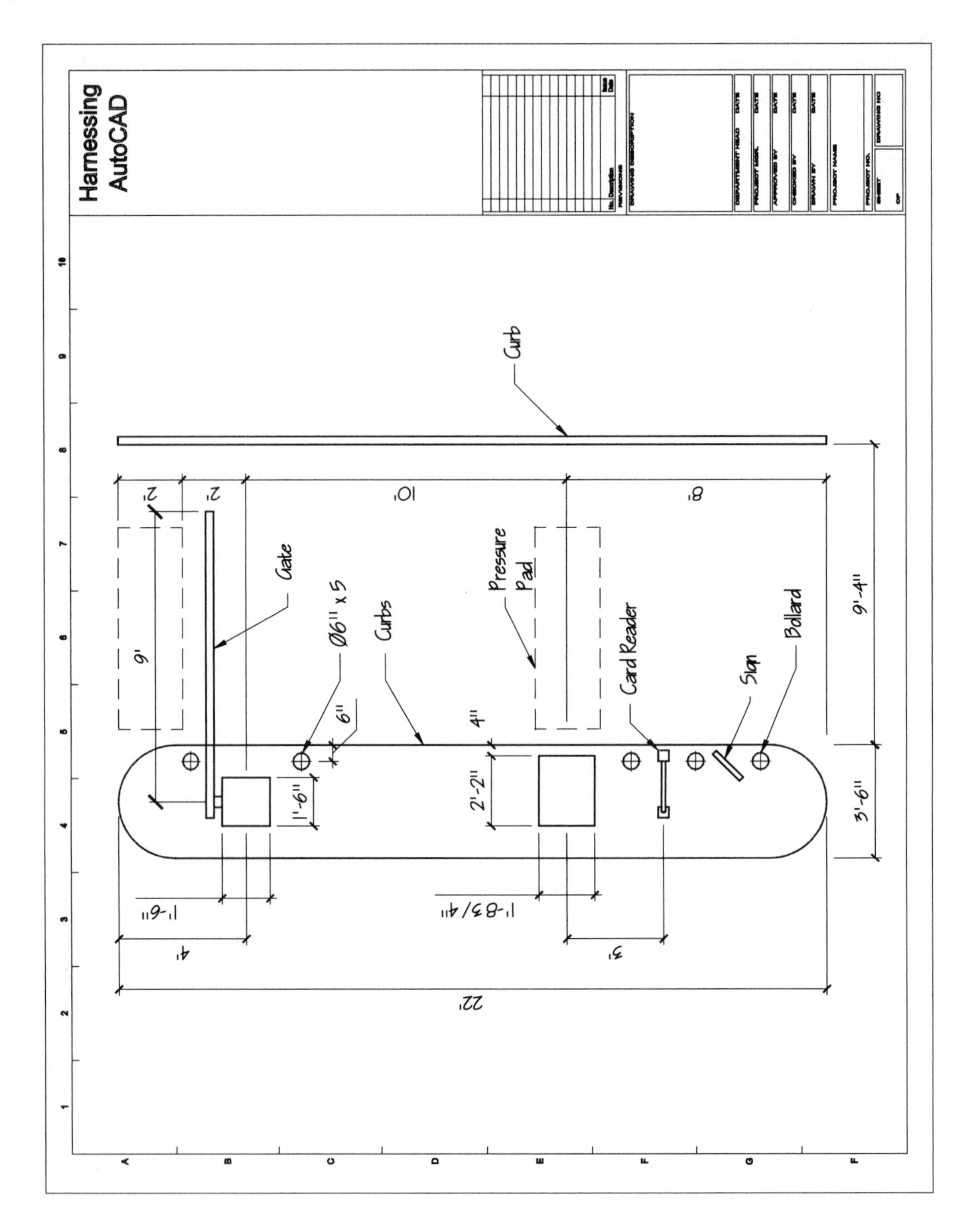


**Figure Ex4–13** *Parking lot security card access entrance*

## EXERCISE 4–14

Create a drawing of the pre-cast concrete plank used for a typical parking structure, as shown in Figure Ex4–14, according to the settings given in the following table:

| Settings | Value | | |
|---|---|---|---|
| 1. Units | Architectural | | |
| 2. Limits | Lower left corner: 0,0 | | |
| | Upper right corner: 86',68' | | |
| 3. Grid | 1' | | |
| 4. Snap | 6" | | |
| 5. Layers | *NAME* | *COLOR* | *LINETYPE* |
| | Border | Red | Continuous |
| | Cline | Green | Continuous |
| | Pcpanel | Magenta | Continuous |
| | Notes | Blue | Continuous |

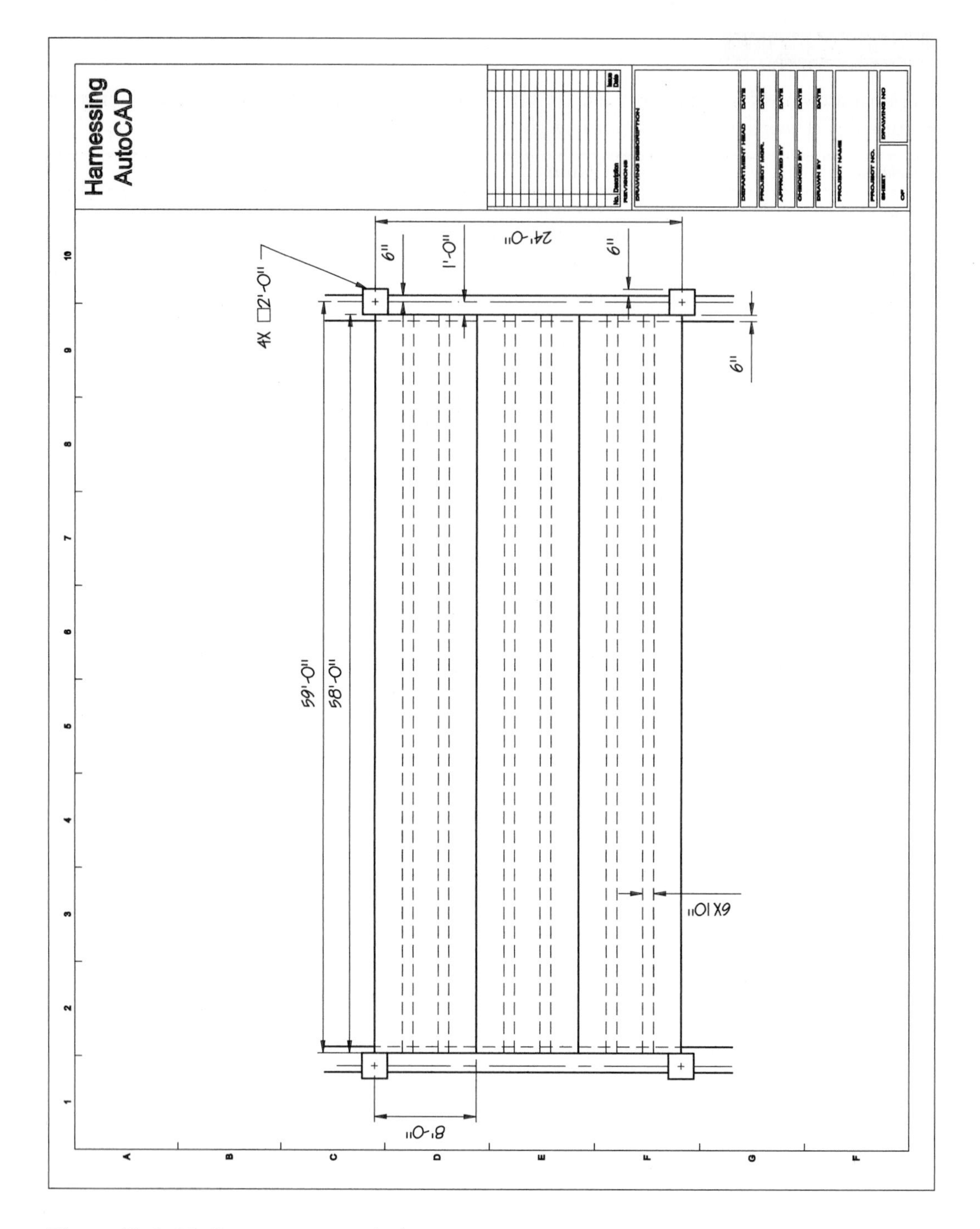


**Figure Ex4–14** *Pre-cast concrete plank*

## EXERCISE 4–15

Create a drawing of the pile cap layout shown in Figure Ex4–15 according to the settings given in the following table:

| Settings | Value | | |
|---|---|---|---|
| 1. Units | Architectural | | |
| 2. Limits | Lower left corner: 0,0<br>Upper right corner: 66'.50' | | |
| 3. Grid | 1' | | |
| 4. Snap | 6" | | |
| 5. Layers | *NAME* | *COLOR* | *LINETYPE* |
| | Border | Red | Continuous |
| | Cline | Green | Continuous |
| | column | Blue | Continuous |
| | pileandpilecap | Magenta | Continuous |
| | Notes | Blue | Continuous |

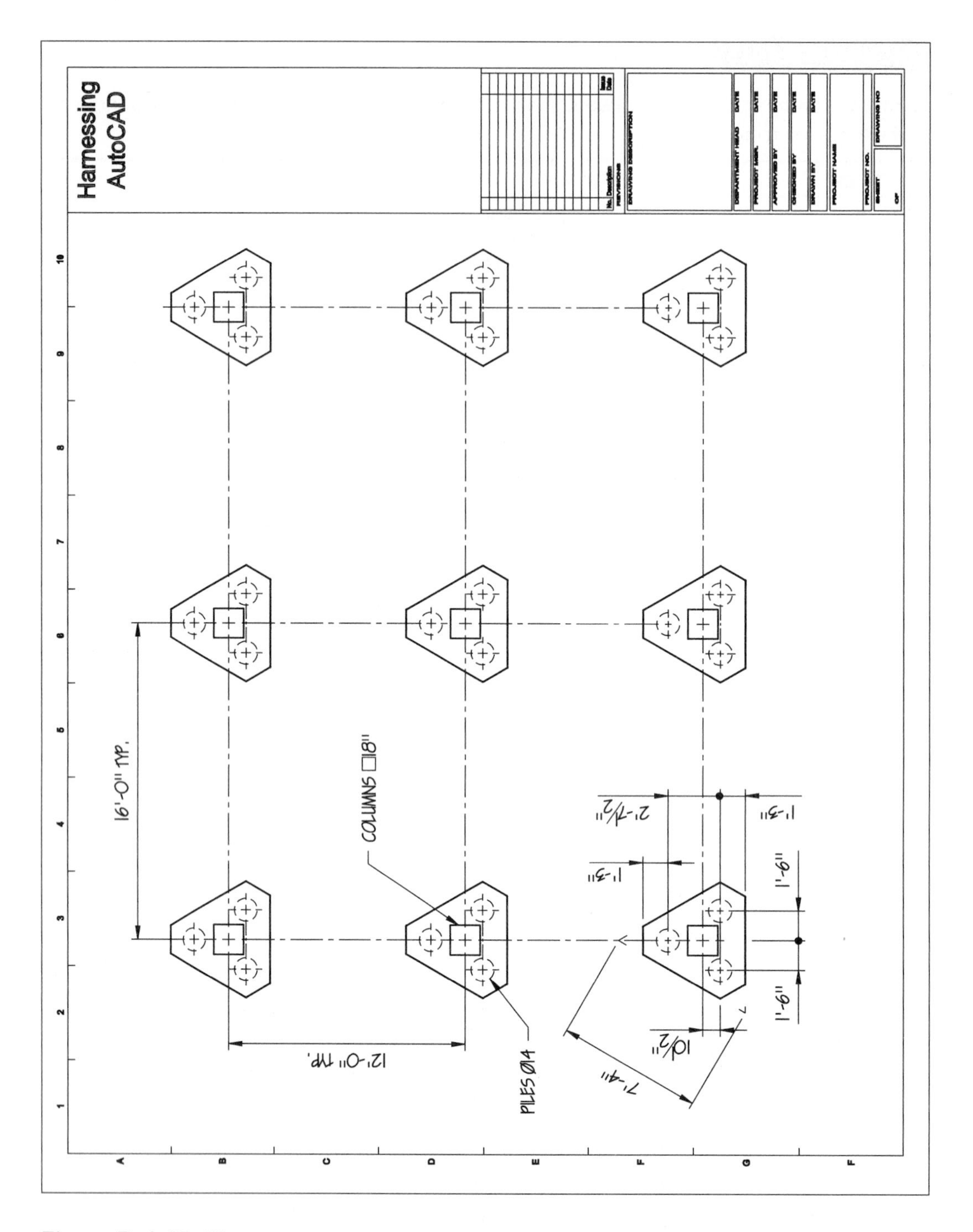


**Figure Ex4–15** *Pile cap*

## EXERCISE 4–16

Create a drawing of the input-output card shown in Figure Ex4–16 according to the settings given in the following tables.

| Settings | Value | | |
|---|---|---|---|
| 1. Units | Decimal | | |
| 2. Limits | Lower left corner: 0,0 | | |
| | Upper right corner: 12,9 | | |
| 3. Grid | 0.25 | | |
| 4. Snap | 0.25 | | |
| 5. Layers | *NAME* | *COLOR* | *LINETYPE* |
| | Object | Green | Continuous |
| | Center | Red | Center |
| | Hidden | Magenta | Hidden |
| | Text | Blue | Continuous |

**Hints:** The input-output card can be created using the PLINE, LINE, and CIRCLE commands, and the text objects can be drawn with the TEXT or MTEXT commands. Create a circle with radius 0.01 in the lower left corner of the array of dots, and then, using the ARRAY command, create an array of 15 rows and 25 columns.

The internal circuits can be drawn using the PLINE command, with the narrower segments set at 0.04 and the wider, "connection" end set at 0.10. This change should be done without exiting the PLINE command.

The wide-line circles at the ends of the circuits can be created by using the PLINE command with the Arc option, drawing two half-circles with a line width of 0.02 and a radius of 0.0625. The snap resolution will have to be set to 0.03125.

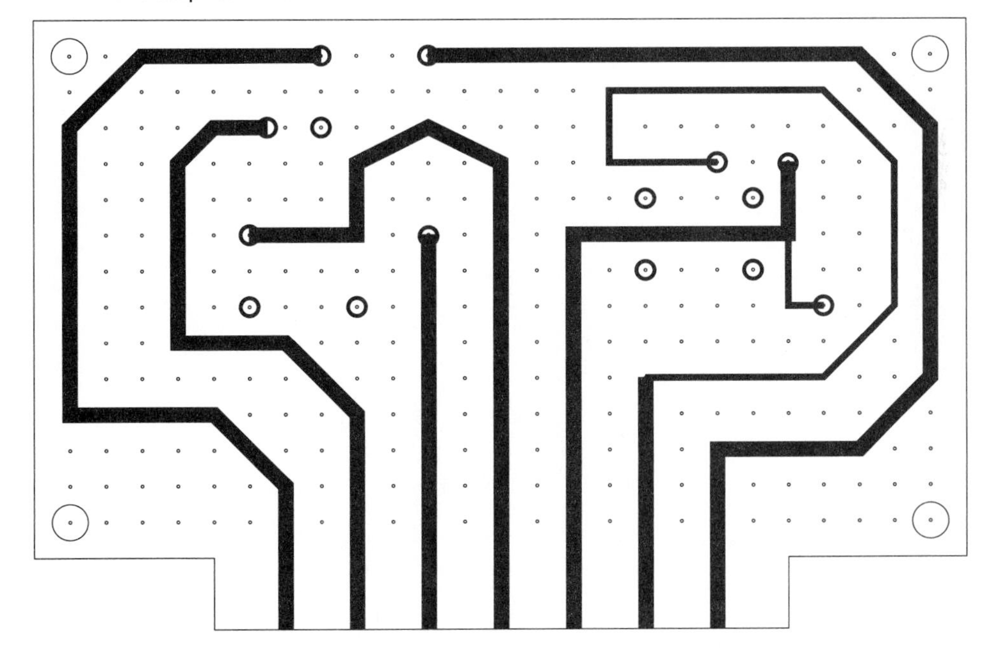

**Figure Ex4–16** *Completed drawing*

## EXERCISE 4–17

Create a drawing of the exterior ornamental lighting layout shown in Figure Ex4–17, according to the settings given in the following table:

| Settings | Value | | |
|---|---|---|---|
| 1. Units | Architectural | | |
| 2. Limits | Lower left corner: 0,0<br>Upper right corner: 85',66' | | |
| 3. Grid | 12" | | |
| 4. Snap | 6" | | |
| 5. Layers | *NAME* | *COLOR* | *LINETYPE* |
| | Border | Red | Continuous |
| | Bldgbackground | Green | Continuous |
| | lights | Blue | Continuous |
| | wiring | Magenta | Continuous |
| | Notes | Blue | Continuous |

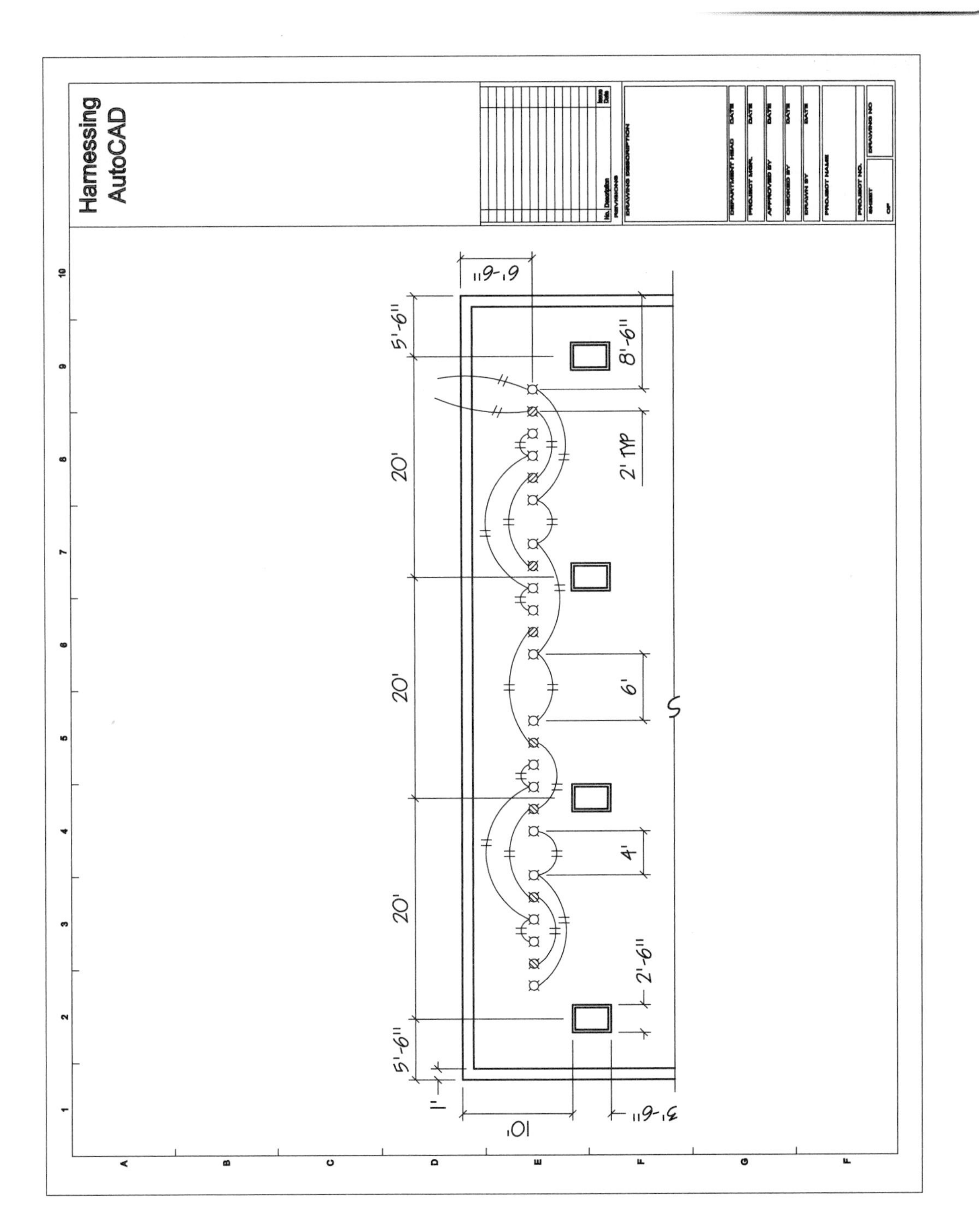


**Figure Ex4–17** *Exterior ornamental lighting*

## EXERCISE 4–18

Create a drawing of the circuit diagram shown in Figure Ex4–18, according to the settings given in the following table:

| Settings | Value | | |
|---|---|---|---|
| 1. Units | Decimals | | |
| 2. Limits | Lower left corner: 0,0 | | |
| | Upper right corner: 12,9 | | |
| 3. Grid | 1" | | |
| 4. Snap | .5" | | |
| 5. Layers | *NAME* | *COLOR* | *LINETYPE* |
| | Border | Red | Continuous |
| | Switches | Green | Continuous |
| | Wire | Magenta | Continuous |

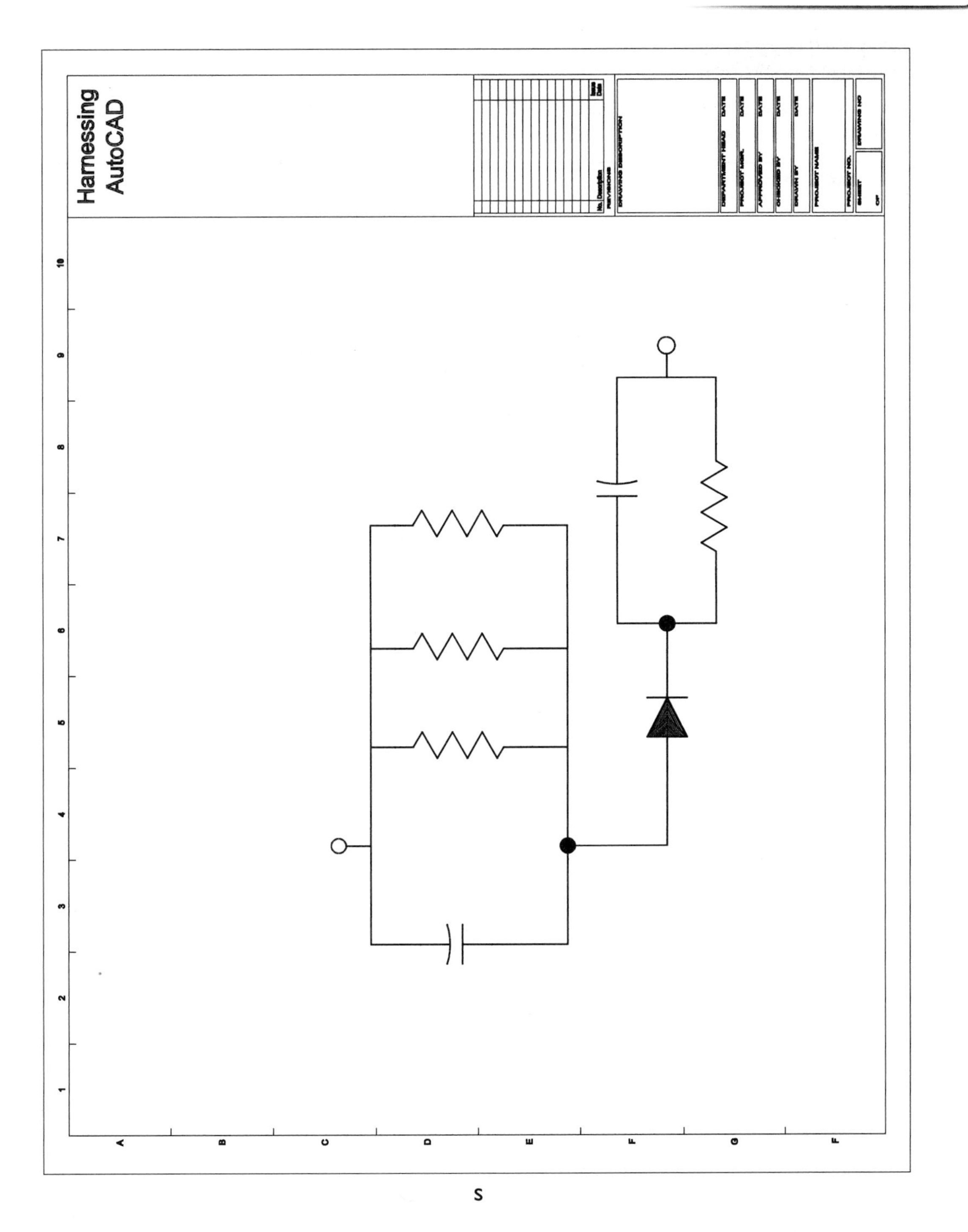


**Figure Ex4–18** *Circuit diagram*

## EXERCISE 4–19

Create a drawing of the lighting switch circuit and wiring diagram shown in Figure Ex4–19, according to the settings given in the following table:

| Settings | Value |
|---|---|
| 1. Units | Architectural |
| 2. Limits | Lower left corner: 0,0<br>Upper right corner: 28',22' |
| 3. Grid | 12" |
| 4. Snap | 6" |

| Settings | Value | | |
|---|---|---|---|
| 5. Layers | *NAME* | *COLOR* | *LINETYPE* |
| | Border | Red | Continuous |
| | Bldgbackground | Green | Continuous |
| | lights | Blue | Continuous |
| | duplexoutlets | White | Continuous |
| | wiring | Magenta | Continuous |
| | Notes | Blue | Continuous |



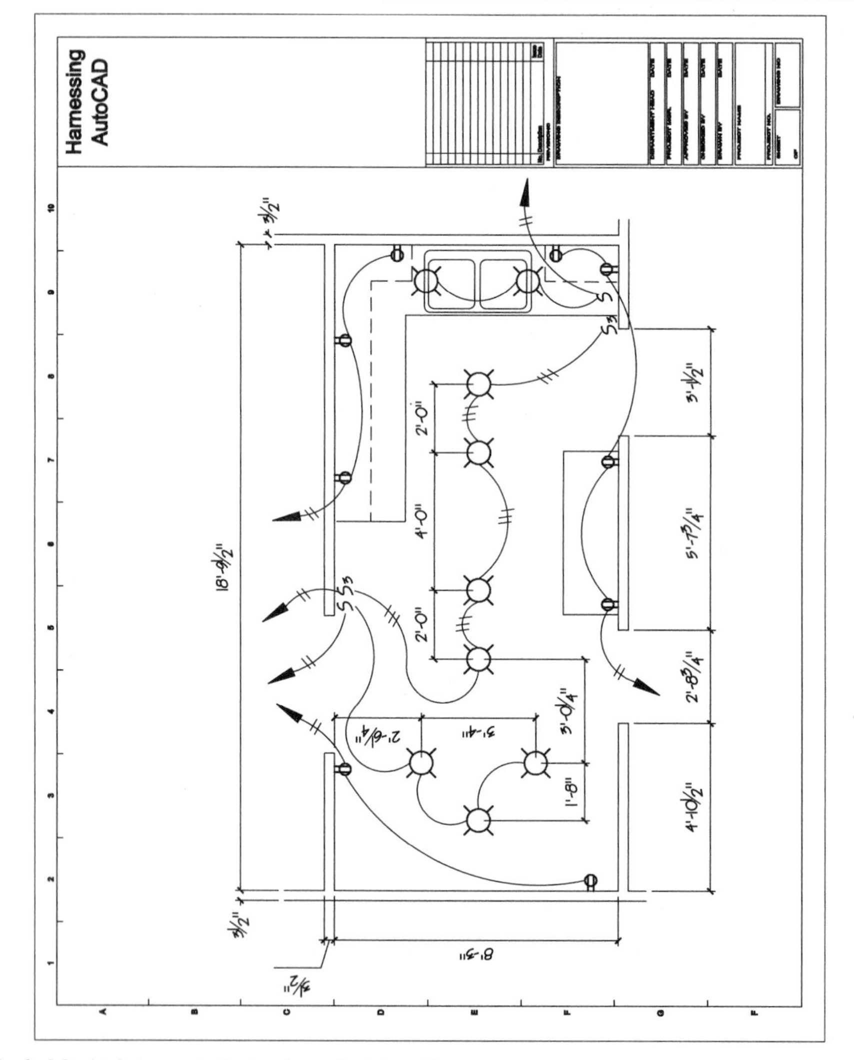

**Figure Ex4–19** *Lighting switch circuit and wiring diagram*

CHAPTER 5

# Fundamentals IV

## PROJECT EXERCISE

This project exercise provides point-by-point instructions for setting up the drawing with layers and then creating the objects shown in the accompanying figure.

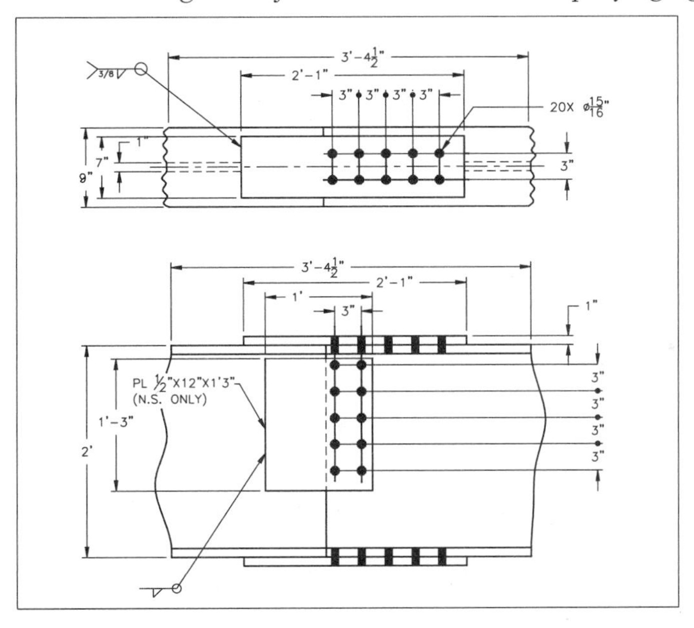

**Figure P5–1** *Completed project drawing (dimensions added for reference)*

In this project you will learn how to do the following:

- Set up the drawing, including Limits and Layers.
- Use the SKETCH, SOLID, and DONUT commands to create objects.
- Use the COPY command to create objects from other objects.
- Use the ROTATE and STRETCH commands to modify objects.
- Use the TEXT command to create text objects.

## SET UP THE DRAWING AND DRAW A BORDER

**Step 1:** Start the AutoCAD program.

**Step 2:** Make sure system variable STARTUP is set to 1, that allows AutoCAD to create a new drawing using a wizard. To create a new drawing, invoke the NEW command from the Standard toolbar or select New from the File menu.

AutoCAD displays the Create New Drawing dialog box. Select the **Use a Wizard** button and AutoCAD lists available wizards. Select Quick Setup wizard from the Select a Wizard list box and choose the **OK** button.

AutoCAD displays the Quick Setup dialog box with the listing of the options available for unit of measurement. Select the "Architectural" radio button, and choose the **Next>** button.

AutoCAD displays the Quick Setup dialog box with the display for Area setting. Under the **Width:** text box enter **9'**, and under the **Length:** text box enter **7'**. Choose the **Finish** button to close the Quick Step dialog box.

**Step 3:** Invoke the LAYER command from the Layers toolbar, or select Layer... from the Format menu. AutoCAD displays the Layer Properties Manager dialog box.

Create six layers, and rename them as shown in the following table, assigning the appropriate color and linetype.

| Layer Name | Color | Linetype | Lineweight |
|---|---|---|---|
| Border | Red | Continuous | Default |
| Object | White | Continuous | Default |
| Solid-Donut | White | Continuous | Default |
| Center | Red | Center | Default |
| Hidden | Magenta | Hidden | Default |
| Const | Blue | Continuous | Default |

Set Border as the current layer, and close the Layer Properties Manager dialog box.

## CREATING OBJECTS

**Step 4:** Draw the border using the PLINE command, with a width of 0.5" from 0'-2",0-2" to 0'-2",6'-10" to 8'-10",6'-10" to 8'-10",0'-2", and then close back to 0'-2",0-2".

**Step 5:** Make the Object layer current and draw the bottom plate using the RECTANGLE command from 3'-4",1'-6" to 5'-5",1'-7". Your drawing should look like Figure P5–2.

**Step 6:** Copy the rectangle twice, with displacements of 0'-0",2'-1" and 0'-0",3'-1". See Figure P5–3.

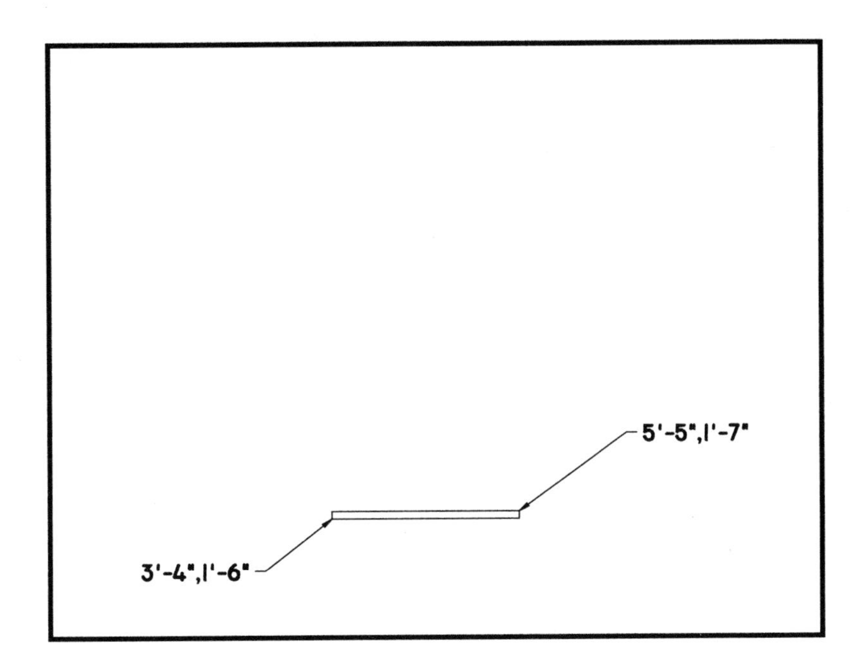


**Figure P5–2** *Bottom plate*

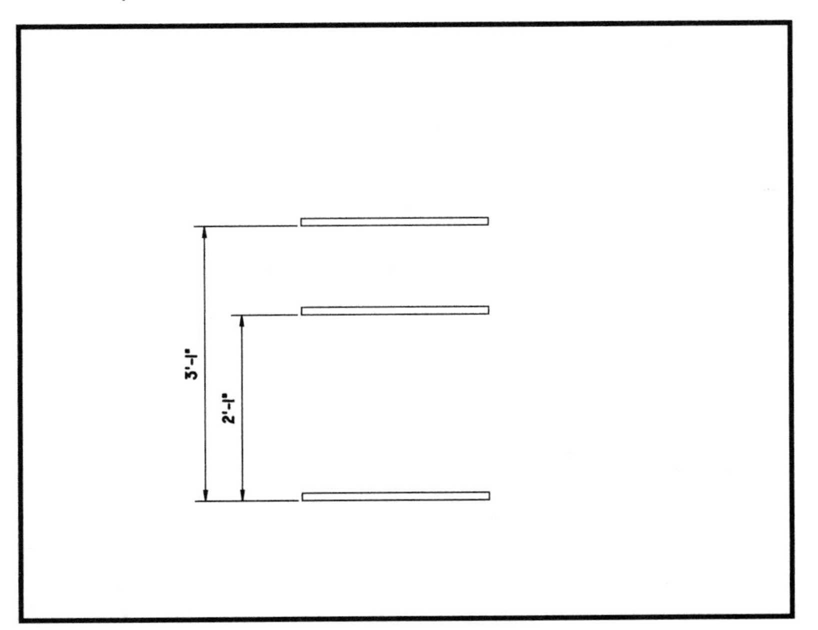


**Figure P5–3** *Setting of the base*

**Step 7:** Stretch the top rectangle with a displacement of 0",6" to change it from a 2'-1" x 1" rectangle to a 2'-1" x 7" rectangle. See Figures P5–4 and P5–5.

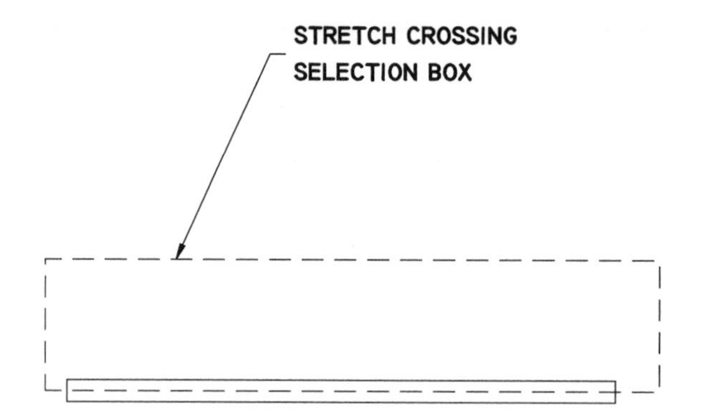


**Figure P5–4** *Display with the stretch crossing*

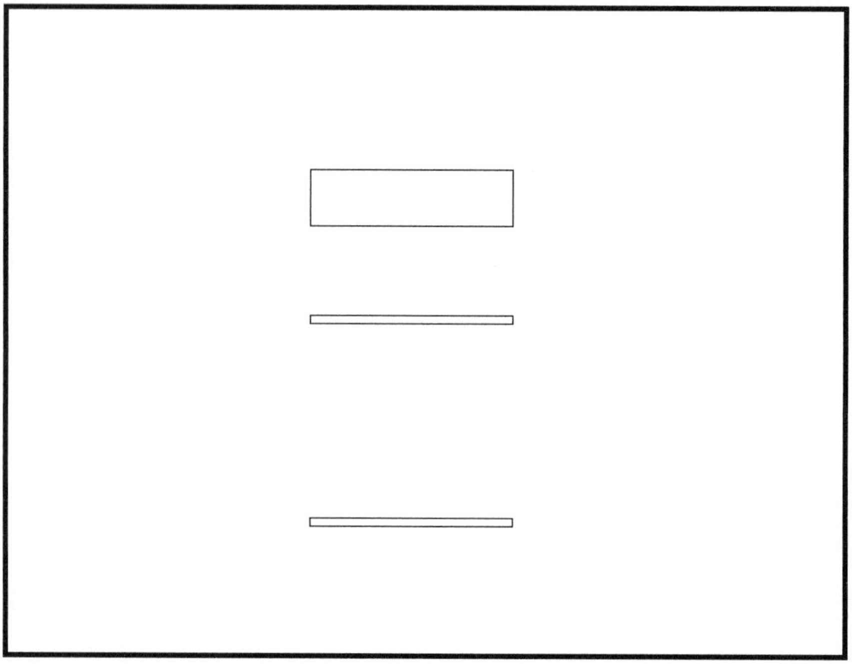

**Figure P5–5** *Result after stretching the top rectangle*

**Step 8:** Draw a horizontal line through the top of the bottom rectangle, and then copy it five times up 1", up 1'-11", up 2'-0", up 2'-11", and up 3'-8". Your drawing should look like Figure P5–6.

**Step 9:** Use the SKETCH command to create the end "breaks" of the top and side views of the beam. Your drawing should look like Figure P5–7.

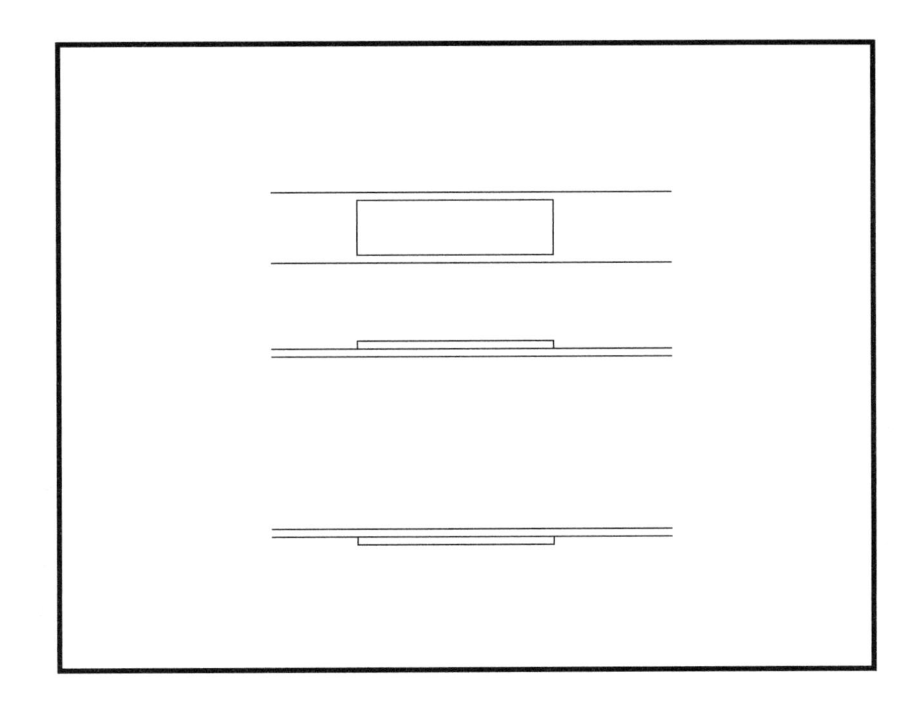

**Figure P5–6** *Basic layout*

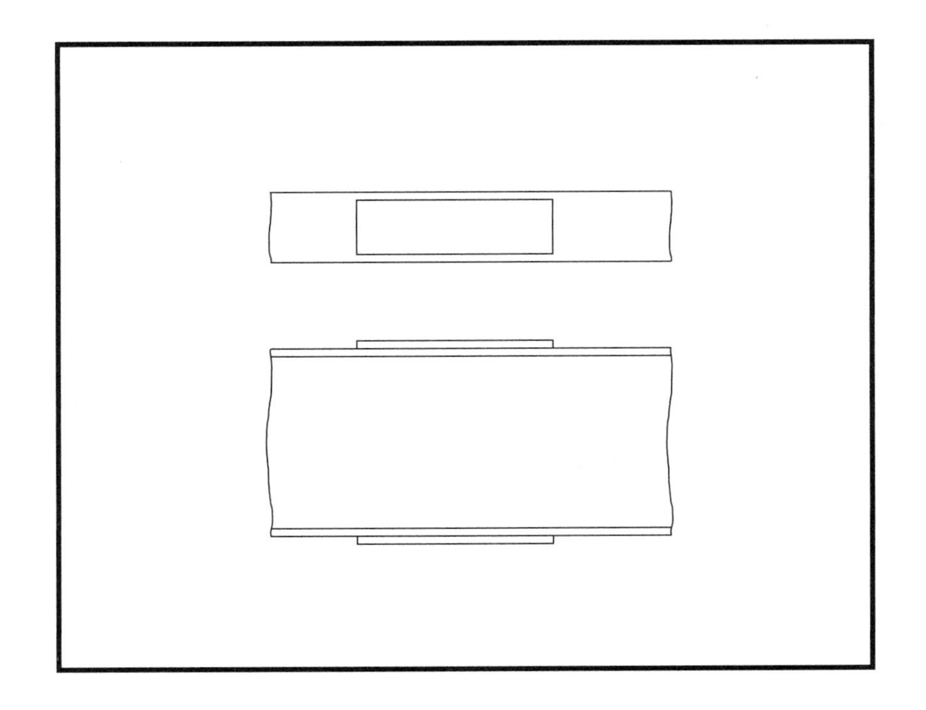

**Figure P5–7** *Drawing with sketch lines*

**Step 10:** While the Object layer is still current, draw the added lines shown in Figure P5–8. Refer to Figure P5–1 for the size of the objects, and make sure to draw in appropriate layers (Center and Hidden). When this step is completed, your drawing should look like Figure P5–9.

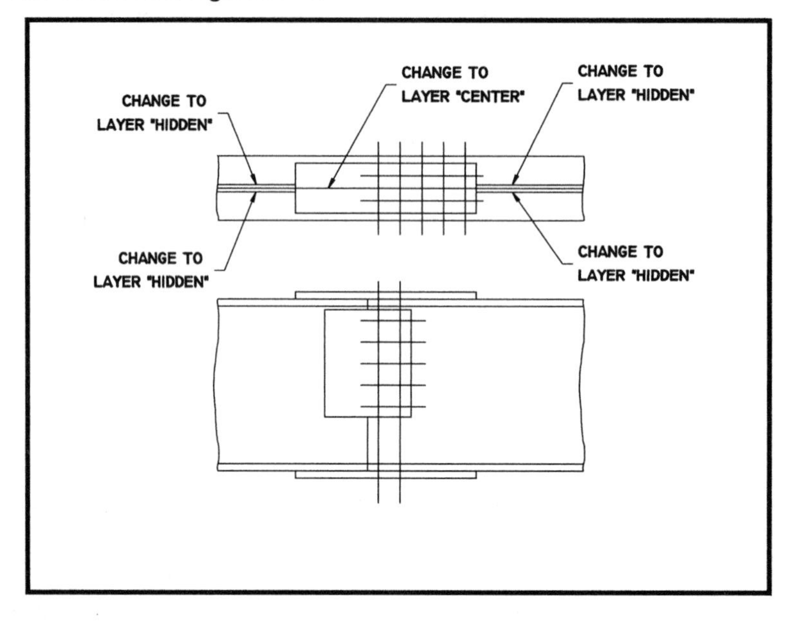


**Figure P5–8** *Additional objects added to the layout*

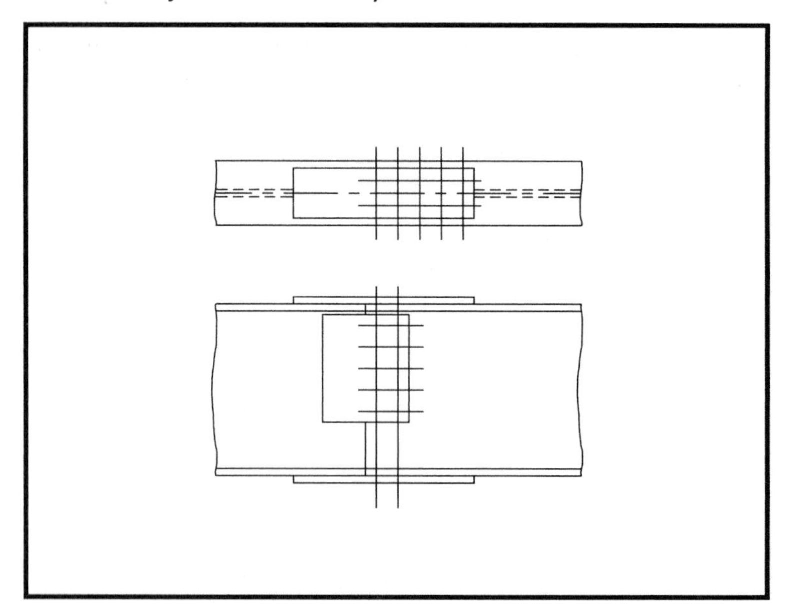

**Figure P5–9** *Drawing after changing the linetypes*

**Step 11:** Set the Solid-Donut layer as the current layer. Create the solid-filled rectangles with the SOLID command. You can create each one individually, or you can create one and then COPY it to the required locations. Refer to Figure P5–1 for location of solid-filled rectangles. Your drawing should then look like Figure P5–10.

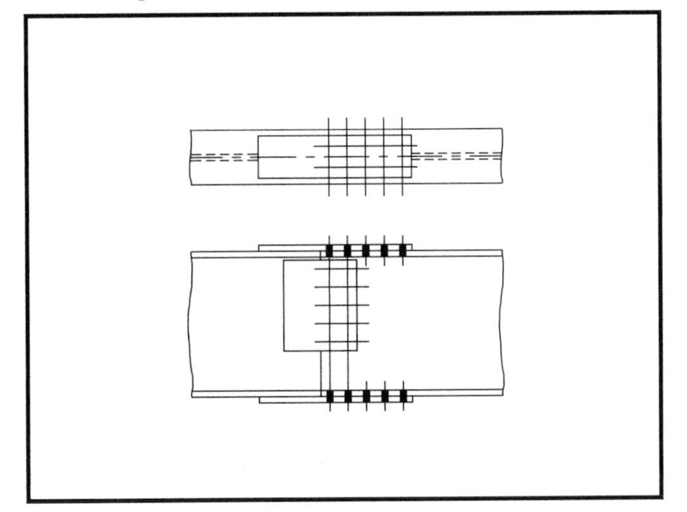

**Figure P5–10** *Drawing with the solid objects drawn*

**Step 12:** Create the solid-filled circles with the DONUT command. The DONUT command permits you to place multiple created objects after you have specified the inside and outside diameters. These (the donuts) might be more easily created one by one simply by picking their individual insertion points without exiting the DONUT command. Refer to Figure P5–1 for location of solid-filled rectangles. Your drawing should now look like Figure P5–11.

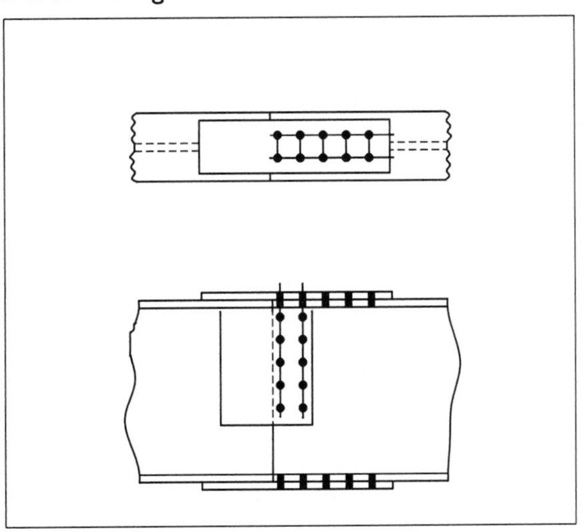

**Figure P5–11** *Drawing with the donut objects drawn*

**Step 13:** Set the Text layer as your current layer. Add the annotations as shown in Figure P5–12. Save the drawing as PROJ5.DWG.

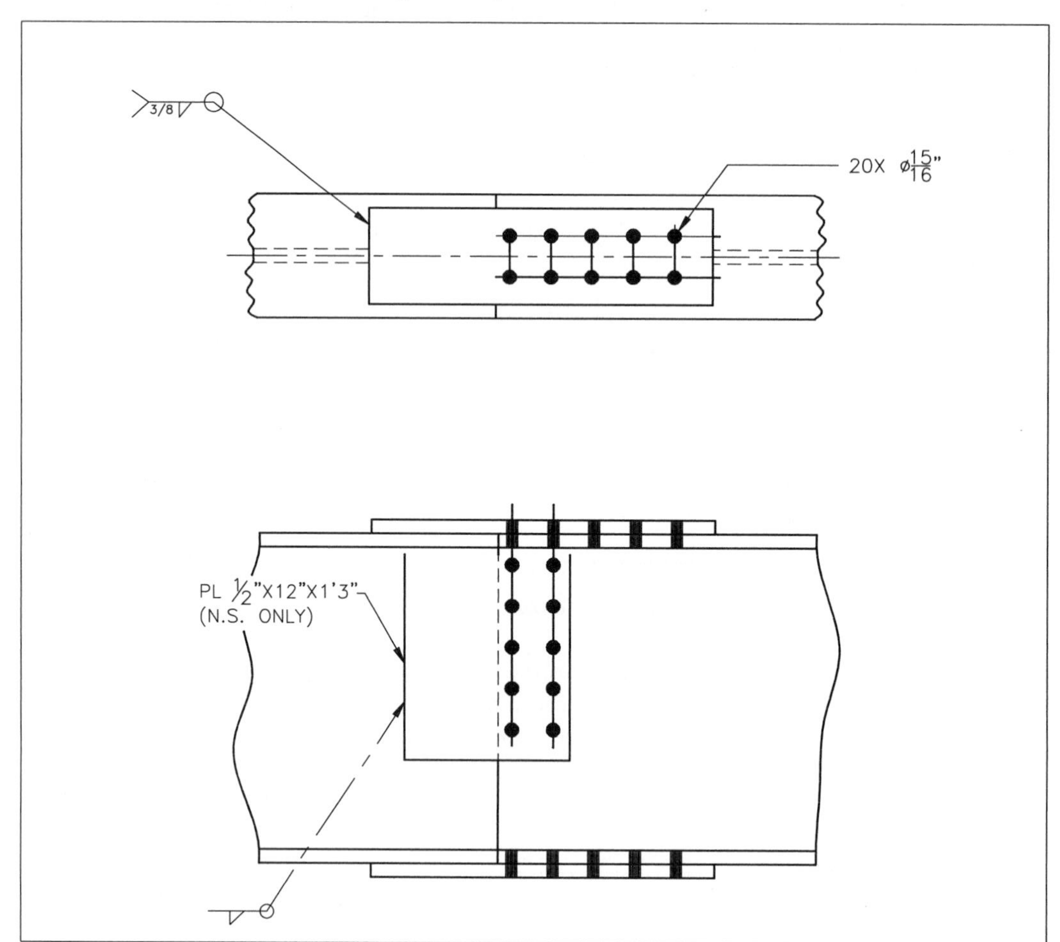


**Figure P5–12** *Completed drawing*

## EXERCISE 5–1

Create the plan and section of transition ductwork drawing as shown in figure according to the settings given in the following table:

| Settings | Value | | |
|---|---|---|---|
| 1. Units | Architectural | | |
| 2. Limits | Lower left corner: 0,0 | | |
| | Upper right corner: 22',17' | | |
| 3. Grid | 6" | | |
| 4. Snap | 3" | | |
| 5. Layers | *NAME* | *COLOR* | *LINETYPE* |
| | Centerline | Cyan | Center |
| | Border | Red | Continous |
| | Structure | Green | Continuous |
| | Ceiling | Magenta | Hidden |
| | Duct | Blue | Continuous |
| | Notes | Blue | Continuous |

**Hint:** The duct will have a clear inside area of 20" x 12" and will have a thickness of ½".The transition must be take place within a 2'=0" length. Do not dimension.

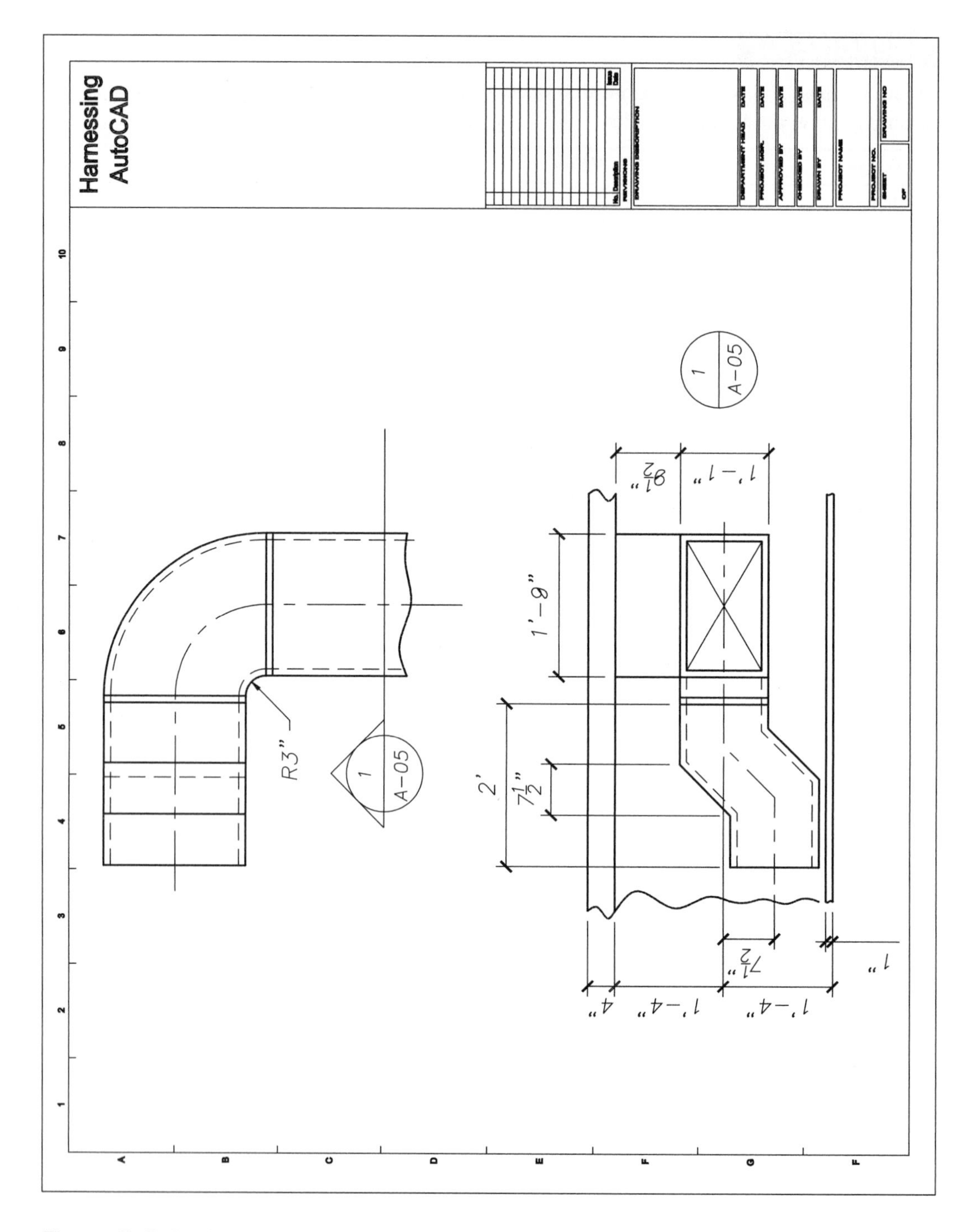


**Figure Ex5–1** *Plan and Section of transition ductwork*

## EXERCISE 5–2

Create the elevation and section of a 6" 600# raised face and ring joint ANSI B 16.5 flanges drawing as shown in Figure Ex5–2 according to the settings/specifications given in the following table. Do not dimension.

| Settings | Value | | |
|---|---|---|---|
| 1. Units | Decimal | | |
| 2. Limits | Lower left corner: 0,0<br>Upper right corner: 40,26 | | |
| 3. Grid | 1 | | |
| 4. Snap | 0.5 | | |
| 5. Layers | *NAME* | *COLOR* | *LINETYPE* |
| | Center | Cyan | Center |
| | Border | Red | Continuous |
| | Raised face | Green | Continuous |
| | Ring joint | Magenta | Hidden |
| 6. Specifications | *Dimensions of ANSI B16.5 Class 600 Flanges and Flange Bolts* | | |

| Nominal Pipe Size | O.D. of Flange | Flange Thickness | Dia. of Hub | Slip-on Socket | Dia. of Bold Circle | Dia. of Bolt Holes | No. of Bolts | Size of Bolts | Raized Face | Ring Joint |
|---|---|---|---|---|---|---|---|---|---|---|
| 6 | 14.00 | 1.88 | 8.75 | 2.62 | 11.50 | 1.12 | 12 | 1 | 8.5 | See Dwg |

mechanical EXERCISE

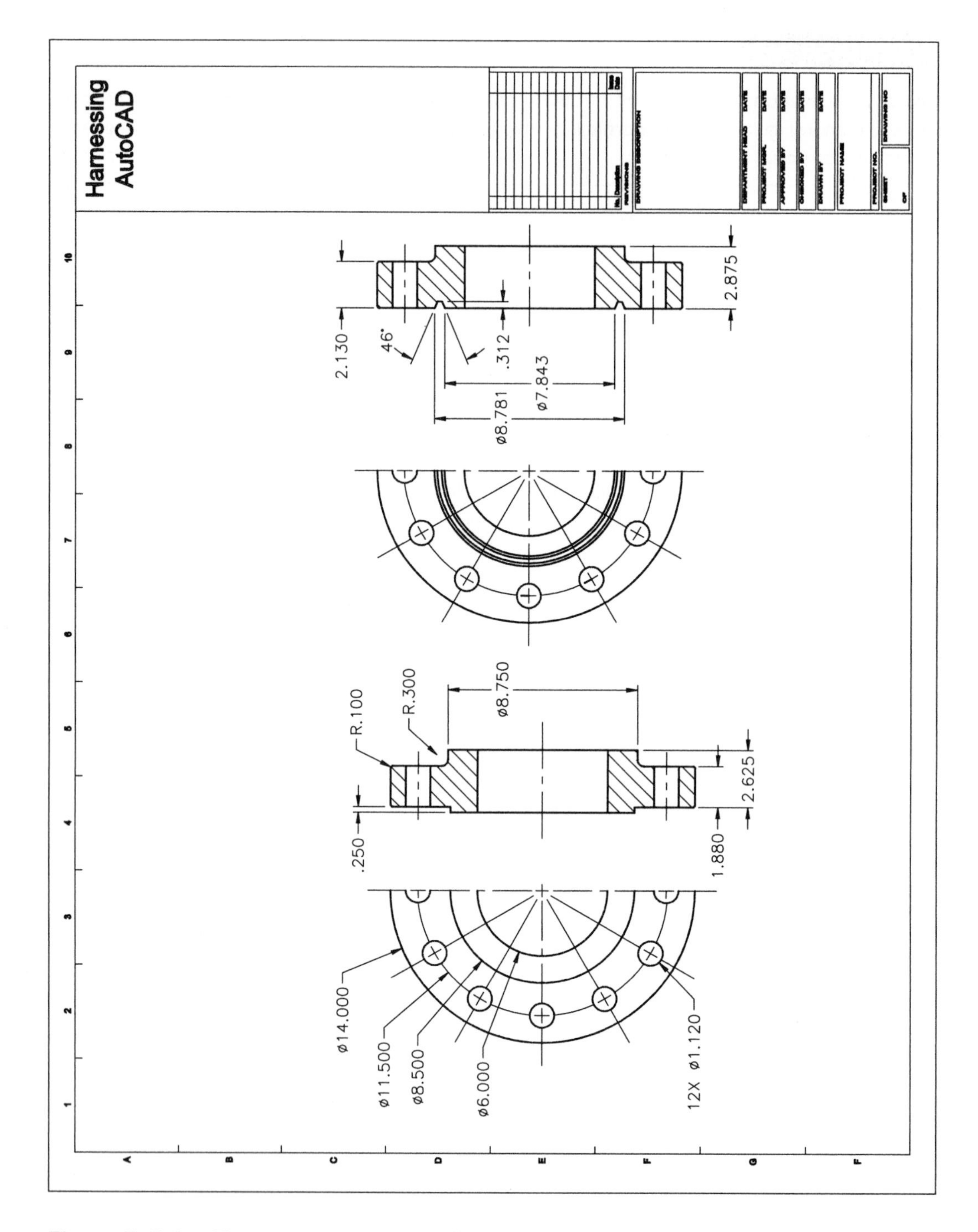


**Figure Ex5–2** *Elevation and section of a 6" 600# raised face and ring joint flange*

## EXERCISE 5–3

Create the top view and section of a multi-groove compressor pulley drawing, as shown in Figure Ex5–3, according to the dimensions and settings given in the following table. Do not dimension.

| **Settings** | **Value** | | |
|---|---|---|---|
| 1. Units | Decimal | | |
| 2. Limits | Lower left corner: 0,0 | | |
| | Upper right corner: 40,26 | | |
| 3. Grid | | | |
| 4. Snap | 0.5 | | |
| 5. Layers | *NAME* | *COLOR* | *LINETYPE* |
| | Center | Cyan | Center |
| | Border | Red | Continuous |
| | Oject | Green | Continuous |
| | Notes | Green | Continuous |

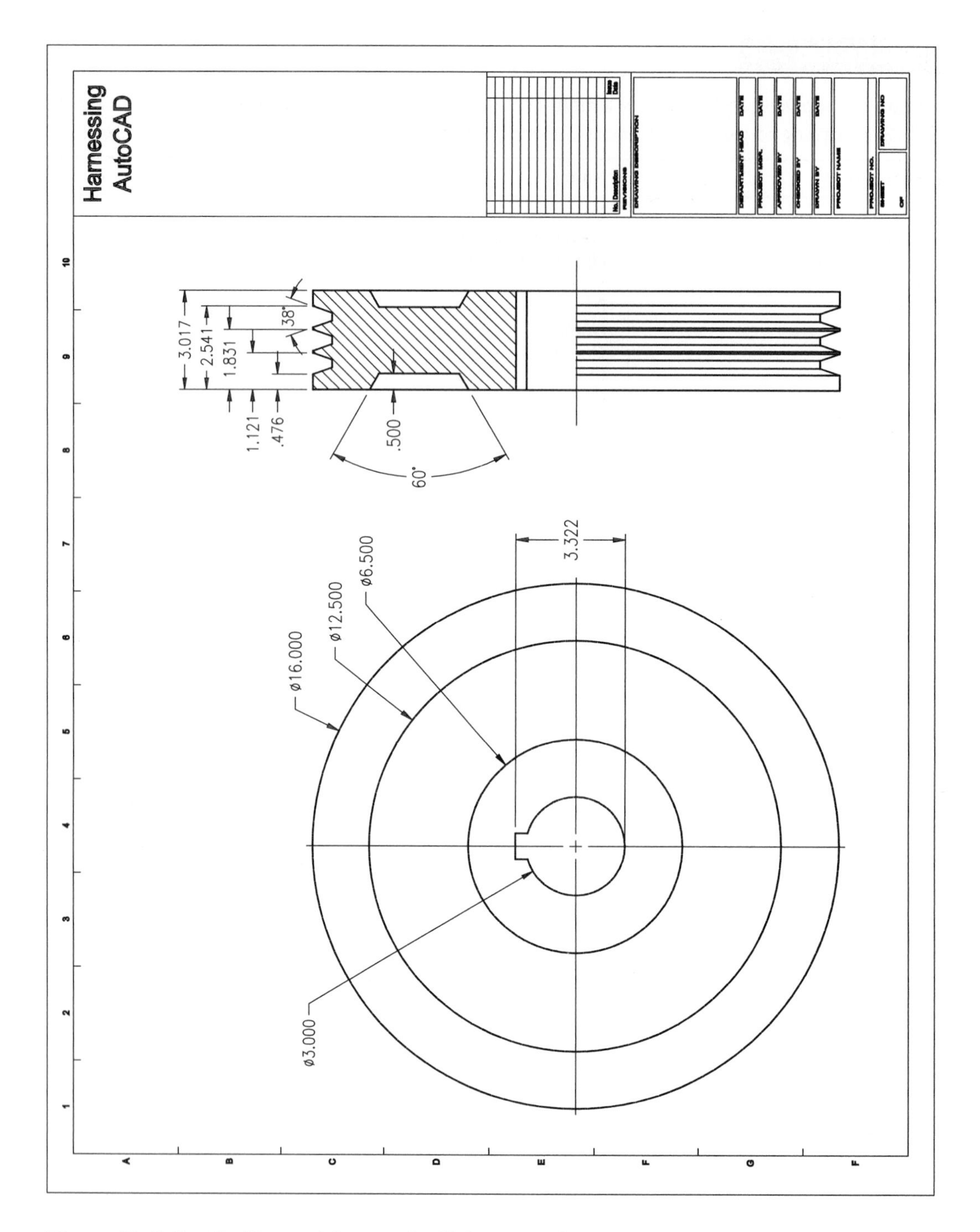


**Figure Ex5–3** *Top View and Section of a Multi-groove Compressor Pulley*

## EXERCISE 5–4

Create the elevation and section of a surface mounted flagpole bracket drawing, as shown in Figure Ex5–4, according to the settings given in the following table. Do not dimension.

| Settings | Value | | |
|---|---|---|---|
| 1. Units | Decimal | | |
| 2. Limits | Lower left corner: 0,0 | | |
| | Upper right corner: 50,32 | | |
| 3. Grid | 1 | | |
| 4. Snap | 0.5 | | |
| 5. Layers | *NAME* | *COLOR* | *LINETYPE* |
| | Center | Cyan | Center |
| | Border | Red | Continuous |
| | Object | Green | Continuous |

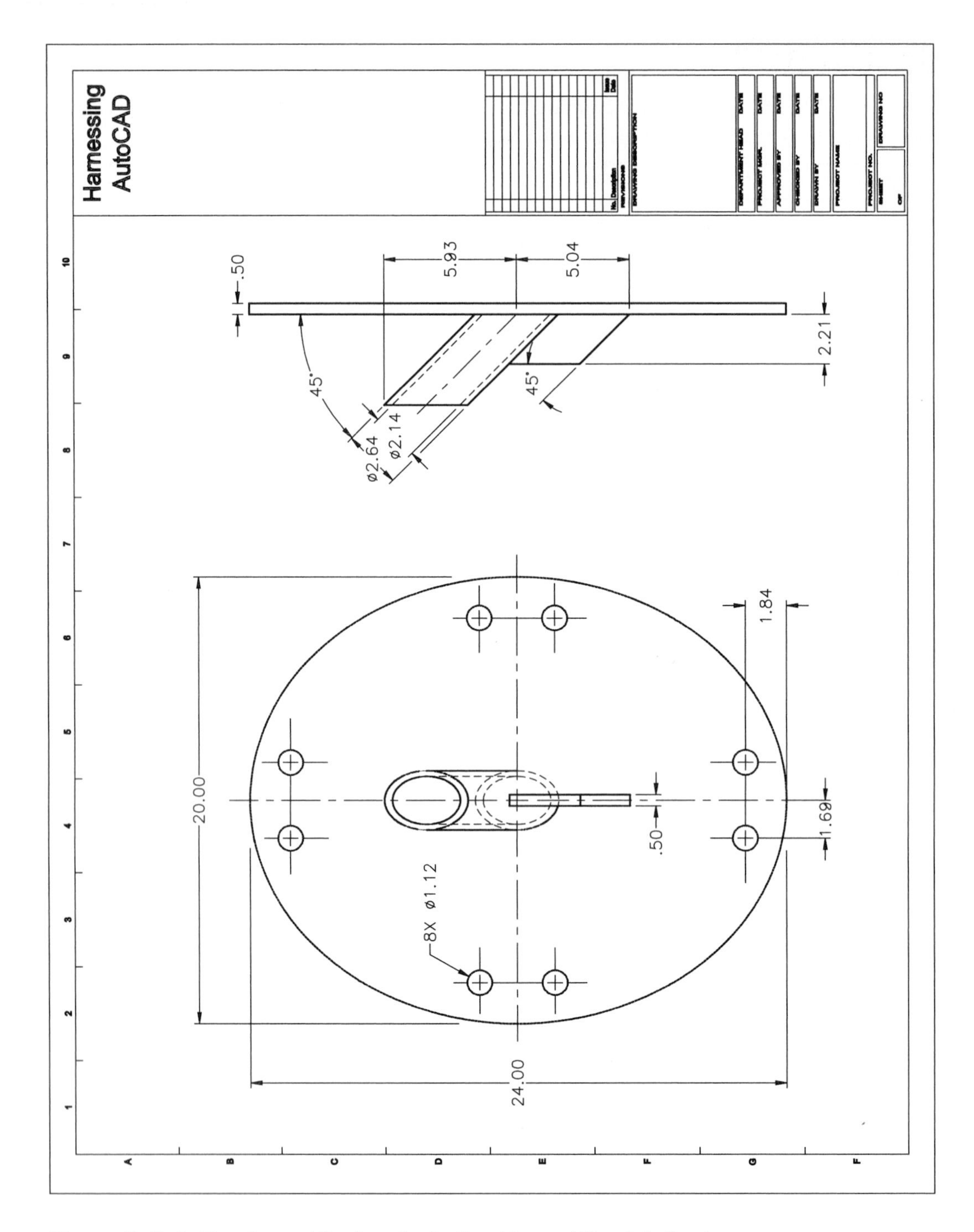


**Figure Ex5–4** *Elevation and Section of a Surface Mounted Flag Pole Bracket*

## EXERCISE 5–5

Create the drawing shown according to the settings given in the following table:

| Settings | Value |
|---|---|
| 1. Units | Architectural |
| 2. Limits | Lower left corner: 0',0'<br>Upper right corner: 9',6' |
| 3. Grid | 4" |
| 4. Snap | 2" |
| 5. Ltscale | 4 |

| Settings | Value | | |
|---|---|---|---|
| 6. Layers | *NAME* | *COLOR* | *LINETYPE* |
| | Center | Cyan | Center |
| | Border | Red | Continous |
| | Backboard | Green | Continuous |
| | Rim | Magenta | Hidden |
| | Text | Blue | Continuous |

**Hints:** Make sure to make each layer current before beginning to draw any objects that are on it.

The border can be drawn from coordinates 3",3" to 8'-9",5'-6" using the RECTANGLE command.

Draw a vertical construction line (xline) through a point whose X coordinate is 4'-0".

Offset the vertical xline 2'-3" to each side of the line just created.

Draw a horizontal construction line (xline) through a point whose *Y* coordinate is 2'-0".

Offset the horizontal construction line 3" above and below the line just created. Offset the lower line 1'-2½" toward the bottom of the drawing.

The construction lines just created will guide you in drawing the circles and lines required for completing the outline of the object.

Using the intersections of the lines and/or circles, draw the sides and sloped bottom lines of the backboard on the Backboard layer.

Using the TRIM command, trim the lines and circles to complete the shape.

After the drawing is finished, either the Construction Layer can be turned OFF or the construction lines can be erased. Do not dimension.

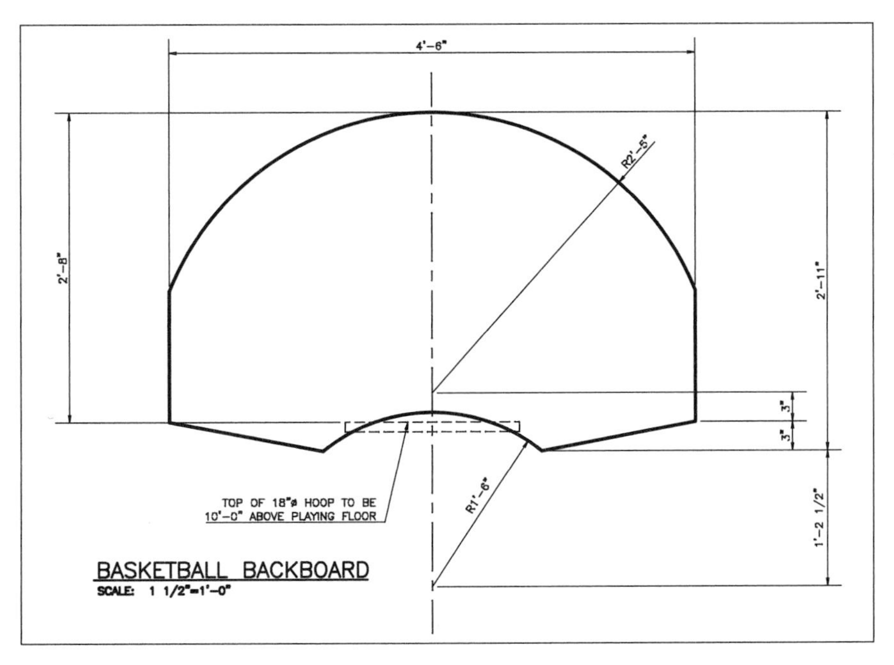

**Figure Ex5–6**

architectural EXERCISE

## EXERCISE 5–6

Create the drawing shown according to the settings given in the following table:

| Settings | Value |
|---|---|
| 1. Units | Architectural |
| 2. Limits | Lower left corner: 0',0'<br>Upper right corner: 12',9' |
| 3. Grid | 6" |
| 4. Snap | 3" |

| Settings | Value | | |
|---|---|---|---|
| 5. Layers | *NAME* | *COLOR* | *LINETYPE* |
| | Construction | Cyan | Continuous |
| | Border | Red | Continuous |
| | Object | Green | Continuous |
| | Center | Magenta | Center |
| | Text | Blue | Continuous |
| | Hidden | White | Hidden |

**Hints:** Make sure to make each layer current before beginning to draw any objects that are on it..

The front view of the wheel handle can be drawn using the PLINE command. The top of the handle is 8" above the top of the closure plate.

The circle part of the top view of the wheel handle can be drawn using either the DONUT command or the PLINE command with the arc option, creating two 180-degree polyarcs with a width of 1". Do not dimension.

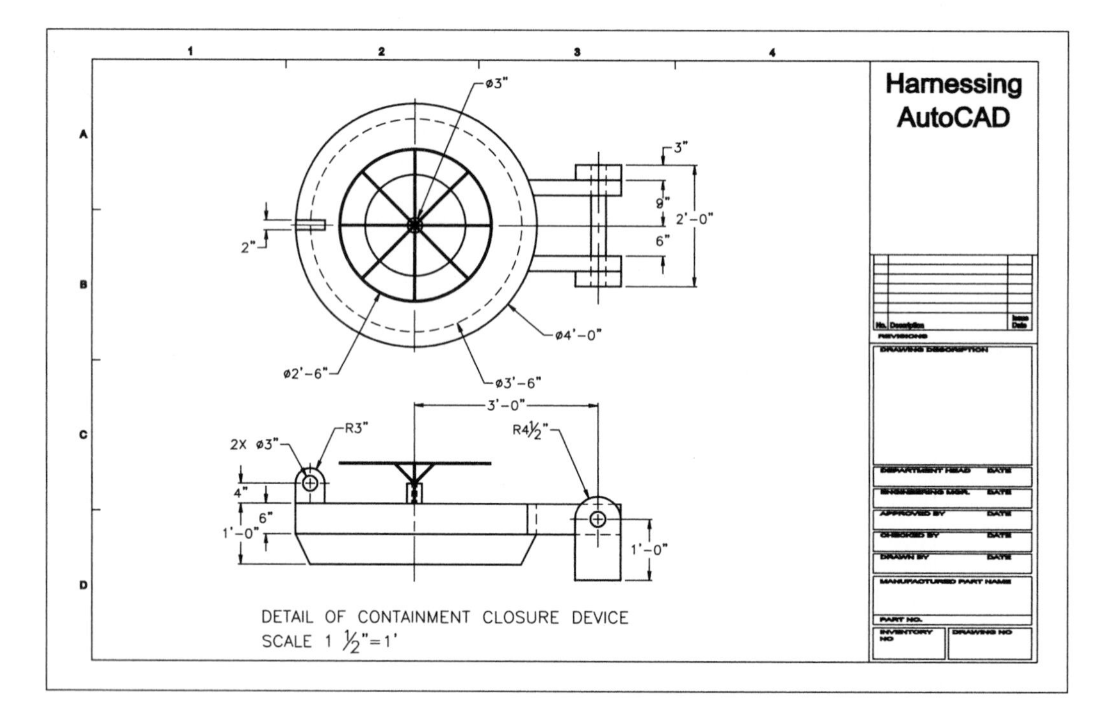


**Figure Ex5–6**

## EXERCISE 5–7

Create the drawing shown in Figure Ex5–7 according to the settings given in the following table:

| Settings | Value | | |
|---|---|---|---|
| 1. Units | Architectural | | |
| 2. Limits | Lower left corner: 0',0' | | |
| | Upper right corner: 60',45' | | |
| 3. Grid | 12" | | |
| 4. Snap | 6" | | |
| 5. Layers | *NAME* | *COLOR* | *LINETYPE* |
| | Construction | Cyan | Continuous |
| | Border | Red | Continuous |
| | Object | Green | Continuous |
| | Text | Blue | Continuous |

**Hints:** Be sure to make each layer current before beginning to draw any objects that are on it.

The border can be drawn from coordinates 2',2" to 58',43" using the RECTANGLE command.

The outline of the foundation wall can be drawn with the PLINE command. Then, by using the OFFSET command (set at 6"), create the additional four lines, each one from the previous.

The Detail of Stair Riser and Sections "A" and "B" can be drawn to true dimensions. Then you can use the SCALE command to enlarge the view of the Detail by a factor of 1.5 and to enlarge the view of the Sections by a factor of 2.0. Do not dimension.

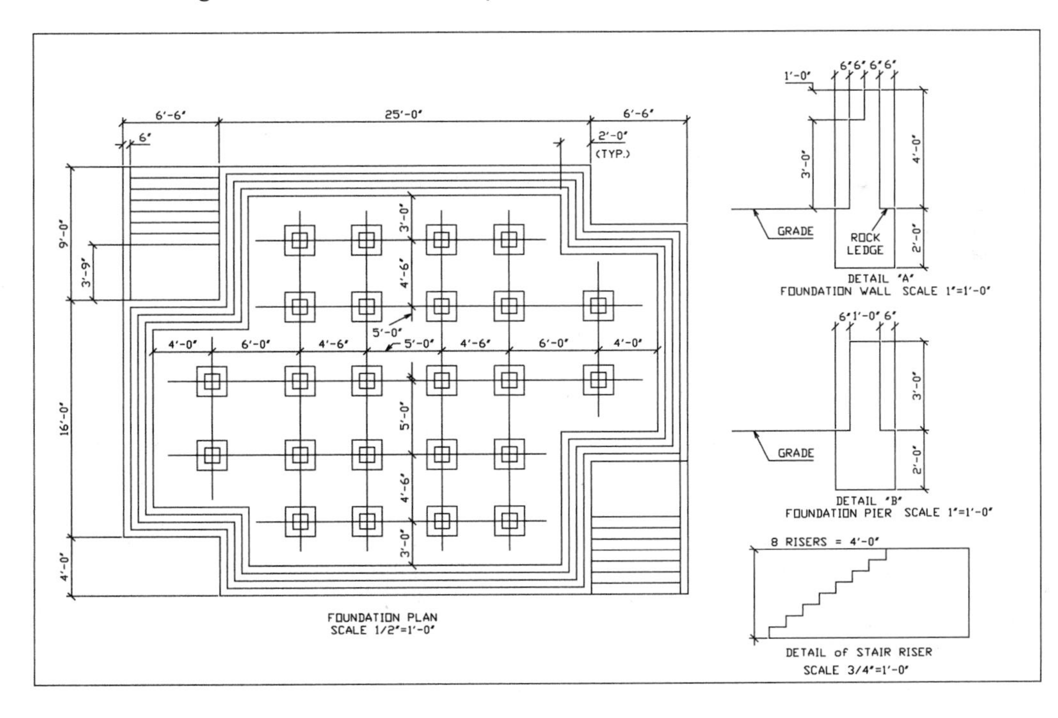


**Figure Ex5–7**

## EXERCISE 5–8

Create the drawing shown in Figure Ex5–8 according to the settings given in the following table:

| Settings | Value |
|---|---|
| 1. Units | Architectural |
| 2. Limits | Lower left corner: 0',0'<br>Upper right corner: 60',45' |
| 3. Grid | 12" |
| 4. Snap | 6" |

| Settings | Value | | |
|---|---|---|---|
| 5. Layers | *NAME* | *COLOR* | *LINETYPE* |
| | Construction | Cyan | Continuous |
| | Border | Red | Continuous |
| | Object | Green | Continuous |
| | Text | Blue | Continuous |
| | Hidden | White | Hidden |

**Hints:** Open the drawing that you created in Exercise 5–7 and save it as EX5–8. Erase the Detail of Stair Riser and Sections "A" and "B."
Erase the internal double squares and the inside line of the foundation wall in the plan.
Change the foundation wall lines to be on the Hidden layer.
The roof lines can be drawn with polylines with a width of 1/2".

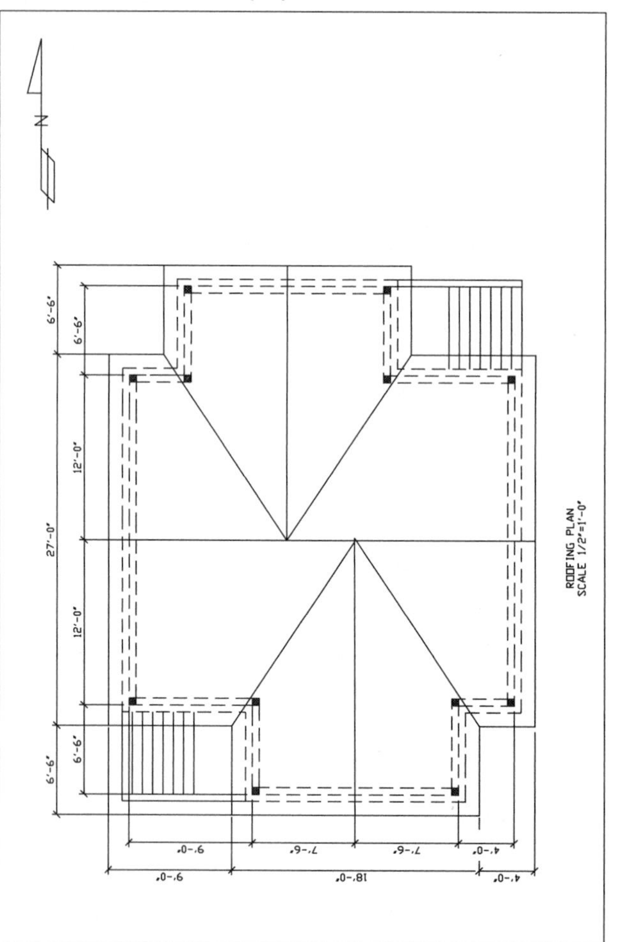


Figure Ex5–8

## EXERCISE 5–9

Create the floor entry pattern drawing shown in Figure Ex5–9 according to the settings given in the following table:

| Settings | Value | | |
|---|---|---|---|
| 1. Units | Architectural | | |
| 2. Limits | Lower left corner: 0,0<br>Upper right corner: 12',9' | | |
| 3. Grid | 12 | | |
| 4. Snap | 6 | | |
| 5. Layers | *NAME* | *COLOR* | *LINETYPE* |
| | Border | Red | Continuous |
| | Holes | Green | Continuous |

**Hint:** Use DONUT and ARRAY commands. Make sure set FILLMODE to ON. Do not dimension.

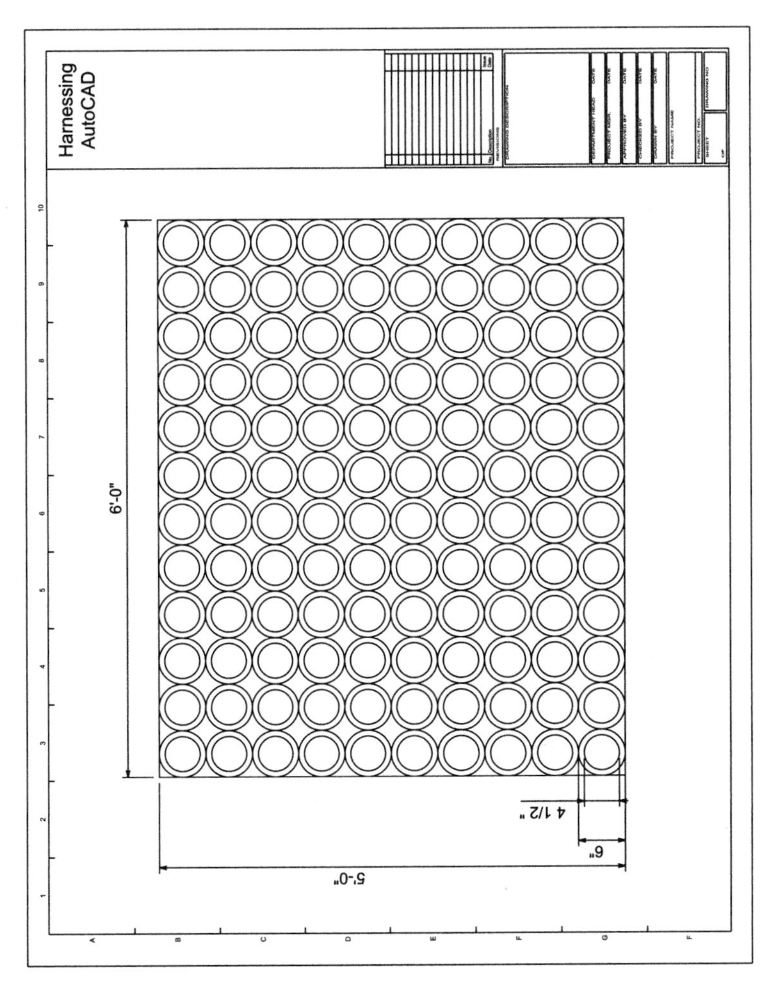


**Figure Ex5–9** *Floor Entry Pattern*

## EXERCISE 5–10

Create the contour lines for the plot plan drawing shown in Figure Ex5–10 according to the settings given in the following table:

| Settings | Value | | |
|---|---|---|---|
| 1. Units | Architectural | | |
| 2. Limits | Lower left corner: 0,0 | | |
| | Upper right corner: 640',496' | | |
| 3. Layers | *NAME* | *COLOR* | *LINETYPE* |
| | Border | White | Continuous |
| | Propline | Blue | Phantom |
| | Road | White | Continuous |
| | Bldgwall | Blue | Continuous |
| | Drive | Green | Continuous |
| | Existcontours | Green | Hidden |
| | Newcontours | Green | Continuous |

civil EXERCISE

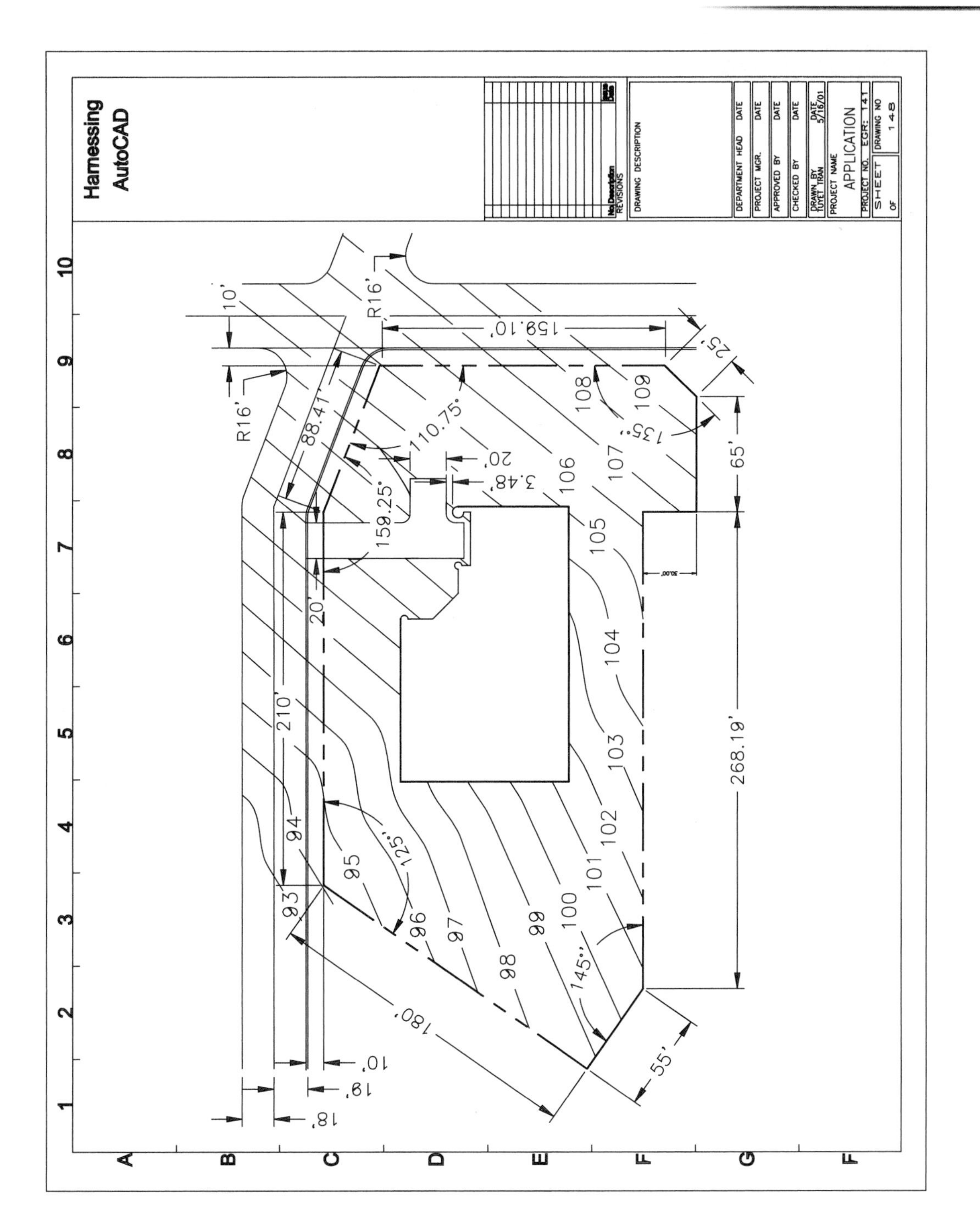


**Figure Ex5–10** *Contour lines for the Plot Plan*

## EXERCISE 5–11

Create the layout of a National Football League playing field drawing shown in Figure Ex5–11 according to the settings given in the following table:

| Settings | Value | | |
|---|---|---|---|
| 1. Units | Architectural | | |
| 2. Limits | Lower left corner: 0,0 | | |
| | Upper right corner: 515',400' | | |
| 3. Layers | *NAME* | *COLOR* | *LINETYPE* |
| | Border | White | Continuous |
| | Playfield | White | Continuous |
| | Yardline | White | Continuous |
| | Notes | Red | Continuous |
| | Dash limit | White | Hidden |

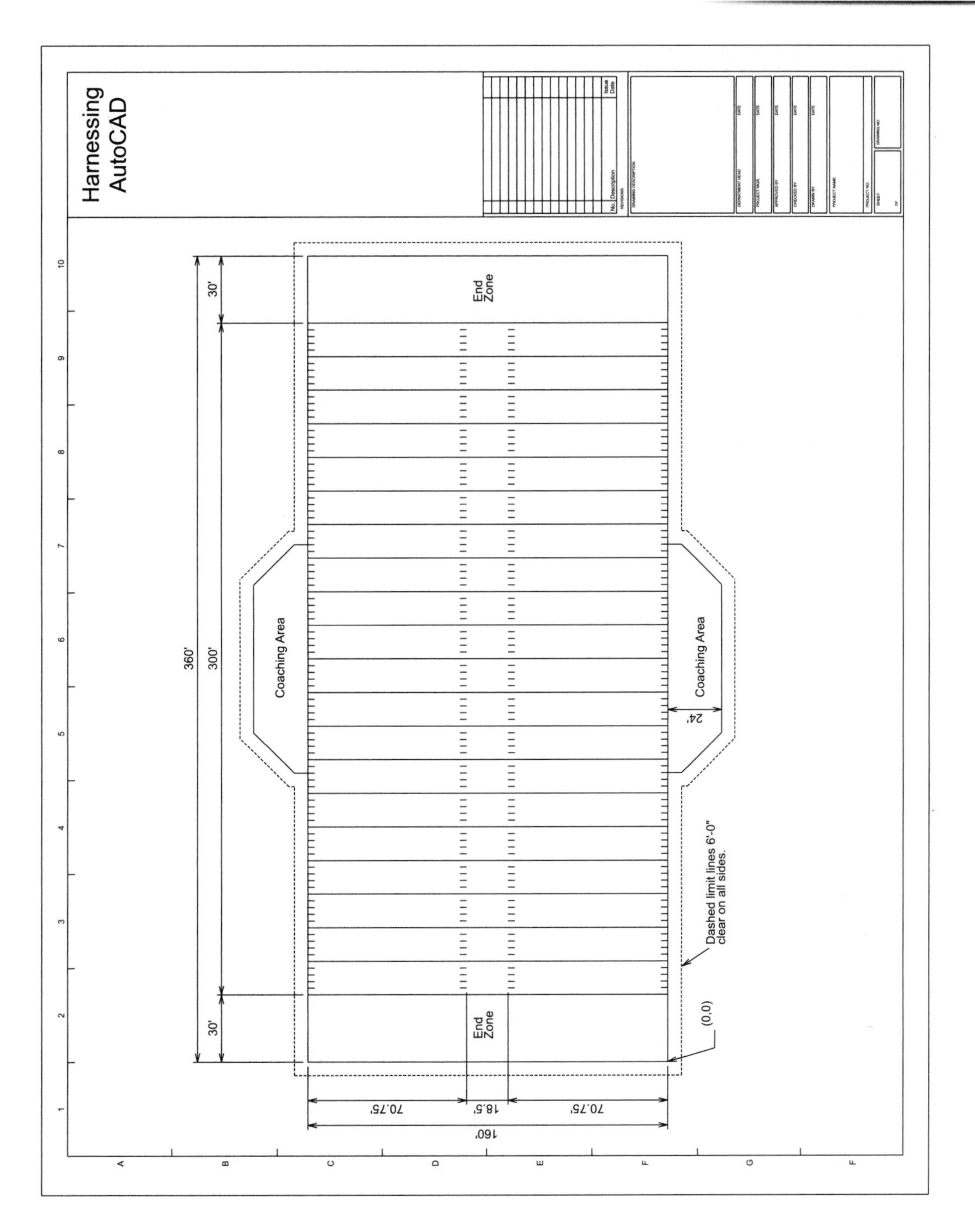


**Figure Ex5–11** *Layout of a National Football League playing field*

## EXERCISE 5–12

Create the layout of property subdivisions in a city block shown in Figure Ex5–12 according to the settings given in the following table:

| Settings | Value | | |
|---|---|---|---|
| 1. Units | Architectural | | |
| 2. Limits | Lower left corner: 0,0, | | |
| | Upper right corner: 460',355' | | |
| 3. Layers | NAME | COLOR | LINETYPE |
| | Cityblock | White | Continuous |
| | Propline | White | Phantom |
| | Bldgperim | Red | Continuous |
| | Bldgwall | Green | Continuous |
| | StrAddr | White | Continuous |

**Hint:** Buildings share a common party wall that is 1 foot wide; except those at the ends of a row.

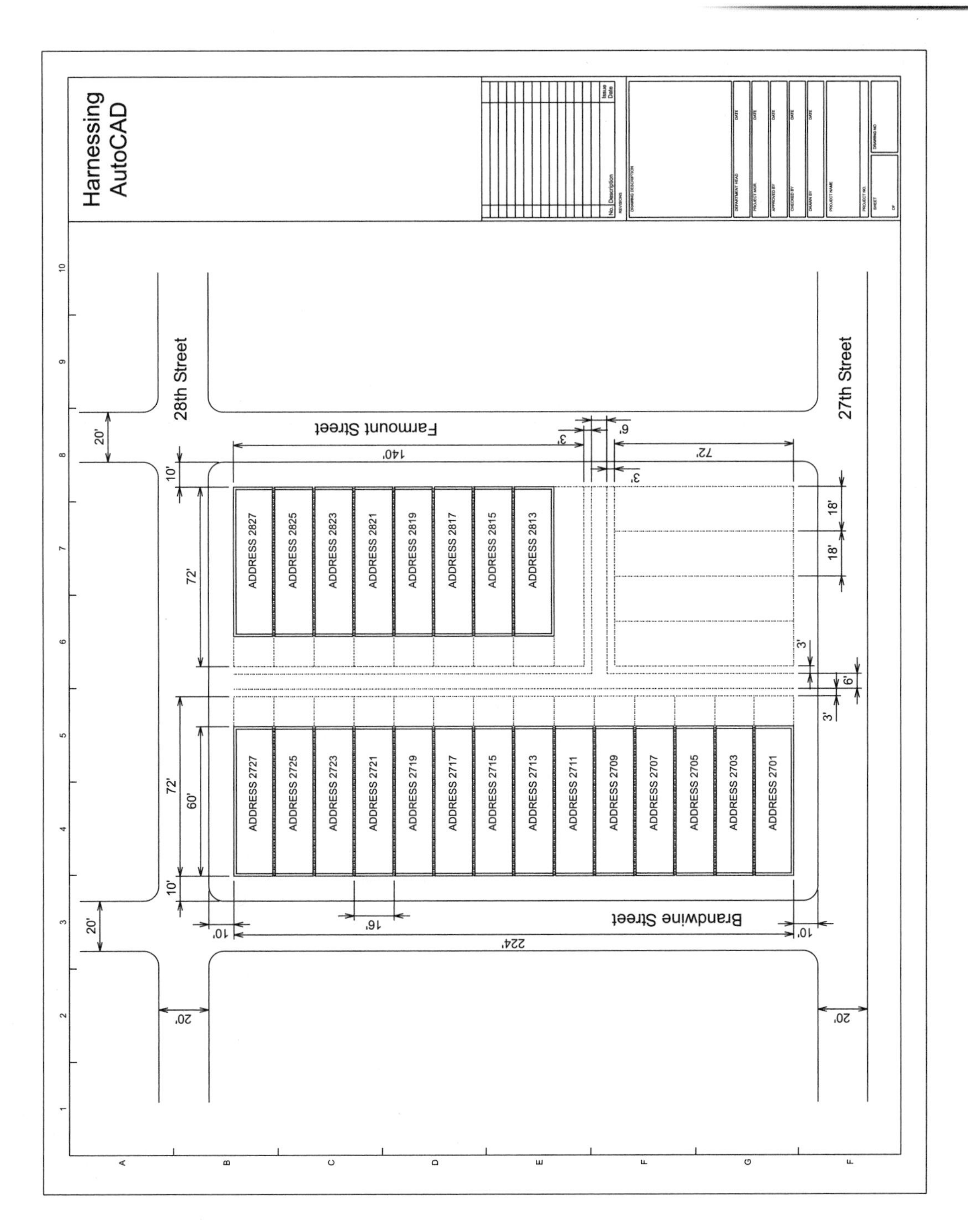


**Figure Ex5–12** *Layout of a Subdivision in a city block*

## EXERCISE 5–13

Create the drawing of a pre-cast concrete panel wall of typical panel size shown in Figure Ex5–13 according to the dimensions given.

| Settings | Value |
|---|---|
| 1. Units | Architectural |
| 2. Limits | Lower left corner: 0,0<br>Upper right corner: 55',40' |

| Settings | Value | | |
|---|---|---|---|
| 3. Layers | NAME | COLOR | LINETYPE |
| | Center | White | Center |
| | Pc panel | White | Continuous |
| | Text | Yellow | Continuous |

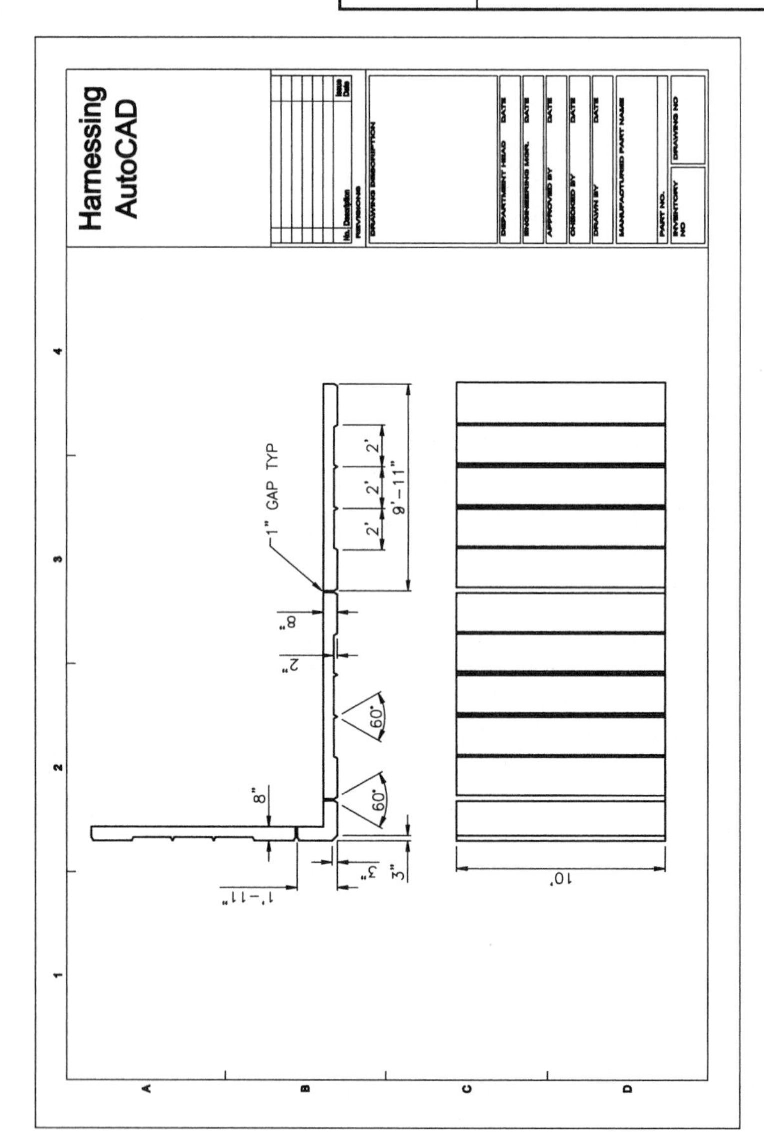


**Figure Ex5–13** *Pre-cast Concrete Panel Wall with typical panel sizes*

## EXERCISE 5–14

Create the schematic layout of an electronic fire alarm and communication diagram shown in Figure Ex5–14 according to the settings given in the following table:

| Settings | Value |
|---|---|
| 1. Units | Architectural |
| 2. Limits | Lower left corner: 0,0<br>Upper right corner: 15',10' |
| 3. Grid | 12 |
| 4. Snap | 6 |

| Settings | Value | | |
|---|---|---|---|
| 5. Layers | *NAME* | *COLOR* | *LINETYPE* |
| | Border | White | Continuous |
| | Module | Blue | Continuous |
| | Symbols | Red | Continuous |
| | Notes | Green | Continuous |

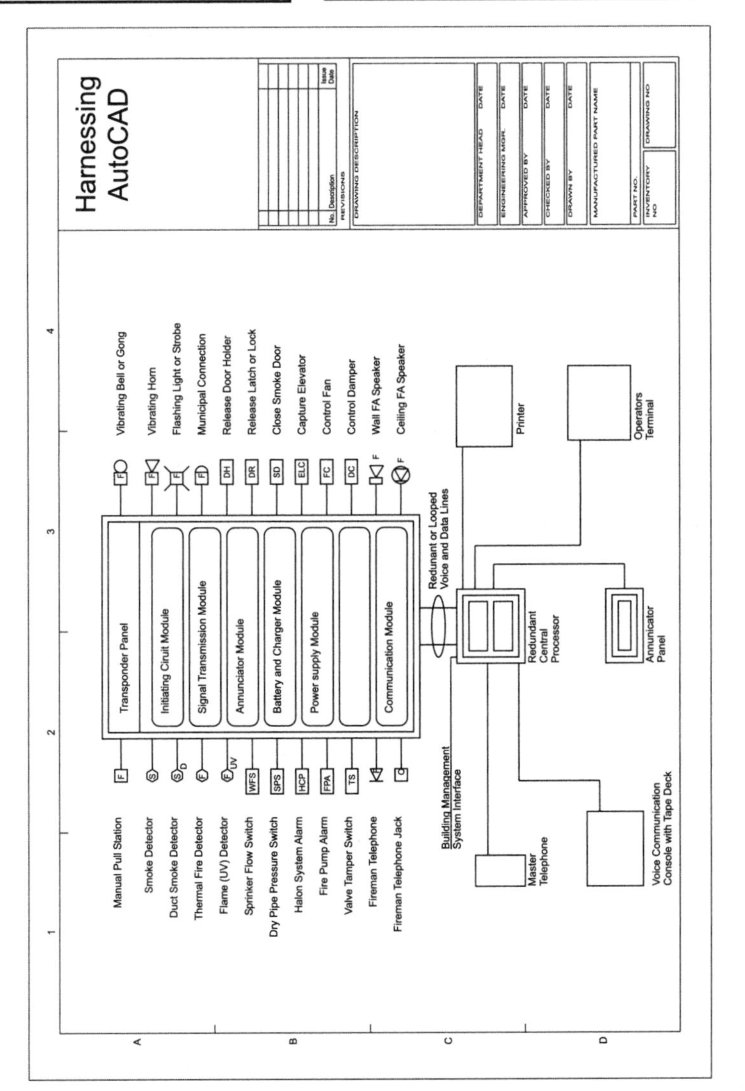


**Figure Ex5–14** *Schematic layout of Electronic Fire Alarm and Communication diagram*

## EXERCISE 5–15

Create the schematic diagram of a simple Ethernet peer to peer network shown in Figure Ex5–15 according to the settings given in the following table:

| Settings | Value | | |
|---|---|---|---|
| 1. Units | Architectural | | |
| 2. Limits | Lower left corner: 0,0 | | |
| | Upper right corner: 5',4' | | |
| 3. Grid | 6 | | |
| 4. Snap | 3 | | |
| 5. Layers | *NAME* | *COLOR* | *LINETYPE* |
| | Border | White | Continuous |
| | Cable | Green | Continuous |
| | Symbols | Red | Continuous |
| | Text | White | Continuous |

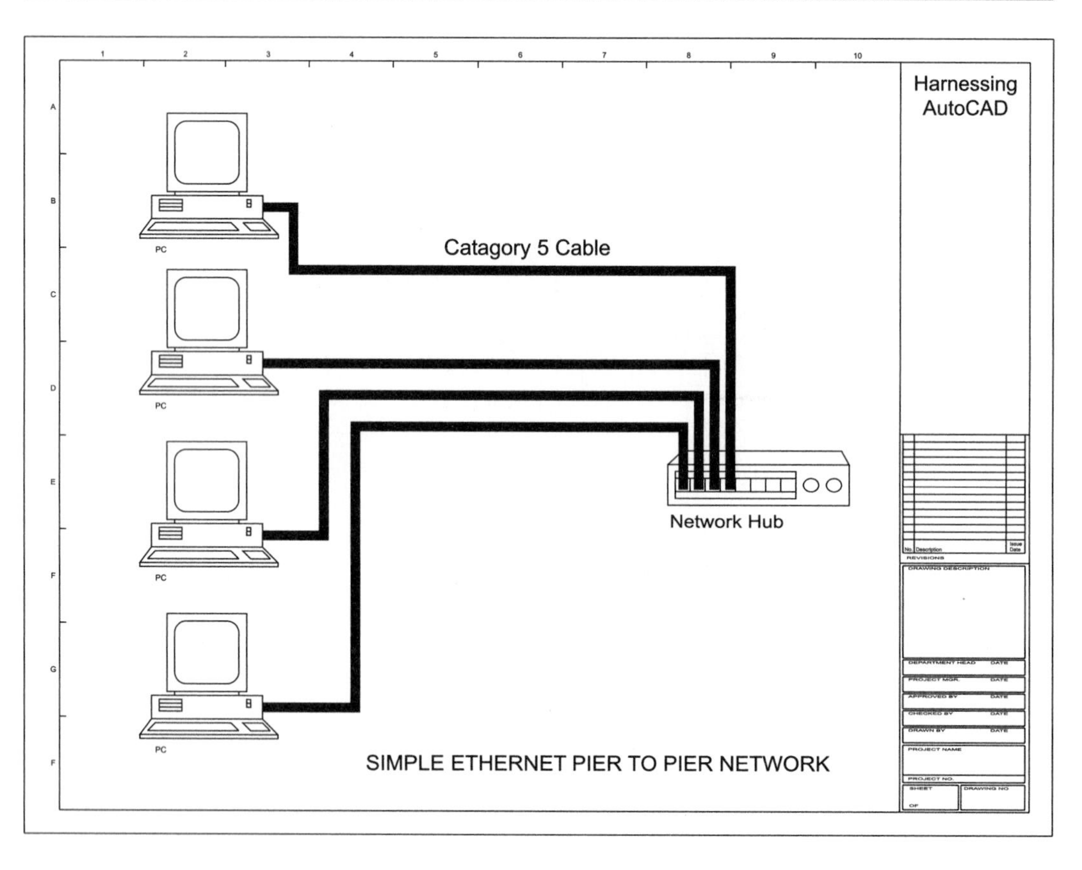


**Figure Ex5–15** *Schematic diagram of a simple Ethernet Pier to Pier Network*

## EXERCISE 5–16

Create the schematic diagram of a two-segment Ethernet network shown in Figure Ex5–16 according to the settings given in the following table:

| Settings | Value |
|---|---|
| 1. Units | Architectural |
| 2. Limits | Lower left corner: 0,0<br>Upper right corner: 12',9' |
| 3. Grid | 12" |
| 4. Snap | 6" |

| Settings | Value | | |
|---|---|---|---|
| 5. Layers | *NAME* | *COLOR* | *LINETYPE* |
| | Border | Red | Continuous |
| | Cable | Green | Continuous |
| | Symbols | Blue | Continuous |
| | Text | White | Continuous |



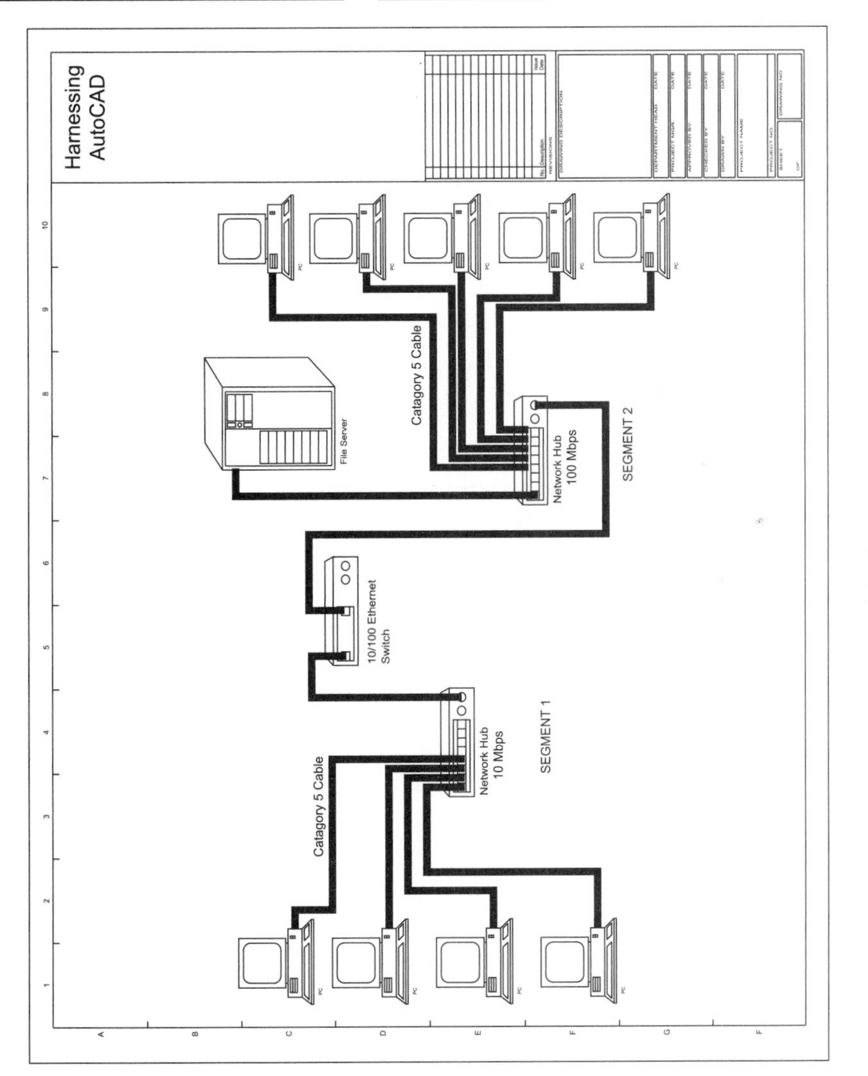


**Figure Ex5–16** *Schematic diagram of a two-segment Ethernet Network diagram*

## EXERCISE 5–17

Create the schematic diagram of residential circuit breakers, as shown in Figure Ex5–17, according to the settings given in the following table:

| Settings | Value |
|---|---|
| 1. Units | Architectural |
| 2. Limits | Lower left corner: 0,0<br>Upper right corner:4',4' |

| Settings | Value | | |
|---|---|---|---|
| 3. Layers | *NAME* | *COLOR* | *LINETYPE* |
| | Border | White | Continuous |
| | Wire | Green | Continuous |
| | Symbols | Blue | Continuous |
| | Text | White | Continuous |

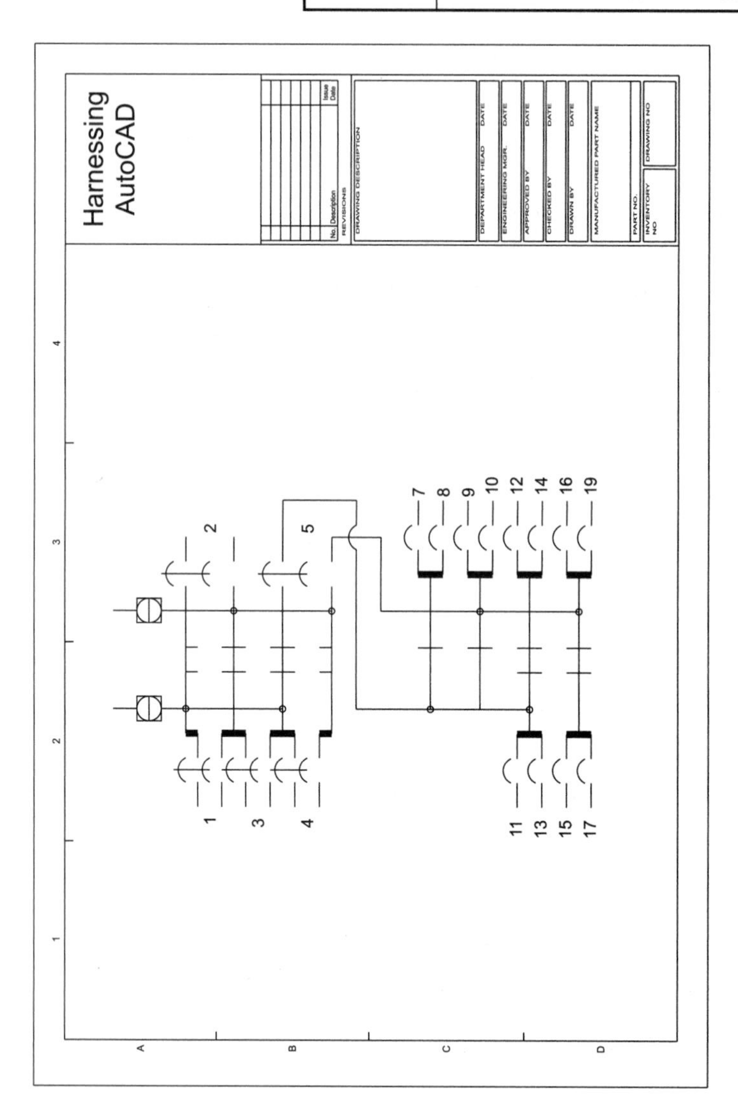


**Figure Ex5–20** *Schematic diagram of a Residential Circuit Breaker*

CHAPTER 6

# Fundamentals V

project EXERCISE

## PROJECT EXERCISE

This project exercise provides point-by-point instructions to draw the objects shown in Figure P6–1. In this exercise you will apply the skills acquired in Chapters 1 through 6.

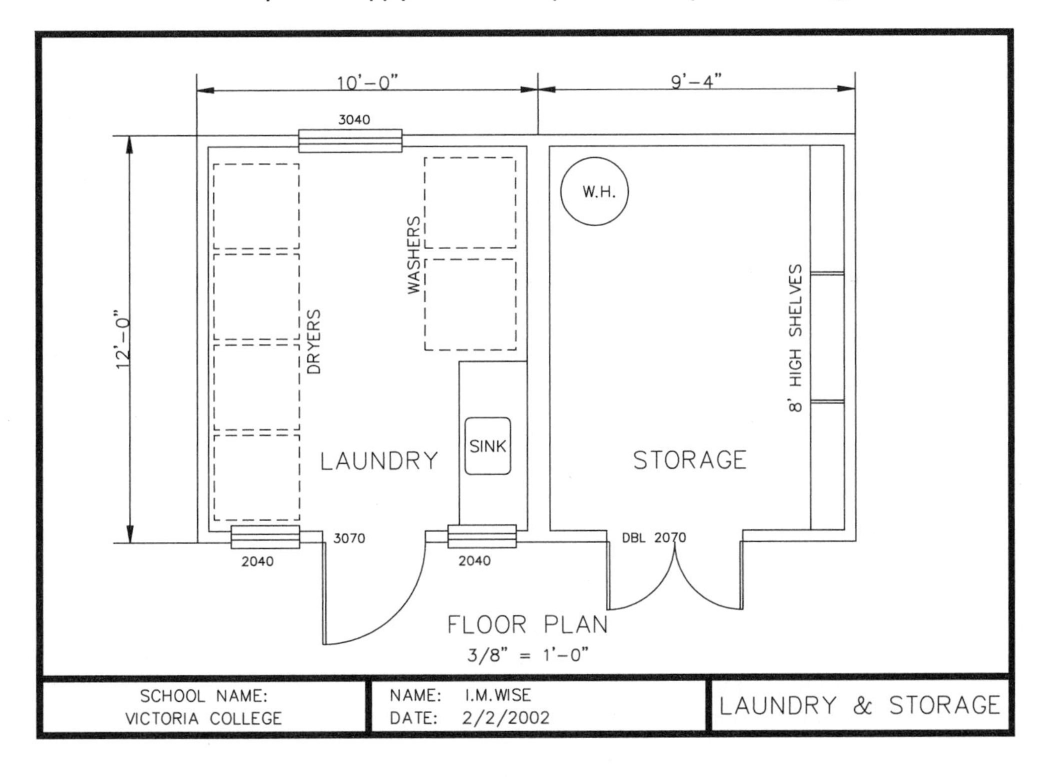


**Figure P6–1** *Completed project drawing (dimensions added for reference)*

In this project you will learn how to do the following:

- Set up the drawing, including limits and layers.
- Use the XLINE, PLINE, and MLINE commands to create objects.
- Use the MLEDIT command for special joining of multilines.
- Use the TRIM and FILLET commands to modify objects.
- Use the TEXT command to create text objects.

## SET UP THE DRAWING AND DRAW A BORDER

**Step 1:** Start the AutoCAD program.

**Step 2 :** Make sure system variable STARTUP is set to 1 that allows AutoCAD to create a new drawing using wizard. To create a new drawing, invoke the NEW command from the Standard toolbar or select New from the File menu.

Select **Start from Scratch** button from Create New Drawing dialog box with the imperial (feet and inches) selection.

**Step 3:** Set drawing units and drawing limits as given in the table. Invoke the LAYER command from the Layers toolbar, or select **Layer...** from the pull-down Format menu. AutoCAD displays the Layer & Properties Manager dialog box. Create five layers, and rename them as shown in the table, assigning appropriate color and linetype.

| SETTINGS | VALUE | | |
|---|---|---|---|
| UNITS | Architectural | | |
| LIMITS | Lower left corner: –5'-0",–6'-0" | | |
| | Upper right corner: 24'-4",15'-4" | | |
| GRID | 4 | | |
| SNAP | 4 | | |
| LAYERS | *NAME* | *COLOR* | *LINETYPE* |
| | Construction | Cyan | Continuous |
| | Border | Red | Continuous |
| | Object | White | Continuous |
| | Hidden | Magenta | Hidden |
| | Text | Blue | Continuous |

## ESTABLISHING CONSTRUCTION LINES

**Step 4:** Set Construction as the current layer, and set ORTHO, GRID, and SNAP all to ON. Invoke the ZOOM ALL command to display the screen to the limits.

**Step 5:** Invoke the XLINE command from the Draw toolbar. AutoCAD prompts:

Command: **xline** (ENTER)
Specify a point or [Hor/Ver/Ang/Bisect/Offset]: **v** (ENTER)
Specify through point: *(place the cursor so that the X coordinate shown in the status bar is at –4'-8" and press the pick button)*
Specify through point: *(continue placing six additional vertical construction lines whose X coordinates are* ***0'-0", 5'-0", 10'-0", 15'-0", 19'-4"****, and* ***24'-0"****, and press* ENTER *to terminate the command sequence)*

Command: *(press* ENTER *to repeat the* XLINE *command)*

Command: XLINE Specify a point or [Hor/Ver/Ang/Bisect/Offset]:
**h** (ENTER)
Specify through point: *(place the cursor so that the Y coordinate shown in the status bar is at –5'-8" and press the pick button)*
Specify through point: *(continue placing four additional horizontal construction lines whose Y coordinates are* ***-4'-0"***, ***0'-0"***, ***12'-0"***, *and* ***15'-0"***, *and press* ENTER *to terminate the command sequence)*

The display should appear as shown in Figure P6–2.

**Figure P6–2** *Drawing with construction lines*

**Step 6:** Set Border as the current layer. Specify Intersection as the Running Object mode.

**Step 7:** Invoke the PLINE command from the Draw toolbar. AutoCAD prompts:

Command: **pline** (ENTER)
Specify start point: *(specify point 1 as shown in Figure P6–3)*
Current line-width is 0'-0"
Specify next point or [Arc/Halfwidth/Length/Undo/Width]:
**w** (ENTER)
Specify starting width <0'-0">: **2** (ENTER)
Specify ending width <0'-2">: (ENTER)
Specify next point or [Arc/Close/Halfwidth/Length/Undo/Width]: *(specify point 2 as shown in Figure P6–3)*
Specify next point or [Arc/Close/Halfwidth/Length/Undo/Width]: *(specify point 3 as shown in Figure P6–3)*
Specify next point or [Arc/Close/Halfwidth/Length/Undo/Width]: *(specify point 4 as shown in Figure P6–3)*
Specify next point or [Arc/Close/Halfwidth/Length/Undo/Width]:
**c** (ENTER)

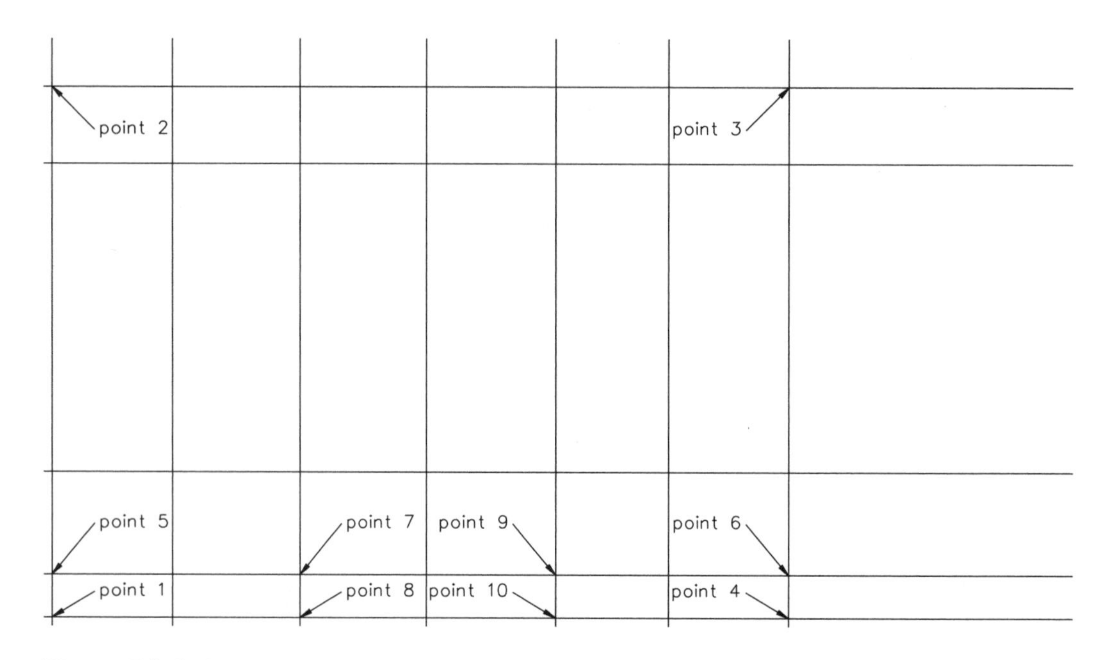


**Figure P6–3** *Points (intersections) for drawing the border*

Command: *(press* ENTER *to repeat the* PLINE *command)*
Specify start point: *(specify point 5 as shown in Figure P6–3)*
Current line-width is 0'-2"
Specify next point or [Arc/Close/Halfwidth/Length/Undo/Width]: *(specify point 6 as shown in Figure P6–3)*
Specify next point or [Arc/Close/Halfwidth/Length/Undo/Width]: *(press* ENTER *to terminate the command sequence)*

Command: *(press* ENTER *to repeat the* PLINE *command)*
Specify start point: *(specify point 7 as shown in Figure P6–3)*
Current line-width is 0'-2"
Specify next point or [Arc/Close/Halfwidth/Length/Undo/Width]: *(specify point 8 as shown in Figure P6–3)*
Specify next point or [Arc/Close/Halfwidth/Length/Undo/Width]: *(press* ENTER *to terminate the command sequence)*

Command: *(press* ENTER *to repeat the* PLINE *command)*
Specify start point: *(specify point 9 as shown in Figure P6–3)*
Current line-width is 0'-2"
Specify next point or [Arc/Close/Halfwidth/Length/Undo/Width]: *(specify point 10 as shown in Figure P6–3)*
Specify next point or [Arc/Close/Halfwidth/Length/Undo/Width]: *(press* ENTER *to terminate the command sequence)*

The display should appear as shown in Figure P6–4.

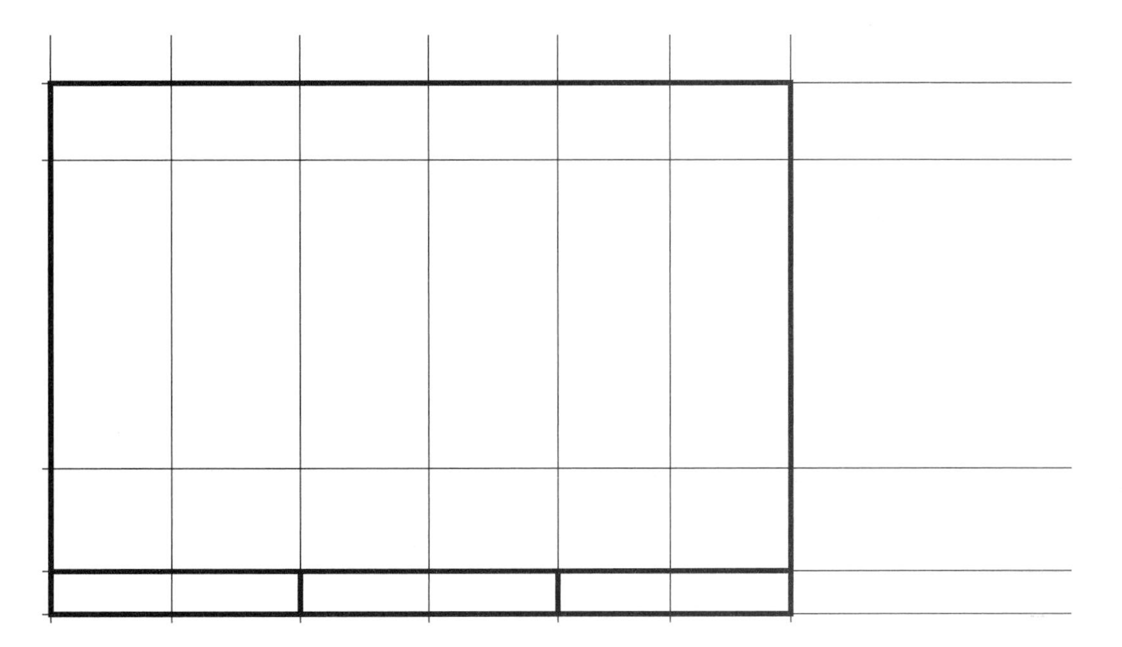

**Figure P6–4** *Drawing with the border*

## DRAWING TEXT IN THE TITLE BLOCK

**Step 8:** Set Construction as the current layer.

**Step 9:** To create additional construction lines for drawing text, invoke the LINE command from the Draw toolbar. AutoCAD prompts:

Command: **line** (ENTER)
Specify first point: *(specify a point on line A, as shown in Figure P6–5, at coordinates –4'8",–4'8")*
Specify next point or [Undo]: *(invoke the Object Snap mode Perpendicular, and select line B, as shown in Figure P6–5)*

Command: *(press* ENTER *to repeat the* LINE *command)*
LINE Specify first point: *(specify a point on line A, as shown in Figure P6–5, at coordinates –4'8",–5',4")*
Specify next point or [Undo]: *(invoke the Object Snap mode Perpendicular, and select line B, as shown in Figure P6–5)*
Specify next point or [Undo]: *(press* ENTER *to terminate the command sequence)*

Command: *(press* ENTER *to repeat the* LINE *command)*
LINE Specify first point: *(invoke the Object Snap mode Midpoint, and select line C, as shown in Figure P6–5)*
Specify next point or [Undo]: *(use the Object Snap mode Perpendicular, and select line D, as shown in Figure P6–5)*
Specify next point or [Undo]: *(press* ENTER *to terminate the command sequence)*

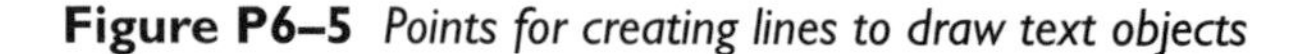


**Figure P6–5** *Points for creating lines to draw text objects*

**Step 10:** Set Text as the current layer.

**Step 11:** Invoke the TEXT command by selecting Text from the Draw toolbar and then selecting Single line text from the flyout menu to draw the text. AutoCAD prompts:

Command: **text** (ENTER)
Specify start point of text or [Justify/Style]: **j** (ENTER)
Enter an option [Align/Fit/Center/Middle/Right/TL/TC/TR/ML/MC/MR/BL /BC/BR]: **c** (ENTER)
Specify center point of text: *(invoke the Object Snap mode Midpoint, and select the first construction line drawn in the previous step to establish point 1, as shown in Figure P6–6)*
Specify height <current>: **4** (ENTER)
Specify rotation angle of text <current>: **0** (ENTER)
Enter text: *(enter* ***SCHOOL NAME:****)*

Enter text: (ENTER)
Command: *(press* ENTER *to repeat the* TEXT *command)*
Specify start point of text or [Justify/Style]: **j** (ENTER)
Enter an option [Align/Fit/Center/Middle/Right/TL/TC/TR/ML/MC/MR/BL /BC/BR]: **c** (ENTER)
Specify center point of text: *(invoke the Object Snap mode Midpoint, and select the second construction line drawn in the previous step to establish point 2 as shown in Figure P6–6)*
Specify height <0'-4">: (ENTER)
Specify rotation angle of text <0>: (ENTER)
Enter text: *(enter the name of your school)*
Enter text: (ENTER)

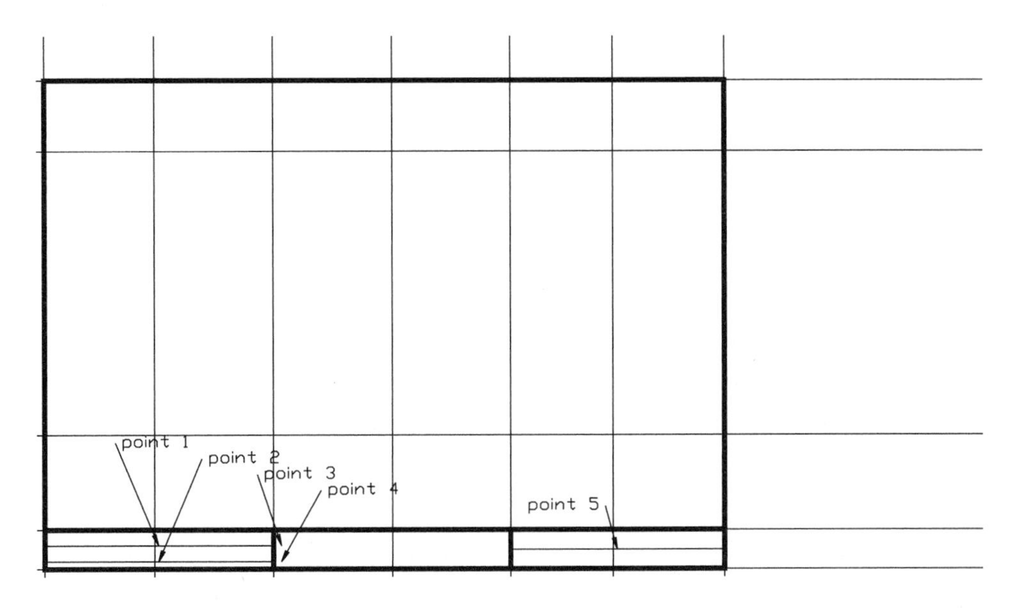


**Figure P6–6** *Points to draw text for the title block*

Command: *(press* ENTER *to repeat the* TEXT *command)*
Specify start point of text or [Justify/Style]: *(specify point 3, as shown in Figure P6–6, at coordinates 5'-4",–4'-8")*
Specify height <0'-4">: (ENTER)
Specify rotation angle of text <0>: (ENTER)
Enter text: **NAME:** *(enter **NAME:** followed by your name)*
Enter text: (ENTER)

Command: *(press* ENTER *to repeat the* TEXT *command)*
Specify start point of text or [Justify/Style]: *(specify point 4, as shown in Figure P6–6 at coordinates 5'-4", -5'-4")*
Specify height <0'-4">: (ENTER)
Specify rotation angle of text <0>: (ENTER)
Enter text: **DATE:** *(enter **DATE:** followed by today's date)*
Enter text: (ENTER)

Command: *(press* ENTER *to repeat the* TEXT *command)*
Specify start point of text or [Justify/Style]: **j** (ENTER)
Enter an option [Align/Fit/Center/Middle/Right/TL/TC/TR/ML/MC/MR/BL/BC/BR]: **m** (ENTER)
Specify middle point of text: *(invoke the Object Snap mode Midpoint, and select the first construction line drawn in the previous step to establish point 5, as shown in Figure P6–6)*
Specify height <0'-4">: **6** (ENTER)
Specify rotation angle of text <0>: (ENTER)
Enter text: (ENTER ***LAUNDRY & STORAGE***)

Enter text: (ENTER)

After drawing the text as indicated, the display should appear as shown in Figure P6–7.

With the Construction layer set to OFF (do *not* turn it OFF at this time), the display would appear as shown in Figure P6–8.

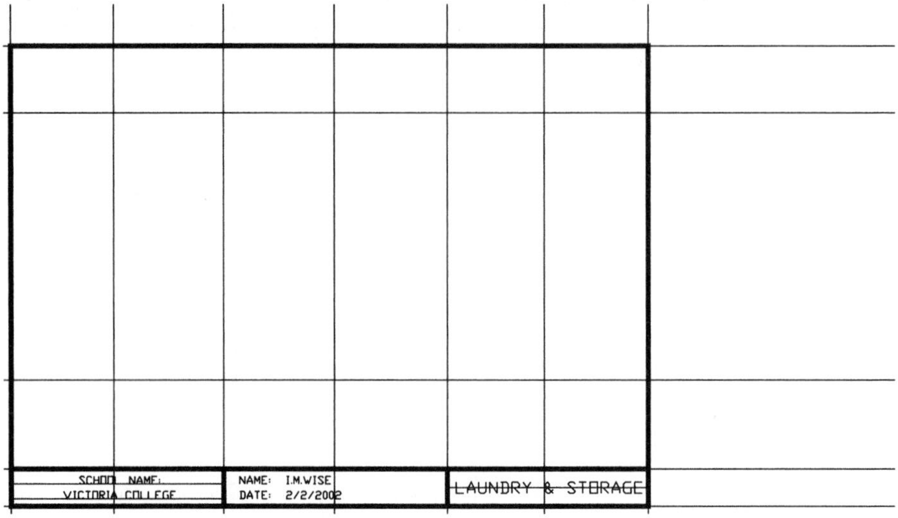

**Figure P6–7** *Drawing with the border and title block (Construction layer ON)*

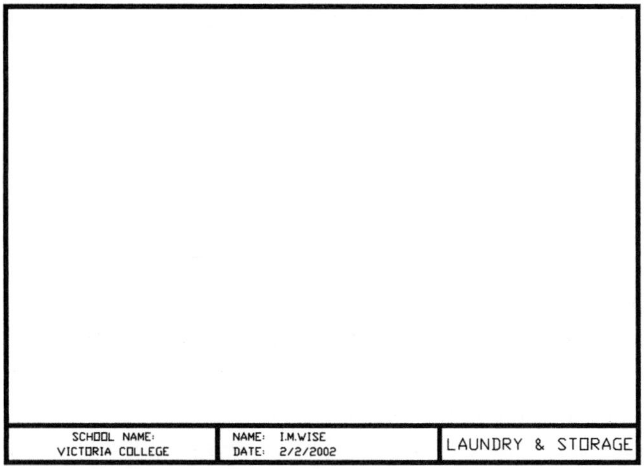

**Figure P6–8** *Drawing with the border and title block (Construction layer OFF)*

**Step 12:** From the Standard toolbar, invoke the ZOOM SCALE command. AutoCAD prompts:

Command: **zoom** (ENTER)
Specify corner of window, enter a scale factor (nX or nXP), or [All/Center/Dynamic/Extents/Previous/Scale/Window] <real time>: **.8** (ENTER)

**Step 13:** Erase the construction lines that are not needed (see Figure P6–9) by invoking the ERASE command from the Modify toolbar. AutoCAD prompts:

Command: **erase** (ENTER)
Select objects: *(select all of the Construction xlines with the designated symbol, plus the Construction xlines created for the title block text, as shown in Figure P6–9).*

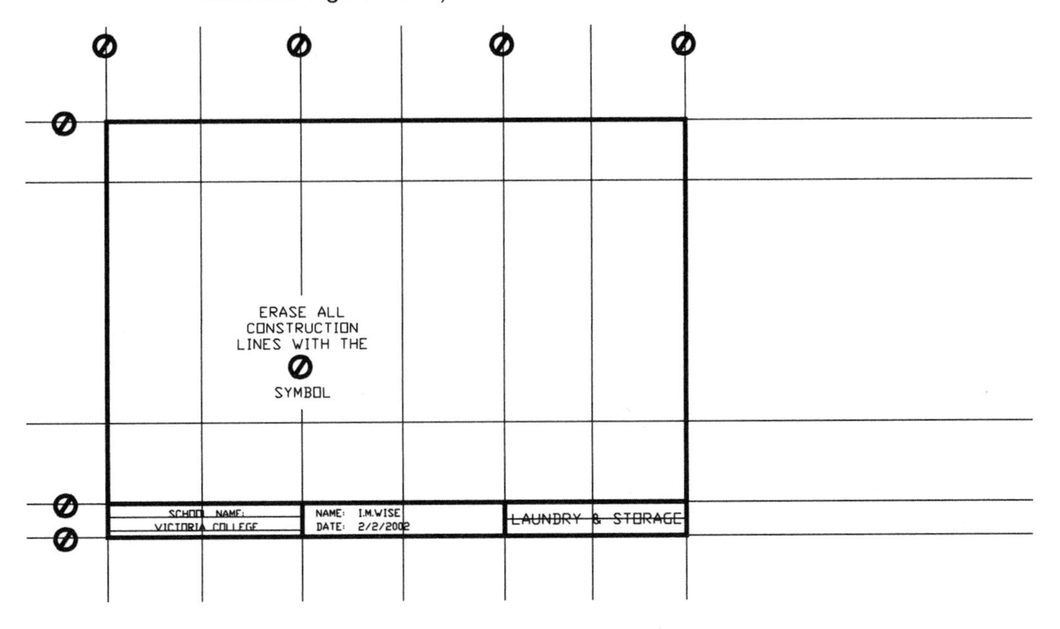


**Figure P6–9** *Drawing with the lines to be erased indicated*

Invoke the ZOOM ALL command to display the whole drawing.

## DRAWING THE WALLS

**Step 14:** Set Object as the current layer. To draw the walls, invoke the MLINE command from the Draw toolbar. Set the Scale factor to 4, and draw the closed rectangle by *specifying* the points indicated in Figure P6–10. AutoCAD prompts:

Command: **mline** (ENTER)
Current settings: Justification = Top, Scale = 1.00, Style = STANDARD
Specify start point or [Justification/Scale/STyle]: **s** (ENTER)
Enter mline scale <1.00>: **4** (ENTER)
Current settings: Justification = Top, Scale = 4.00, Style = STANDARD
Specify start point or [Justification/Scale/STyle]: *(specify point 1 as shown in Figure P6–10)*
Specify next point: *(specify point 2 as shown in Figure P6–10)*
Specify next point or [Undo]: *(specify point 3 as shown in Figure P6–10)*
Specify next point or [Close/Undo]: *(specify point 4 as shown in Figure P6–10)*
Specify next point or [Close/Undo]: **c** (ENTER)

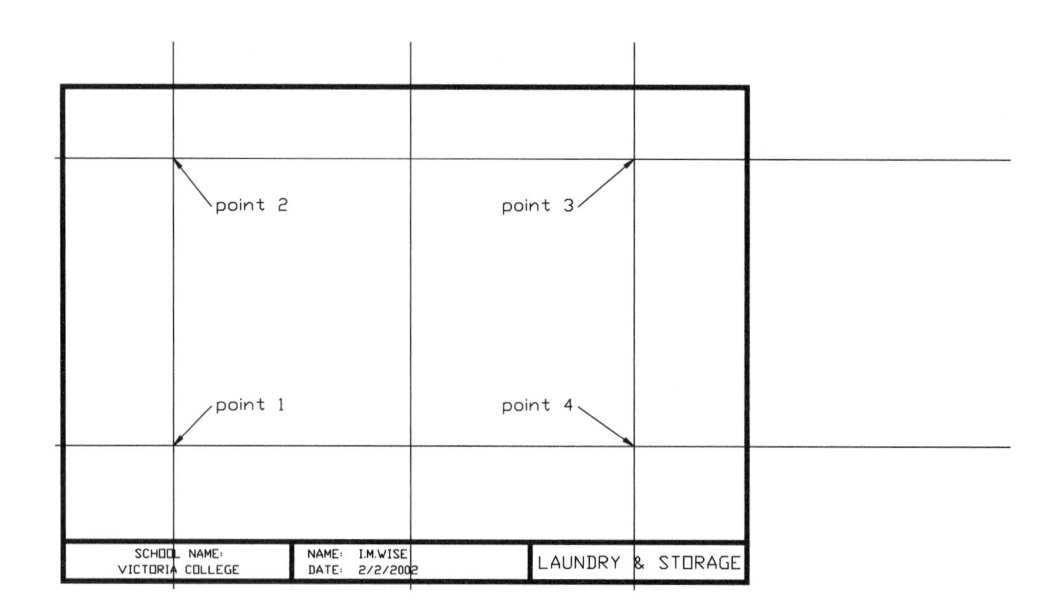

**Figure P6–10** *Points for drawing the walls*

**Step 15:** To draw the interior wall, again invoke the MLINE command from the Draw toolbar. Set the Scale to 8 and the Justification to 0, and draw the interior wall by *specifying* the points indicated in Figure P6–11. AutoCAD prompts:

Command: **mline** (ENTER)
Current settings: Justification = Top, Scale = 4.00, Style = STANDARD
Specify start point or [Justification/Scale/STyle]: **s** (ENTER)
Enter mline scale <4.00>: **8** (ENTER)
Current settings: Justification = Top, Scale = 8.00, Style = STANDARD
Specify start point or [Justification/Scale/STyle]: **j** (ENTER)
Enter justification type [Top/Zero/Bottom] <top>: **z** (ENTER)
Current settings: Justification = Zero, Scale = 8.00, Style = STANDARD
Specify start point or [Justification/Scale/STyle]: *(specify point 1 as shown in Figure P6–11)*
Specify next point: *(specify point 2 as shown in Figure P6–11)*
Specify next point or [Undo]: *(press ENTER to terminate the command sequence)*

**Step 16:** Set the Running Object Snap mode to OFF.

**Step 17:** To edit the multiline as shown in Figure P6–12, invoke the MLEDIT command from the Modify II toolbar. AutoCAD displays the Multiline Edit Tools dialog box. Choose the Open Tee option, and then choose the OK button to close the dialog box. AutoCAD prompts:

Command: **mledit** (ENTER)
Select first mline: *(select the interior wall at point A, as shown in Figure P6–12)*

Select second mline: *(select the exterior wall at point B, as shown in Figure P6–12)*
Select first mline or [Undo]: *(select the interior wall at point C, as shown in Figure P6–12)*
Select second mline: *(select the exterior wall at point D, as shown in Figure P6–12)*
Select first mline or [Undo]: *(press* ENTER *to terminate the command sequence)*

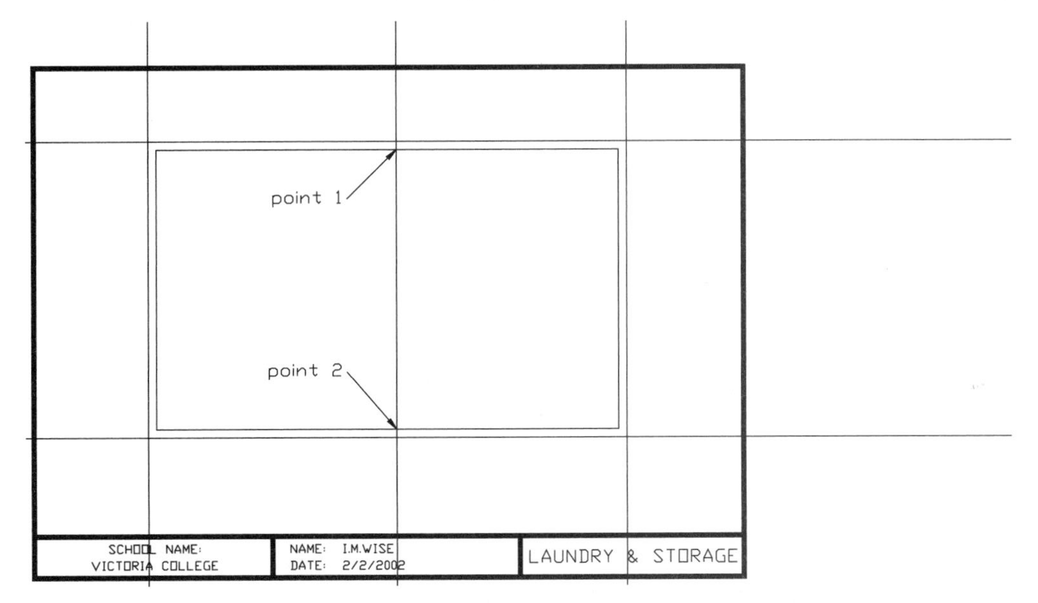


**Figure P6–11** *Points for drawing the interior walls*

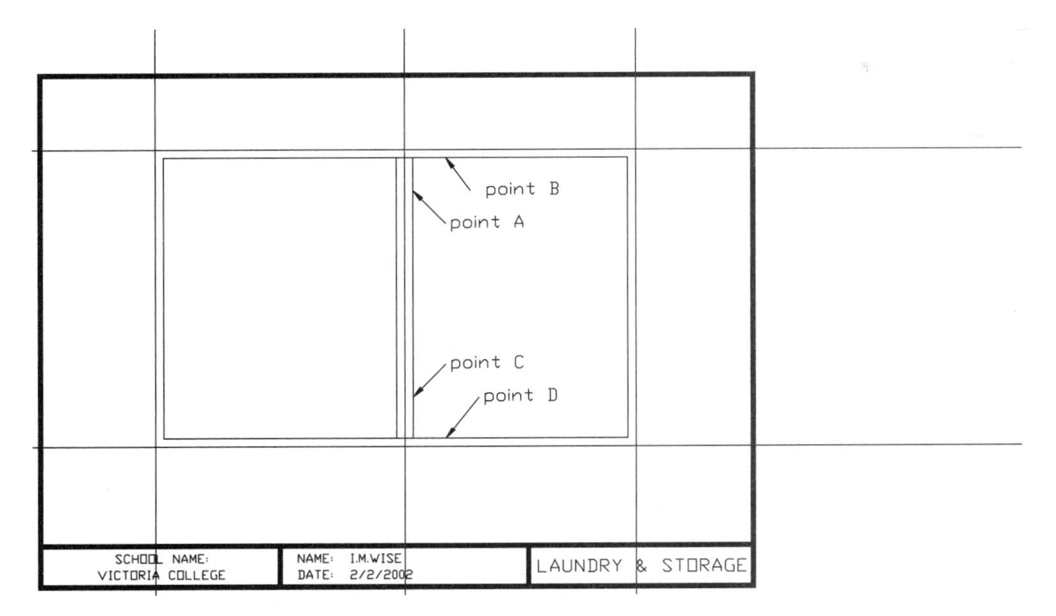


**Figure P6–12** *Points for editing multilines*

After drawing the mline for the interior and exterior walls and "opening" the tees as indicated, the display should appear as shown in Figure P6–13.

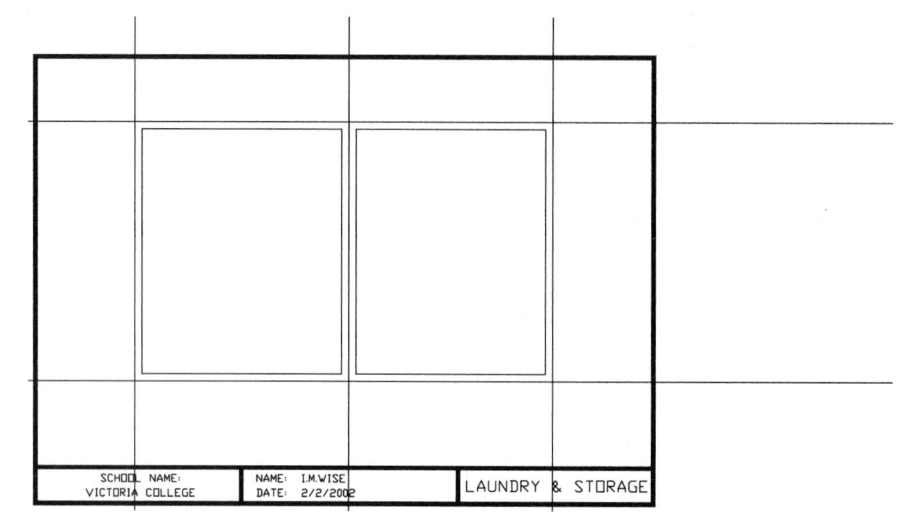


**Figure P6–13** *Drawing with interior and exterior walls*

**Step 18:** Invoke the MOVE command from the Modify toolbar. AutoCAD prompts:

Command: **move** (ENTER)
Select objects: *(select the middle vertical construction line)*
Select objects: *(press ENTER to complete the selection of objects)*
Specify base point or displacement: *(specify a point on the selected line)*
Specify second point of displacement or <use first point as displacement>: *(specify a point that is 2'-4" left of the line)*

Command: *(press ENTER to repeat the MOVE command)*
Select objects: *(select the right vertical construction line)*
Select objects: *(press ENTER to complete the selection of objects)*
Specify base point or displacement: *(specify a point on the selected line)*
Specify second point of displacement or <use first point as displacement>: *(specify a point that is 1'-4" left of the line)*

Command: *(press ENTER to repeat the MOVE command)*
Select objects: *(select the lower horizontal construction line)*
Select objects: *(press ENTER to complete the selection of the objects)*
Specify base point or displacement: *(specify a point on the selected line)*
Specify second point of displacement or <use first point as displacement>: *(specify a point that is 5'-4"above the line)*

## DRAWING THE COUNTER FOR THE SINK

**Step 19:** Set Intersection as the Running Object Snap mode. Invoke the LINE command from the Draw toolbar. AutoCAD prompts:

Command: **line** (ENTER)
Specify first point: *(specify point 1 as shown in Figure P6–14)*
Specify next point or [Undo]: *(specify point 2 as shown in Figure P6–14)*
Specify next point or [Undo]: *(specify point 3 as shown in Figure P6–14)*
Specify next point or [Undo]: *(press* ENTER *to terminate the command sequence)*

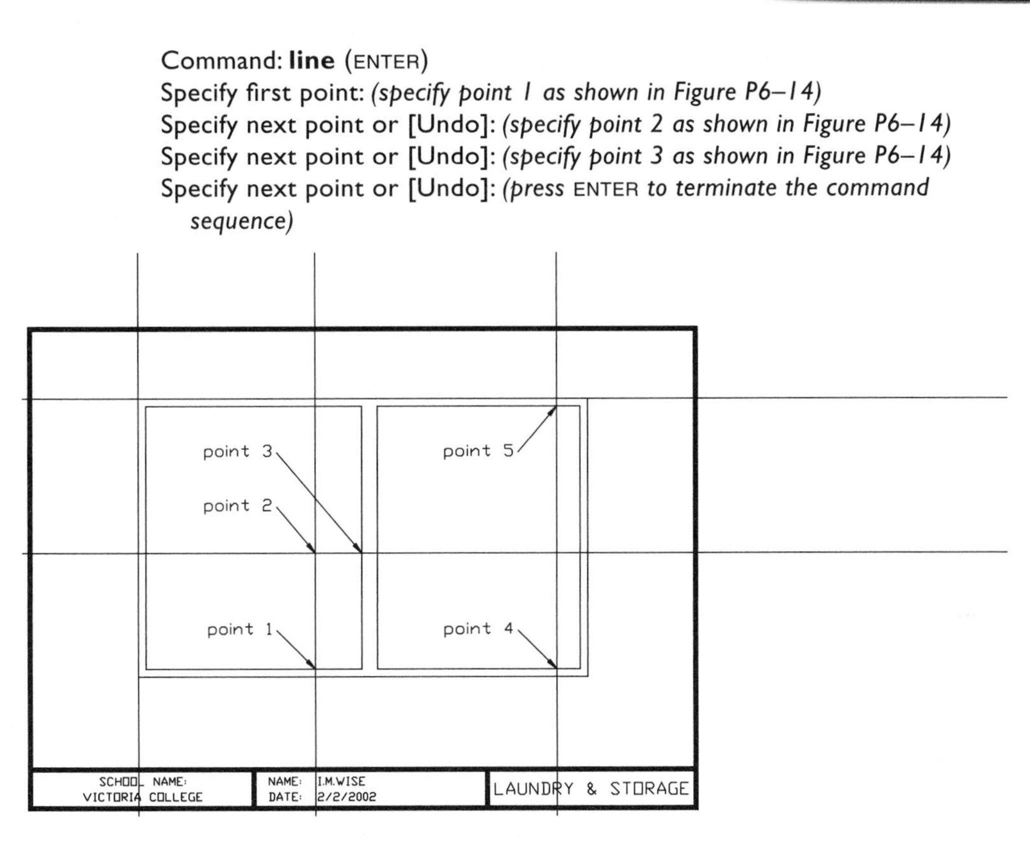


**Figure P6–14** *Points for drawing the counter for the sink*

Command: *(press* ENTER *to repeat the* LINE *command)*
LINE Specify first point: *(specify point 4 as shown in Figure P6–14)*
Specify next point or [Undo]: *(specify point 5 as shown in Figure P6-14)*
Specify next point or [Undo]: *(press* ENTER *to terminate the command sequence)*

**Step 20:** Invoke the ERASE command from the Modify toolbar to erase the remaining construction lines. Set the Running Object Snap mode to OFF.

## OPENING THE WALLS FOR DOORS AND WINDOWS

**Step 21:** To create the opening of the walls, invoke the MLEDIT command from the Modify II toolbar. AutoCAD displays the Multiline Edit Tools dialog box. Choose the Cut All option, and then choose OK to close the dialog box. AutoCAD prompts:

Command: **mledit** (ENTER)
Select mline: *(specify the lower wall at point 1'-0",0'-0")*
Select second point: *(specify the lower wall at point 3'-0",0'-0")*
Select mline or [Undo]: *(specify the lower wall at point 3'-8",0'-0")*
Select second point: *(specify the lower wall at point 6'-8",0'-0")*
Select mline or [Undo]: *(specify the lower wall at point 7'-4",0'-0")*

Select second point: *(specify the lower wall at point 9'-4",0'-0")*
Select mline or [Undo]: *(specify the lower wall at point 12'-0",0'-0")*
Select second point: *(specify the lower wall at point 16'-0",0'-0")*
Select mline or [Undo]: *(specify the upper wall at point 3'-8",12'-0")*
Select second point: *(specify the upper wall at point 6'-8",12'-0")*
Select mline or [Undo]: *(press* ENTER *to terminate the command sequence)*

## DRAWING THE WINDOWS AND THE WATER HEATER

**Step 22:** Invoke the ZOOM WINDOW command to zoom in as shown in Figure P6–15.

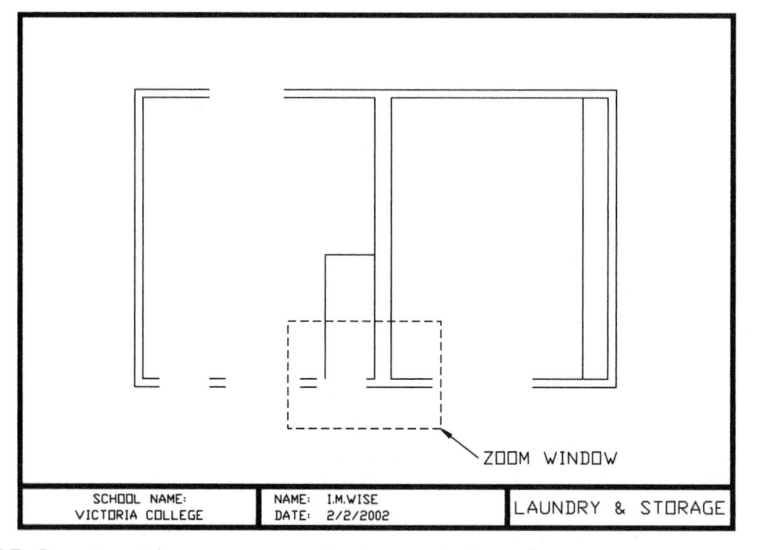


**Figure P6–15** *Drawing with* ZOOM WINDOW *in the area indicated*

**Step 23:** Change the SNAP setting to 2. Invoke the MLINE command from the Draw toolbar to draw the windows. AutoCAD prompts:

Command: **mline** (ENTER)
Current settings: Justification = Zero, Scale = 8.00, Style = STANDARD
Specify start point or [Justification/Scale/STyle]: **s** (ENTER)
Enter mline scale <8.00>: **2** (ENTER)
Current settings: Justification = Zero, Scale = 2.00, Style = STANDARD
Specify start point or [Justification/Scale/STyle]: *(specify point A, as shown in Figure P6–16, at coordinates 7'-4",0'-2")*
Specify next point: *(specify the endpoint at coordinates 9'-4",0'-2")*
Specify next point or [Undo]: *(press* ENTER *to terminate the command sequence)*

Command: *(press* ENTER *to repeat the* MLINE *command)*
Current settings: Justification = Zero, Scale = 2.00, Style = STANDARD
Specify start point or [Justification/Scale/STyle]: **s** (ENTER)
Enter mline scale <2.00>: **8** (ENTER)
Current settings: Justification = Zero, Scale = 8.00, Style = STANDARD

Specify start point or [Justification/Scale/STyle]: *(specify point A at coordinates 7'-4",0'-2")*
Specify next point: *(specify the endpoint at coordinates 9'-4",0'-2")*
Specify next point or [Undo]: *(press* ENTER *to terminate the command sequence)*

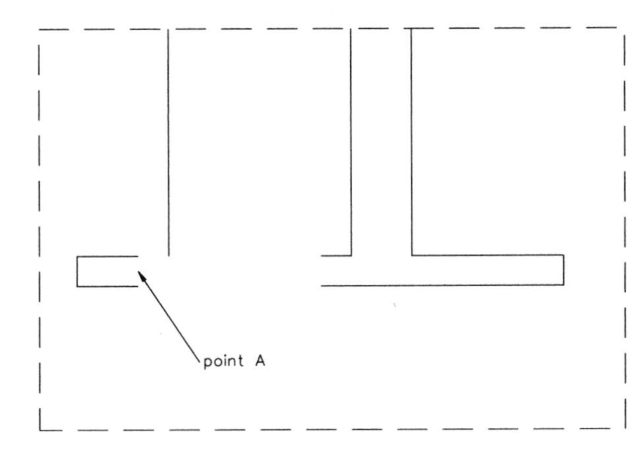


**Figure P6–16** *Points indicating where the starting point of the window has to be drawn*

**Step 24:** Invoke the TRIM command from the Modify toolbar to trim the line for the counter even with the line for the window. AutoCAD prompts:

Command: **trim** (ENTER)
Select cutting edges: *(select the mline for the cutting edge as shown in Figure P6–17)*
Select objects: *(press* ENTER *to complete selection of the cutting edge)*
Select object to trim or shift-select to extend or [Project/Edge/Undo]: *(select the object to trim as shown in Figure P6–17)*
Select object to trim or shift-select to extend or [Project/Edge/Undo]: *(press* ENTER *to complete selection of the object to trim)*

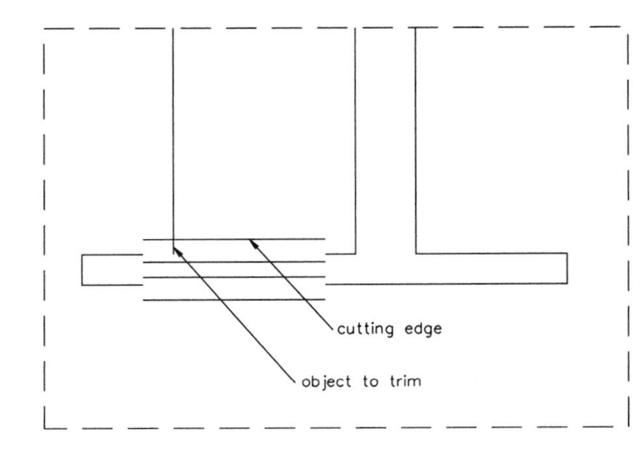


**Figure P6–17** *Indication of the lines to trim*

**Step 25:** Use the LINE command to close the end of the multilines for the window openings and door jambs.

After trimming the line for the cabinet and drawing the lines for closing the ends of the window, the display should appear as shown in Figure P6–18.

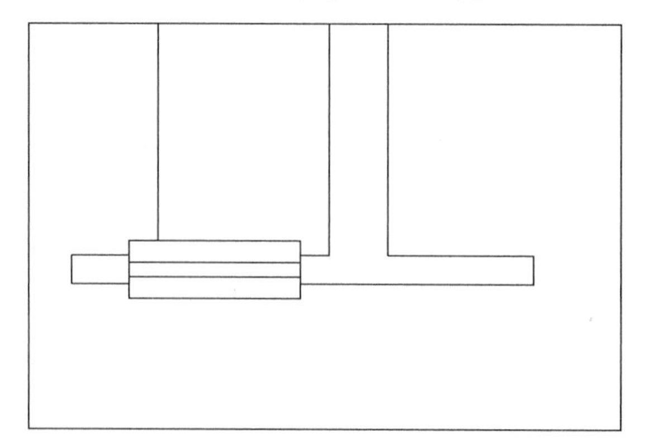

**Figure P6–18** *Drawing showing the window opening*

**Step 26:** Draw two additional windows with the MLINE command with Scales 8 and 2 and Justification set to 0, and close the ends of the other multilines representing window and door jambs.

**Step 27:** Invoke the ZOOM ALL command to display the complete drawing. Draw the water heater with the CIRCLE command, with center at coordinates 11'-8",10'-4" and radius of 1'-0".

After drawing the remaining windows, closing the jambs, and drawing the water heater, the display should appear as shown in Figure P6–19.

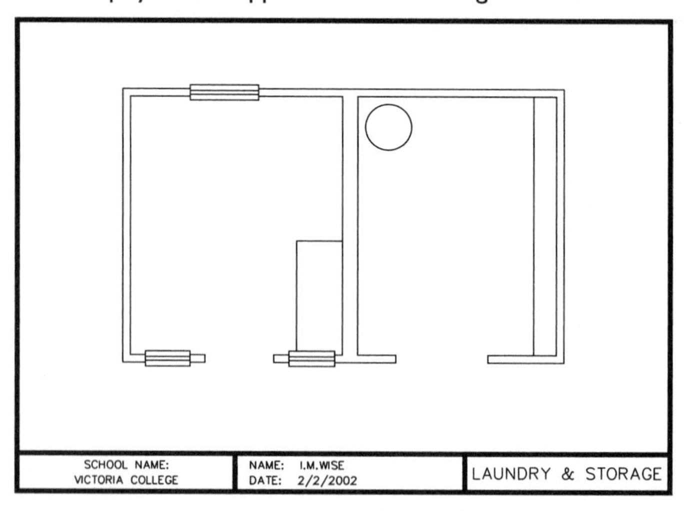

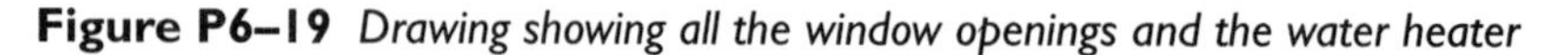


**Figure P6–19** *Drawing showing all the window openings and the water heater*

## DRAWING THE DOORS

**Step 28:** Using the ZOOM WINDOW option, enlarge the area as shown in Figure P6–20 to begin drawing doors.

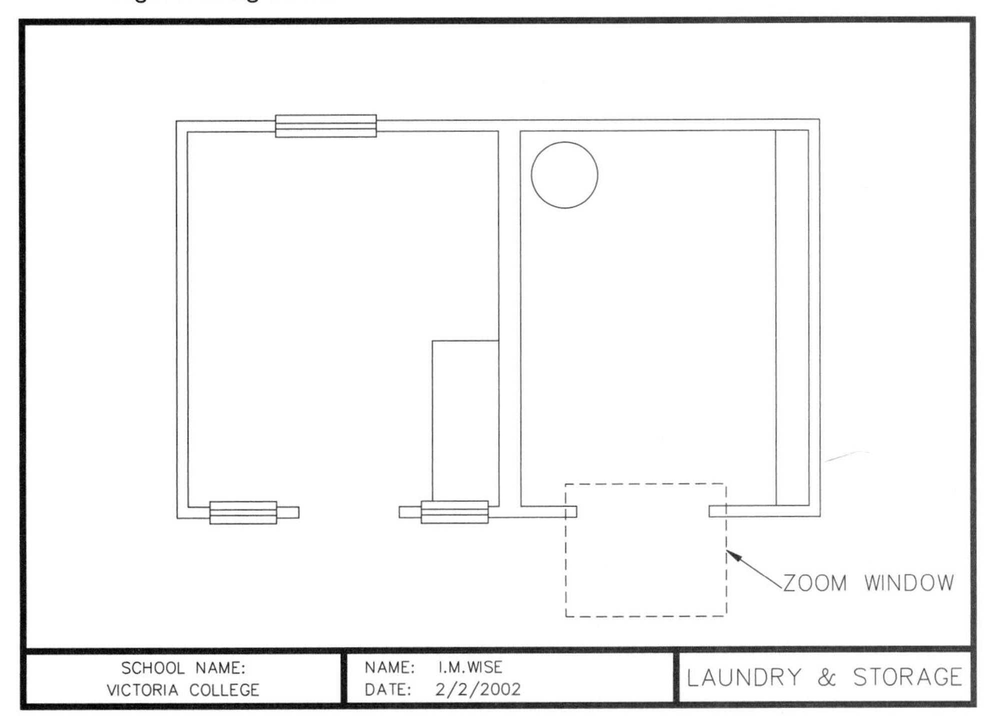


**Figure P6–20** *Points indicating the area to be enlarged by means of the* ZOOM WINDOW *command*

**Step 29** Invoke the MLINE command from the Draw toolbar. Set the Scale to 1, and draw the double line representing the door on the right by specifying the points indicated in Figure P6–21. AutoCAD prompts:

Command: **mline** (ENTER)
Current settings: Justification = Zero, Scale = 8.00, Style = STANDARD
Specify start point or [Justification/Scale/STyle]: **j** (ENTER)
Enter justification type [Top/Zero/Bottom] <top>: **t** (ENTER)
Current settings: Justification = Top, Scale = 8.00, Style = STANDARD
Specify start point or [Justification/Scale/STyle]: **s** (ENTER)
Enter mline scale <2.00>: **1** (ENTER)
Current settings: Justification = Top, Scale = 1.00, Style = STANDARD
Specify start point or [Justification/Scale/STyle]: *(specify point A, as shown in Figure P6–21)*
Specify next point: *(specify point B, as shown in Figure P6–21, at absolute coordinates 16'-0",-2'-0", or use the relative coordinates of @24<270)*
Specify next point or [Undo]: *(press* ENTER *to terminate the command sequence)*

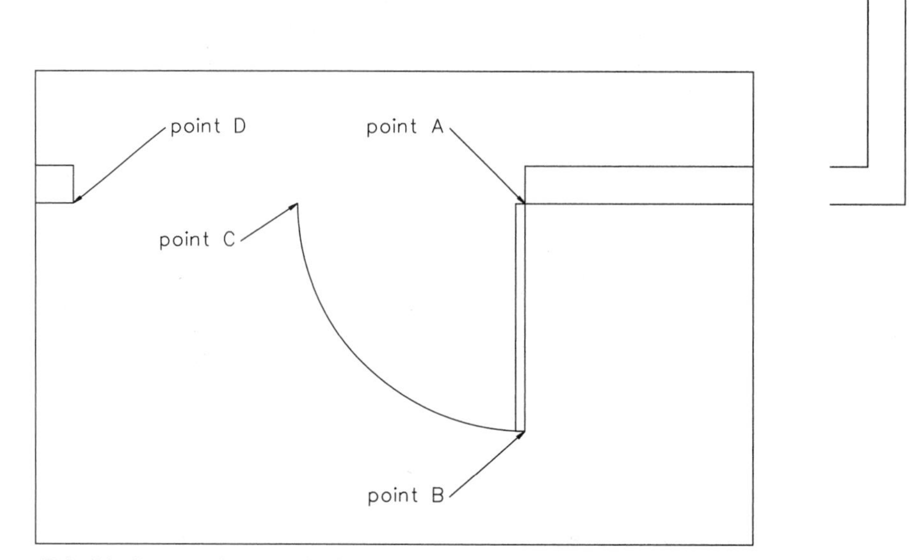


**Figure P6–21** *Points indicating the location of the door*

**Step 30:** Change the SNAP setting to 1. Invoke the LINE command from the Draw toolbar to close the ends of the "door" just drawn.

**Step 31:** Invoke the ARC command with Start/Center/Angle option to draw an arc to represent the swing of the door. AutoCAD prompts:

Command: Specify start point of arc or [CEnter]: *(specify point B, as shown in Figure P6–21)*
Specify second point of arc or [CEnter/ENd]: **c** (right-click and choose Center)
Specify center point of arc: *(specify point A, as shown in Figure P6–21)*
Specify end point of arc or [Angle/chord Length]: *(right-click and select Angle from the shortcut menu)*

Specify included angle: **-90** (ENTER)

**Step 32:** Invoke the MIRROR command from the Modify toolbar. AutoCAD prompts:

Command: **mirror** (ENTER)
Select objects: *(select the multiline, end closing lines, and arc representing the door)*
Select objects: *(press* ENTER *to complete object selection)*
Specify first point of mirror line: *(specify point C, as shown in Figure P6–21)*
Specify second point of mirror line: *(specify a point on the left side of point C)*
Delete source objects? [Yes/No] <N>: (ENTER)

**Step 33:** Use the PAN command to move to the viewing area to include the opening made in the wall on the left, and repeat the preceding method to draw the multiline, end lines, and arc to represent the 3'-0" door on the left, as shown in Figure P6–22.

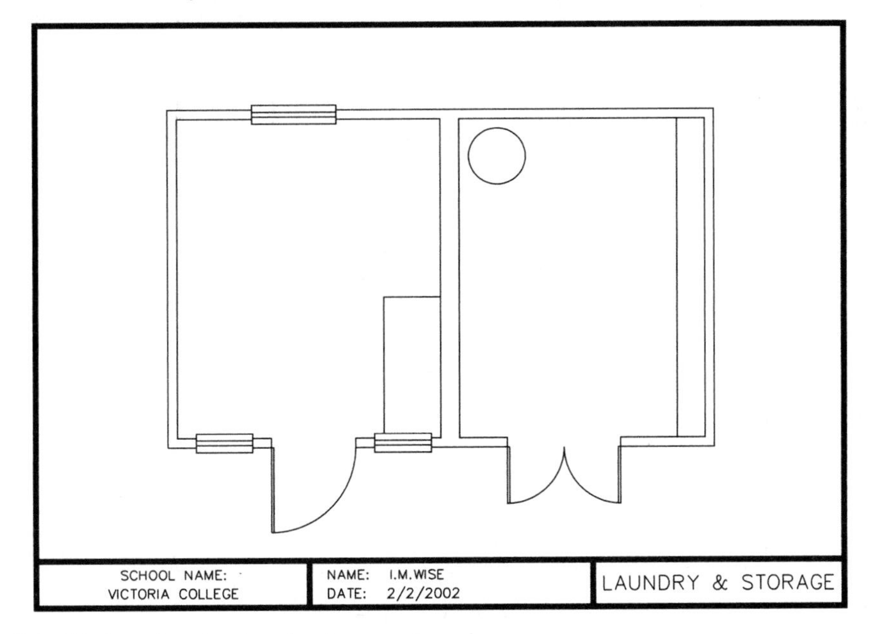


**Figure P6–22** *Drawing showing the doors and windows*

## DRAWING THE SINK

**Step 34:** Use the PAN and/or ZOOM command to move and resize the viewing area to include the counter and to draw a sink as shown in Figure P6–23.

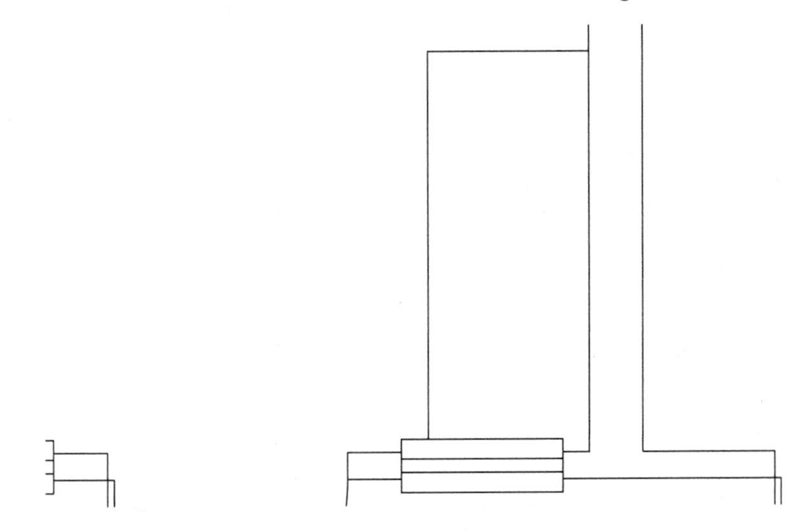

**Figure P6–23** *Drawing showing the closeup of the counter*

**Step 35:** Invoke the RECTANGLE command from the Draw toolbar to draw the sink. AutoCAD prompts:

Command: **rectangle** (ENTER)
Specify first corner point or [Chamfer/Elevation/Fillet/Thickness/Width]: *(specify the lower left point at 7'-10",2'-0")*
Specify other corner point: *(specify the upper right point at 9'-2",3'-8")*

**Step 36:** Invoke the FILLET command from the Modify toolbar. AutoCAD prompts:

Command: **fillet** (ENTER)
Current settings: Mode = TRIM, Radius = 0'-0 1/2"
Select first object or [Polyline/Radius/Trim]: **r** (ENTER)
Specify fillet radius <0'-0 1/2">: **2** (ENTER)
Select first object or [Polyline/Radius/Trim]: **p** (ENTER)
Select 2D polyline: *(select the rectangle drawn in the previous step)*

The results should appear as shown in Figure P6–24.

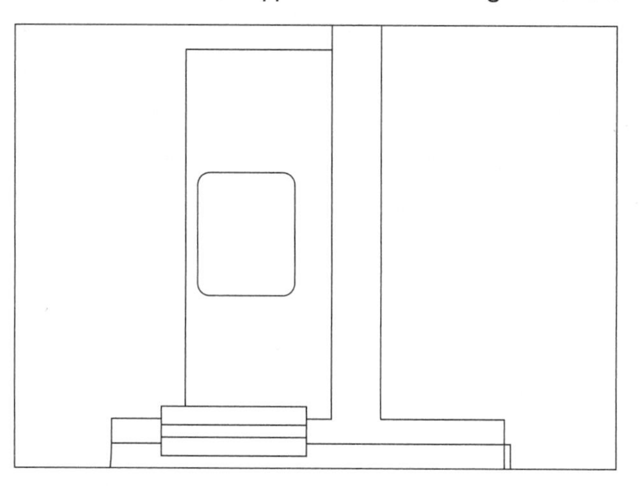

**Figure P6–24** *Drawing with counter and sink*

## DRAWING THE WASHERS AND DRYERS

**Step 37:** Invoke the ZOOM ALL command to display the complete drawing. Set Hidden as the current layer.

**Step 38:** Invoke the RECTANGLE command from the Draw toolbar. AutoCAD prompts:

Command: **rectangle** (ENTER)
Specify first corner point or [Chamfer/Elevation/Fillet/Thickness/Width]: *(specify the lower left point at 0'-6",0'-8")*
Specify other corner point: *(specify the upper right point at 3'-0",3'-2")*

Command: *(press ENTER to repeat the RECTANGLE command)*

Specify first corner point or [Chamfer/Elevation/Fillet/Thickness/Width]: *(specify the lower left point at 6'-8",5'-8")*
Specify other corner point: *(specify the lower left point at 9'-4",8'-4")*

**Step 39:** Invoke the ARRAY command from the Modify toolbar. AutoCAD prompts:

Command: **-array** (ENTER)
Select objects: *(select the first rectangle drawn in step 38)*
Select objects: *(press ENTER to complete selection)*
Enter the type of array [Rectangular/Polar] <R>: **r** (ENTER)
Enter the number of rows (—) <1>: **4**
Enter the number of columns (|||) <1>: (ENTER)
Enter the distance between rows or specify unit cell (—): **2'8**

**Step 40:** Invoke the COPY command from the Modify toolbar. AutoCAD prompts:

Command: **copy** (ENTER)
Select objects: *(select the second rectangle drawn in step 38)*
Select objects: *(press ENTER to complete selection)*
Specify base point or displacement, or [Multiple]: *(specify a point)*
Specify second point of displacement or <use first point as displacement>: *(specify a relative displacement point at 3'-0" at 90 degrees)*

Figure P6–25 shows the result after drawing the washers and dryers.

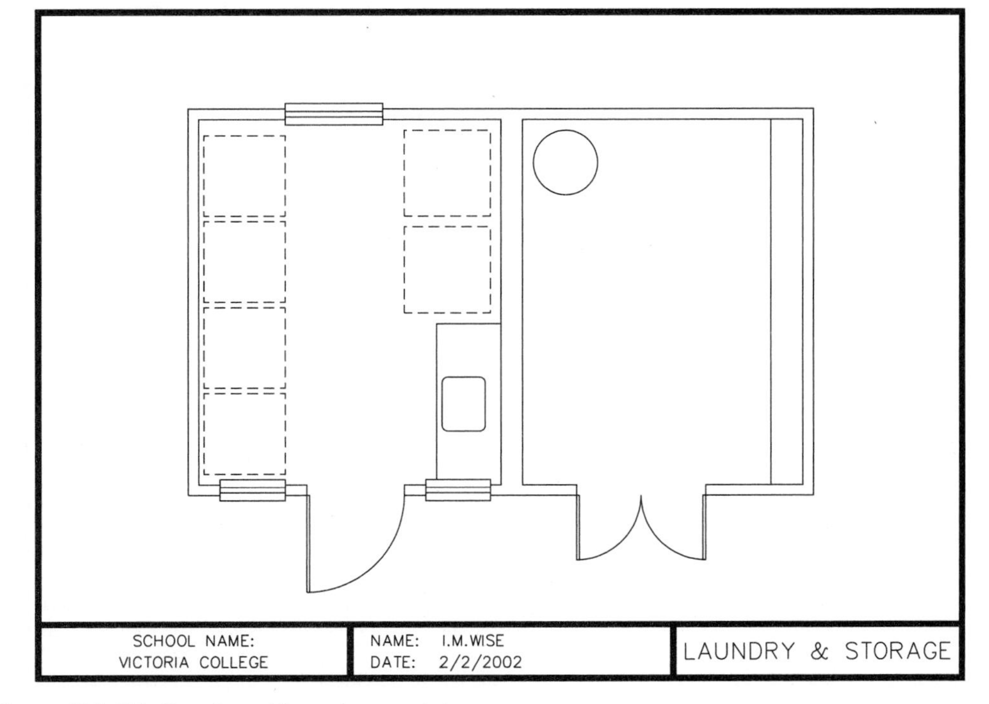


**Figure P6–25** *Drawing with washers and dryers*

## COMPLETING THE SHELVES

**Step 41:** To draw the shelves, invoke the MLINE command from the Draw toolbar. AutoCAD prompts:

Command: **mline** (ENTER)
Current settings: Justification = Top, Scale = 1.00, Style = STANDARD
Specify start point or [Justification/Scale/STyle]: **j** (ENTER)
Enter justification type [Top/Zero/Bottom] <top>: **z** (ENTER)
Current settings: Justification = Zero, Scale = 1.00, Style = STANDARD
Specify start point or [Justification/Scale/STyle]: *(specify a point at 18'-0",8'-0")*
Specify next point:*(specify a point at 19'-0",8'-0")*
Specify next point or [Undo]: *(press* ENTER *to terminate the command sequence)*

Command: *(press* ENTER *to repeat the* MLINE *command)*
Current settings: Justification = Zero, Scale = 1.00, Style = STANDARD
Specify start point or [Justification/Scale/STyle]: *(specify a point at 18'-0",4'-0")*
Specify next point: *(specify a point at 19'-0",4'-0")*
Specify next point or [Undo]: *(press* ENTER *to terminate the command sequence)*

Figure P6–26 shows the result after drawing the shelves.

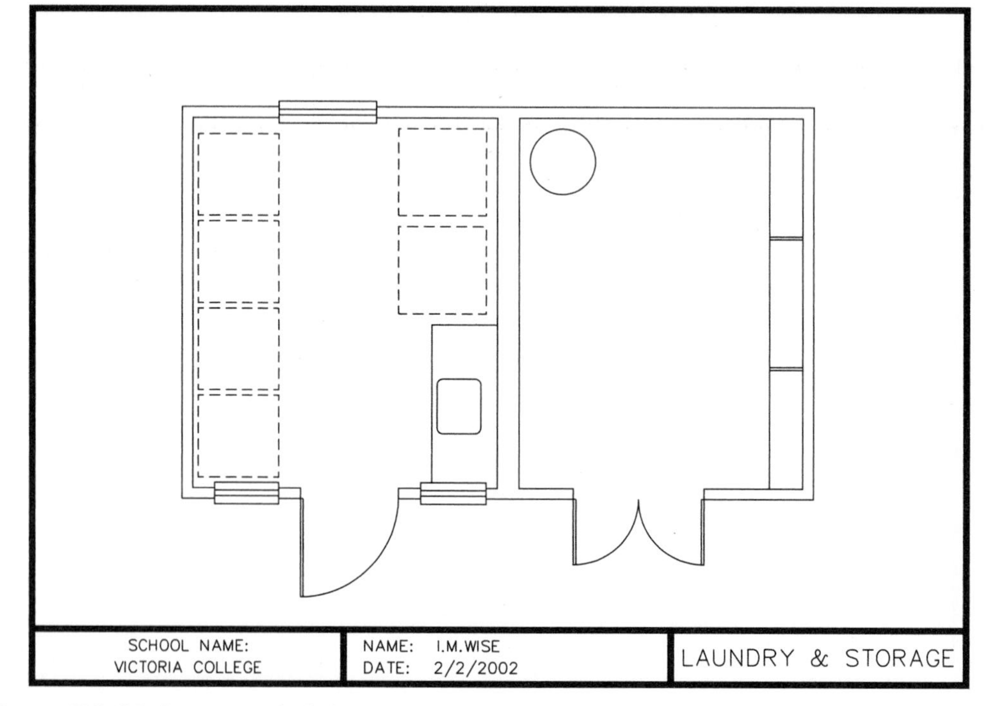


**Figure P6–26** *Drawing with shelves*

## DRAWING TEXT OBJECTS

**Step 42** Set Text as the current layer. Invoke the TEXT command, and draw the text at appropriate text size, as shown in Figure P6–27.

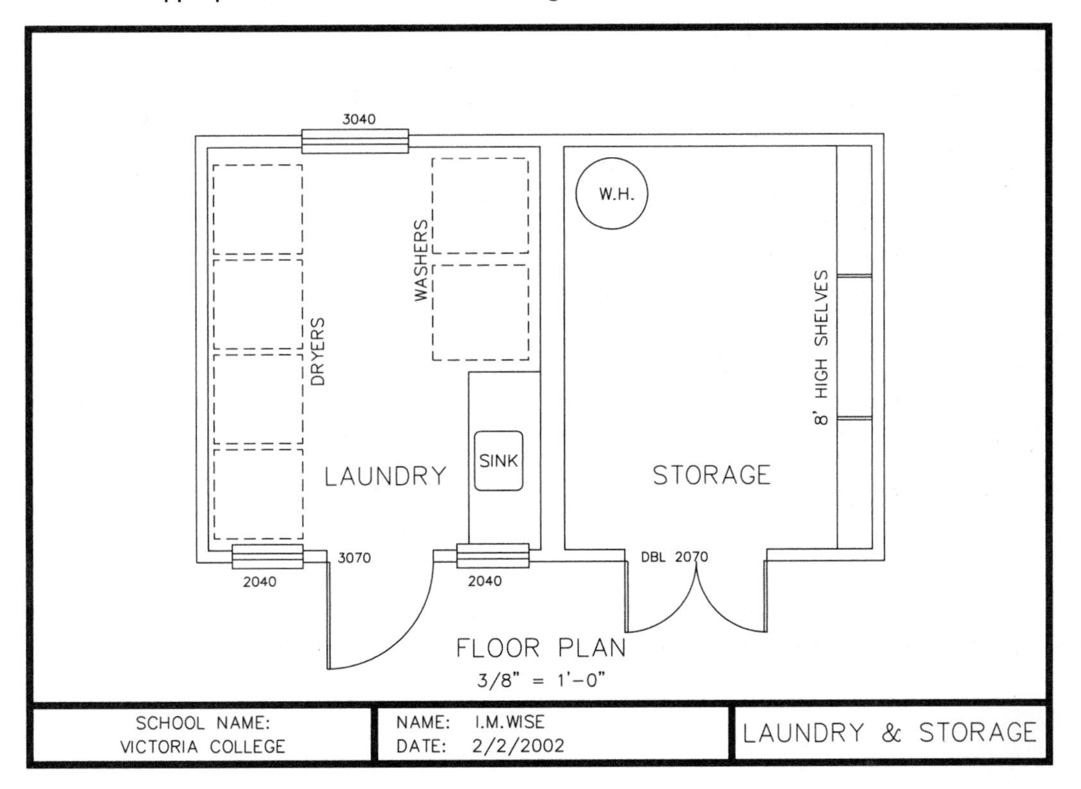


**Figure P6–27** *Drawing with text objects*

Congratulations! You have just successfully applied several AutoCAD concepts in creating the drawing.

## EXERCISE 6–1

Create the section drawing of a built-up beam as shown in Figure Ex6–1, according to the settings given in the following table. You will use plates of ½" thick and four angles 6" x 4" x ½" to create this build-up beam. Do not crosshatch the sections. Do not add dimensions.

| Settings | Value | | |
|---|---|---|---|
| 1. Units | Architectural | | |
| 2. Limits | Lower left corner: 0,0 | | |
| | Upper right corner: 36,24 | | |
| 3. Grid | 1" | | |
| 4. Snap | .5" | | |
| 5. Layers | *NAME* | *COLOR* | *LINETYPE* |
| | Centerline | Cyan | Center |
| | Border | Red | Continous |
| | Steel | Green | Continuous |

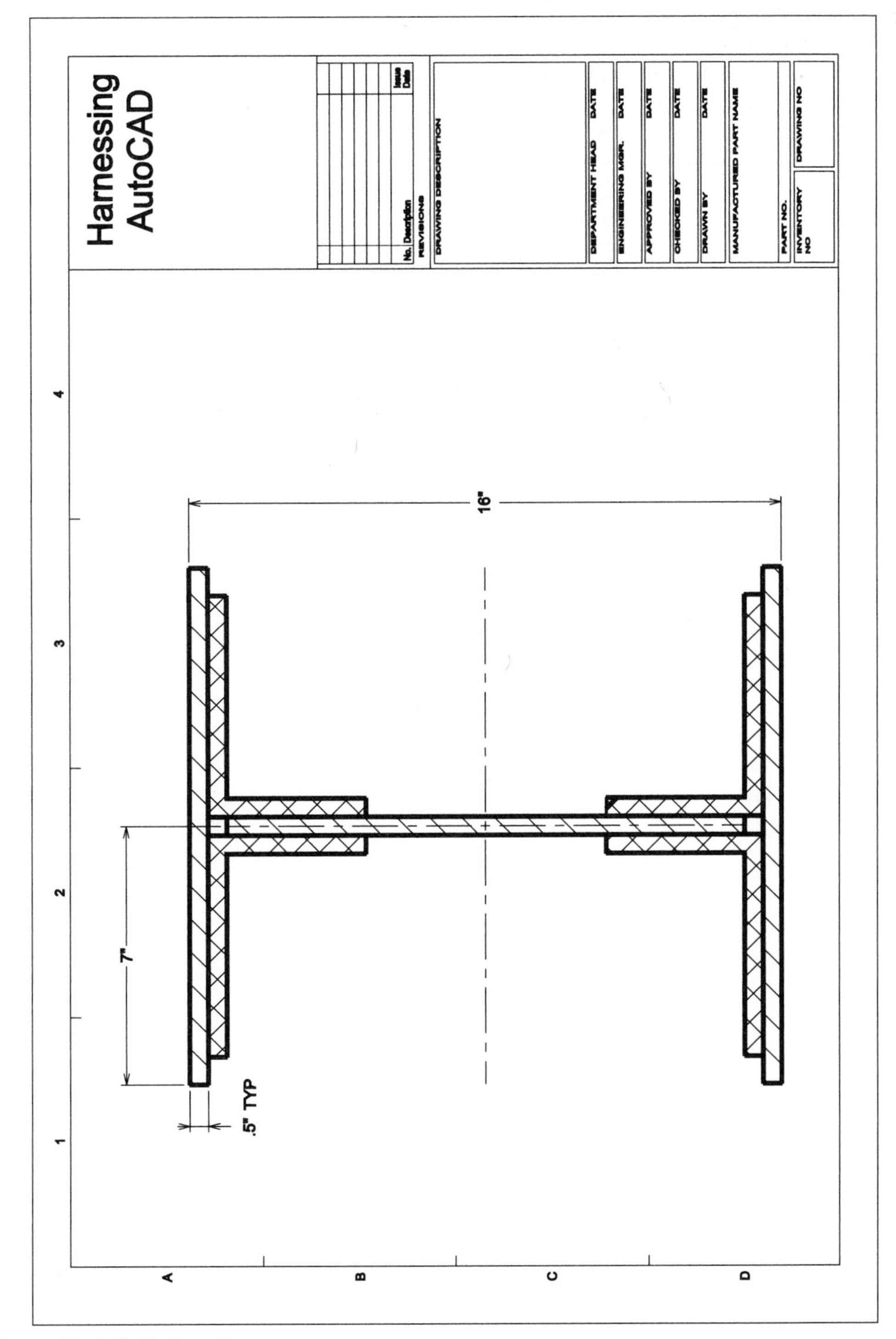


**Figure Ex6–1** *Built-up beam*

## EXERCISE 6–2

Create a section view of the turnbuckle drawing as shown in Figure Ex6–2, according to the dimensions and settings given in the following table:

| Settings | Value | | |
|---|---|---|---|
| 1. Units | Architectural | | |
| 2. Limits | Lower left corner: 0,0 | | |
| | Upper right corner: 38, 24 | | |
| 3. Grid | .125" | | |
| 4. Snap | .0625" | | |
| 5. Layers | *NAME* | *COLOR* | *LINETYPE* |
| | Center | Cyan | Center |
| | Border | Red | Continuous |
| | Object | Green | Continuous |
| | Notes | White | Continuous |

### Specifications

| Diameter D | A | B | C | E | F |
|---|---|---|---|---|---|
| 3" | 15" | 6" | 4 1/2" | 3 5/8" | 6 1/8" |

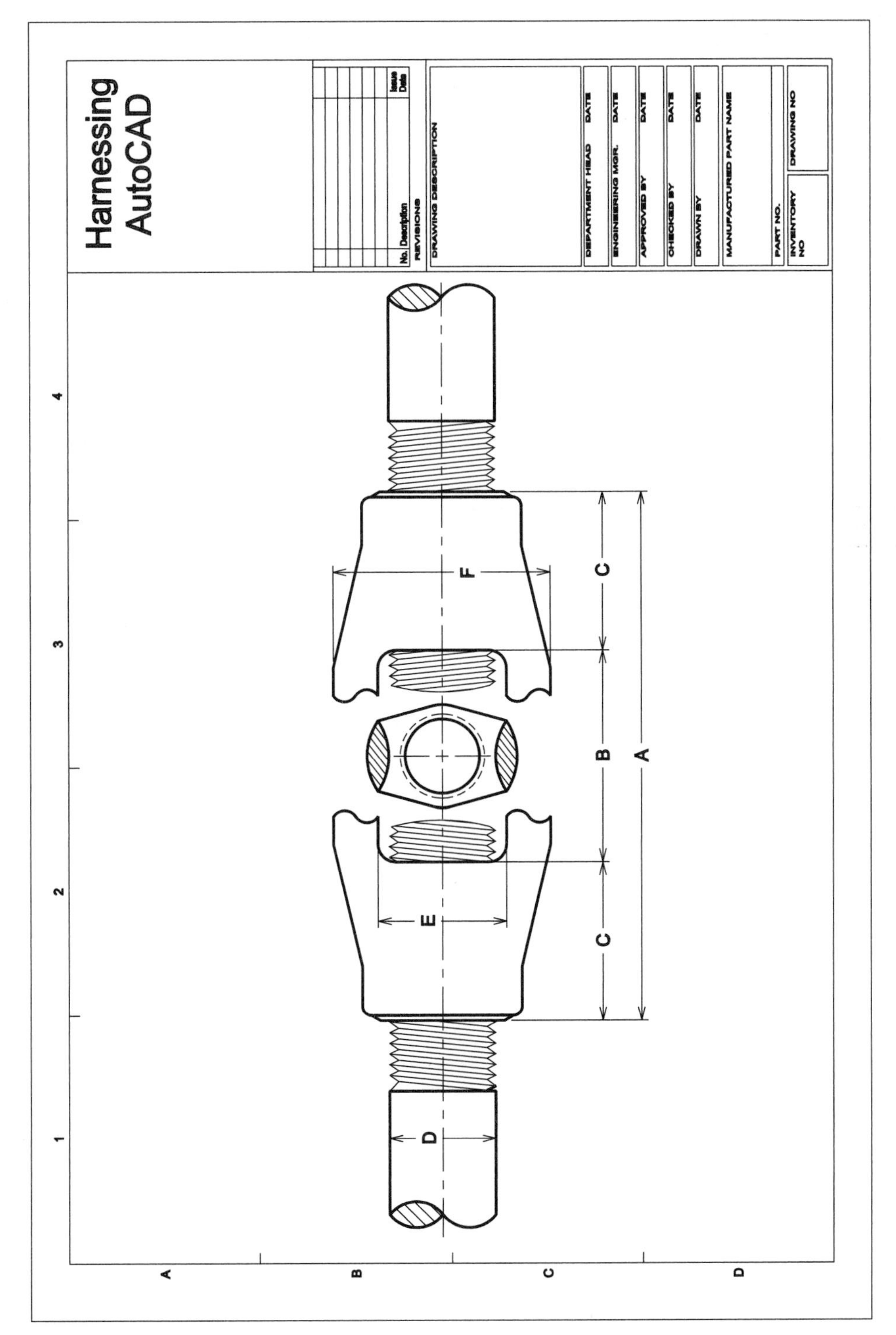


**Figure Ex6–2** *Turnbuckle*

## EXERCISE 6–3

Create the flange cuts at the column web connections drawing as shown in Figure Ex6–3, according to the settings given in the following table:

| Settings | Value |
|---|---|
| 1. Units | Architectural |
| 2. Limits | Lower left corner: 0,0<br>Upper right corner: 6',4' |
| 3. Grid | .125" |
| 4. Snap | .0625" |
| 5. Layers | *NAME* *COLOR* *LINETYPE*<br>Center Cyan Center<br>Border Red Continous<br>Object Green Continuus<br>Web Blue Hidden |

### Table of Steel Sizes

| Designation | A | B | C | D |
|---|---|---|---|---|
| W10x77 | 10 5/8" | 10 ¼" | ½" | 7/8" |
| W10x730 | 22 3/8" | 3 1/16" | 17 7/8" | 4 15/16" |

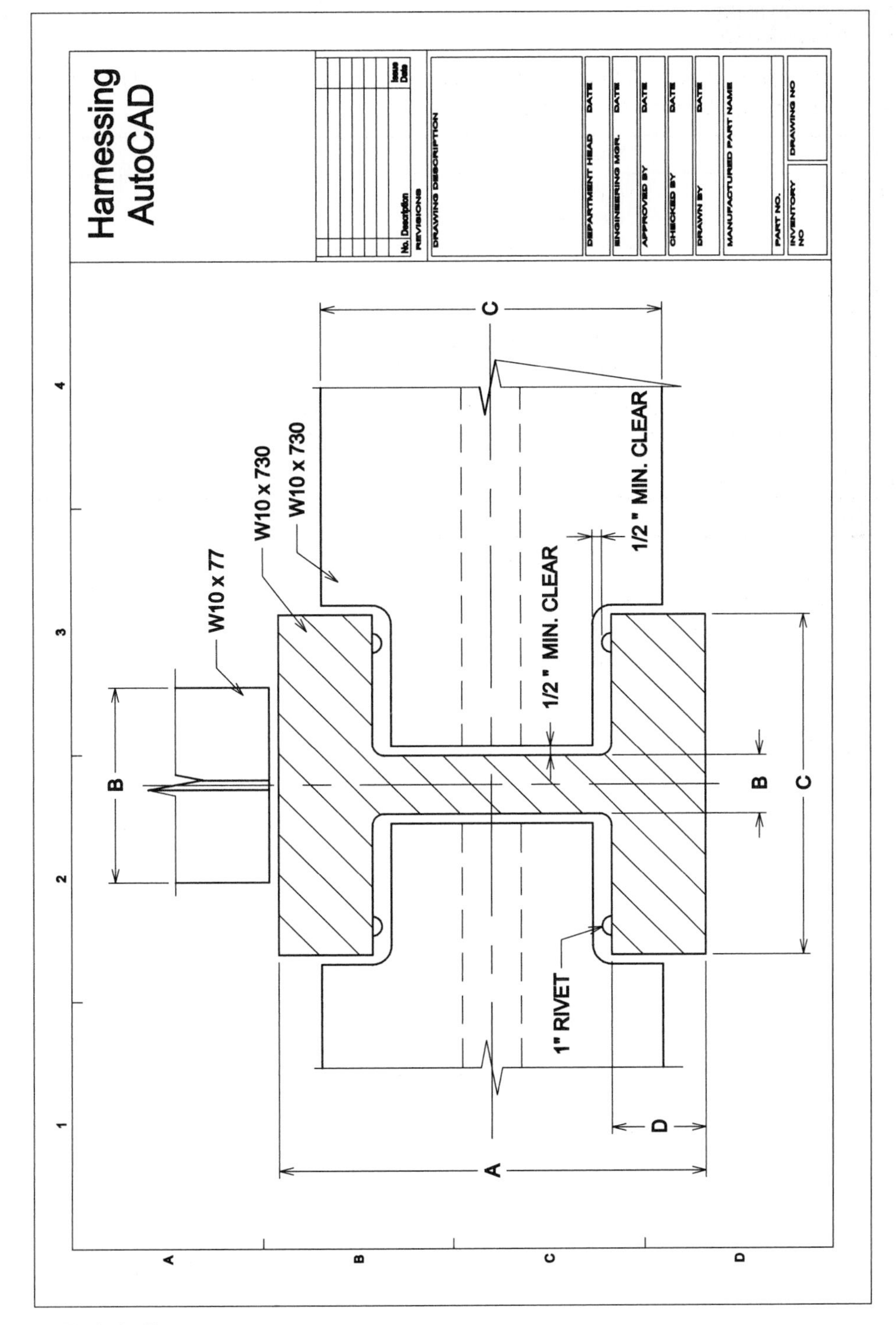


**Figure Ex6–3** *Flange cuts*

## EXERCISE 6–4

Create the boiler flue and stack offset connection drawing as shown in Figure Ex6–4 according to the settings given below:

| Settings | Value | | |
|---|---|---|---|
| 1. Units | Architectural | | |
| 2. Limits | Lower left corner: 0,0 | | |
| | Upper right corner: 48',36' | | |
| 3. Grid | 1' | | |
| 4. Snap | 6 | | |
| 5. Layers | *NAME* | *COLOR* | *LINETYPE* |
| | Center | Cyan | Center |
| | Border | Red | Continuous |
| | Concretepad | Green | Continuous |
| | Boiler | Magenta | Hidden |
| | Flue | Blue | Continuous |
| | Transition | White | Continuous |
| | Stack | Green | Continuous |
| | Notes | White | Continuous |

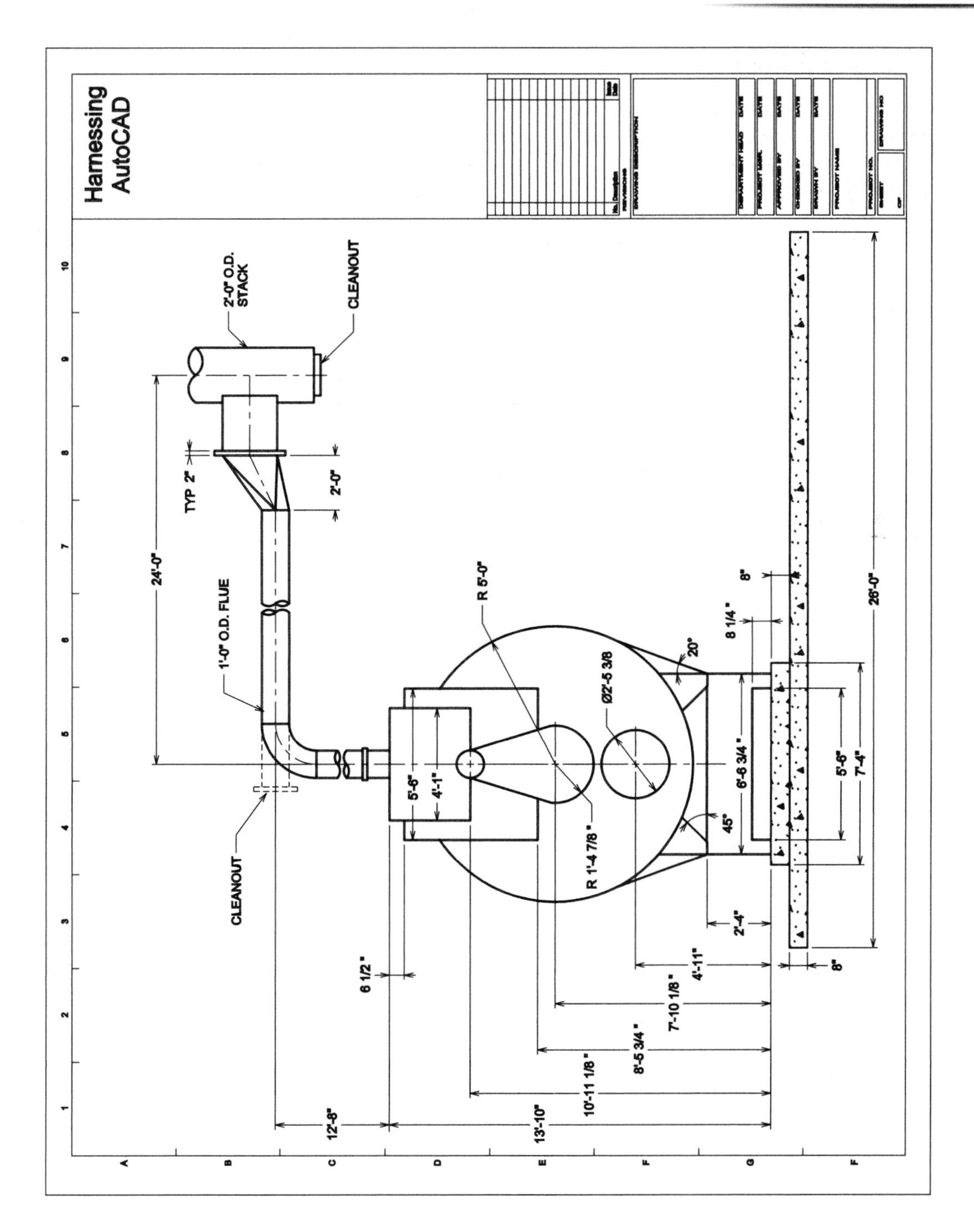


**Figure Ex6–4** *Boiler Flue and Stack Offset Connection*

## EXERCISE 6–5

Create the drawing shown in Figure Ex6–5, according to the settings given in the following table. (Do *not* dimension this drawing.)

| Settings | Value |
|---|---|
| 1. Units | Architectural |
| 2. Limits | Lower left corner: 0',0'<br>Upper right corner: 20',20' |
| 3. Grid | 6" |
| 4. Snap | 3" |

| Settings | Value | | |
|---|---|---|---|
| 5. Layers | *NAME* | *COLOR* | *LINETYPE* |
| | Border | Red | Continuous |
| | Object | Green | Continuous |
| | Text | Blue | Continuous |

**Hints:** Use the MLINE command to draw the door panel perimeters with a style that displays the joints. One of the top panels of the front door can be copied down. Then, using the STRETCH command, you can lengthen the copy and then copy the newly created longer panel. Repeat for the lower panels of all the doors.

Draw the door handles with a style that closes the ends of the mline.

The window inside the back door will have to be drawn using a style that does not display the joints or close the ends.

Use the MLEDIT command to connect the window mullions in open tees.

The MTEXT command is used to create text objects.

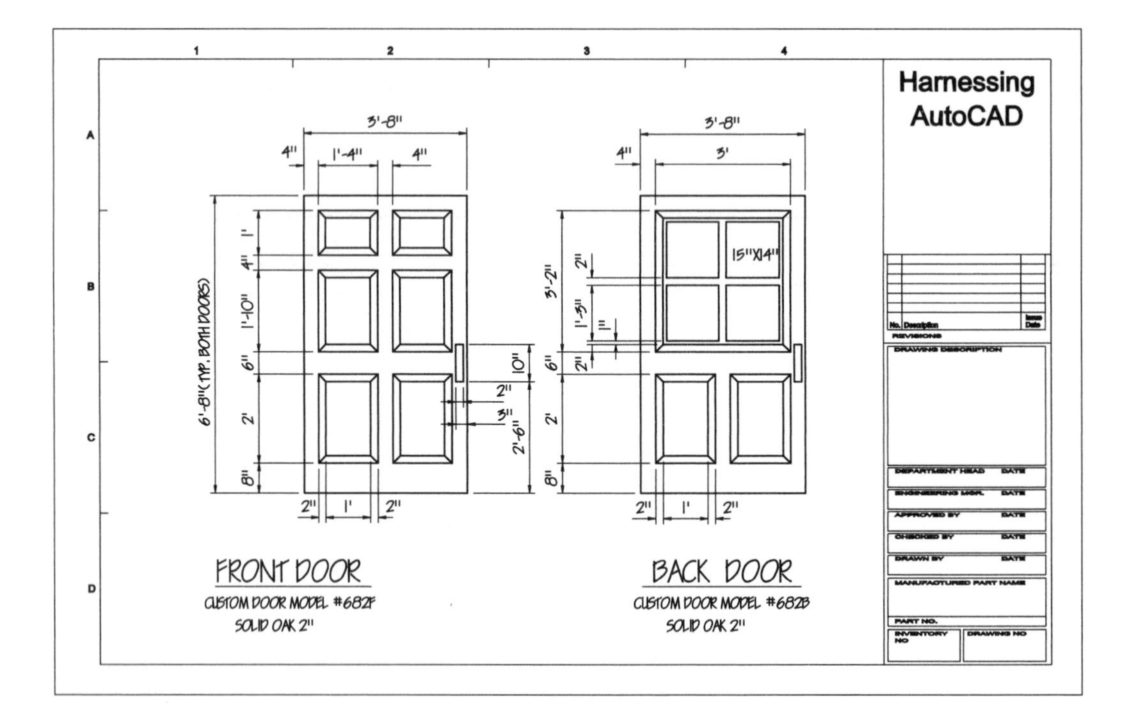


**Figure Ex6–5** *Detailed drawing of a front door and a back door*

## EXERCISE 6–6

Create a layout of a building drawing as shown in Figure Ex6–6, according to the settings given in the following table:

| Settings | Value | | |
|---|---|---|---|
| 1. Units | Architectural | | |
| 2. Limits | Lower left corner: 0,0 | | |
| | Upper right corner: 83',64' | | |
| 3. Grid | 2' | | |
| 4. Snap | 1' | | |
| 5. Layers | *NAME* | *COLOR* | *LINETYPE* |
| | Border | Red | Continous |
| | Holes | Green | Continuous |
| | Center | Blue | Phantom |
| | Extwall | Blue | Continuous |
| | Extdoor | White | Continuous |
| | Partition | Yellow | Continuous |
| | Intdoors | White | Continuous |
| | Notes | White | Continuous |

### Exterior Wall Dimensions

| Brick | Insulation | Air Space | Concrete Masonry Unit | Total Wall Thickness |
|---|---|---|---|---|
| 3 5/8" | 2" | 1" | 7 5/8" | 14 ¼" |

### Interior Wall Dimensions

| Metal Stud Width | Gypsum Board Thickness | Total Wall Thickness |
|---|---|---|
| 3 5/8" | 5/8" | 4 7/8" |

architectural EXERCISE

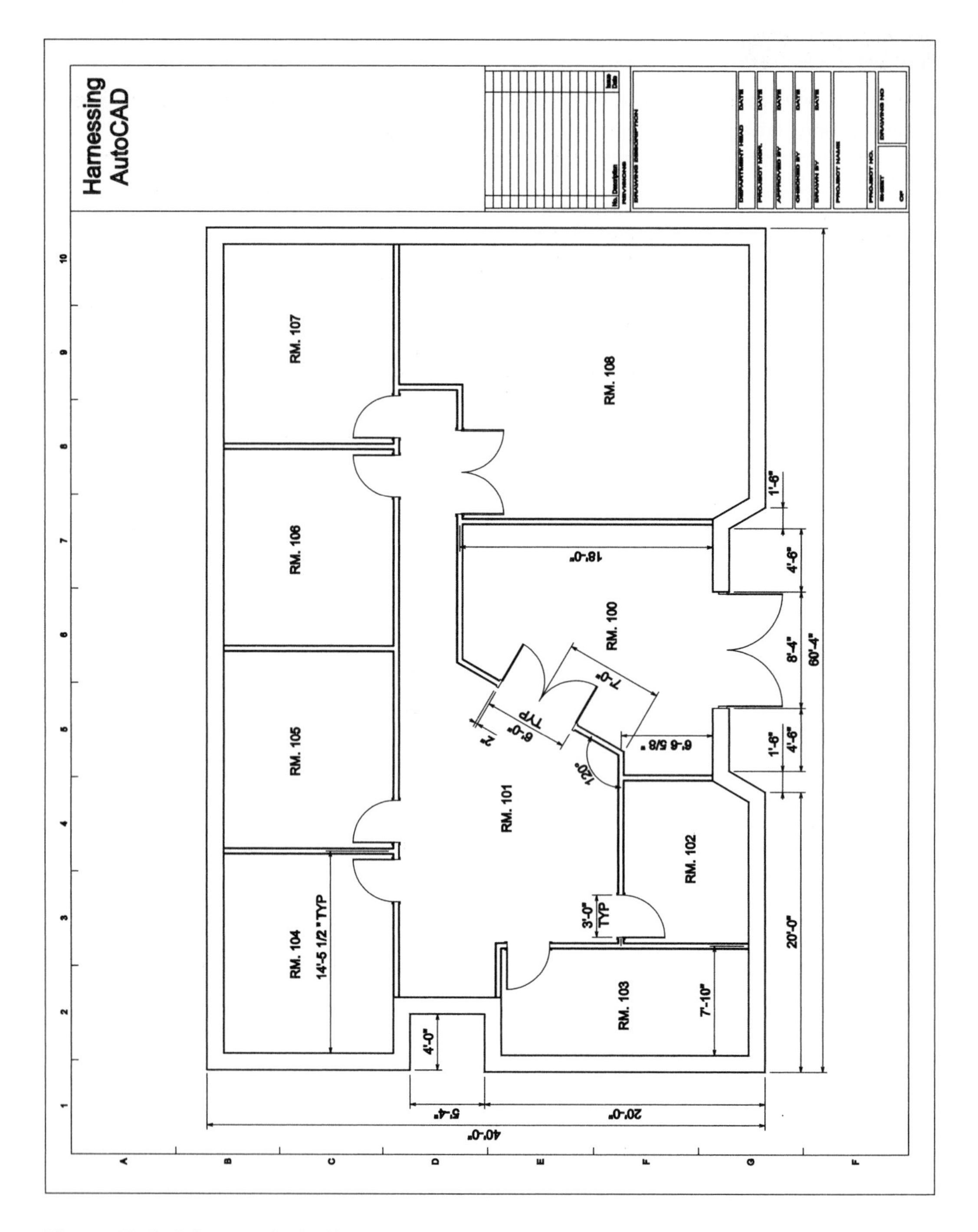


**Figure Ex6–6** *Layout of a Building*

## EXERCISE 6–7

Create the fire rated shaft wall around elevators drawing as shown in Figure Ex6–7, according to the settings given in the following table:

| Settings | Value | | |
|---|---|---|---|
| 1. Units | Architectural | | |
| 2. Limits | Lower left corner: 0,0 | | |
| | Upper right corner: 60',48' | | |
| 3. Layers | *NAME* | *COLOR* | *LINETYPE* |
| | Border | White | Continuous |
| | Shear wall | Blue | Phantom |
| | Elevators | White | Continuous |
| | Partition | Blue | Continuous |
| | Frames | Green | Continuous |
| | Notes | Green | Continuous |

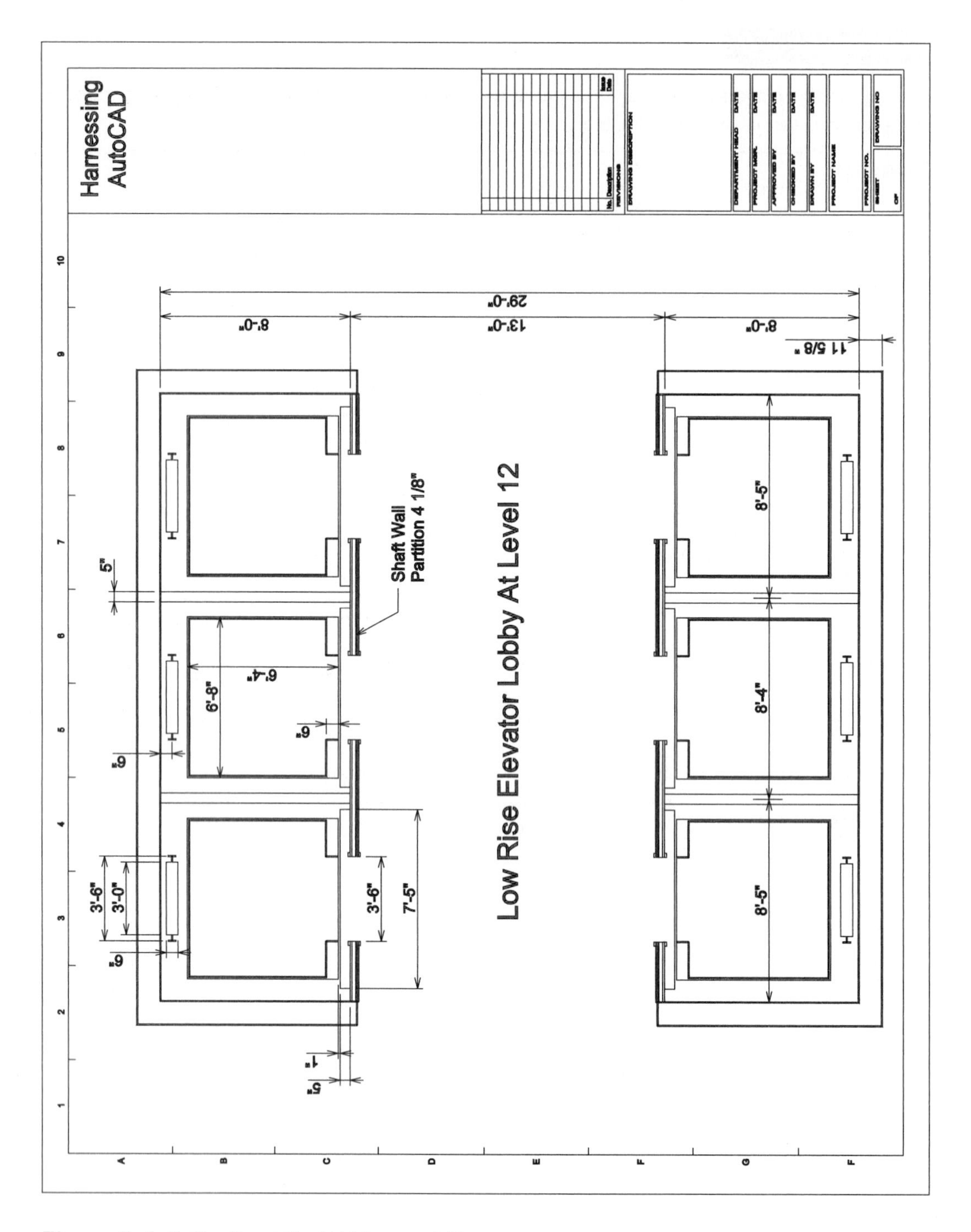


**Figure Ex6–7** *Fire Rated Shaft Wall around Elevators*

## EXERCISE 6–8

Create the elevation of a skylight drawing of equally sized glass panels as shown in Figure Ex6–8, according to the settings given in the following table:

| Settings | Value | | |
|---|---|---|---|
| 1. Units | Architectural | | |
| 2. Limits | Lower left corner: 0,0 | | |
| | Upper right corner: 52',40' | | |
| 3. Grid | 2' | | |
| 4. Snap | 1' | | |
| 5. Layers | *NAME* | *COLOR* | *LINETYPE* |
| | Border | White | Continous |
| | Amullion | White | Continuous |
| | Bmullion | White | Continuous |
| | Cmuillion | Red | Continuous |
| | Notes | White | Continuous |

**Hint:** The skylight mullions are three different sizes. The largest two are vertical and they are 10" wide and 6" wide. The smallest is horizontal and is 3" wide. The 10" mullions are equally spaced on centers. The 6" mullions are equally spaced between the 10" mullions. The 3" horizontal mullions are equally spaced. Create equally sized glass panels as shown in the drawing.

architectural EXERCISE

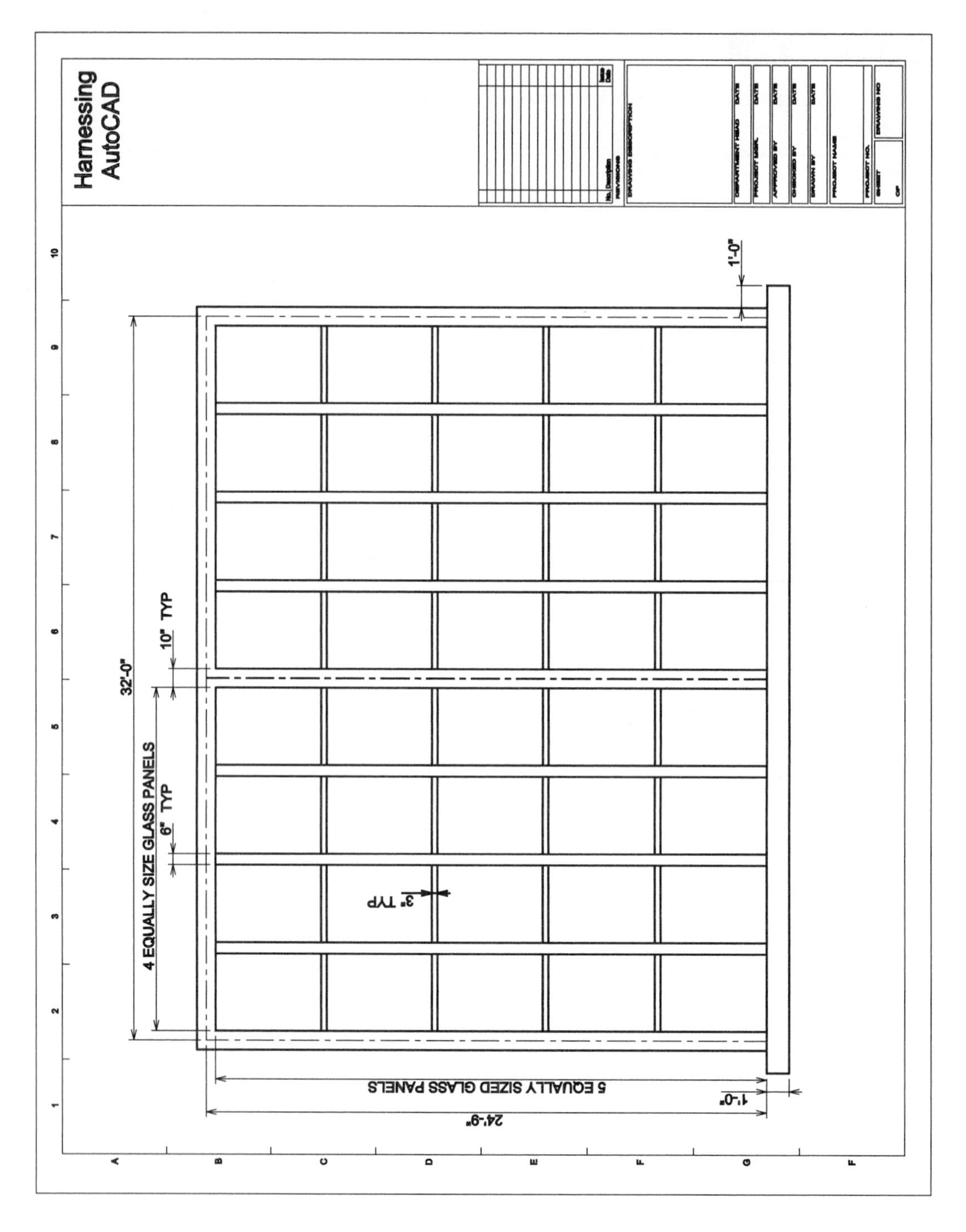


**Figure Ex6–8** *Skylight*

## EXERCISE 6–9

Create the threshold as shown in Figure Ex6–9, according to the settings given in the following table:

| Settings | Value | | |
|---|---|---|---|
| 1. Units | Architectural | | |
| 2. Limits | Lower left corner: 0,0, | | |
| | Upper right corner: 14,10 | | |
| 3. Layers | *NAME* | *COLOR* | *LINETYPE* |
| | Border | White | Continuous |
| | Left profile | White | Continuous |
| | Right profile | Red | Continuous |

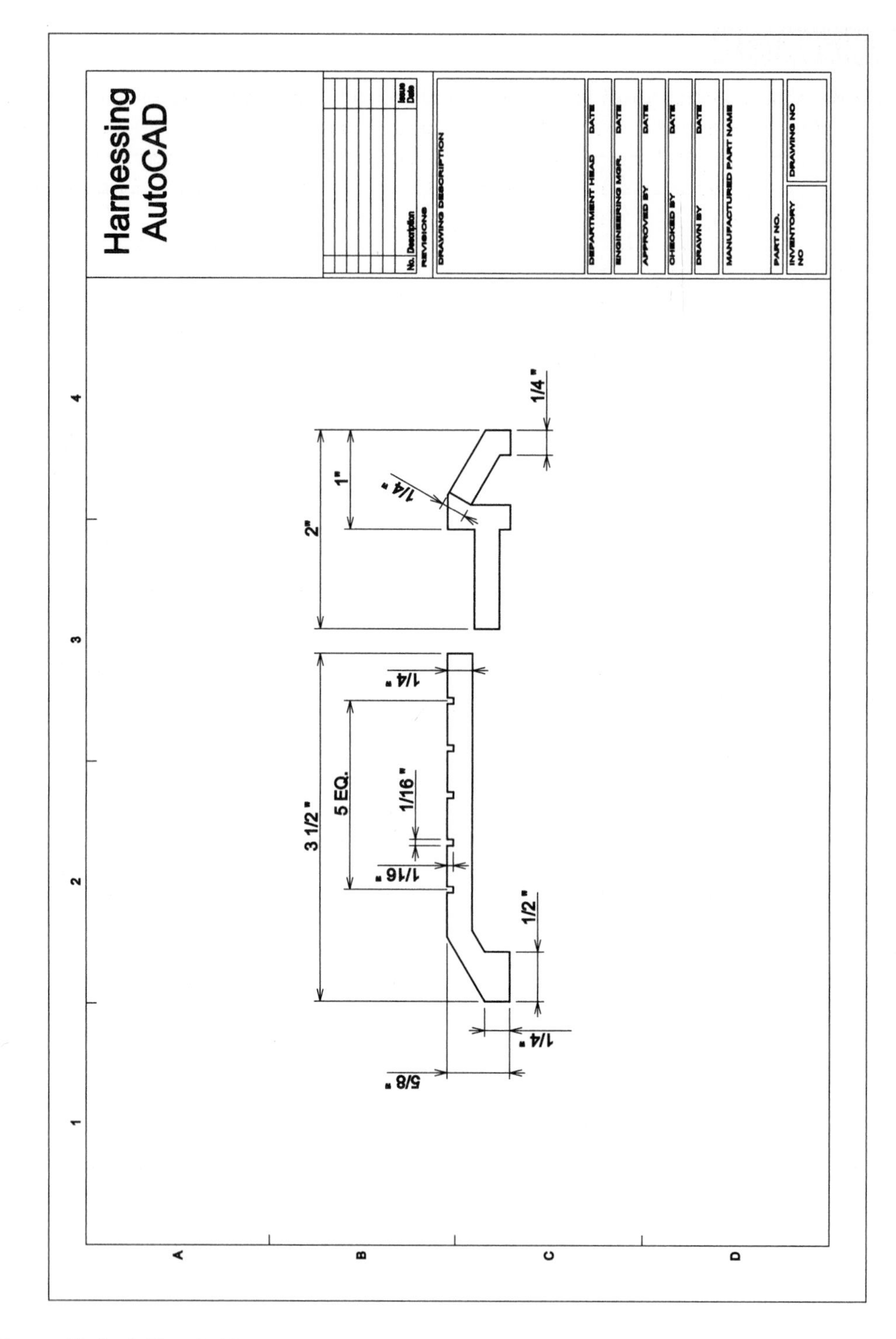


**Figure Ex6–9** *Threshold*

## EXERCISE 6–10

The drawing shown here is typical of informational graphics/text drawings whose objects are not dimensioned but parametric in nature. This cross-section of a reinforced concrete column is typical of those with varying dimensions and number of reinforcing bars. It is for reference only. The important information is given in the written text. Create the drawing shown in Figure Ex6–10, according to the settings given in the following table.

| Settings | Value | | |
|---|---|---|---|
| 1. Units | Decimal | | |
| 2. Limits | Lower left corner: 0,0 | | |
| | Upper right corner: 12,9 | | |
| 3. Grid | .5 | | |
| 4. Snap | .5 | | |
| 5. Layers | *NAME* | *COLOR* | *LINETYPE* |
| | Center | Cyan | Continuous |
| | Border | Red | Continuous |
| | Object | Green | Continuous |
| | Text | Blue | Continuous |
| | Hidden | Cyan | Continuous |
| | Pattern | White | Continuous |

**Hints:** Make sure you are in the appropriate layer before you draw the objects.

Use the DONUT command to draw two circles with large OD/ID and multiple circles with small OD and zero ID.

The patterning in the cross section can be generated by drawing one small triangle with the PLINE command and then copying and rotating it randomly throughout the area. The stipples that are conventional in representing concrete can also be randomly placed, by means of the POINT command. One of the conventions of patterning for concrete is that the stippling is usually more concentrated near the edges of the cross-section than in the center.

The MTEXT command is used to create blocks of text within prescribed rectangles. The justification determines where the block of text anchors and where the overage spills out of the rectangle.

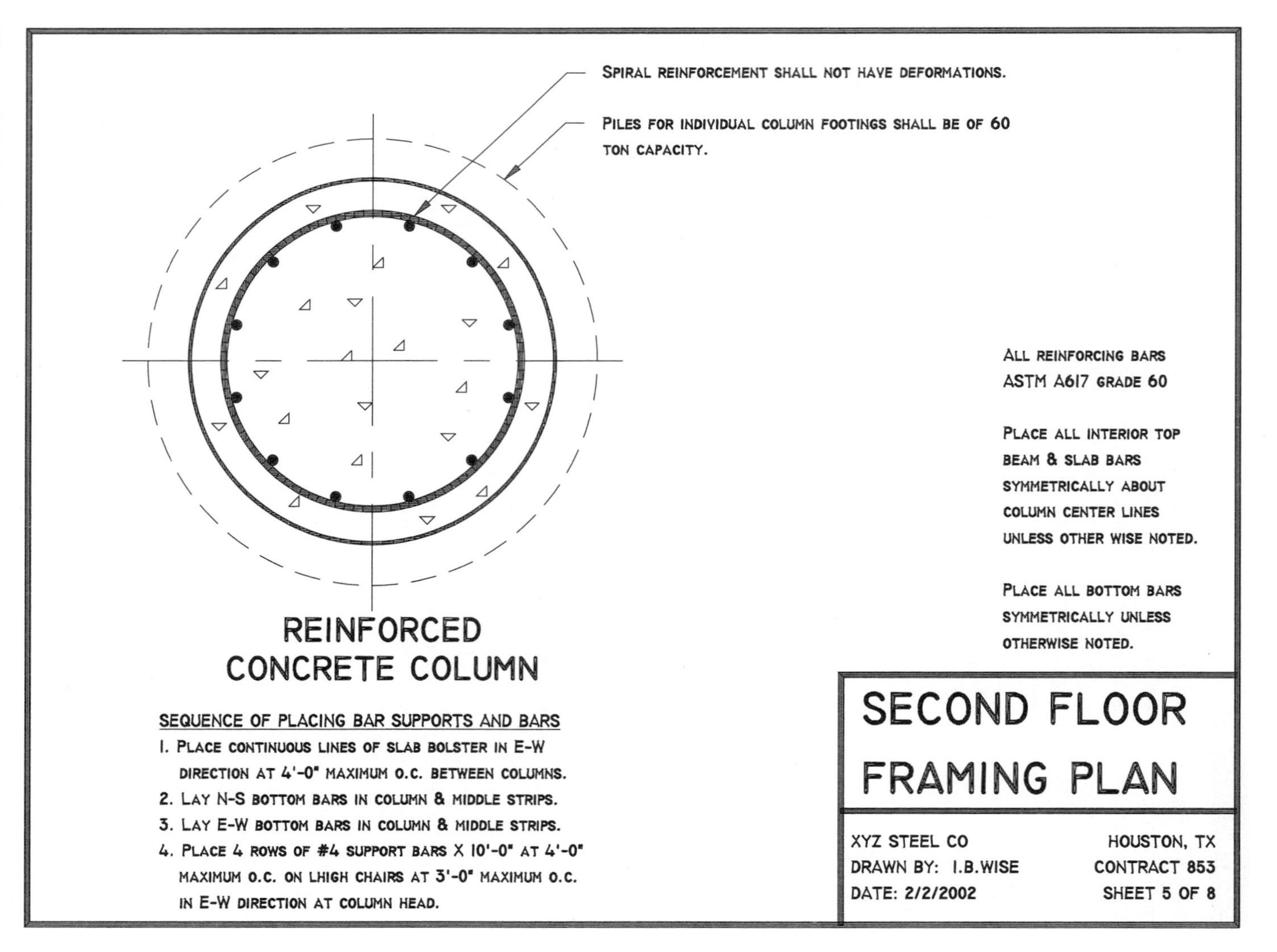


**Figure Ex6–10** *Completed drawing*

## EXERCISE 6–11

Create the drawing shown in Figure Ex6–11, according to the settings given in the following table. (Do *not* dimension this drawing.)

| Settings | Value |
|---|---|
| 1. Units | Architectural |
| 2. Limits | Lower left corner: 0'-0,0'-0<br>Upper right corner: 65'-0,54'-0 |
| 3. Grid | 12 |
| 4. Snap | 6 |

| Settings | Value | | |
|---|---|---|---|
| 5. Layers | *NAME* | *COLOR* | *LINETYPE* |
| | Center | Cyan | Continuous |
| | Border | Red | Continuous |
| | Object | Green | Continuous |
| | Text | Blue | Continuous |
| | Hidden | Cyan | Hidden |

**Hints:** Make sure you are in the appropriate layer before you draw the objects.

Use the MLINE command to draw the beams in continuous linetype and the footings in hidden linetype.

Use the MLEDIT command to connect the beams in open tees.

The MTEXT command is used to create blocks of text within prescribed rectangles. The justification determines where the block of text anchors and where the overage spills out of the rectangle.

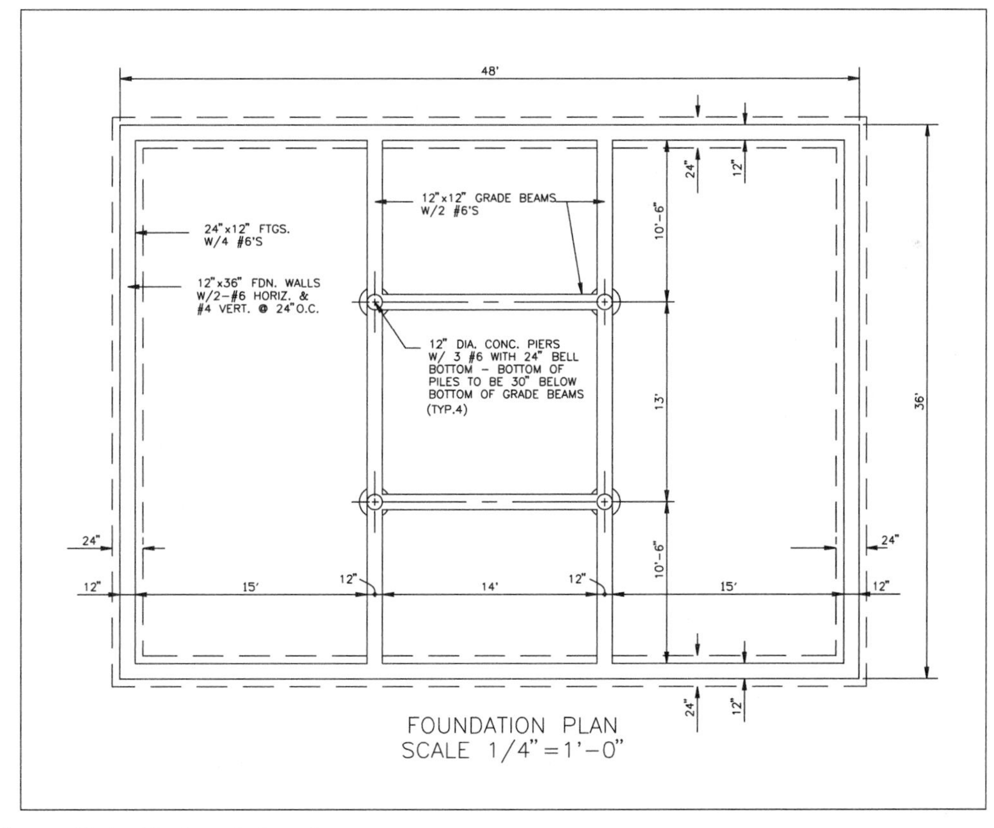


**Figure Ex6–11** *Foundation plan*

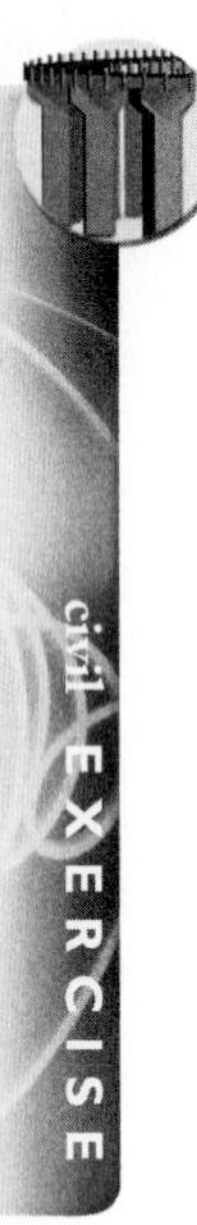


## EXERCISE 6–12

Create a glued laminate truss as shown in Figure Ex6–12, according to the settings given in the following table:

| Settings | Value | | |
|---|---|---|---|
| 1. Units | Architectural | | |
| 2. Limits | Lower left corner: 0,0 | | |
| | Upper right corner: 64', 48' | | |
| 3. Grid | 2' | | |
| 4. Snap | 1' | | |
| 5. Layers | *NAME* | *COLOR* | *LINETYPE* |
| | Center | White | Phantom |
| | Walls | White | Continuous |
| | Truss | Red | Continuous |
| | Text | Green | Continuous |

**Hint:** The truss is composed of ten 3 1/8" x 1 ½" wood members.

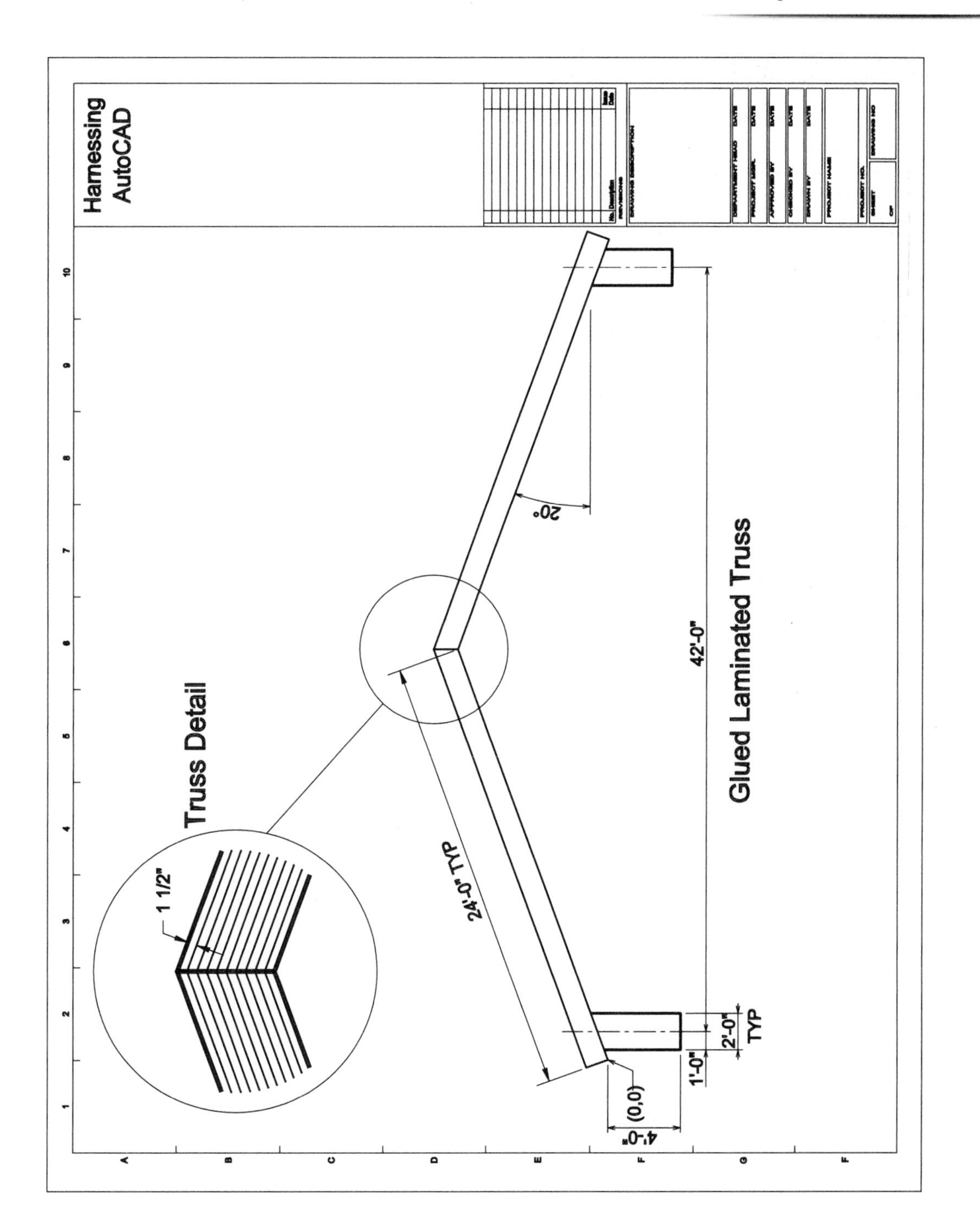


**Figure Ex6–12** *Glued Laminate Truss*

## EXERCISE 6–13

Create the drawing of reinforcing (elevation view) of the concrete beam as shown in Figure Ex6–13, according to the dimensions given.

| Settings | Value | | |
|---|---|---|---|
| 1. Units | Architectural | | |
| 2. Limits | Lower left corner: 0,0 | | |
| | Upper right corner: 48', 36' | | |
| 3. Grid | 2' | | |
| 4. Snap | 1' | | |
| 5. Layers | *NAME* | *COLOR* | *LINETYPE* |
| | Border | White | Continuous |
| | Center | White | Center |
| | Beam | Blue | Continuous |
| | Column | Red | Continuous |
| | Reinforcing | White | Continuous |
| | Notes | White | Continuous |

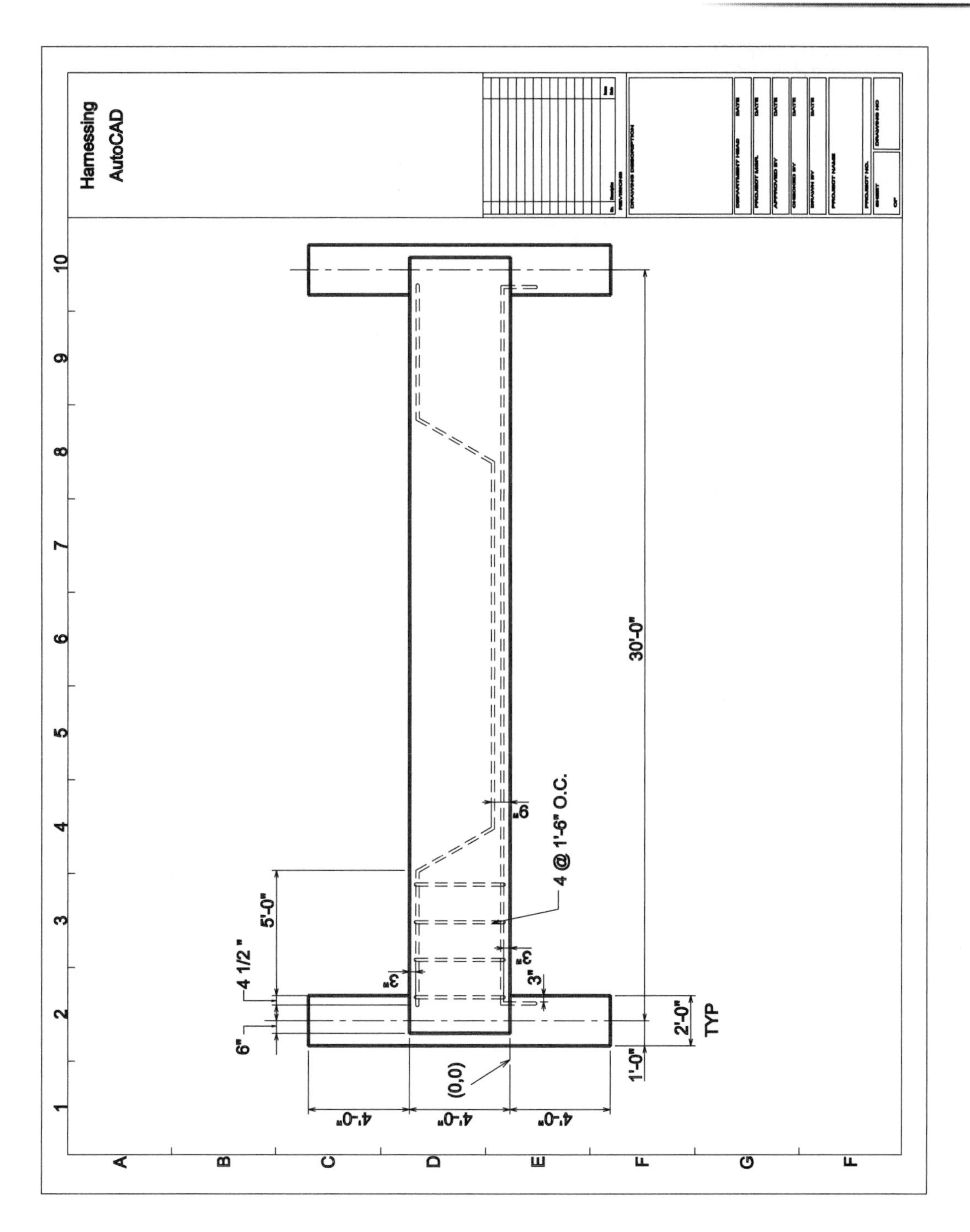


**Figure Ex6–13** *Reinforcing (elevation view) of the Concrete Beam*

## EXERCISE 6–14

Create the fluorescent light tubes drawings as shown in Figure Ex6–14, according to the settings given in the following table:

| Settings | Value | | |
|---|---|---|---|
| 1. Units | Architectural | | |
| 2. Limits | Lower left corner: 0,0<br>Upper right corner: 6',6' | | |
| 3. Grid | .25 | | |
| 4. Snap | .125" | | |
| 5. Layers | *NAME* | *COLOR* | *LINETYPE* |
| | Border | White | Continuous |
| | Type1 | Blue | Continuous |
| | Type 2 | Red | Continuous |
| | Type 3 | Yellow | Continuous |
| | Notes | White | Continuous |

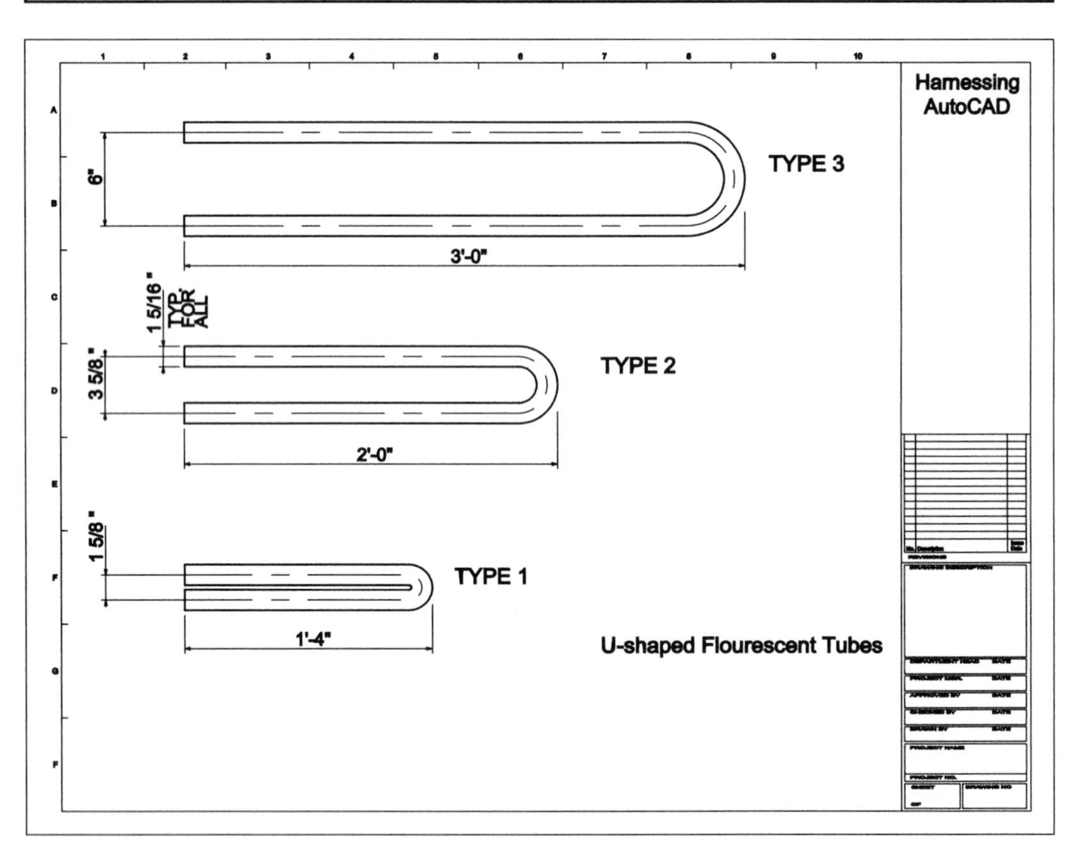


**Figure Ex6–14** *Fluorescent Light Tubes*

## EXERCISE 6–15

Create the high power electrical conduits encased in a concrete vault as shown in Figure Ex6–15, according to the settings given in the following table.

| Settings | Value | | |
|---|---|---|---|
| 1. Units | Architectural | | |
| 2. Limits | Lower left corner: 0,0 | | |
| | Upper right corner:30',24' | | |
| 3. Grid | .25" | | |
| 4. Snap | .125" | | |
| 5. Layers | *NAME* | *COLOR* | *LINETYPE* |
| | Border | White | Continuous |
| | Vault wall | Yellow | Hidden |
| | Conduit | Blue | Continuous |
| | Panels | Green | Continuous |
| | Main switch | Red | Continuous |
| | Ladder | White | Continuous |
| | Access panel | White | Hidden |
| | Notes | White | Continuous |

electrical EXERCISE

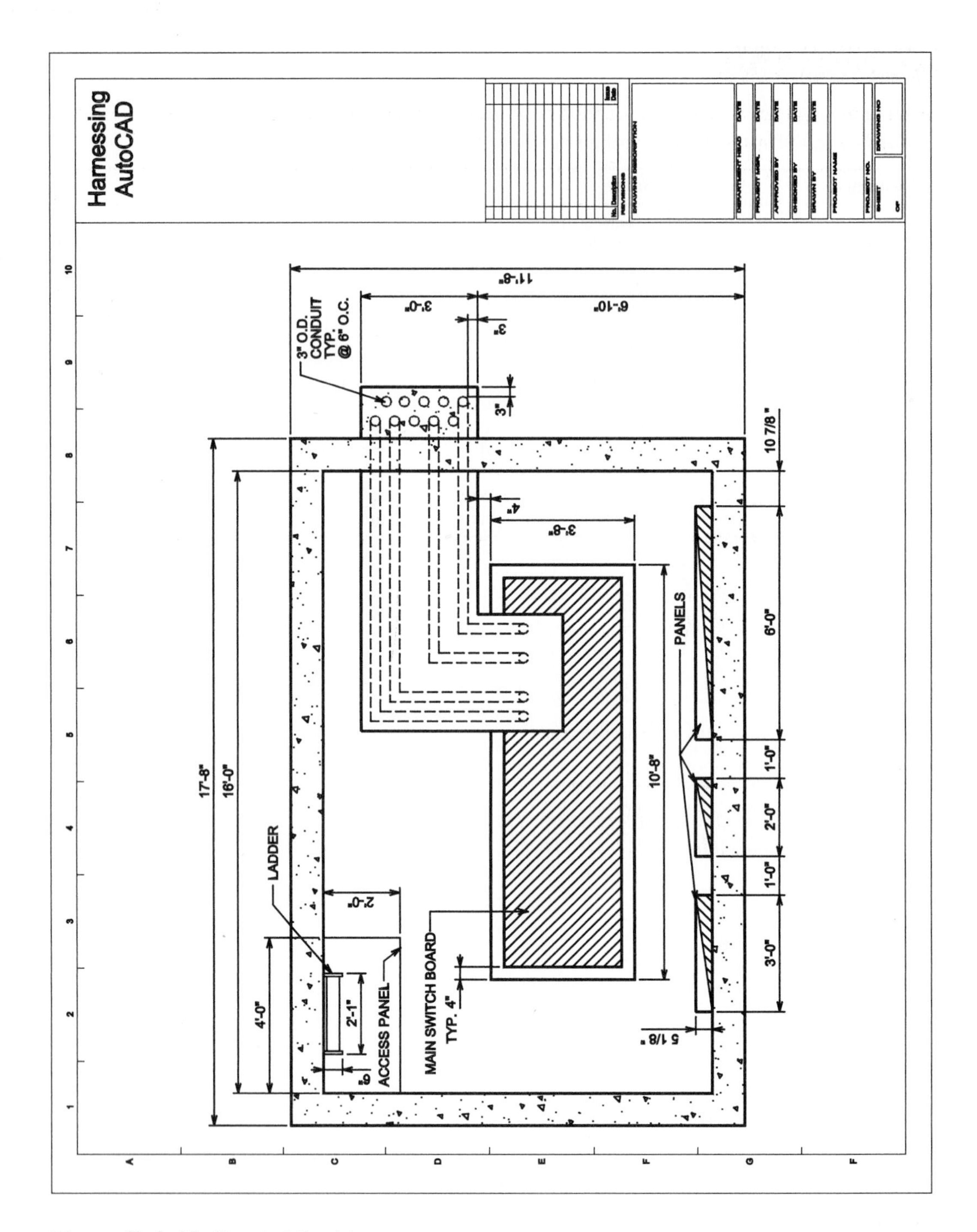


**Figure Ex6–15** *Electrical Conduits*

## EXERCISE 6–16

Create the elevation of a surface mounted electrical tombstone with its section through a concrete slab as shown in Figure Ex6–16, according to the settings given in the following table.

| Settings | Value | | |
|---|---|---|---|
| 1. Units | Architectural | | |
| 2. Limits | Lower left corner: 0,0 | | |
| | Upper right corner: 3',2' | | |
| 3. Grid | .125" | | |
| 4. Snap | .0625" | | |
| 5. Layers | *NAME* | *COLOR* | *LINETYPE* |
| | Border | White | Continuous |
| | Concrete | Green | Continuous |
| | Reinforcing | Red | Hidden |
| | Center | Green | Center |
| | Elec Box | Red | Continuous |
| | Conduit | Blue | Continuous |
| | Tombstone | White | Continuous |
| | Notes | White | Continuous |

electrical EXERCISE

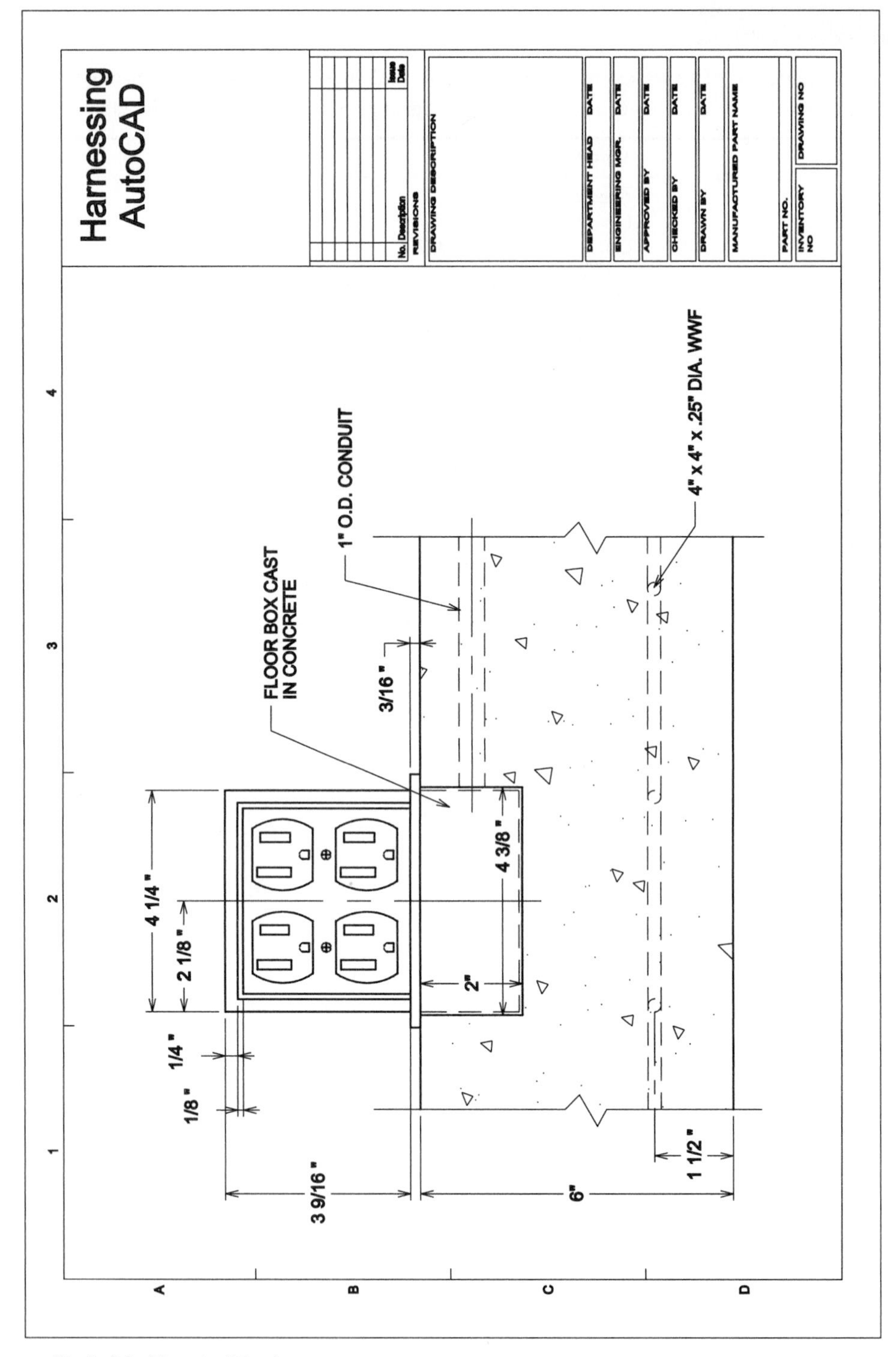


**Figure Ex6–16** *Electrical Tombstone*

## EXERCISE 6–17

Create the schematic drawing shown in Figure Ex6–17, according to the settings given in the following table:

| Settings | Value | | |
|---|---|---|---|
| 1. Units | Decimal | | |
| 2. Limits | Lower left corner: 0,0 | | |
| | Upper right corner: 12,9 | | |
| 3. Grid | .25 | | |
| 4. Snap | .25 | | |
| 5. Layers | *NAME* | *COLOR* | *LINETYPE* |
| | Border | Red | Continuous |
| | Cable | Green | Continuous |
| | Symbols | Blue | Continuous |
| | Notes | White | Continuous |

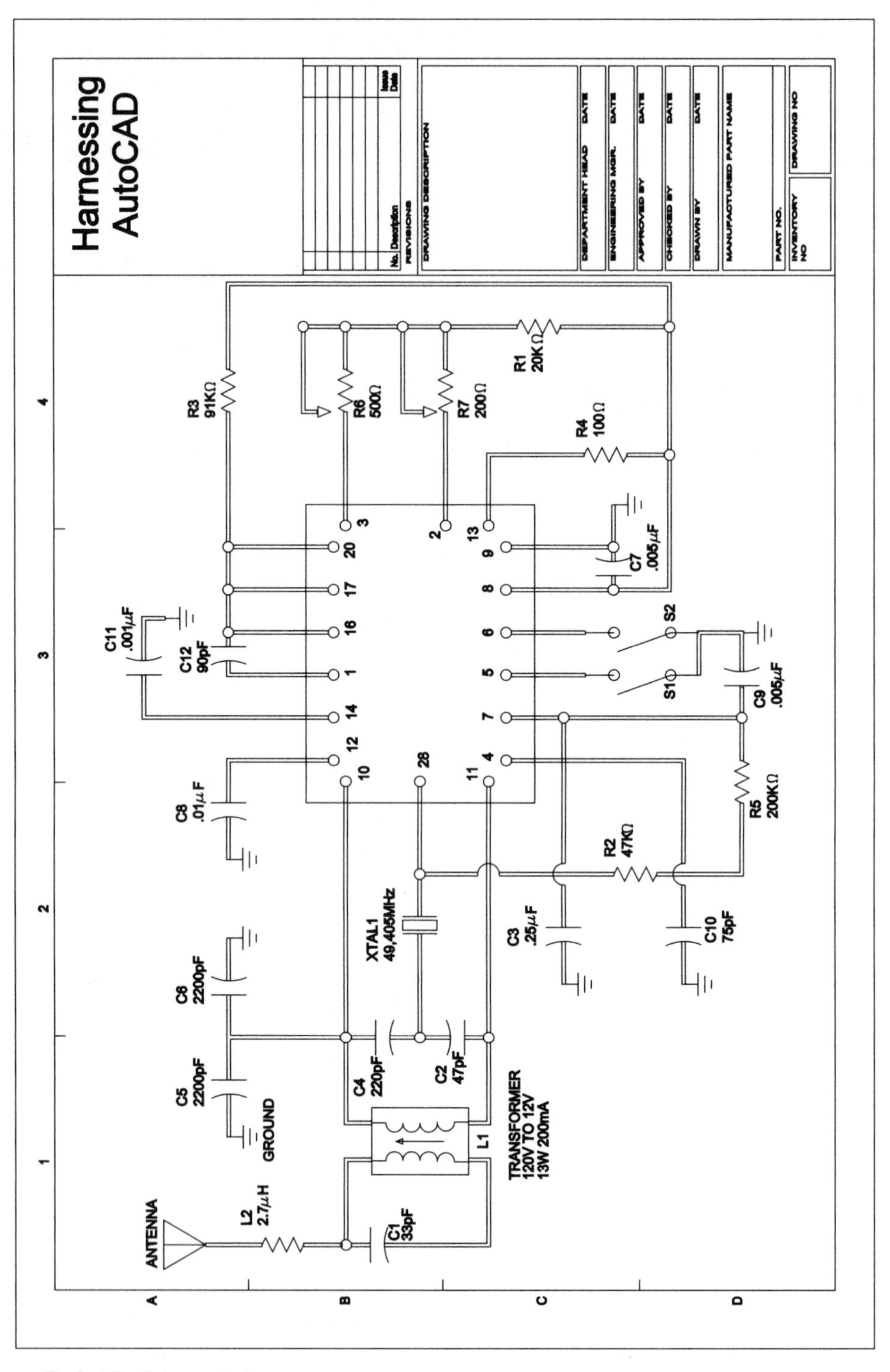


**Figure Ex6–17** *Schematic layout*

CHAPTER 7

# Dimensioning

Project EXERCISE

## PROJECT EXERCISE

This project exercise provides point-by-point instructions for dimensioning the objects in the drawing shown in Figure P7–1.

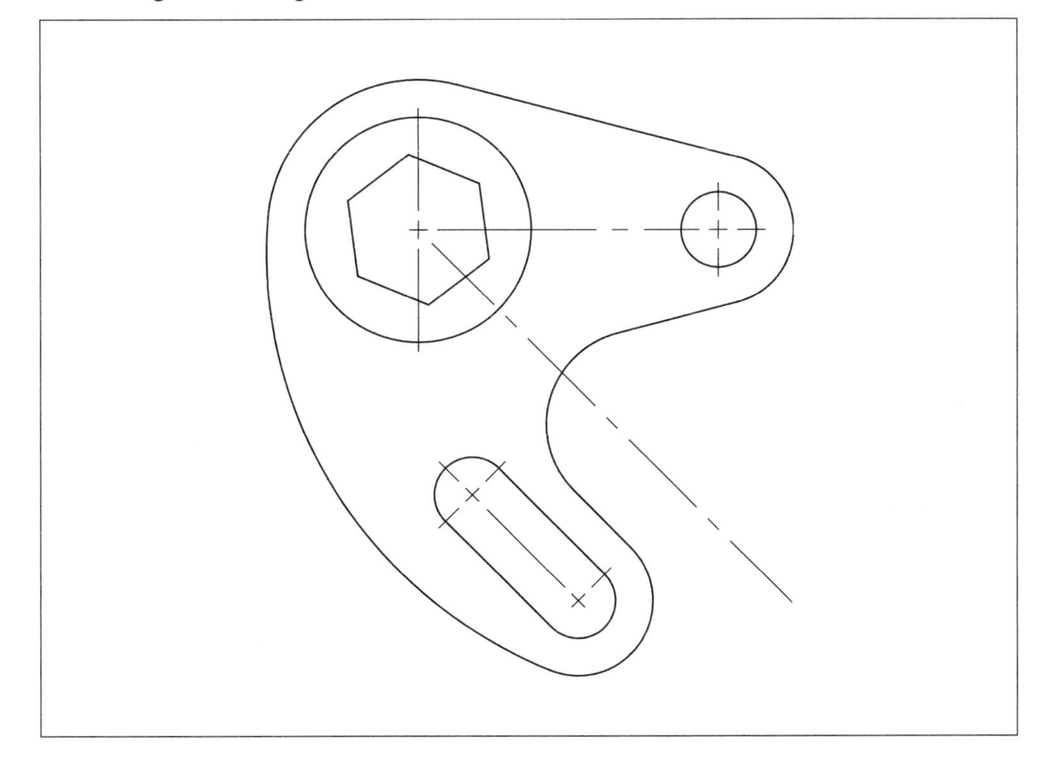

**Figure P7–1** *Completed project drawing*

In this project you will use Dimension commands such as DIMLINEAR, DIMSTYLE, DIMALIGNED, DIMCONTINUE, DIMANGULAR, DIMRADIUS, and DIMDIAMETER to dimension objects.

**Step 1:** Start the AutoCAD program.

**Step 2:** Invoke the OPEN command from the Standard toolbar and select the drawing completed in the Chapter 4 Project Exercise.

**Step 3:** Invoke the LAYER command from the Layers toolbar, or select Layer... from the Format menu. AutoCAD displays the Layer Properties Manager dialog box.

Create one new layer, and rename it as shown in the following table, assigning an appropriate color and linetype.

| Layer Name | Color | Linetype |
|---|---|---|
| Dimension | Blue | Continuous |

Set *dimension* as the current layer, and close the Layer Properties Manager dialog box.

**Step 4:** Select Style from the Dimension menu, AutoCAD displays the Dimension Style Manager dialog box.

Choose the **Modify** button to open the Modify Dimension Style dialog box. Then choose the Lines and Arrows tab. In the Dimension Line section enter **0.50** in the **Baseline spacing** text box. In the Arrowheads section enter **0.125** in the **Arrow Size:** text box, and in the Extension Line section enter **0.125** in the **Extend beyond dim lines:** text box.

Choose the Primary Units tab in the Modify Dimension Style dialog box. From the **Precision:** list box (in Linear and Angular),, select **0.00** for two decimal place precision.

Choose the Text tab in the Modify Dimension Style dialog box. In the Text Appearance section enter **0.125** in the **Text Height:** text box.

Choose **OK** to close the Modify Dimension Style dialog box.

Choose **Close** to close the Dimension Style Manager dialog box.

**Step 5:** From the Object Snap toolbar choose the Object Snap Settings button. The Drafting Settings dialog box will appear. In the Object Snap section set Center object snap mode to ON and the remaining object snap modes to OFF. Also make sure OSNAP toggle button is set to ON in the status bar.

**Step 6:** Invoke the DIMLINEAR command from the Dimension toolbar. Draw the linear dimension, as shown in Figure P7–2.

**Step 7:** Invoke the ALIGNED DIMENSION command from the Dimension toolbar (see Figure P7–3) to add the aligned dimensions to the drawing.

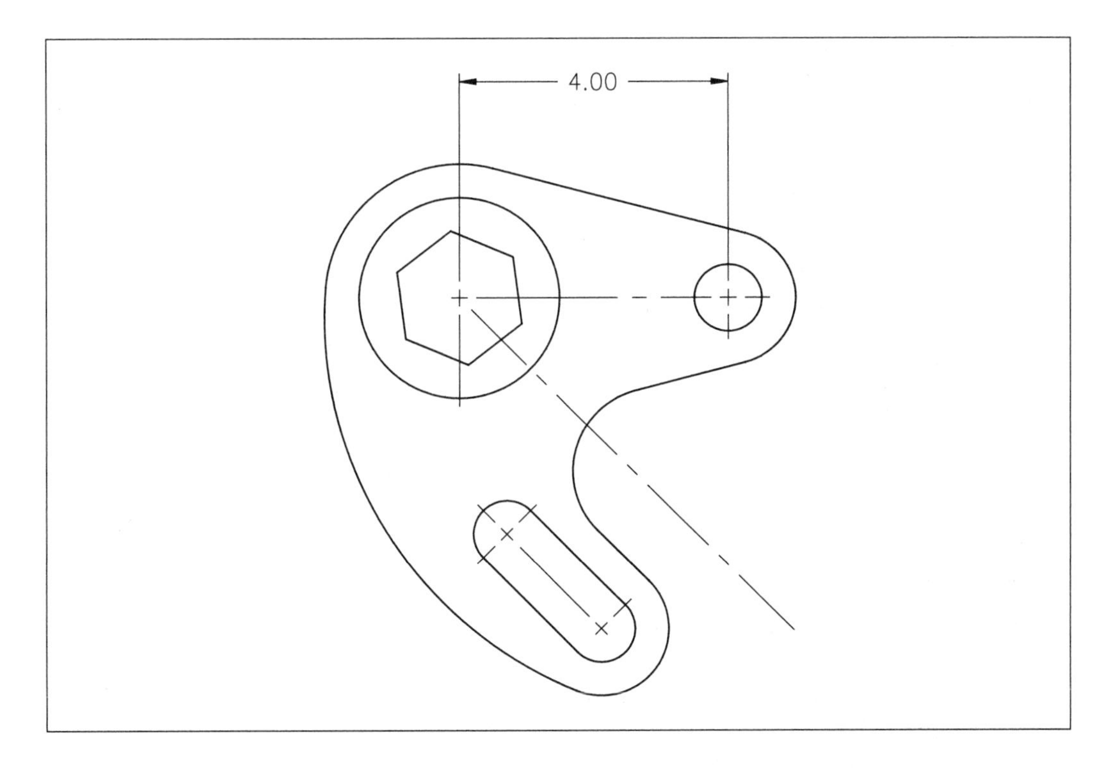


**Figure P7–2** *Completed linear dimension*

**Figure P7–3** *Invoking the* DIMALIGNED *command from the Dimension toolbar*

Command: **dimaligned** (ENTER)
Specify first extension line origin or <select object>: *(select the arc at PT. 1, as shown in Figure P7–4)*
Specify second extension line origin: *(select the arc at PT. 2, as shown in Figure P7–4)*
Specify dimension line location or [Mtext/Text/Angle]: *(select PT. 3 to locate the dimension, as shown in Figure P7–4)*

AutoCAD draws the dimension as shown in Figure P7–5.

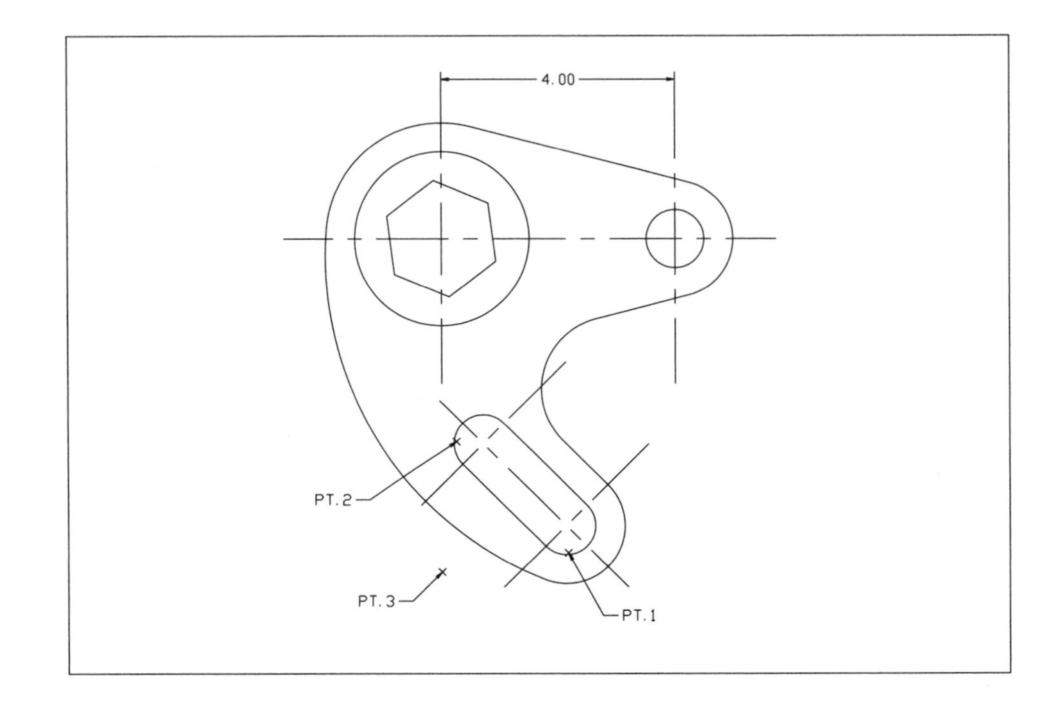


**Figure P7–4** *Specifying points for the aligned dimension*

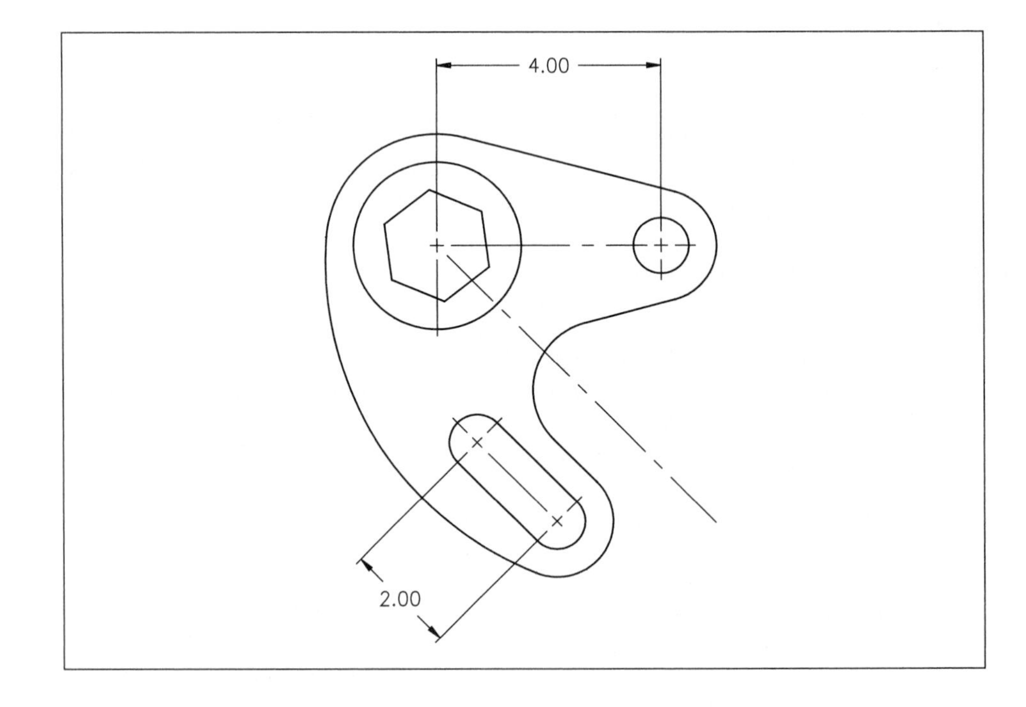


**Figure P7–5** *Completed aligned dimension*

Invoke the DIMCONTINUE command from the Dimension toolbar.

Command: **dimcontinue** (ENTER)
Specify a second extension line origin or [Undo<Select>]: *(select the large circle to continue the dimension, as shown in Figure P7–6 and press* ENTER *to complete the selection).*

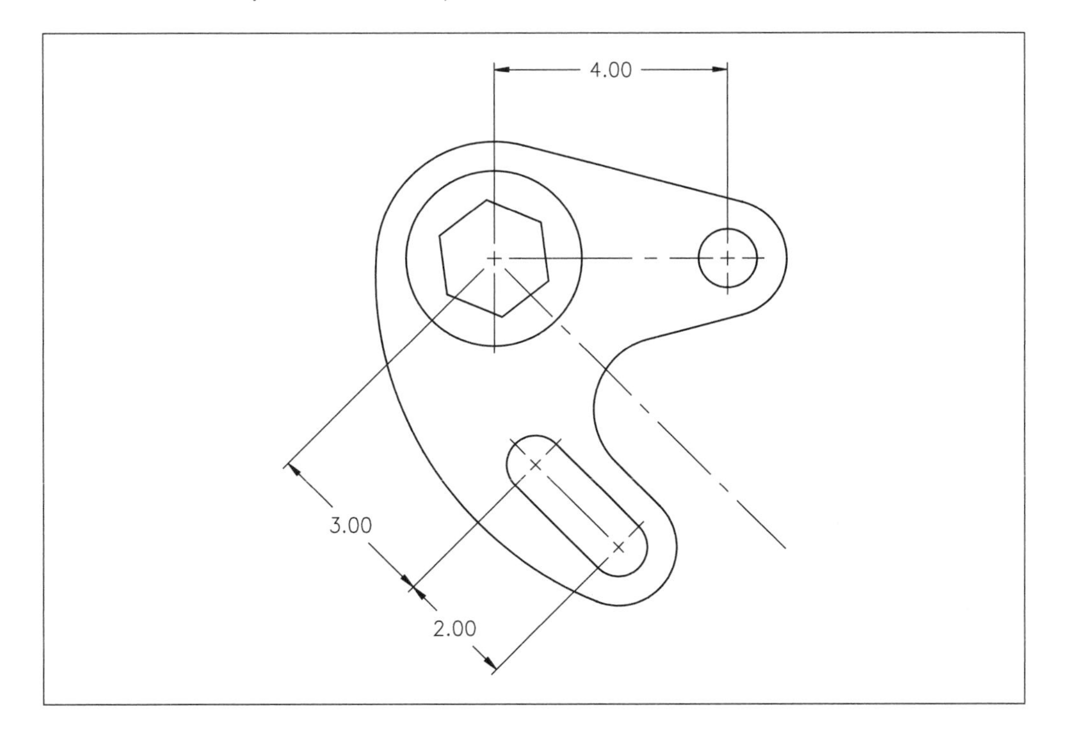


**Figure P7–6** *Aligned dimension*

**Step 8:** Draw centerlines to assist in completing the dimensioning exercise. Set Centerline as the current layer. Invoke the LINE command from the Draw toolbar, and draw lines 1 and 2 as shown in Figure P7–7.

Command: **line** (ENTER)
Specify first point: **7,8** *(start point for line 1)*
Specify next point or [Undo]: **@6.5<315**
Specify next point or [Undo]: (ENTER)

Command: (ENTER)
Specify first point: 7,8 *(start point for line 2)*
Specify next point or [Undo]: *(use the object snap tool "intersection" to draw the line to the intersection of the polygon, as shown in Figure P7–7)*
*Specify next point or [Undo]: (enter)*

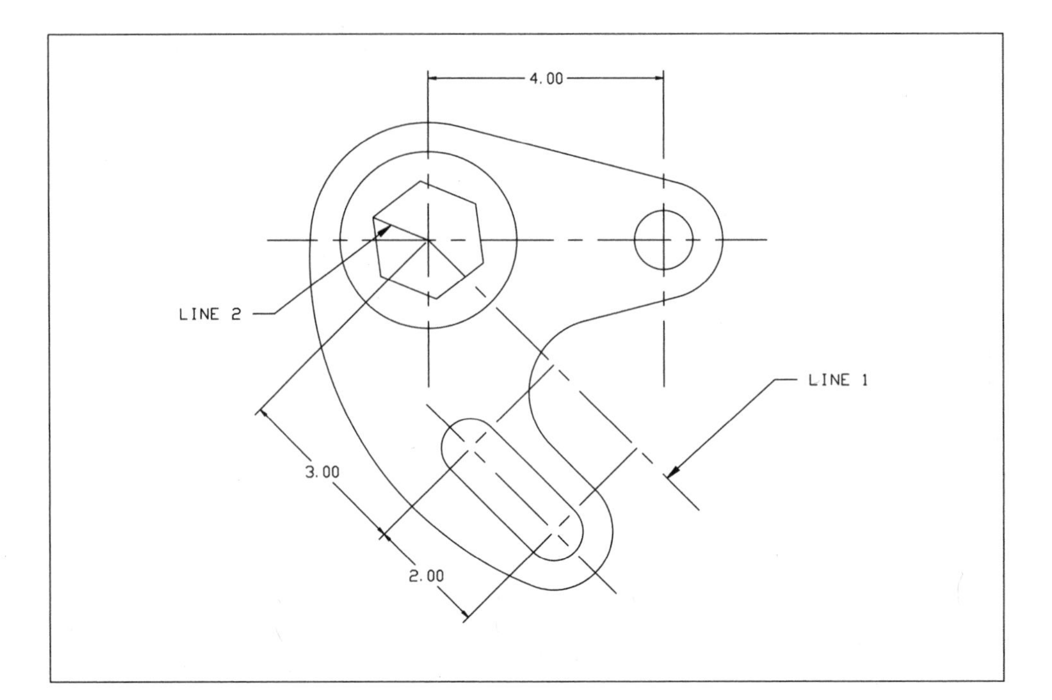

**Figure P7–7** *Lines 1 and 2*

**Step 9:** Set Dimension as the current layer. Invoke the DIMALIGNED command from the Dimension toolbar to draw the remaining aligned dimensions for the drawing.

> Command: **dimaligned** (ENTER)
> Specify first extension line origin or <select object>: *(select the arc at PT. 1, as shown in Figure P7–8)*
> Specify second extension line origin: *(use the object snap tool "perpendicular" to specify PT. 2, as shown in Figure P7–8)*
> Specify dimension line location or [Mtext/Text/Angle]: *(specify PT. 3 to locate the dimension, as shown in Figure P7–8)*

AutoCAD draws the dimension as shown in Figure P7–9.

> Command: **dimaligned** (ENTER)
> Specify first extension line origin or <select object>: *(use the object snap tool "intersection" to locate PT. 1, as shown in Figure P7–10)*
> Specify second extension line origin: *(use the object snap tool "perpendicular" to specify PT. 2, as shown in Figure P7–10)*
> Specify dimension line location or [Mtext/Text/Angle]: *(specify PT. 3 to locate the dimension, as shown in Figure P7–10)*

AutoCAD draws the dimension as shown in Figure P7–11.

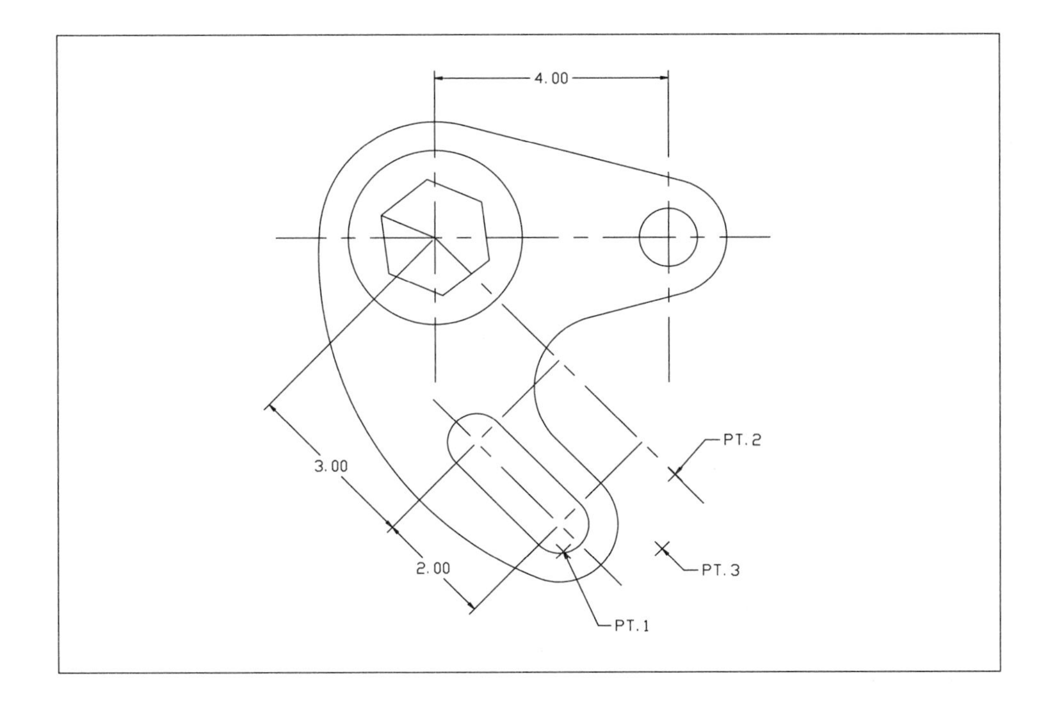


**Figure P7–8** *Drawing the aligned dimension*

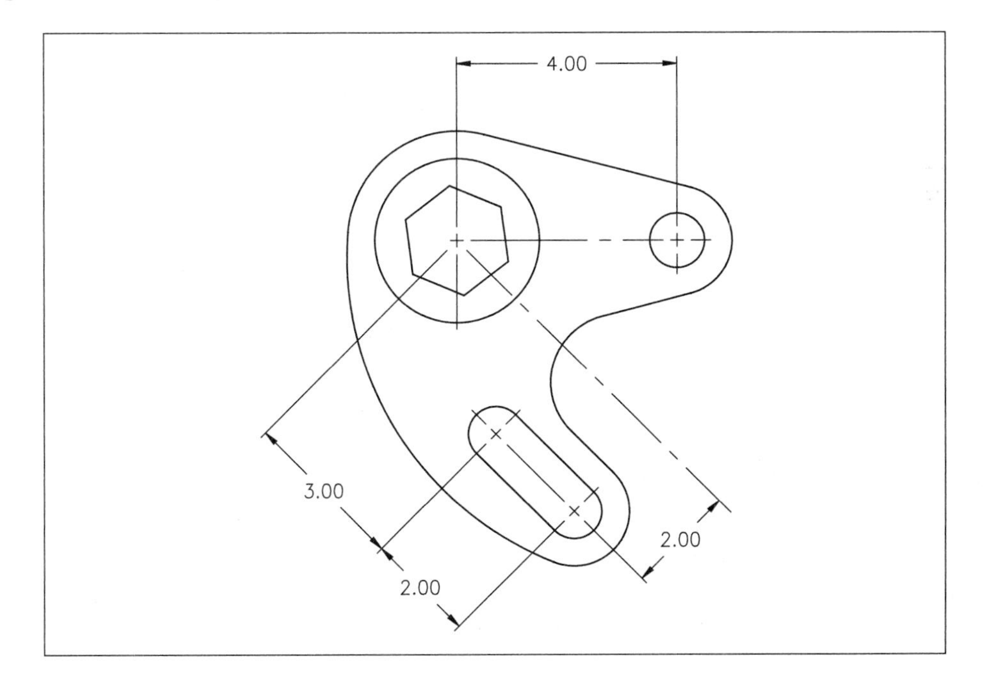


**Figure P7–9** *Completed dimension*

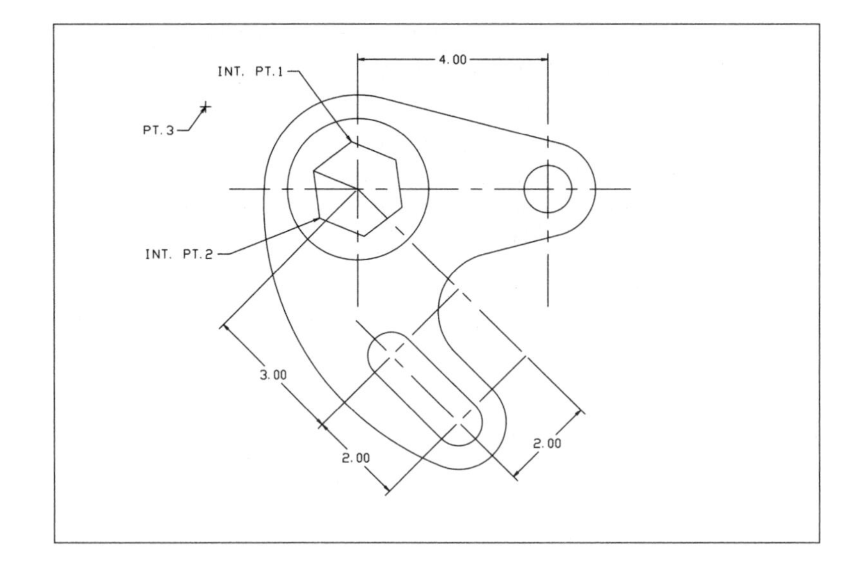


**Figure P7–10** *Drawing the aligned dimension*

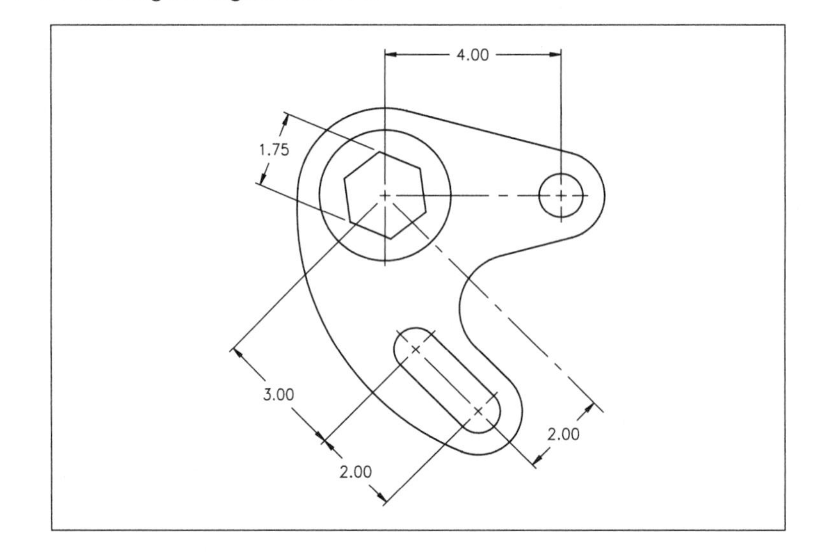


**Figure P7–11** *Completed dimension*

**Step 10:** Set the OSNAP toggle button to OFF in the status bar.

**Step 11:** Invoke the DIMANGULAR command from the Dimension toolbar to draw the angular dimensions.

> Command: **dimangular** (ENTER)
> Select arc, circle, line, or <specify vertex>: *(select line 1 as shown in Figure P7–12)*
> Select second line: *(select line 2 as shown in Figure P7–12)*

Specify dimension arc line location or [Mtext/Text/Angle]: *(locate the dimension as at PT. 1, as shown in Figure P7–12)*

Command: **dimangular** (ENTER)
Select arc, circle, line or <specify vertex>: *(select line 4 as shown in Figure P7–12)*
Select second line: *(select line 5 as shown in Figure P7–12)*
Specify dimension arc line location or [Mtext/Text/Angle]: *(locate the dimension as at PT. 2, as shown in Figure P7–12)*

AutoCAD draws the dimensions as shown in Figure P7–13.

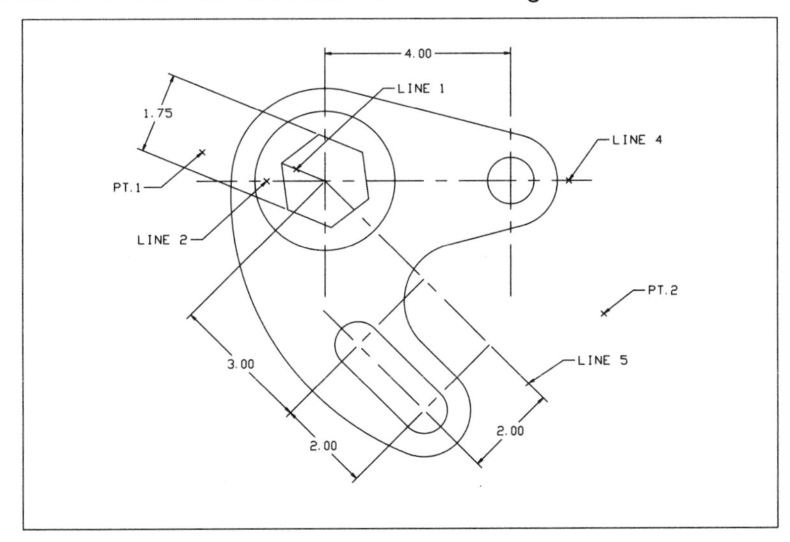


**Figure P7–12** *Selecting points for angular dimensions*

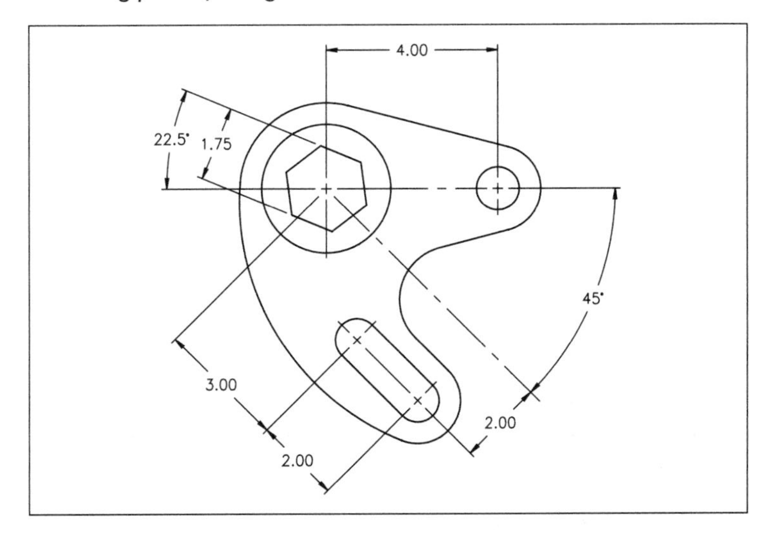


**Figure P7–13** *Completed angular dimensions*

**Step 12:** Invoke the DIMRADIUS and DIMDIAMETER commands to draw the radius and diameter dimensions as shown in Figure P7–14.

**Note:** Before using DIMRADIUS and DIMDIAMETER, open the Dimension Style Manager dialog box from the Dimension menu. Choose the Modify button, then select the Fit sub-dialog box and choose **Place text manually when dimensioning** in the Fine Tuning section of the dialog box.

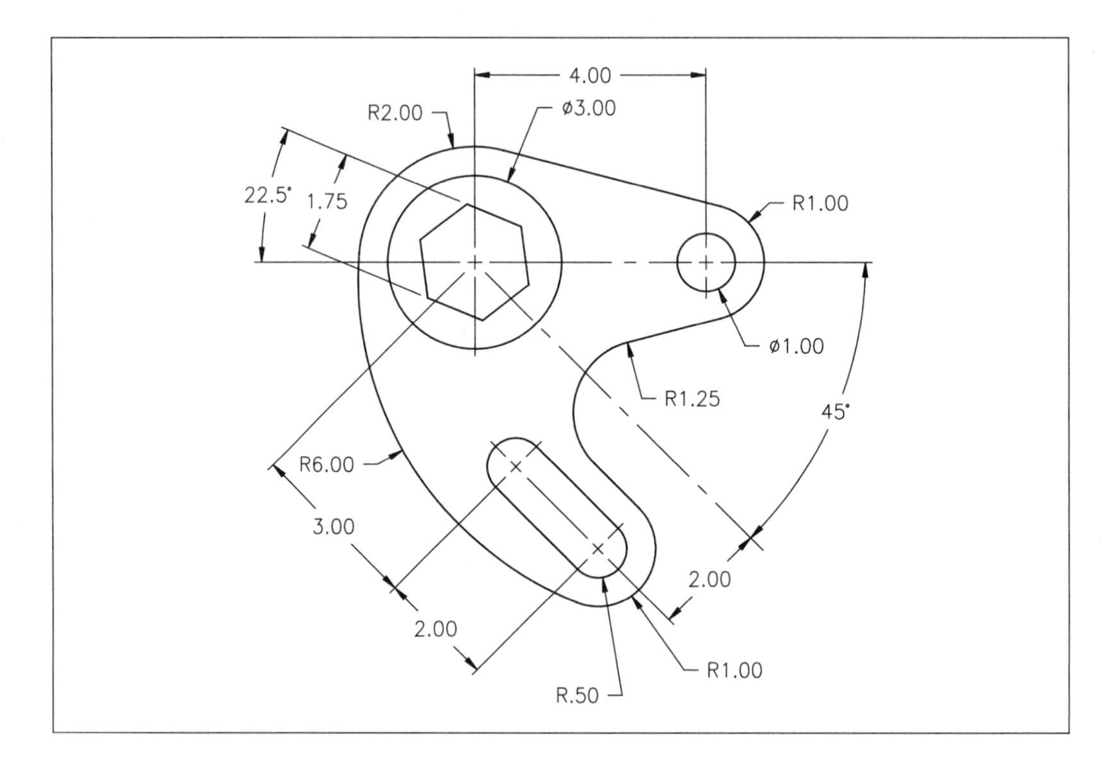


**Figure P7–14** *Completed dimensions*

**Step 13:** Save and close the drawing.

## EXERCISE 7–1

Create the drawing shown in Figure Ex7–1, with all the dimensions (including both lateral tolerancing and geometric tolerancing symbology).

| Settings | Value | | |
|---|---|---|---|
| 1. Units | Decimal | | |
| 2. Limits | Lower left corner: 0,0 | | |
| | Upper right corner: 420,300 | | |
| 3. Grid | 10 | | |
| 4. Snap | 5 | | |
| 5. Layers | *NAME* | *COLOR* | *LINETYPE* |
| | Dimension | White | Continuous |
| | Border | Red | Continuous |
| | Object | Green | Continuous |
| | Center | Magenta | Center |
| | Text | Blue | Continuous |
| | Hidden | White | Hidden |

mechanical EXERCISE

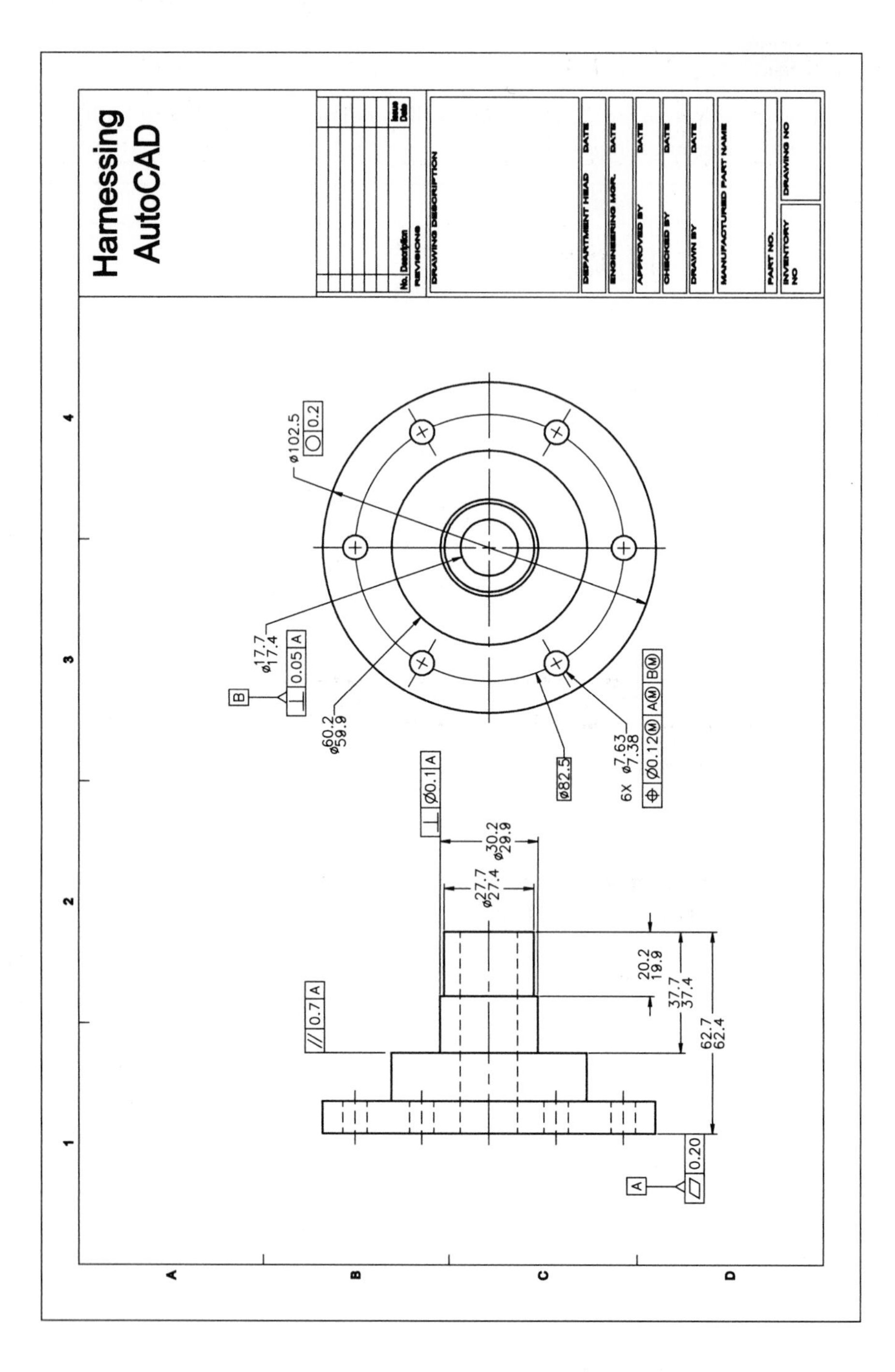


**Figure Ex7–1** *Completed drawing*

## EXERCISE 7–2

Create the drawing of the Lock Ring as shown in Figure Ex7–2, with all the dimensions.

| Settings | Value | | |
|---|---|---|---|
| 1. Units | Decimal | | |
| 2. Limits | Lower left corner: 0,0 | | |
| | Upper right corner: 11,8.5 | | |
| 3. Dimscale | 1 | | |
| 4. Ltscale | 0.5 | | |
| 5. Layers | *NAME* | *COLOR* | *LINETYPE* |
| | Border | Red | Continuous |
| | Object | Green | Continuous |
| | Text | Blue | Continuous |
| | Dimension | White | Continuous |
| | Center | Red | Center |
| | Hidden | Magenta | Hidden |

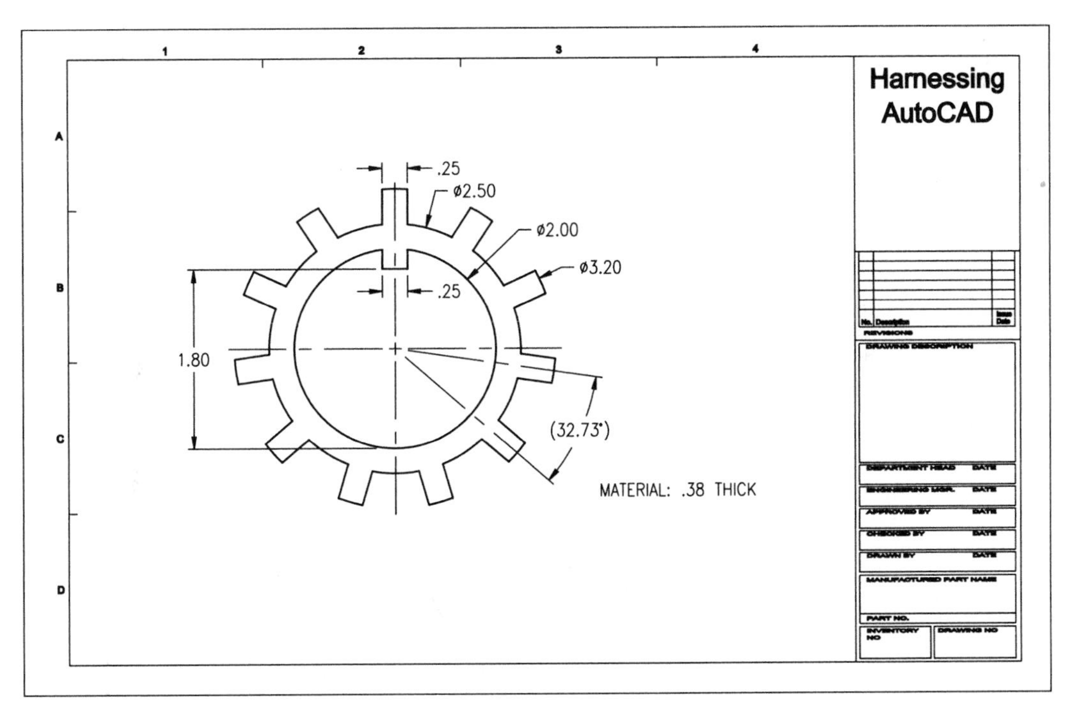


**Figure Ex7–2** *Lock Ring*

## EXERCISE 7–3

Open the Top View and Section of a Multi-groove Compressor Pulley drawing that was completed in Exercise 5–3 and add the dimensions shown in Figure Ex7–3. (Make sure to create a layer for dimensions and draw all dimensions on that layer).

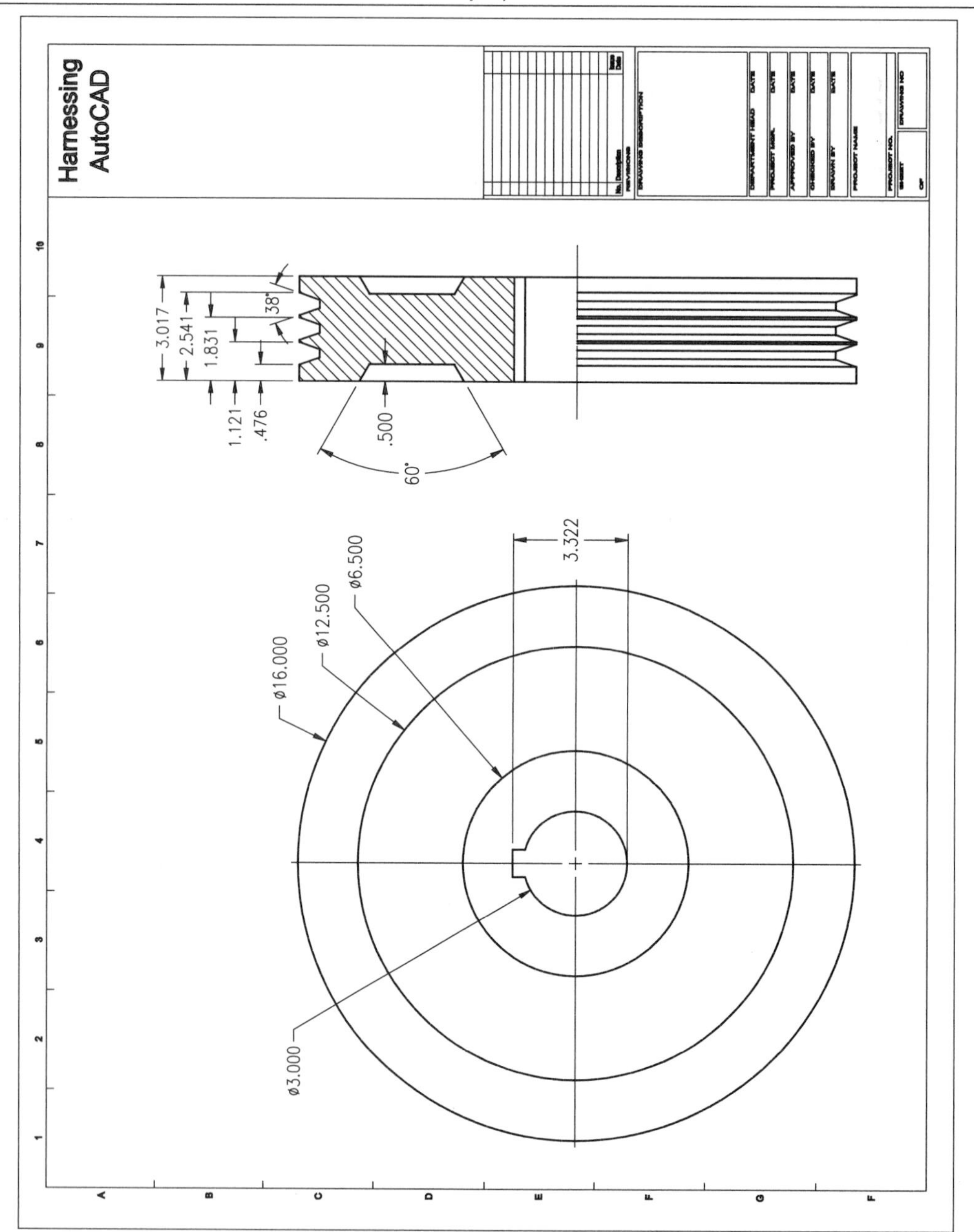


**Figure Ex7–3** *Top View and Section of a Multi-groove Compressor Pulley*

## EXERCISE 7–4

Open the Elevation and Section of a Surface Mounted Flag Pole Bracket drawing that was completed in Exercise 5–4 and add the dimensions shown in Figure Ex7–4. (Make sure to create a layer for dimensions and draw all dimensions on that layer).

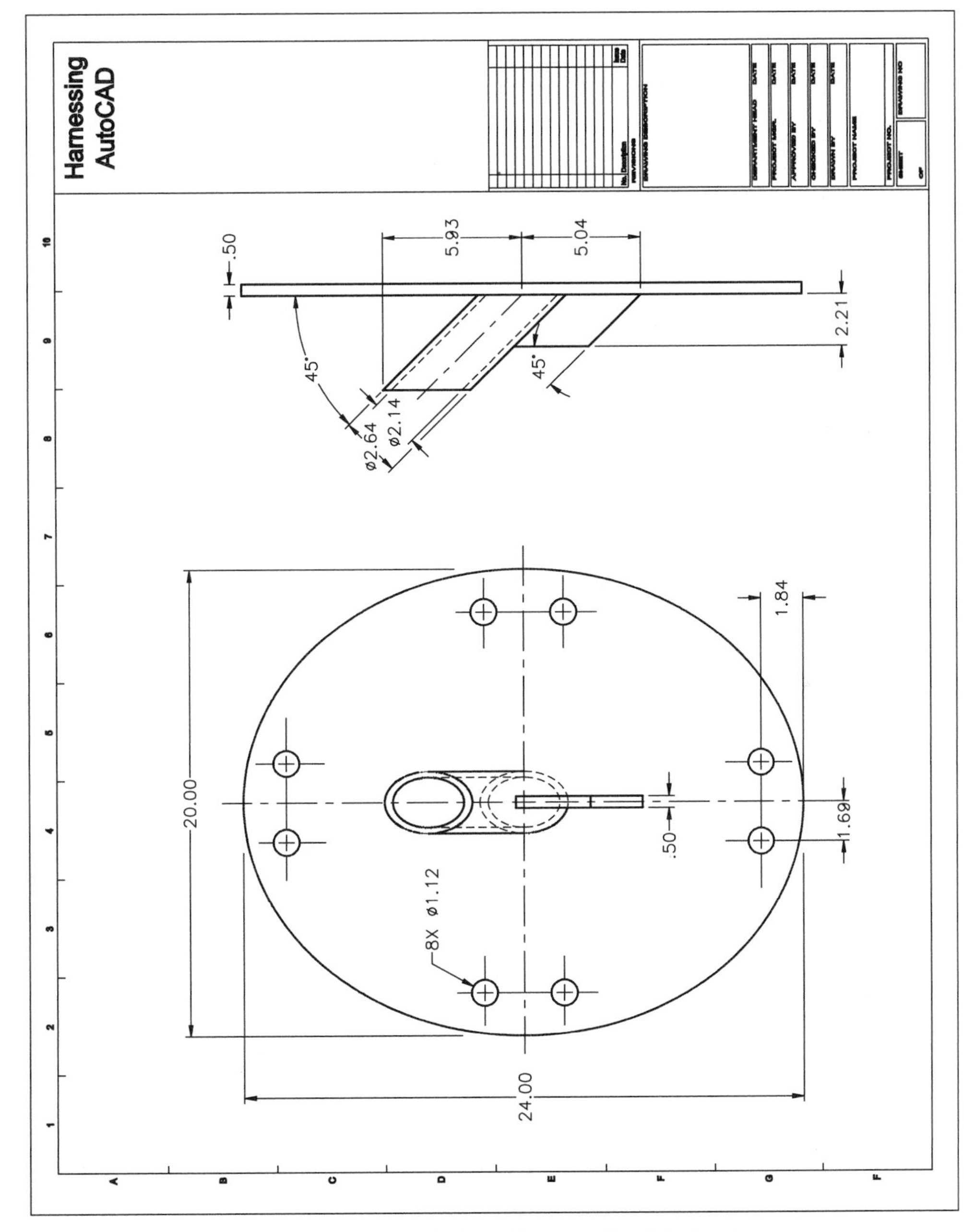


**Figure Ex7–4** *Elevation and Section of a Surface Mounted Flag Pole Bracket*

mechanical EXERCISE

## EXERCISE 7–5

Open the drawing that was completed in Exercise 6–5 and add the dimensions shown in Figure Ex7–5. (Make sure to create a layer for dimensions and draw all dimensions on that layer).

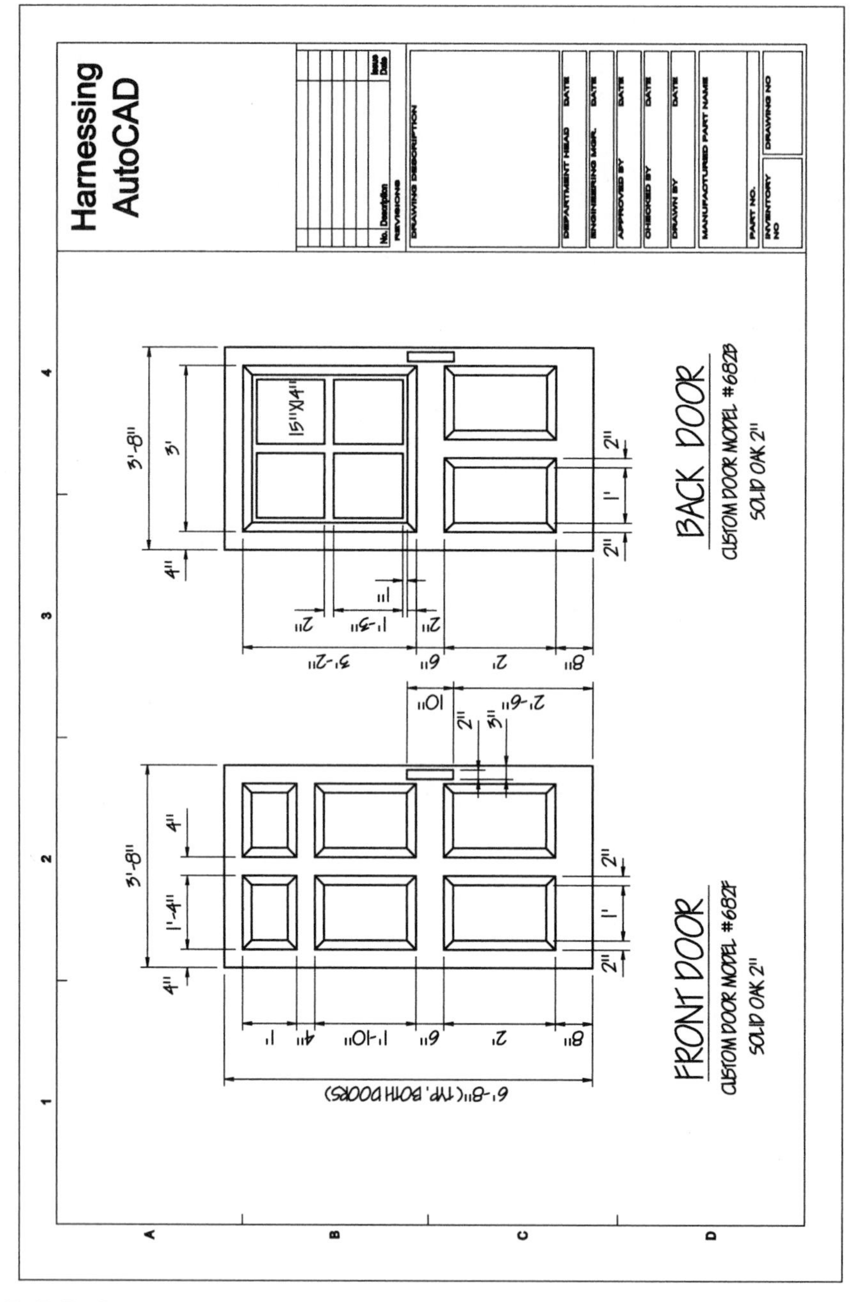


**Figure Ex7–5** *Completed drawing with dimensions*

## EXERCISE 7–6

The drawing of the instrument panel in Figure Ex7–6 has all the information needed to create your drawing. Add necessary dimensions as shown in the figure.

| Settings | Value | | |
|---|---|---|---|
| 1. Units | Decimal | | |
| 2. Limits | Lower left corner: 0,0 | | |
| | Upper right corner: 12,10.5 | | |
| 3. Grid | 0.5" | | |
| 4. Snap | 0.25" | | |
| 5. Layers | *NAME* | *COLOR* | *LINETYPE* |
| | Border | Red | Continuous |
| | Object | Green | Continuous |
| | Text | Blue | Continuous |
| | Dimension | White | Continuous |

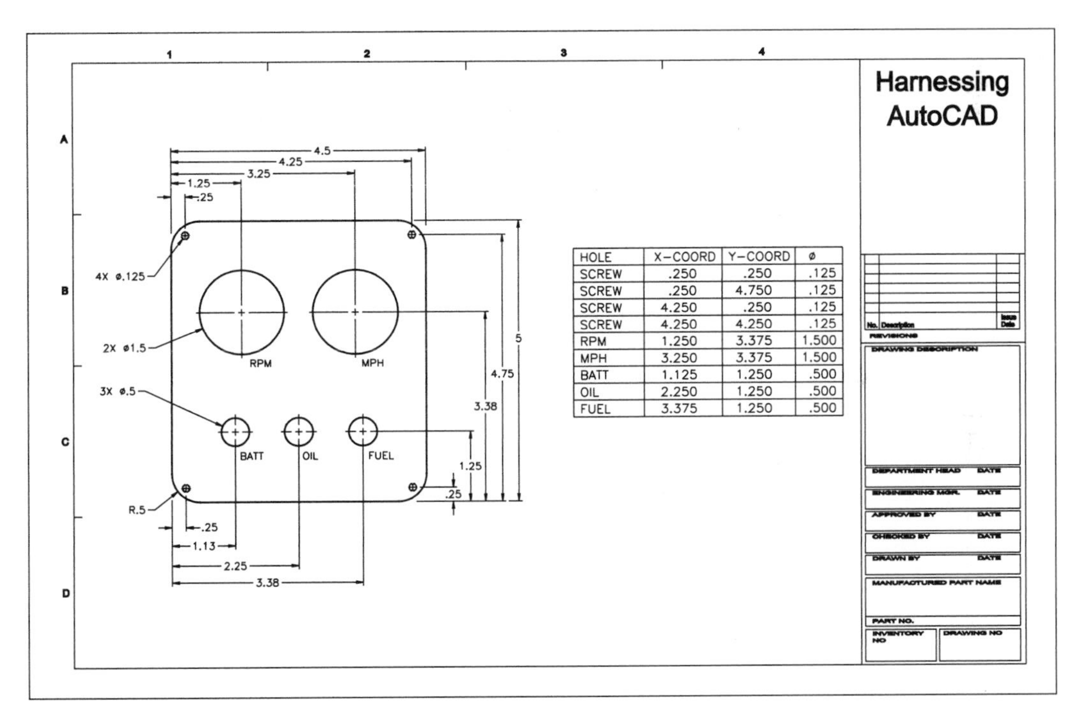

| HOLE | X–COORD | Y–COORD | ø |
|---|---|---|---|
| SCREW | .250 | .250 | .125 |
| SCREW | .250 | 4.750 | .125 |
| SCREW | 4.250 | .250 | .125 |
| SCREW | 4.250 | 4.250 | .125 |
| RPM | 1.250 | 3.375 | 1.500 |
| MPH | 3.250 | 3.375 | 1.500 |
| BATT | 1.125 | 1.250 | .500 |
| OIL | 2.250 | 1.250 | .500 |
| FUEL | 3.375 | 1.250 | .500 |



**Figure Ex7–6** *Layout of the drawing*

## EXERCISE 7–7

Open the Stair Plan drawing that was completed in Exercise 3–7 and add the dimensions shown in Figure Ex7–7. (Make sure to create a layer for dimensions and draw all dimensions on that layer).

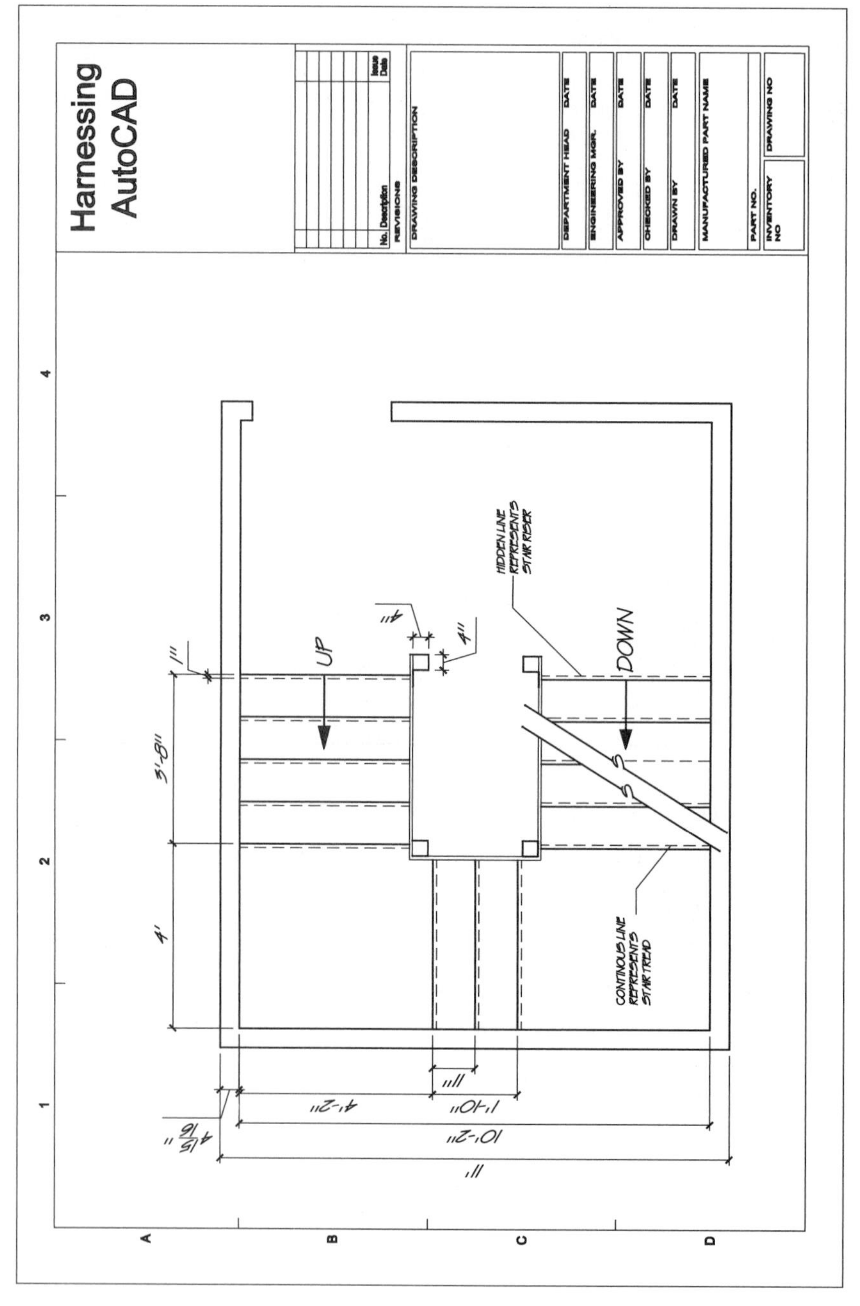


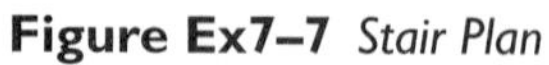
**Figure Ex7–7** *Stair Plan*

## EXERCISE 7–8

Open the Ornamental Fence drawing that was completed in Exercise 4–7 and add the dimensions shown in Figure Ex7–8. (Make sure to create a layer for dimensions and draw all dimensions on that layer).

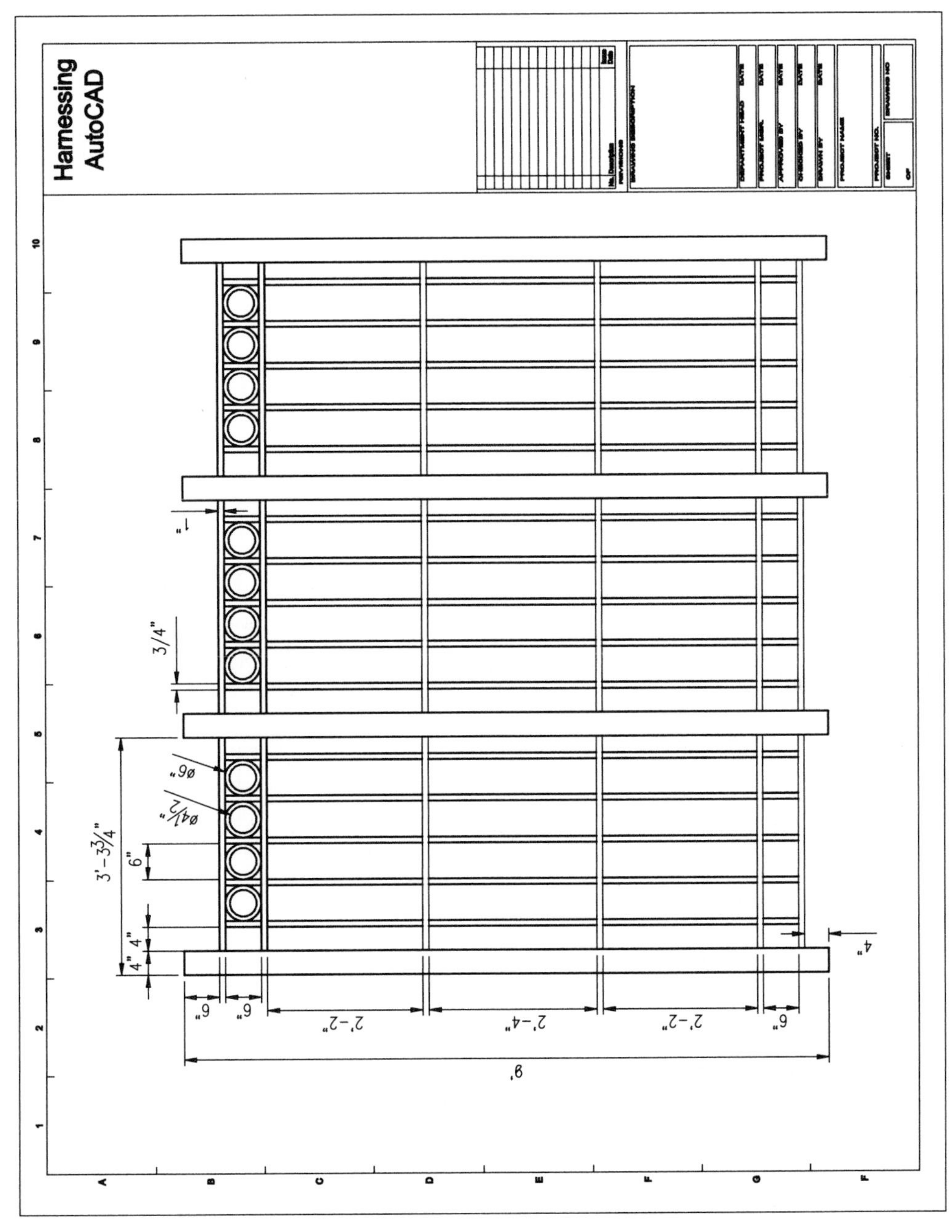


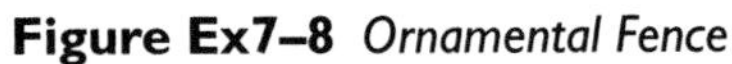
**Figure Ex7–8** *Ornamental Fence*

## EXERCISE 7–9

Open the Recessed Store Front drawing that was completed in Exercise 4–9 and add the dimensions shown in Figure Ex7–9. (Make sure to create a layer for dimensions and draw all dimensions on that layer).

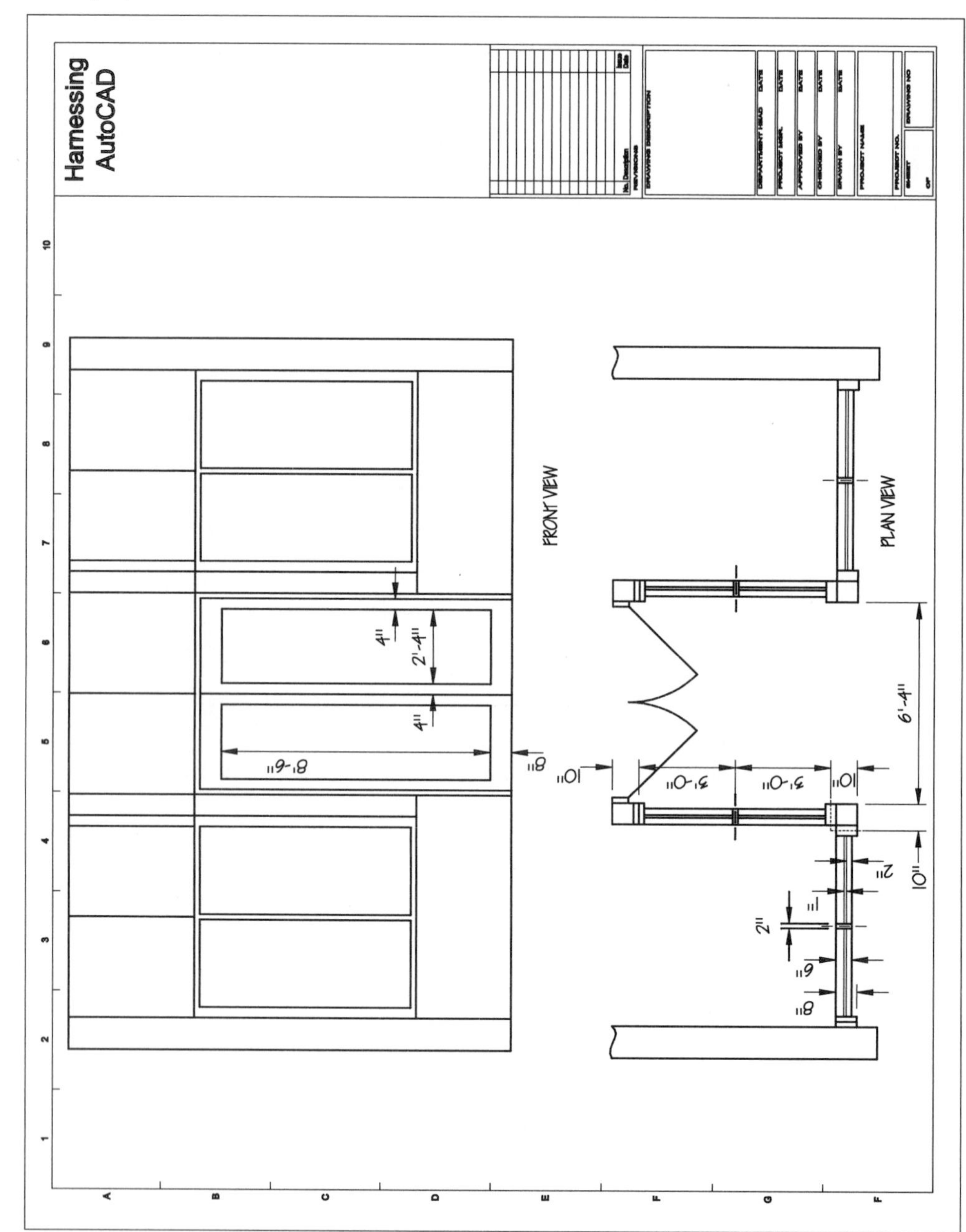


**Figure Ex7–9** *Recessed Store Front*

## EXERCISE 7–10

Open the drawing that was completed in Exercise 4–6 and add the dimensions in Figure Ex7–10.

| Settings | Value |
|---|---|
| 1. Units | Architectural |
| 2. Limits | Lower left corner:<br>10'-0",-10',-0"<br>Upper right corner:<br>50'-0",35',-0" |
| 3. Grid | 12" |
| 4. Snap | 6" |

| Settings | Value | | |
|---|---|---|---|
| 5. Layers | *NAME* | *COLOR* | *LINETYPE* |
| | Beam | White | Continuous |
| | Column | White | Continuous |
| | Dimension | White | Continuous |
| | Border | Red | Continuous |
| | Construction | Green | Continuous |
| | Text | Blue | Continuous |

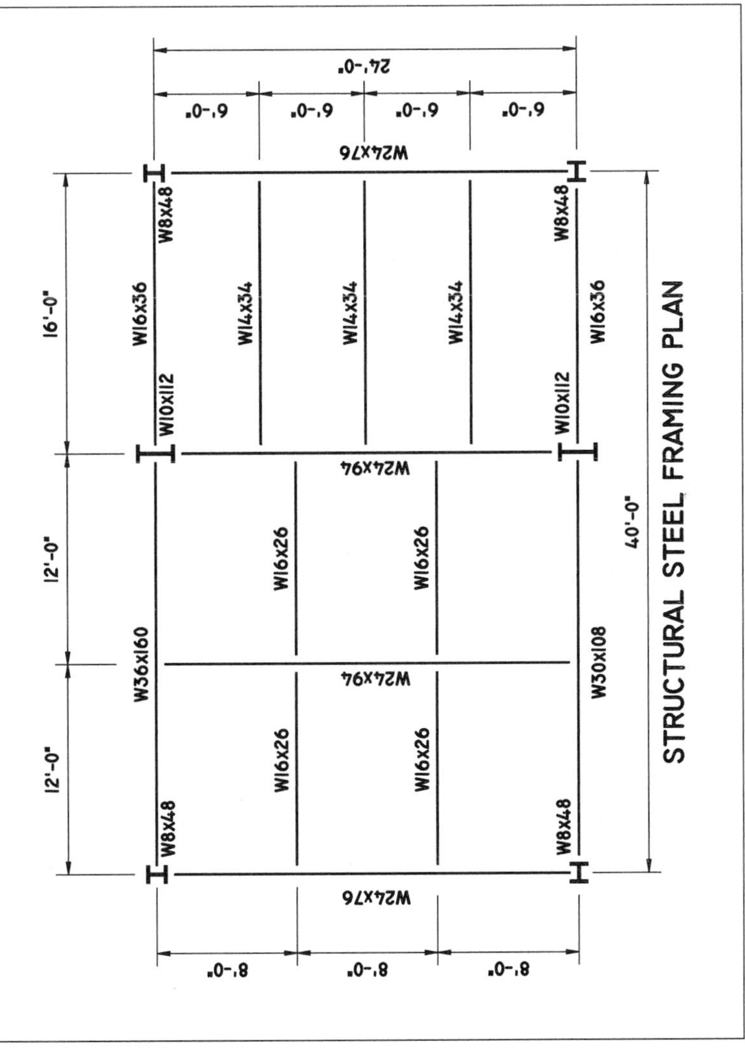


**Figure Ex7–10** *Completed drawing*

## EXERCISE 7–11

Open the Security Retaining Wall drawing that was completed in Exercise 4–12 and add the dimensions shown in Figure Ex7–11. (Make sure to create a layer for dimensions and draw all dimensions on that layer).

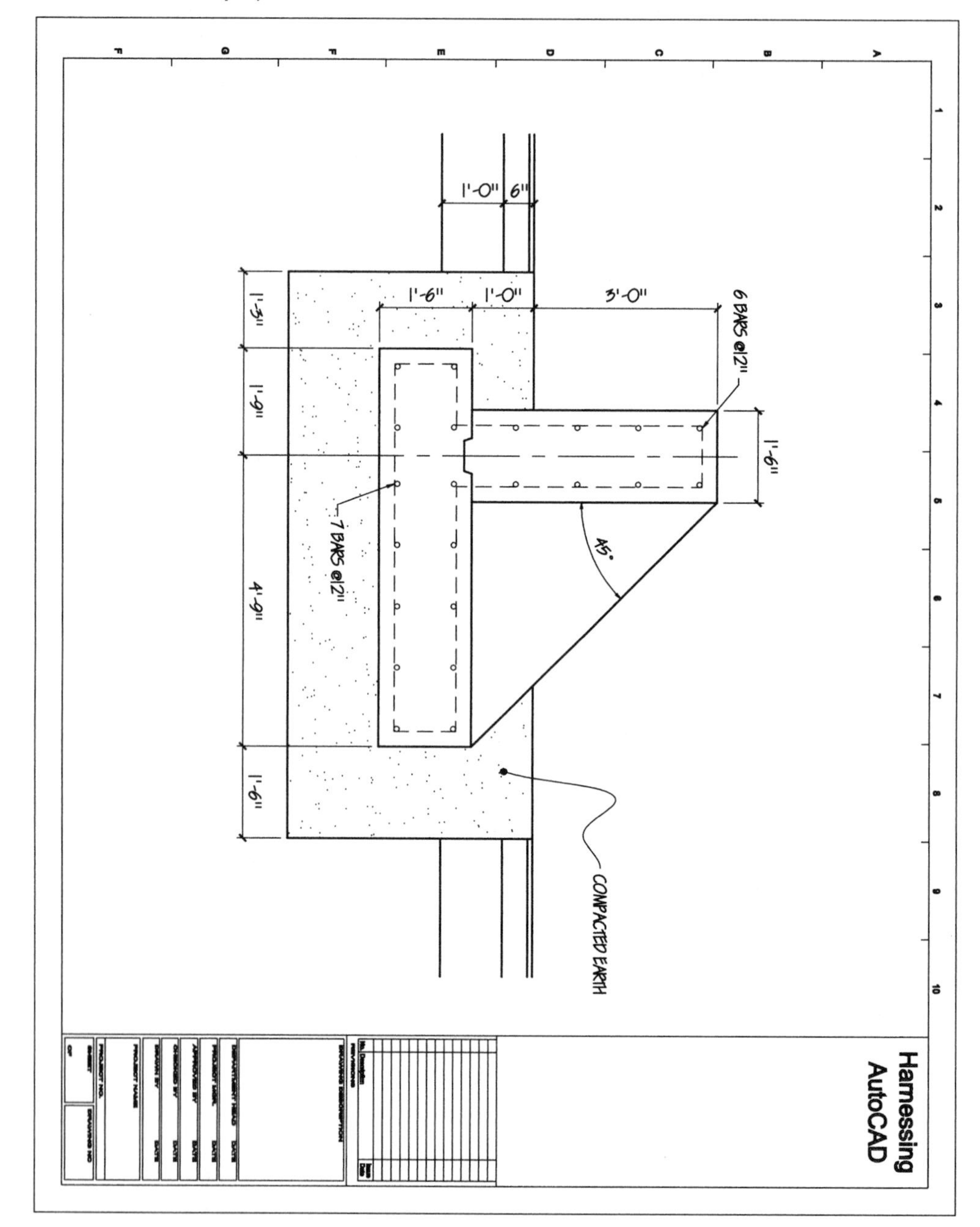


**Figure Ex7–11** *Security Retaining Wall*

## EXERCISE 7–12

Open the Pile Cap drawing that was completed in Exercise 4–15 and add the dimensions shown in Figure Ex7–12. (Make sure to create a layer for dimensions and draw all dimensions on that layer).

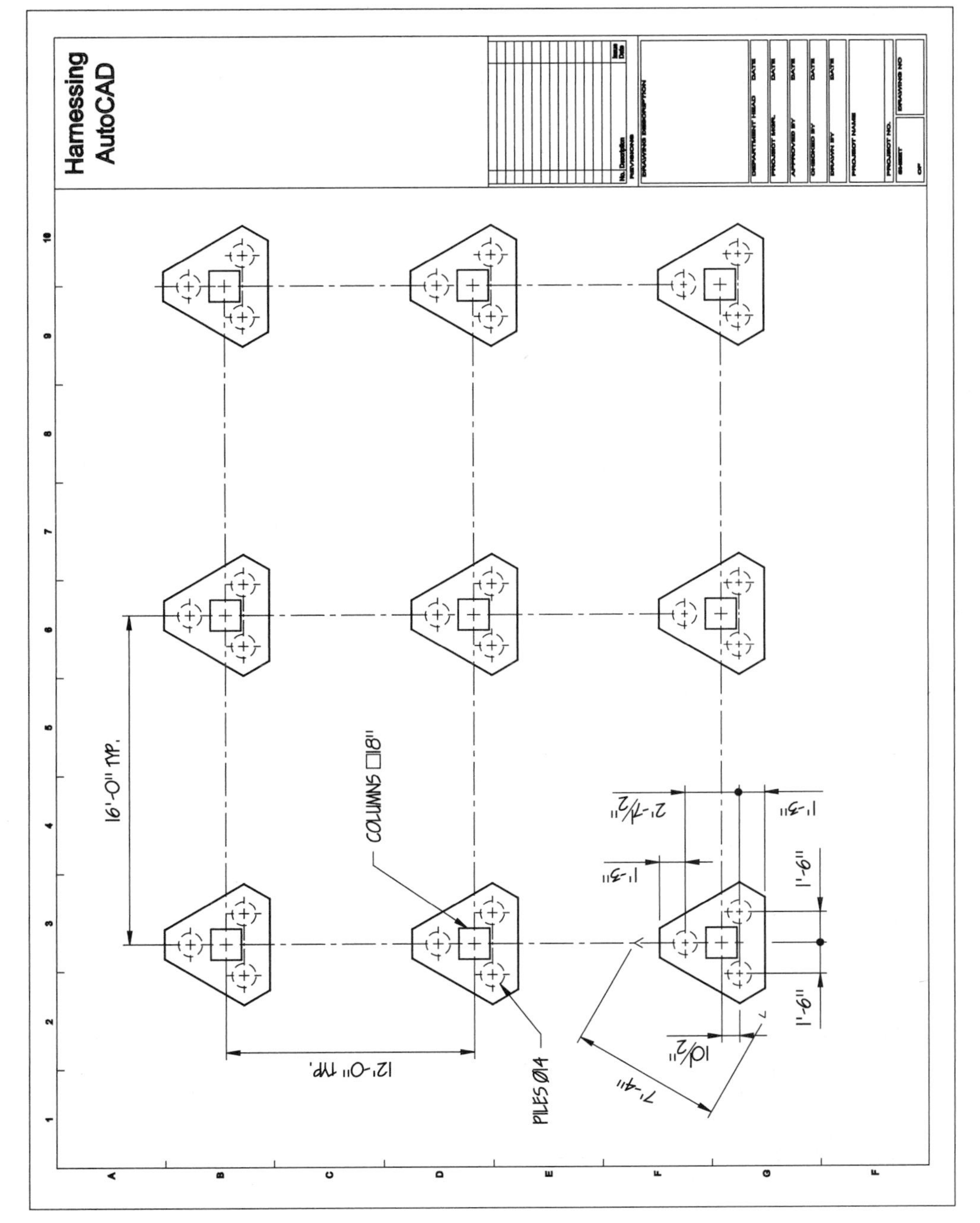


**Figure Ex7–12** *Pile Cap*

## EXERCISE 7–13

The drawing of the instrument panel in Figure Ex7–13 has all the information needed to create your drawing. Add the dimensions shown in Figure with the ORDINATE DIMENSION command.

| Settings | Value | | |
|---|---|---|---|
| 1. Units | Decimal | | |
| 2. Limits | Lower left corner: –1.5,–2.5<br>Upper right corner: 10.5,6.5 | | |
| 3. Grid | 0.5" | | |
| 4. Snap | 0.25" | | |
| 5. Layers | *NAME* | *COLOR* | *LINETYPE* |
| | Border | Red | Continuous |
| | Object | Green | Continuous |
| | Text | Blue | Continuous |
| | Dimension | White | Continuous |

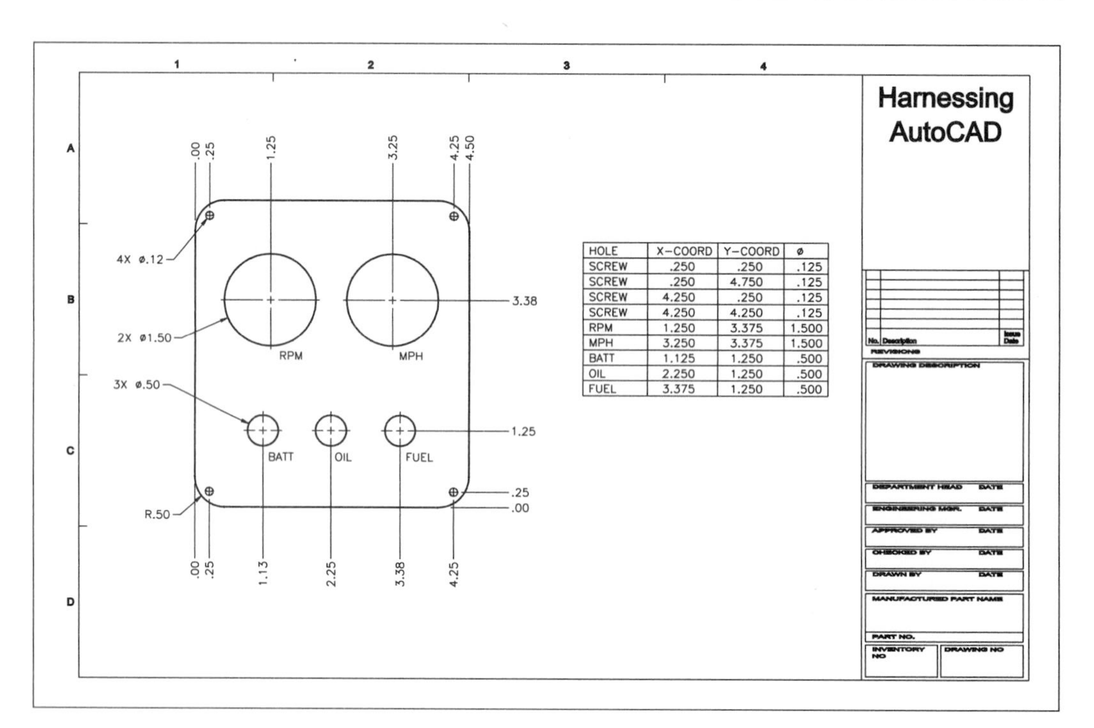

| HOLE | X–COORD | Y–COORD | ø |
|---|---|---|---|
| SCREW | .250 | .250 | .125 |
| SCREW | .250 | 4.750 | .125 |
| SCREW | 4.250 | .250 | .125 |
| SCREW | 4.250 | 4.250 | .125 |
| RPM | 1.250 | 3.375 | 1.500 |
| MPH | 3.250 | 3.375 | 1.500 |
| BATT | 1.125 | 1.250 | .500 |
| OIL | 2.250 | 1.250 | .500 |
| FUEL | 3.375 | 1.250 | .500 |

**Figure Ex7–13** *Layout of the drawing*

## EXERCISE 7–14

Open the drawing that was completed in Exercise 2–25 and add the dimensions in Figure Ex7–14. (Make sure to create a layer for dimensions and draw all dimensions on that layer).

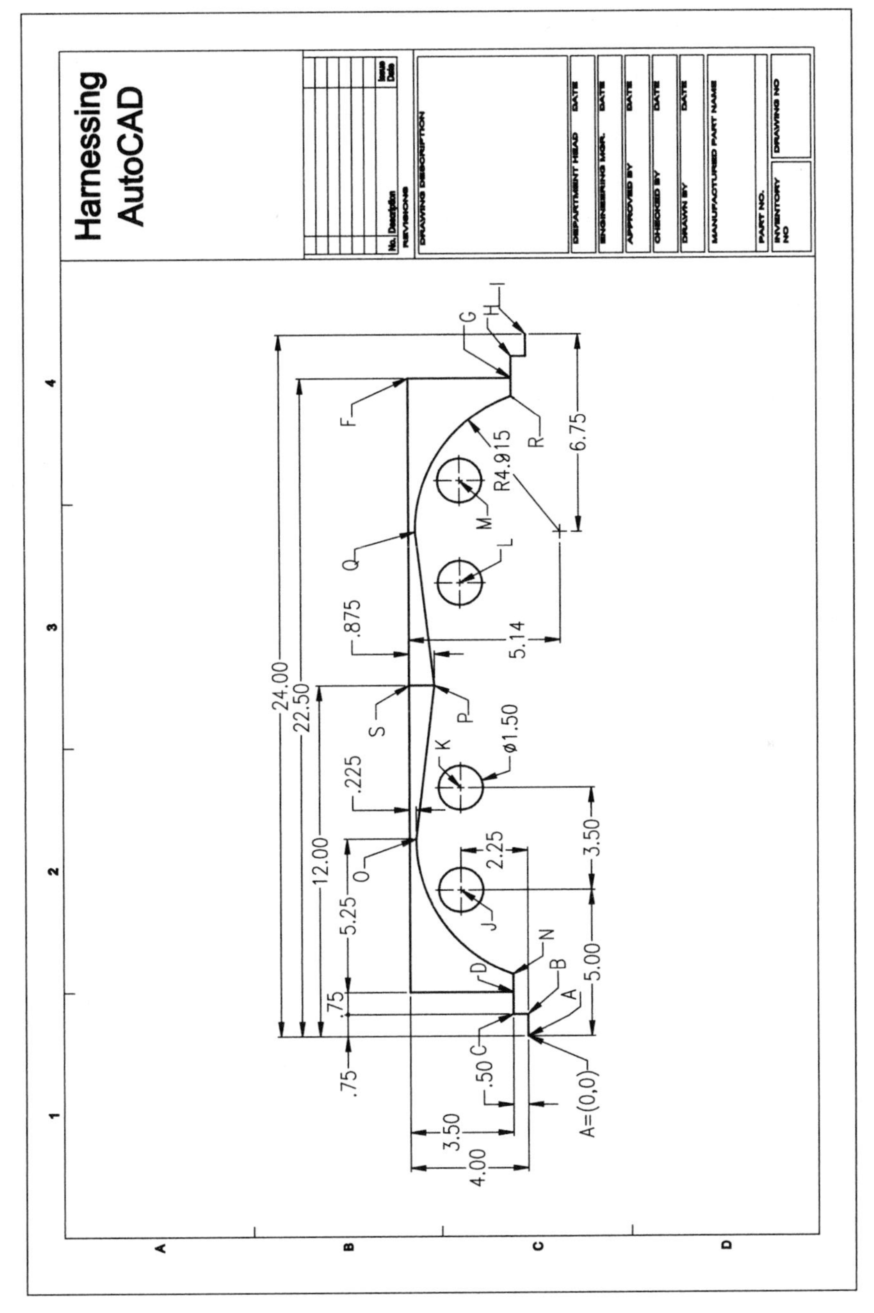


**Figure Ex7–14** *Completed Drawing*

## EXERCISE 7–15

Open the Exterior Ornamental Lighting drawing that was completed in Exercise 4–17 and add the dimensions shown in Figure Ex7–15. (Make sure to create a layer for dimensions and draw all dimensions on that layer).

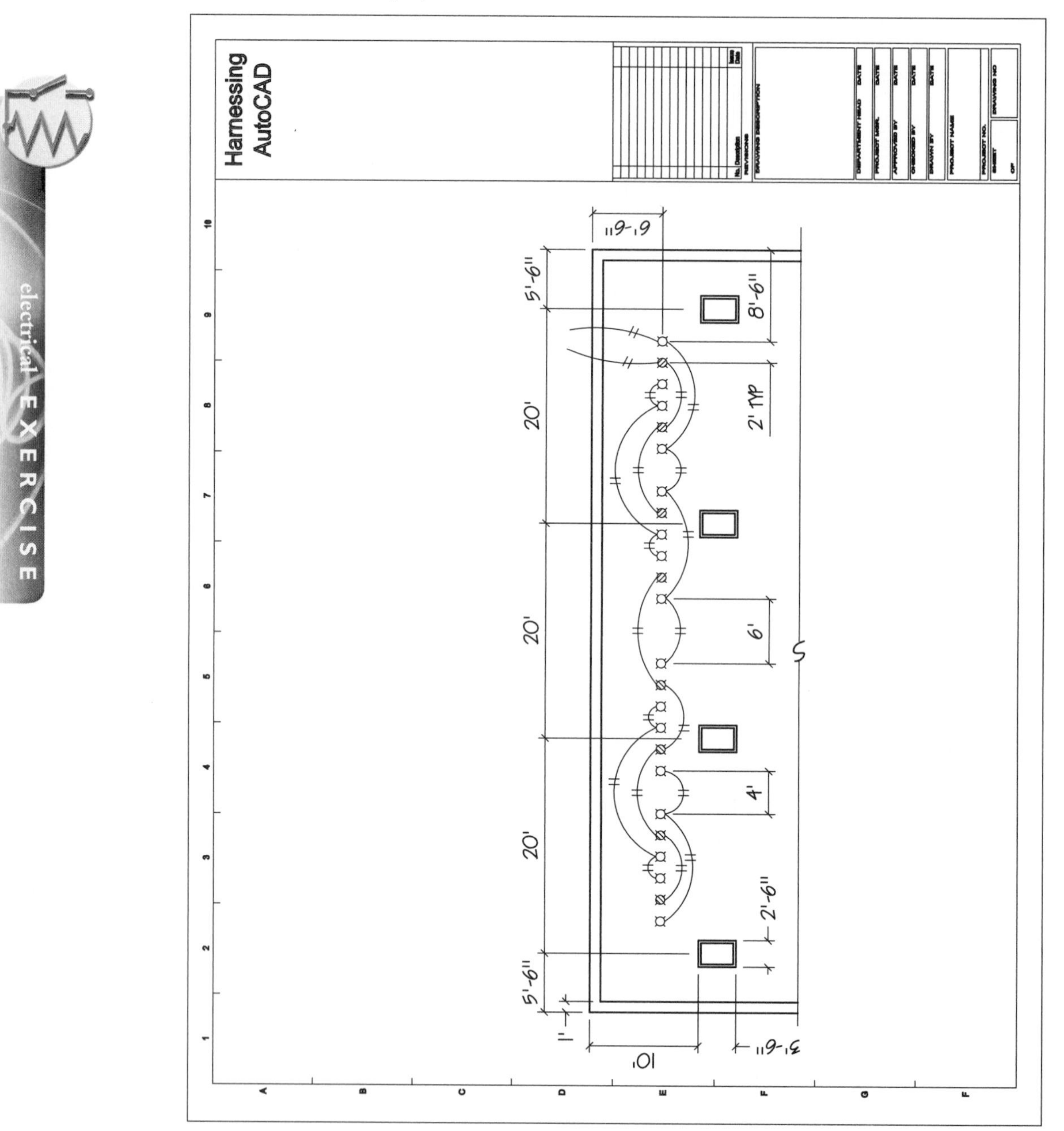


**Figure Ex7–15** *Exterior Ornamental Lighting*

## EXERCISE 7–16

Open the Lighting Switch Circuit and Wiring Diagram drawing that was completed in Exercise 4–19 and add the dimensions shown in Figure Ex7–16. (Make sure to create a layer for dimensions and draw all dimensions on that layer).

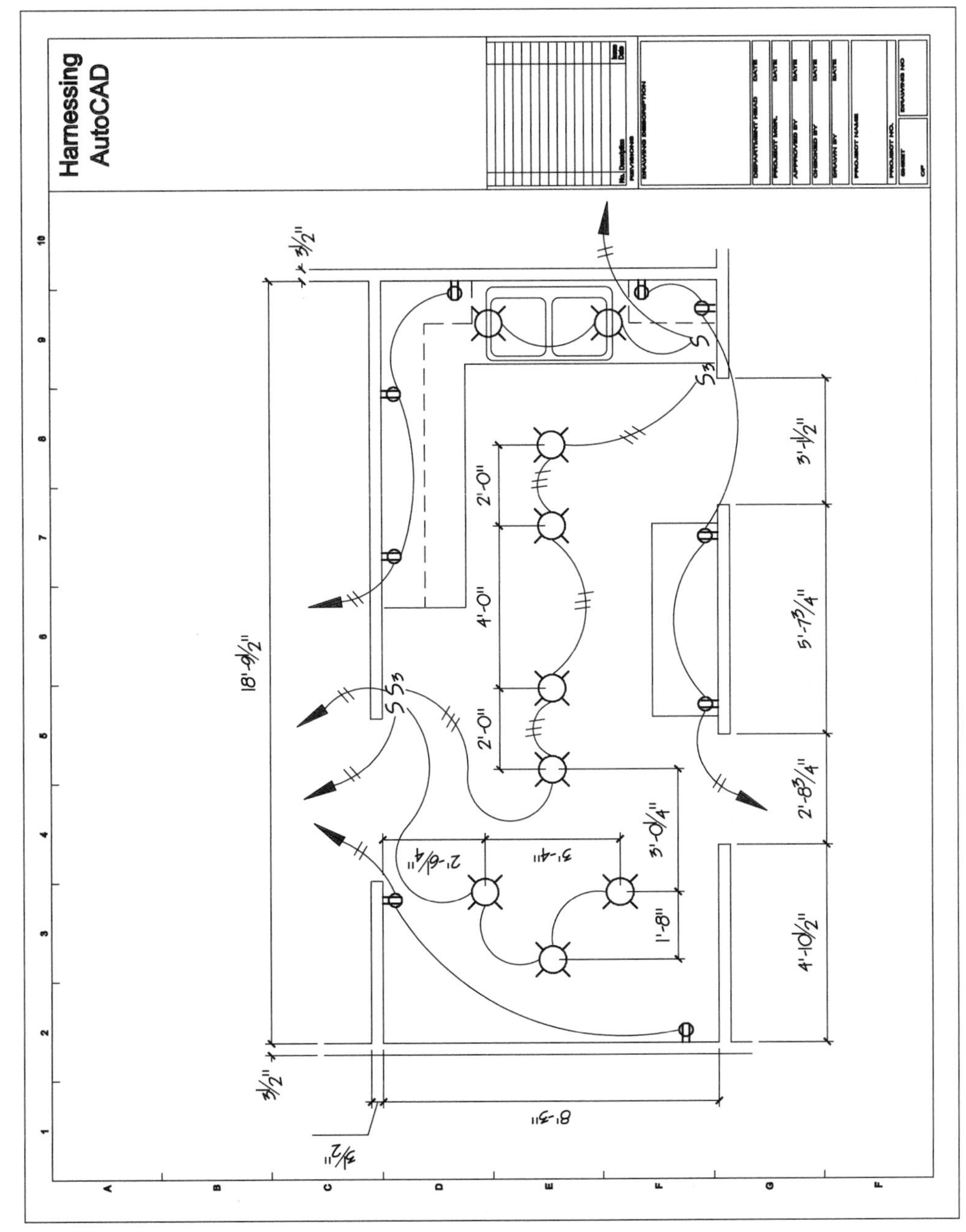


**Figure Ex7–16** *Lighting Switch Circuit and Wiring Diagram*

CHAPTER 8

# Plotting/Printing

Project EXERCISE

## PROJECT EXERCISES

### PROJECT 8–1

This project exercise provides point-by-point instructions for creating a plot style table, changing plot style for selected objects and plotting the drawing.

**Step 1:** Start the AutoCAD program.

**Step 2:** Open drawing PROJ2.DWG that was completed in the Chapter 2 project exercise.

**Step 3:** Activate Layout1 tab. By default, AutoCAD displays Page Setup dialog box similar to Figure P8–1.

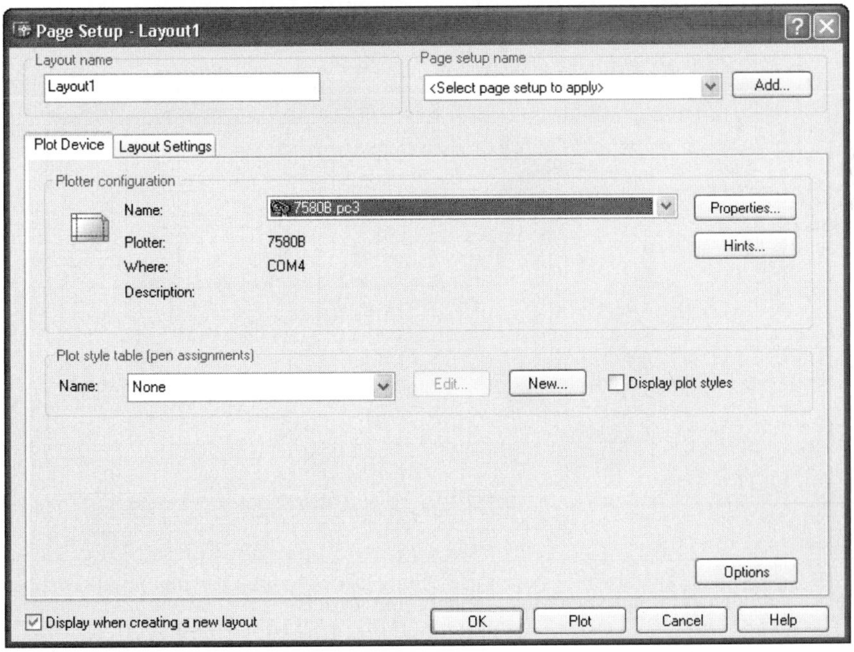

Page Setup - Layout1

Layout name: Layout1

Page setup name: <Select page setup to apply> — Add...

Plot Device | Layout Settings

Plotter configuration

- Name: 7580B.pc3 — Properties...
- Plotter: 7580B — Hints...
- Where: COM4
- Description:

Plot style table (pen assignments)

- Name: None — Edit... — New... — Display plot styles

Options

Display when creating a new layout — OK — Plot — Cancel — Help

**Figure P8–1** *Page Setup dialog box – Plot Device tab*

**Step 4:** From the Plot Device tab, select one of the available plotters in the **Plotter Configuration** section.

**Step 5:** Create a new plot style table by choosing the **New...** button from the Plot Style table section of the Page Setup dialog box. AutoCAD displays **Add Named Plot Style Table – Begin** wizard as shown in Figure P8–2.

Add Named Plot Style Table - Begin

Begin
Browse File
File Name
Finish

This wizard provides you with the ability to create a named plot style table from scratch, or by importing pen settings that were saved either in PCP, PC2, or Release 14 CFG files. Once you create a plot style table, you can use it in the current drawing or any other named drawings.

Start from scratch

Create a new plot style table from scratch.

Use a CFG file

Import R14 pen settings that were saved automatically to a CFG file.

Use a PCP or PC2 file

Import R14 pen settings that were saved in a PCP or PC2 file.

< Back | Next > | Cancel

**Figure P8–2** *Add Named Plot Style Table – Begin wizard*

Select the **Start from scratch** radio button to create a new plot style table. Choose the **Next** button. AutoCAD displays the **Add Named Plot Style Table – File Name** wizard as shown in Figure P8–3.

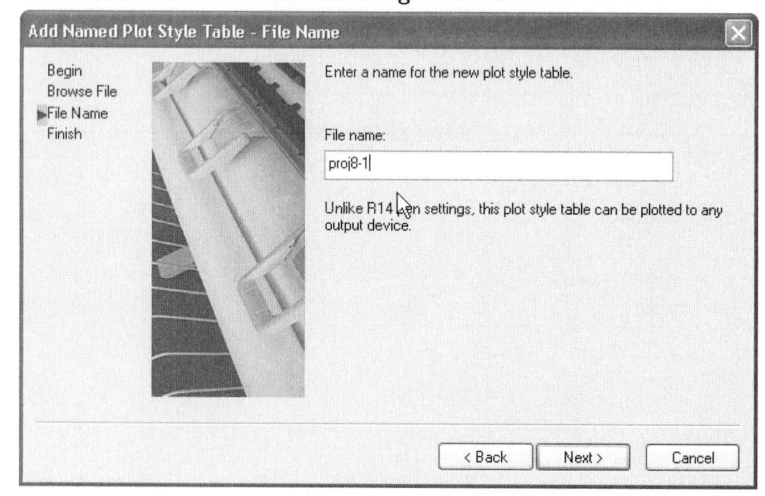

**Figure P8–3** *Add Named Plot Style Table – File Name wizard*

Enter **proj8-1** as the file name for the plot style table in the **File name:** edit field. Choose the **Next** button. AutoCAD displays **Add Named Plot Style Table – Finish** wizard as shown in Figure P8–4.

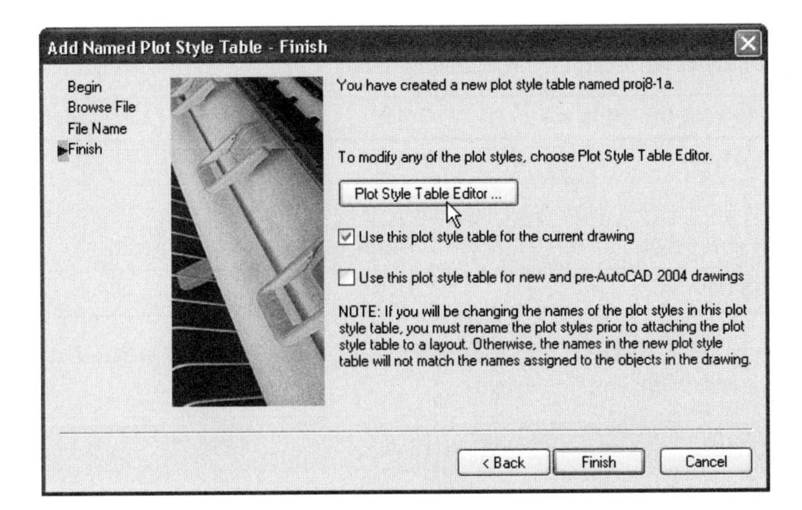


**Figure P8–4** *Add Named Plot Style Table – Finish wizard*

To create new plot styles, choose the **Plot Style Table Editor** button. AutoCAD displays the Plot Style Table Editor as shown in Figure P8–5.

**Note:** Assumption is made that the PROJ2.DWG was created with Use named plot styles as the default plot style behavior for new drawings.

Plot Style Table Editor - proj8-1.stb

General | Table View | Form View

| | |
|---|---|
| Name | Normal |
| Description | Description_1 |
| Color | Use object color |
| Enable dithering | ☑ |
| Convert to grayscale | ☐ |
| Use assigned pen # | Automatic |
| Virtual pen # | Automatic |
| Screening | 100 |
| Linetype | Use object linetype |
| Adaptive adjustment | ☑ |
| Lineweight | Use object lineweight |
| Line End Style | Use object end style |
| Line Join Style | Use object join style |
| Fill Style | Use object fill style |

Add Style | Delete Style | Edit Lineweights... | Save As...

Save & Close | Cancel | Help

**Figure P8–5** *Plot Style Table Editor*

Create four new styles, rename and assign color and lineweight for the newly created styles as listed in the following table:

| Name of the style table | Color | Lineweight |
|---|---|---|
| COIL | RED | 0.0059" |
| 90 DEG ELL | YELLOW | 0.0059" |
| DOOR JAMB | GREEN | 0.0059" |
| INST PANEL | CYAN | 0.0059" |

Figure P8–6 shows the Plot Style Table Editor with newly created styles with the assigned color and lineweight.

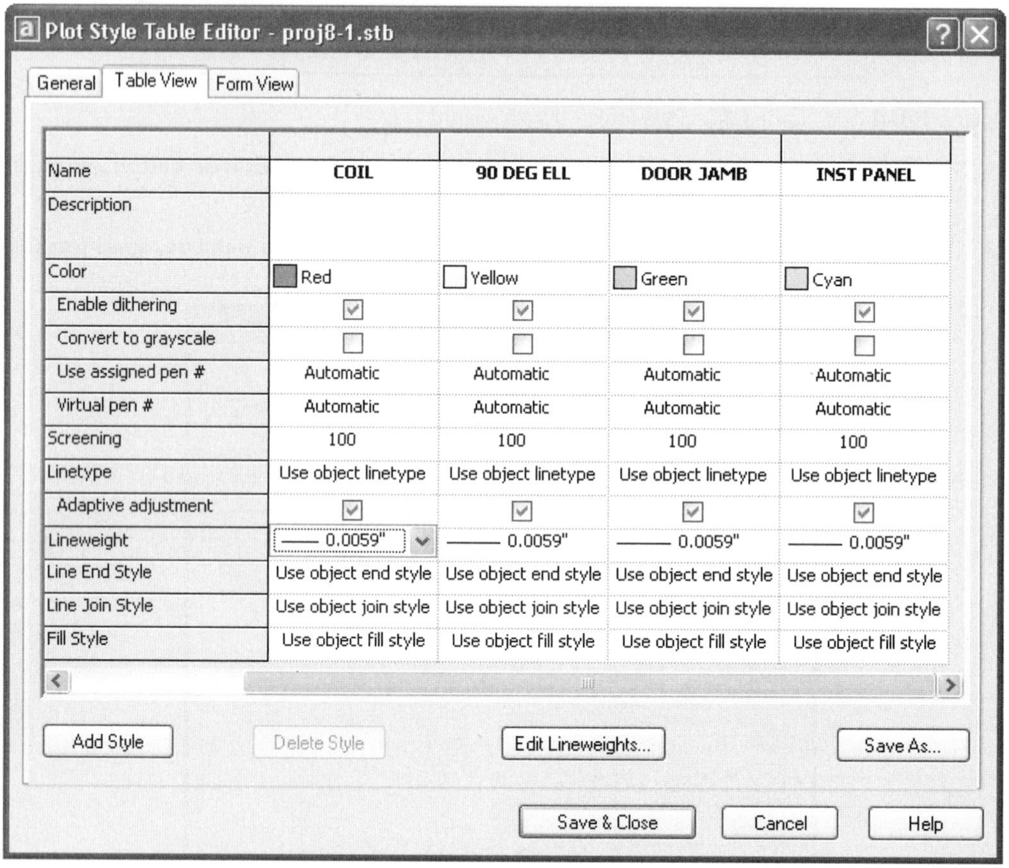

Plot Style Table Editor - proj8-1.stb

General | Table View | Form View

| | | | | |
|---|---|---|---|---|
| Name | COIL | 90 DEG ELL | DOOR JAMB | INST PANEL |
| Description | | | | |
| Color | Red | Yellow | Green | Cyan |
| Enable dithering | ☑ | ☑ | ☑ | ☑ |
| Convert to grayscale | ☐ | ☐ | ☐ | ☐ |
| Use assigned pen # | Automatic | Automatic | Automatic | Automatic |
| Virtual pen # | Automatic | Automatic | Automatic | Automatic |
| Screening | 100 | 100 | 100 | 100 |
| Linetype | Use object linetype | Use object linetype | Use object linetype | Use object linetype |
| Adaptive adjustment | ☑ | ☑ | ☑ | ☑ |
| Lineweight | 0.0059" | 0.0059" | 0.0059" | 0.0059" |
| Line End Style | Use object end style | Use object end style | Use object end style | Use object end style |
| Line Join Style | Use object join style | Use object join style | Use object join style | Use object join style |
| Fill Style | Use object fill style | Use object fill style | Use object fill style | Use object fill style |

Add Style | Delete Style | Edit Lineweights... | Save As...

Save & Close | Cancel | Help

**Figure P8–6** *Plot Style Table Editor with newly created plot styles.*

Choose the **Save & Close** button to save the newly created styles and close the Plot Style Table Editor. Choose the **Finish** button to close the Add Named Plot Style Table.

**Step 6:** Select the **Layout Settings** tab of the Page Setup dialog box and choose the paper size, drawing orientation, plot area and scale, plot offset, and plot options as shown in Figure P8–7.

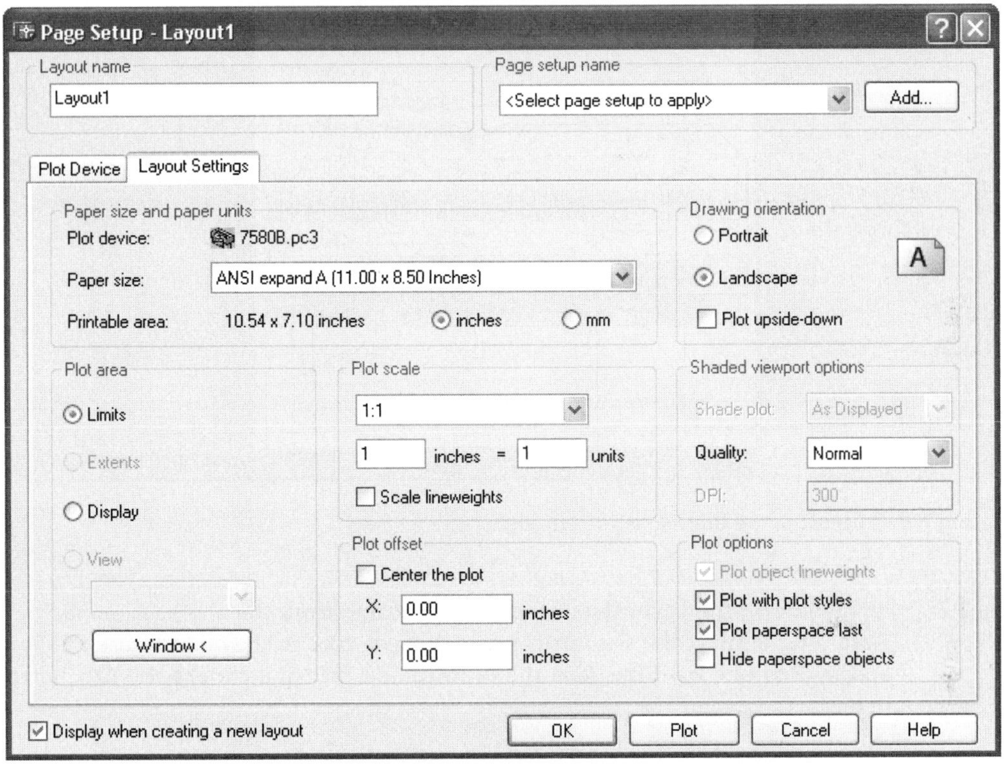

**Figure P8–7** *Page Setup dialog box – Layout Settings tab.*

Choose the **OK** button to close the Page Setup dialog box. AutoCAD displays the drawing in the Layout1 as shown in Figure P8–8.

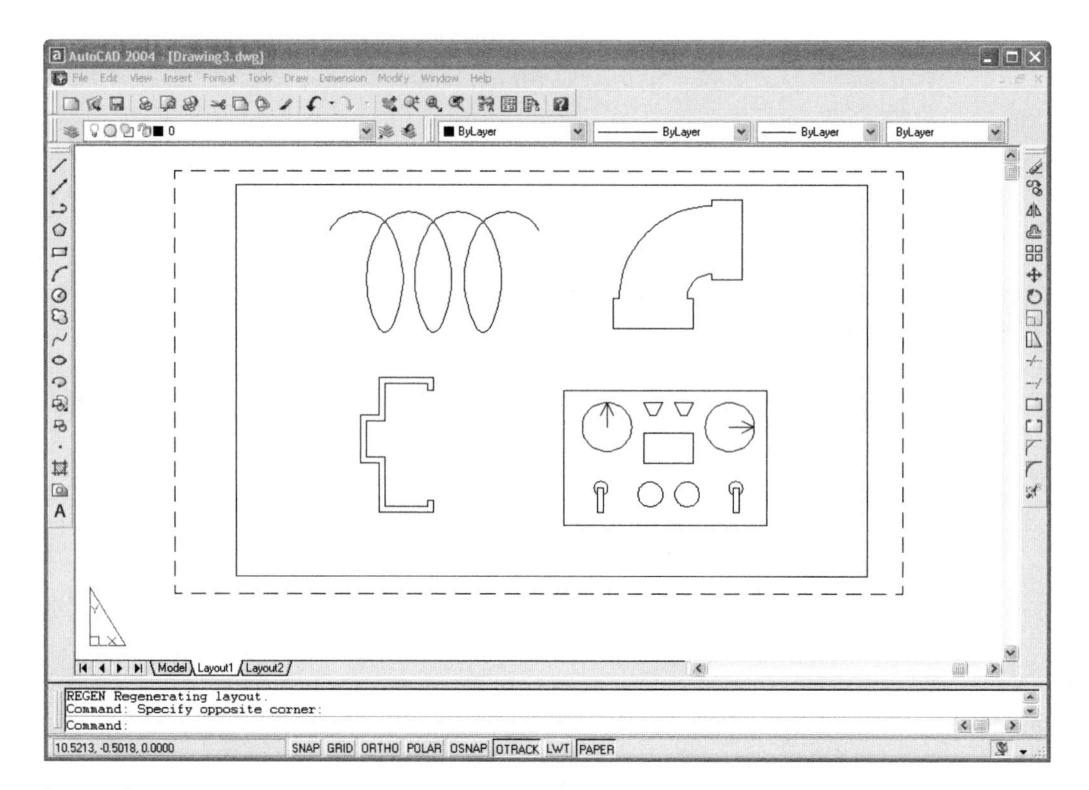


**Figure P8–8** *Layout of the drawing in the Layout1.*

**Step 7:** Right-click on the Layout1 tab and select Rename from the shortcut menu. AutoCAD displays the **Rename Layout** dialog box as shown in Figure P8–9. Specify **Project 2 – Size A** in the **Name:** edit field and choose the **OK** button.

Rename Layout
Name
Project 2 - Size A
OK
Cancel

**Figure P8–9** *Rename Layout dialog box.*

**Step 8:** Choose the **PAPER** button in the status bar to toggle to Model space.

**Step 9:** Select all the objects that comprise the Coil to change the plot style from the default plot style (ByLayer). Right-click and select Properties from the shortcut menu. AutoCAD displays Properties dialog box as shown in Figure P8–10.

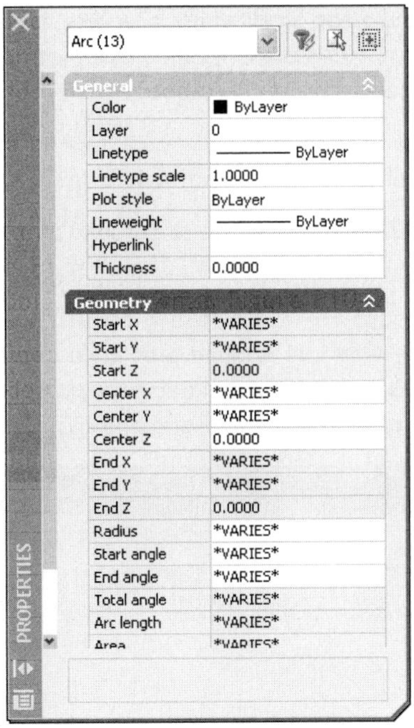

Arc (13)

General

| | |
|---|---|
| Color | ByLayer |
| Layer | 0 |
| Linetype | ByLayer |
| Linetype scale | 1.0000 |
| Plot style | ByLayer |
| Lineweight | ByLayer |
| Hyperlink | |
| Thickness | 0.0000 |

Geometry

| | |
|---|---|
| Start X | *VARIES* |
| Start Y | *VARIES* |
| Start Z | 0.0000 |
| Center X | *VARIES* |
| Center Y | *VARIES* |
| Center Z | 0.0000 |
| End X | *VARIES* |
| End Y | *VARIES* |
| End Z | 0.0000 |
| Radius | *VARIES* |
| Start angle | *VARIES* |
| End angle | *VARIES* |
| Total angle | *VARIES* |
| Arc length | *VARIES* |
| Area | *VARIES* |

PROPERTIES

**Figure P8–10** *Properties dialog box*

Choose **Other** from the list box in the Plot Style section of the Properties dialog box. AutoCAD displays the Select Plot Style dialog box as shown in Figure P8–11.

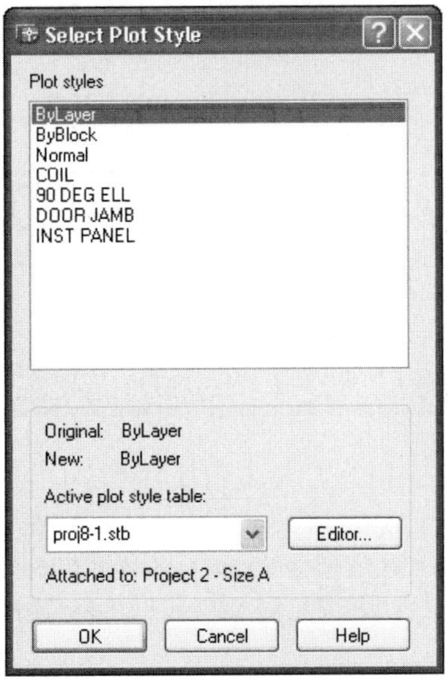

Select Plot Style

Plot styles

ByLayer
ByBlock
Normal
COIL
90 DEG ELL
DOOR JAMB
INST PANEL

Original: ByLayer
New: ByLayer
Active plot style table:
proj8-1.stb
Editor...
Attached to: Project 2 - Size A

OK Cancel Help

**Figure P8–11** *Select Plot Style dialog box*

Select **Coil** from the **Plot Styles:** list box and choose the **OK** button to close the dialog box. AutoCAD assigns Coil as the plot style for the selected objects. Clear the current selection of the objects.

Similarly, change the plot style for $90^0$ ell, Door Jamb and Instrument Panel from bylayer to 90 deg ell, Door Jamb and Inst Panel plot style respectively.

**Step 10:** Choose the **MODEL** button in the status bar to toggle to Paper space.

**Step 11:** Invoke the PLOT command from the File menu. AutoCAD displays the Plot dialog box as shown in Figure P8-12. Choose the **Full Preview** button. AutoCAD displays the drawing with the applied plot styles. Right-click and select **Exit** from the shortcut menu. If necessary, make appropriate changes in the Plot dialog box. Choose the **OK** button to plot the drawing from the selected Layout.

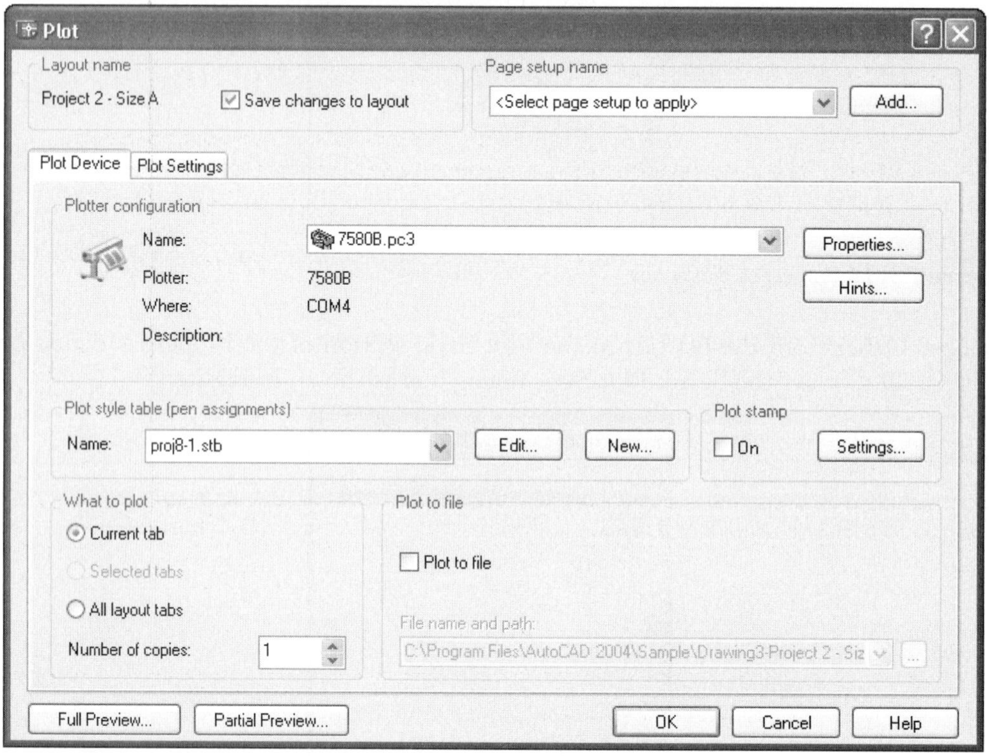

**Figure P8–12** *Plot dialog box*

AutoCAD starts plotting and reports its progress as it converts the drawing into the plotter's graphics language by displaying the number of vectors processed.

## PROJECT 8–2

This project exercise provides point-by-point instructions for creating a plot style table, changing plot style for selected layers and plotting the drawing.

**Step 1:** Start the AutoCAD program.

**Step 2:** Open drawing PROJ6.DWG that was completed in the Chapter 6 project exercise.

**Step 3:** Activate Layout1 tab. By default, AutoCAD displays the Page Setup dialog box.

**Step 4:** From the Plot Device tab, select one of the available plotters that can plot on a Size D paper in the **Plotter configuration** section.

**Step 5:** Create a new plot style table by choosing the **New** button from the Plot Style table section of the Page Setup dialog box. AutoCAD displays the **Add Named Plot Style Table – Begin** wizard. Select the **Start from scratch** radio button to create a new plot style table. Choose the **Next** button. AutoCAD displays the **Add Named Plot Style Table – File Name** wizard. Specify **proj8–2** as the file name for the plot style table in the **File name:** edit field. Choose **Next** button. AutoCAD displays the **Add Named Plot Style Table – Finish** wizard.

To create new plot styles, choose the **Plot Style Table Editor** button. AutoCAD displays the Plot Style Table Editor.

**Note:** Assumption is made that the PROJ6.DWG was created with Use named plot styles as the default plot style behavior for new drawings.

Create four new styles, rename and assign color and lineweight for the newly created styles as listed in the following table:

| Name of the style table | Color | Lineweight |
|---|---|---|
| BORDER | BLUE | 0.012" |
| OBJECT | YELLOW | 0.0059" |
| HIDDEN | GREEN | 0.0059" |
| TEXT | CYAN | 0.0059" |

Figure P8–13 shows the Plot Style Table Editor with newly created styles with the assigned color and lineweight.

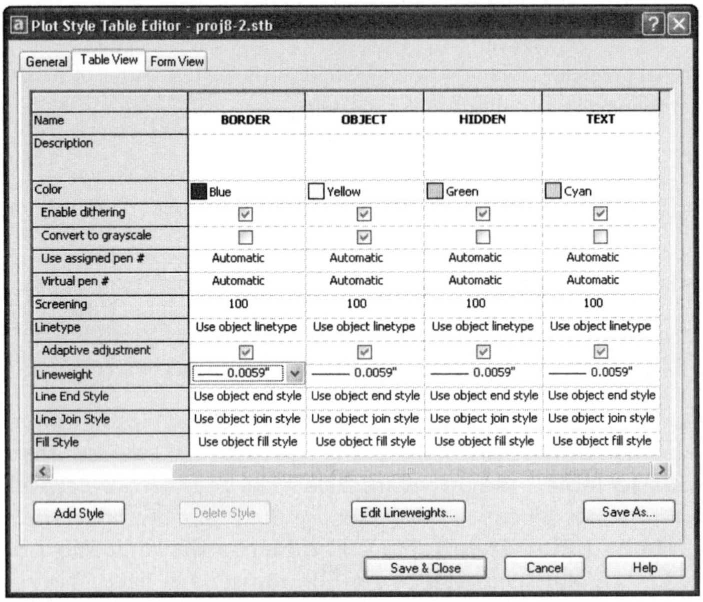

| Name | BORDER | OBJECT | HIDDEN | TEXT |
|---|---|---|---|---|
| Description | | | | |
| Color | Blue | Yellow | Green | Cyan |
| Enable dithering | ☑ | ☑ | ☑ | ☑ |
| Convert to grayscale | ☐ | ☑ | ☐ | ☐ |
| Use assigned pen # | Automatic | Automatic | Automatic | Automatic |
| Virtual pen # | Automatic | Automatic | Automatic | Automatic |
| Screening | 100 | 100 | 100 | 100 |
| Linetype | Use object linetype | Use object linetype | Use object linetype | Use object linetype |
| Adaptive adjustment | ☑ | ☑ | ☑ | ☑ |
| Lineweight | 0.0059" | 0.0059" | 0.0059" | 0.0059" |
| Line End Style | Use object end style | Use object end style | Use object end style | Use object end style |
| Line Join Style | Use object join style | Use object join style | Use object join style | Use object join style |
| Fill Style | Use object fill style | Use object fill style | Use object fill style | Use object fill style |

**Figure P8–13** *Plot Style Table Editor with newly created plot styles.*

Choose the **Save & Close** button to save the newly created styles and close the Plot Style Table Editor. Choose the **Finish** button to close the Add Named Plot Style table.

**Step 6:** Select the **Layout Settings** tab of the Page Setup dialog box and choose paper size, drawing orientation, plot area and scale, plot offset, and plot options as shown in Figure P8–14.

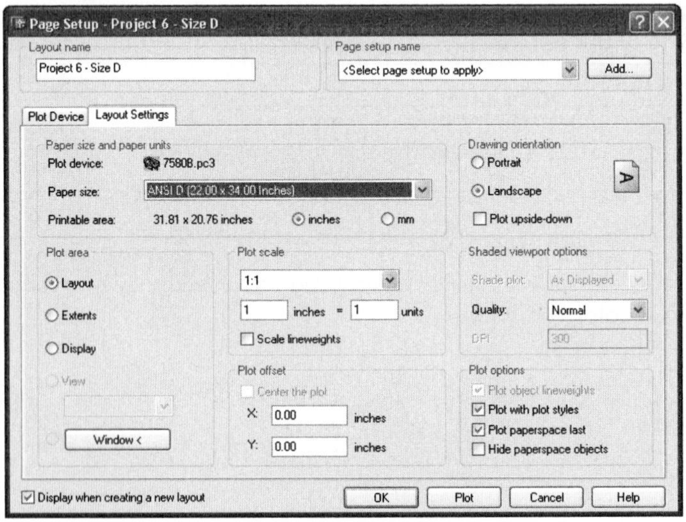

**Figure P8–14** *Page Setup dialog box – Layout Settings tab.*

Choose the **OK** button to close the Page Setup dialog box. AutoCAD by default creates a viewport and displays the drawing in the Layout1.

**Step 7:** Right-click on the Layout1 tab and select Rename from the shortcut menu. AutoCAD displays the **Rename Layout** dialog box. Specify **Project 6 – Size D** in the **Name:** text field and choose the **OK** button.

**Step 8:** Invoke the LAYER command. AutoCAD displays the Layer Properties Manager dialog box. Create a new layer and rename the layer to Viewports and assign color yellow. Set Viewports as the current layer. Change the plot style from the default to the one shown Figure P8–15 for Border, Object, Hidden, and Text layers.

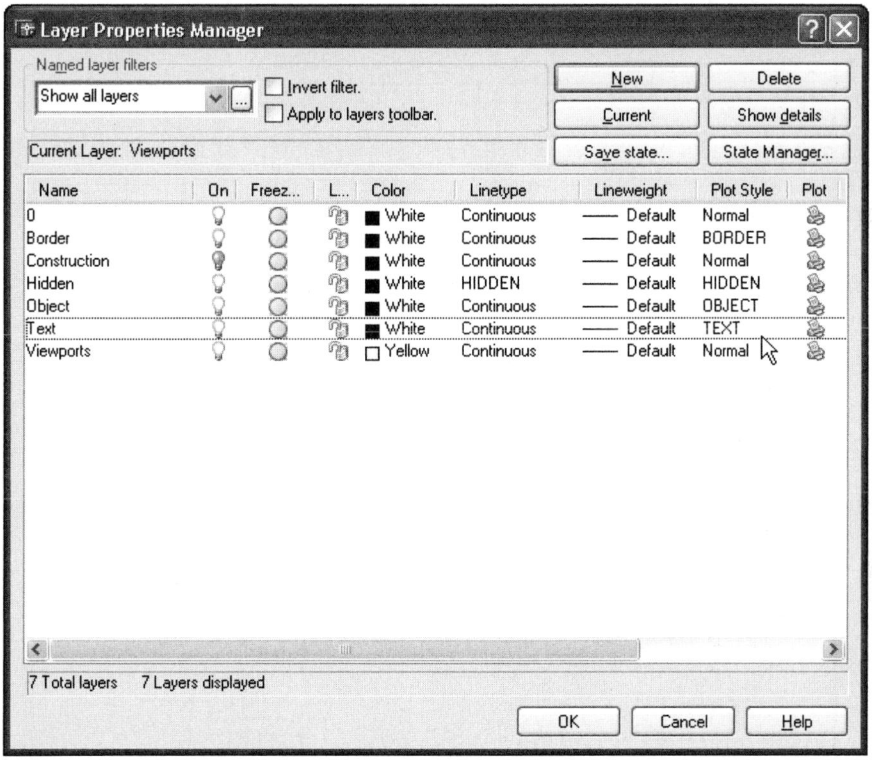

| Name | Color | Linetype | Lineweight | Plot Style |
|---|---|---|---|---|
| 0 | White | Continuous | Default | Normal |
| Border | White | Continuous | Default | BORDER |
| Construction | White | Continuous | Default | Normal |
| Hidden | White | HIDDEN | Default | HIDDEN |
| Object | White | Continuous | Default | OBJECT |
| Text | White | Continuous | Default | TEXT |
| Viewports | Yellow | Continuous | Default | Normal |

**Figure P8–15** *Layer Properties Manager dialog box.*

Choose the OK button to close the Layer Properties Manager dialog box.

**Step 9:** Invoke the ERASE command and delete the existing viewport. Invoke the -VPORTS command and create two new viewports. The first one to a size of 20 x 20 and the second to a size of 8 x 15 as shown in Figure P8–16.

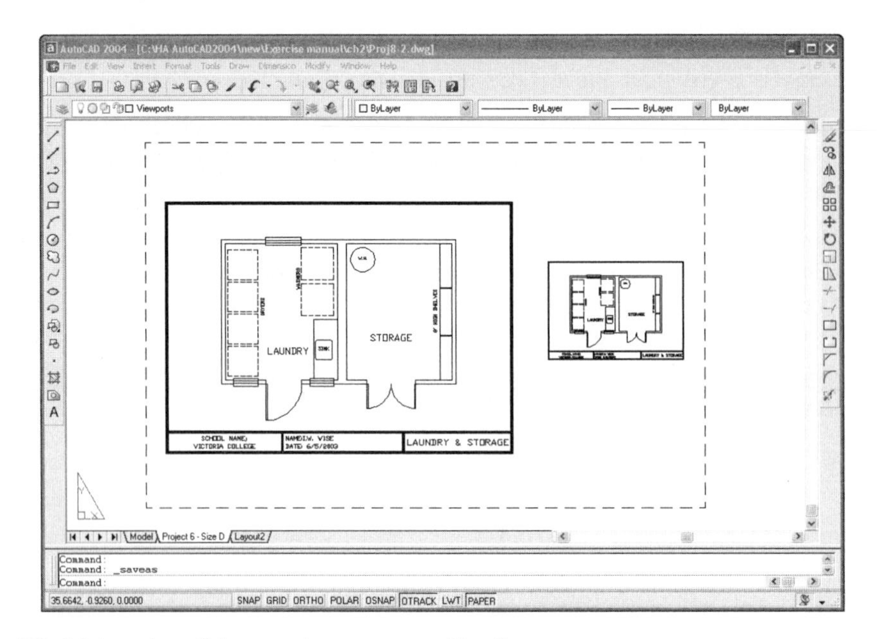


**Figure P8–16** *Location of the two viewports on a Size D paper.*

**Step 10:** Choose the **PAPER** button in the status bar to toggle to Model space.

**Step 11:** Activate the left viewport. Open the Viewports tool bar. To scale the drawing inside the floating viewport, select ½"=1' from Viewport Scale Control as shown in Figure P8–17.

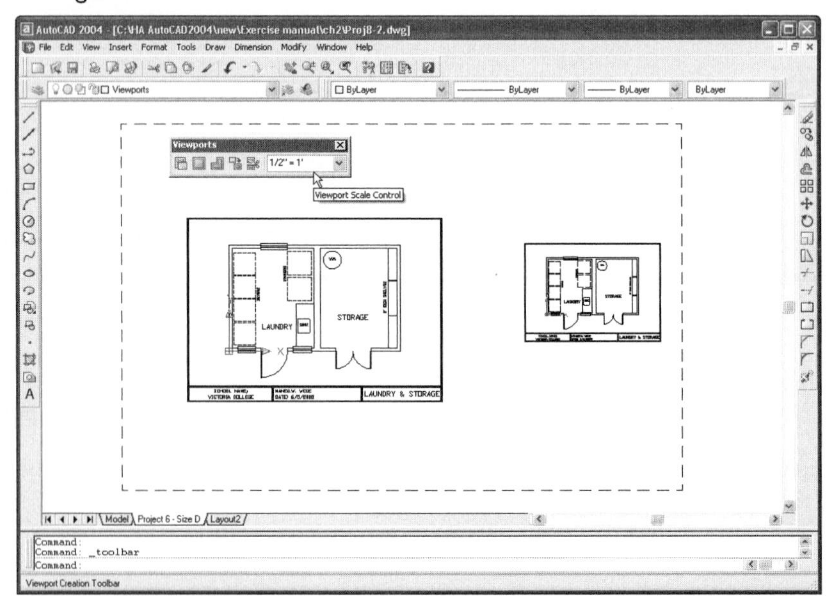


**Figure P8–17** *Viewports tool bar*

Activate the right viewport and set the Viewport Scale Control to 1:16. Pan the drawing in the viewport so that it displays only the Laundry floor plan as shown in Figure P8–18.

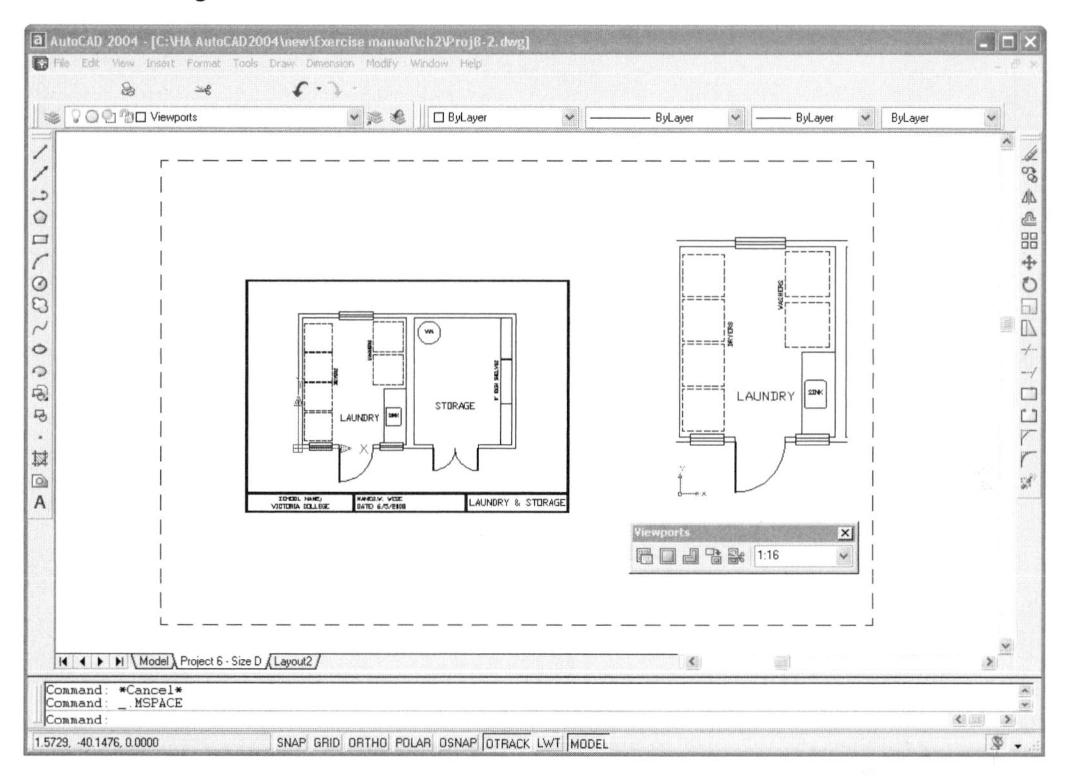


**Figure P8–18** *Display of the drawing in the viewports*

**Step 12:** Choose the **MODEL** button in the status bar to toggle to Paper space. Set Layer 0 as the current layer and turn OFF the Viewports layer. The drawing will be displayed as shown in Figure P8–19.

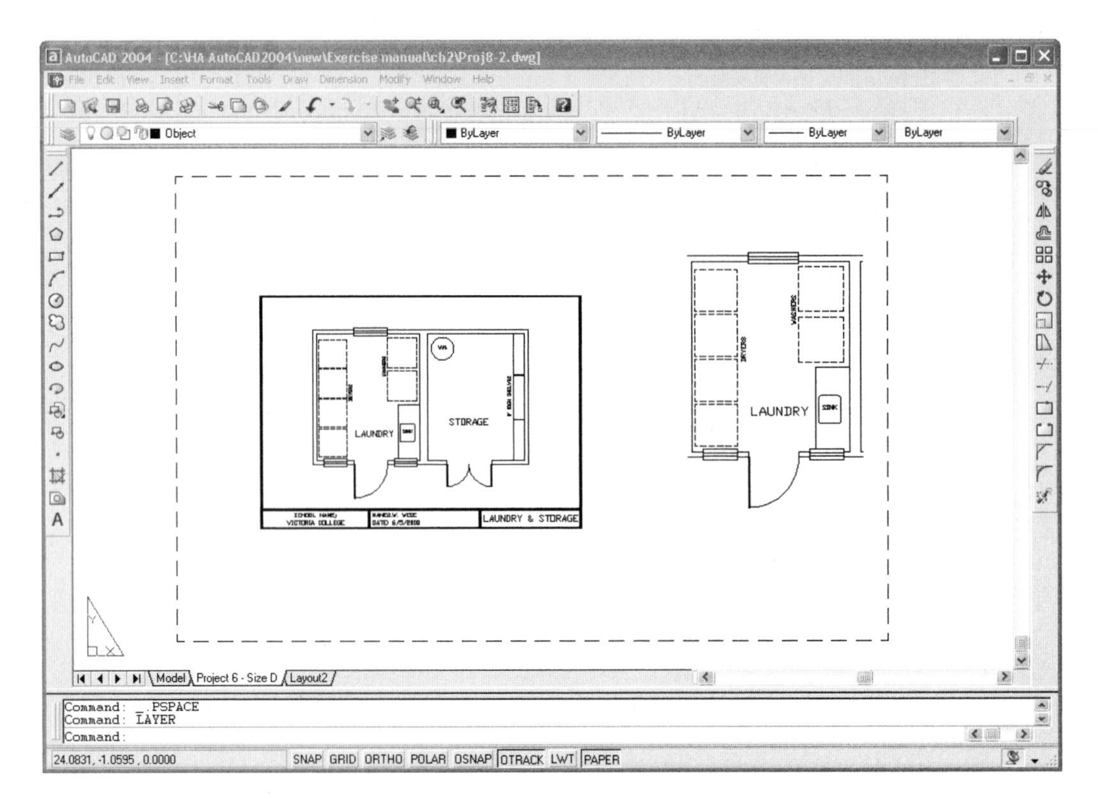

**Figure P8–19** *Display of the drawing in the viewports with the Viewports layer set to OFF.*

**Step 13:** Invoke the PLOT command from the File menu. AutoCAD displays the Plot dialog box. Choose the **Full Preview** button. AutoCAD displays the drawing with the applied plot styles. Right-click and select **Exit** from the shortcut menu. If necessary, make appropriate changes in the Plot dialog box. Choose the **OK** button to plot the drawing from the selected Layout.

AutoCAD starts plotting and reports its progress as it converts the drawing into the plotter's graphics language by displaying the number of vectors processed.

## EXERCISE 8–1

Open EX7–4 drawing that was completed in Chapter 7. Plot the drawing on a size C paper to a scale of 1:2.

## EXERCISE 8–2

Open EX7–5 drawing that was completed in Chapter 7. Select appropriate paper size to plot the drawing to a scale of 1:4.

## EXERCISE 8–3

Open EX7–8 drawing that was completed in Chapter 7. Create a layout to plot the drawing on a size C paper to a scale of 1"=1'0". Add the title block and plot the drawing.

## EXERCISE 8–4

Open EX7–9 drawing that was completed in Chapter 7. Select appropriate paper size to plot the drawing to a scale of ½"=1'0"

## EXERCISE 8–5

Open EX7–13 drawing that was completed in Chapter 7. Select appropriate paper size to plot the drawing to a scale of ½"=1'0".

## EXERCISE 8–6

Open EX7–17 drawing that was completed in Chapter 7. Plot the drawing on size B to Scaled to Fit.

## EXERCISE 8–7

Open EX7–20 drawing that was completed in Chapter 7. Plot the drawing on size C to Scaled to Fit.

## EXERCISE 8–8

Open EX7–20 drawing that was completed in Chapter 7. Plot the drawing on size A to Scaled to Fit.

# Hatching and Boundaries

## PROJECT EXERCISE

This project exercise provides point-by-point instructions for the objects shown in Figure P9–1. In this exercise you will apply the skills acquired in Chapters 1 through 9.

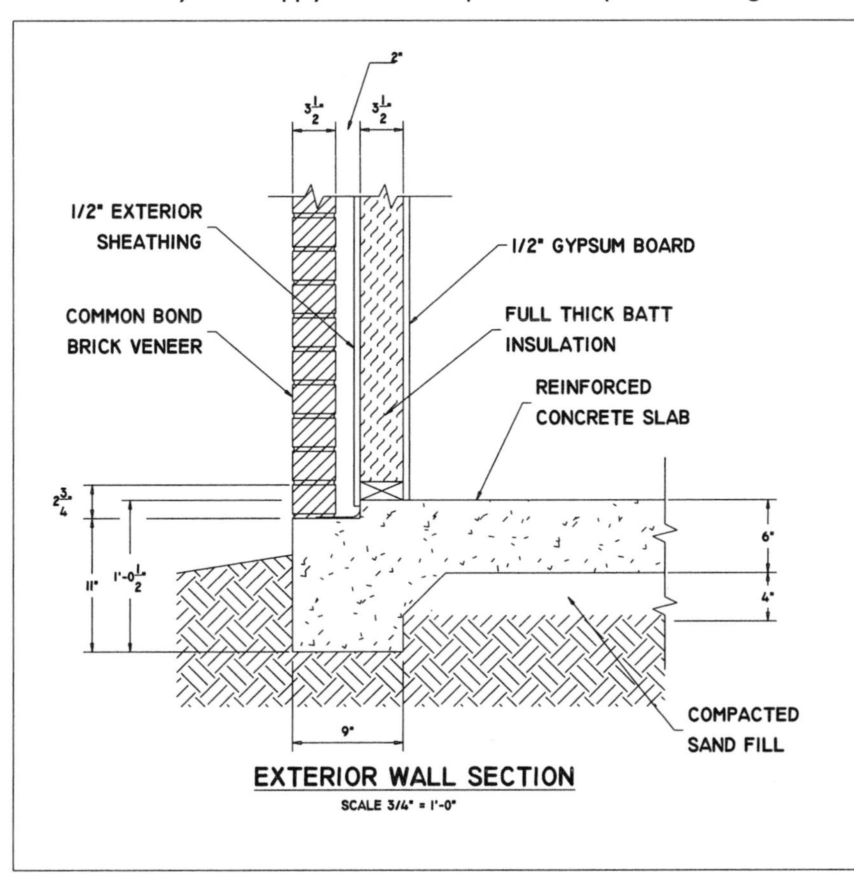


**Figure P9–1** *Completed project drawing*

In this project you will do the following:

- Set up the drawing, including Limits, Units, and Layers.
- Use the LINE, PLINE, and BHATCH commands to create objects.
- Use the ARRAY command to create objects from existing objects.
- Use the FILLET and TRIM commands to modify objects.

**SET UP THE DRAWING**

**Step 1:** Start the AutoCAD program.

**Step 2:** To create a new drawing, invoke the NEW command from the Standard toolbar or select New from the File menu.

AutoCAD displays the Create New Drawing dialog box. Select the Start from Scratch button, and then set the Units, Limits, Snap, and Grid to the values shown in the SETTINGS/VALUES table.

**Step 3:** Invoke the LAYER command from the Layer toolbar, or select Layer from the Format menu. AutoCAD displays the Layer Properties Manager dialog box. Create 10 layers, and rename them as shown in the table, and assign appropriate colors and linetypes.

| SETTINGS | VALUE | | |
|---|---|---|---|
| UNITS | Architectural | | |
| LIMITS | Lower left corner: -2'-0",-2'-6" | | |
| | Upper right corner: 4'-0",3'-6" | | |
| GRID | 1" | | |
| SNAP | 1/2" | | |
| LAYERS | NAME | COLOR | LINETYPE |
| | Border | Cyan | Continuous |
| | Slab | White | Continuous |
| | Text | Blue | Continuous |
| | Brick section | White | Continuous |
| | Concrete | White | Continuous |
| | Earth | White | Continuous |
| | Hatch boundary | Red | Continuous |
| | Insulation | Green | Continuous |
| | Sand | White | Continuous |
| | Wall | White | Continuous |

| HATCH PATTERNS | NAME | LAYER | SCALE | ANGLE |
|---|---|---|---|---|
| | AR-CONC | Concrete | 0'-0 1/2" | 0 |
| | FLEX | Insulation | 0'-3" | 45 |
| | ANSI31 | Brick section | 0'-6" | 0 |
| | EARTH | Earth | 0'-9" | 45 |
| | AR-SAND | Sand | 0'-0 1/2" | 0 |

## DRAWING THE CONCRETE SLAB, BRICK WALL, AND FRAME WALL

**Step 4:** Set Slab as the current layer. Invoke the PLINE command with a width of 0.125 to draw the concrete perimeter and a width of 0.0 to draw the "break" line, as shown in Figure P9–2. Figure P9–1 shows all the required dimensions to draw the layout. After you draw the layout, the drawing should look like Figure P9–2.

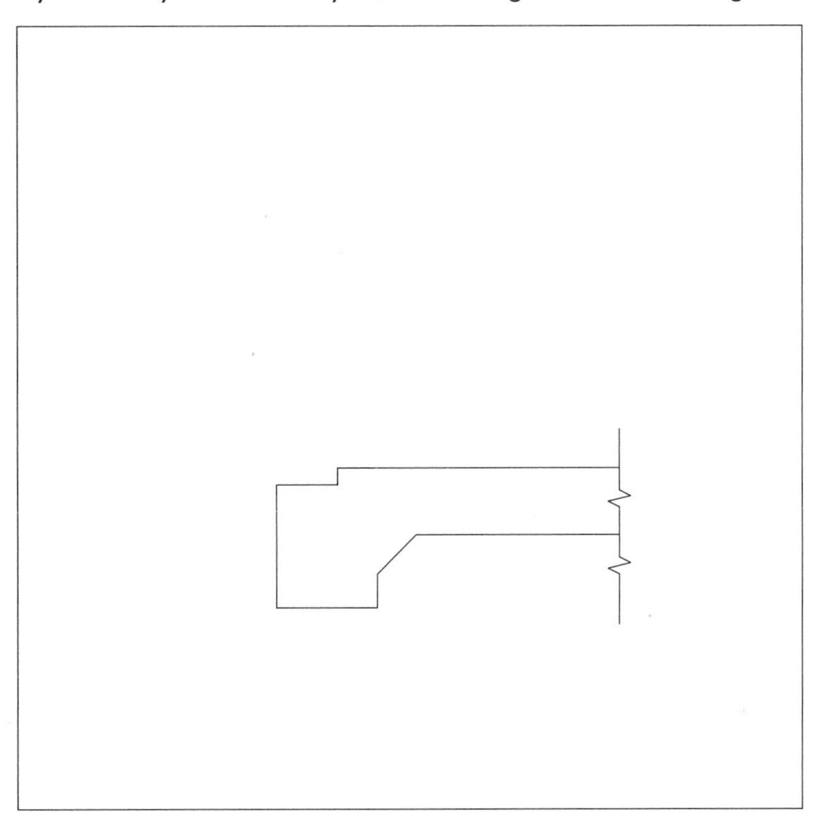

**Figure P9–2** *Layout of the drawing*

**Step 5:** Set Wall as the current layer. Invoke the PLINE command to draw the bricks. The bottom brick is 2-3/8" x 3-1/2" and is 3/8" above the brick ledge.

**Step 6:** In order to create the arc that represents the mortar joints, invoke the CIRCLE command with the Three point option. Select the corners of the concrete and the brick that the desired arc touches, and then drag the cursor to a point that gives you a radius slightly larger than the joint gap. Then, using the TRIM command with the concrete and the brick as the objects to trim to, remove the outer part of the circle, leaving the desired arc. Invoke the MIRROR command to create an arc on the opposite edge of the brick, using the Midpoint Osnap on the bottom line of the brick as the basepoint about which to mirror. The completed drawing should look like Figure P9–3.

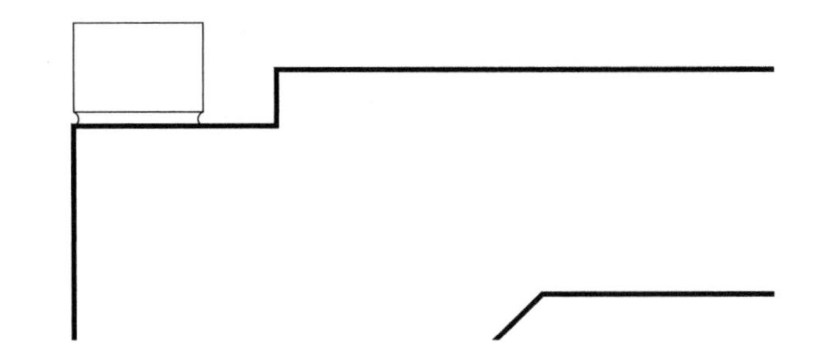

**Figure P9–3** *Layout of a brick to draw the wall*

**Step 7:** Invoke the ARRAY command to duplicate the bricks and mortar joints and create a stack of 10 bricks. Draw a "break" line through the top brick, and use the TRIM command to trim off any brick extending past the "break" line, as shown in Figure P9–4.

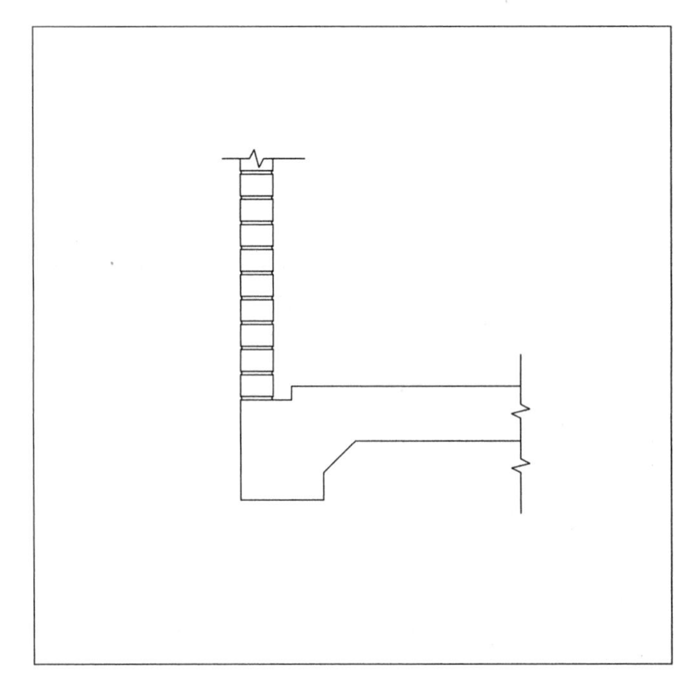

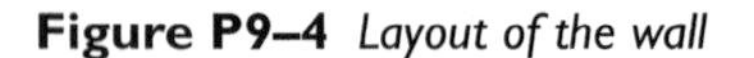

**Figure P9–4** *Layout of the wall*

**Step 8:** Invoke the LINE command to draw the 1/2" gypsum board and the 1/2" exterior sheathing comprising the frame wall. The 2 x 4 sole plate is 1-1/2" thick and is shown by crossed lines. The heavy flashing at the bottom can be drawn with the PLINE command, with 0.125 width; then use the FILLET command to round the corner, as shown in Figure P9–5.

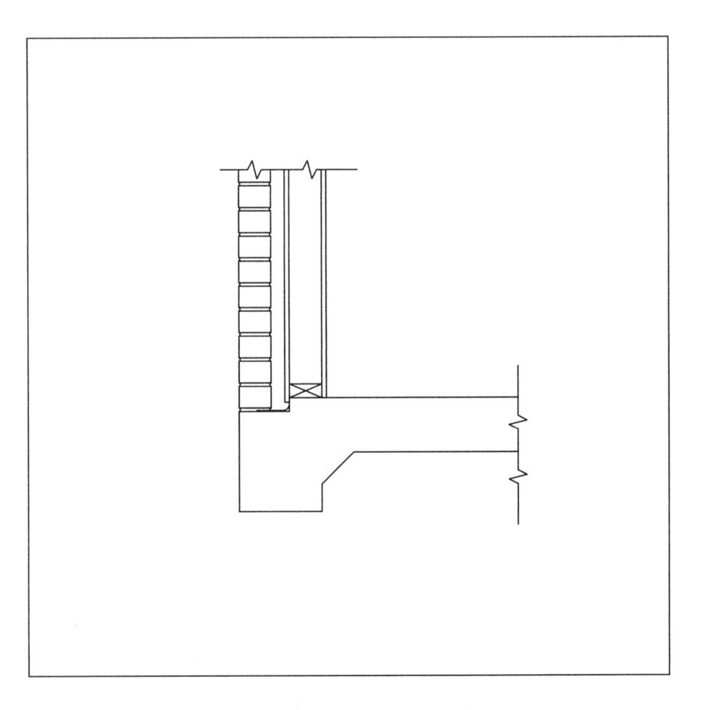

**Figure P9–5** *Layout of the gypsum board and exterior sheathing*

**Step 9:** Set Hatchboundary as the current layer. Draw copies of the two faces of the bottom brick and the two bottom arcs representing the mortar joint. Set the Wall layer to OFF. The layout should look like Figure P9–6.

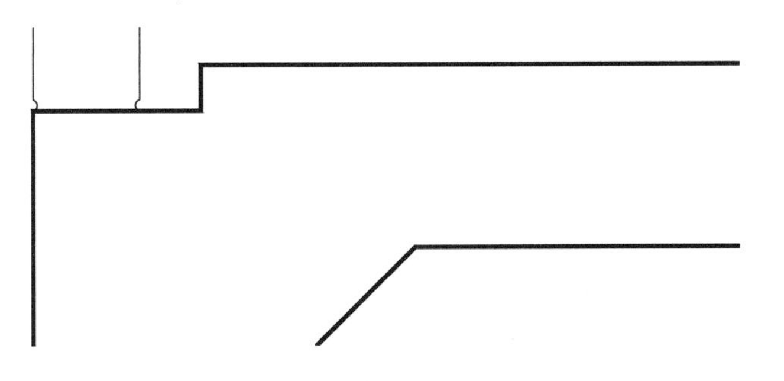

**Figure P9–6** *Layout of the brick and mortar joint*

**Step 10:** Invoke the ARRAY command to duplicate the faces of the bricks and mortar joints into a stack of 10 double faces and double arcs. Set the Wall layer to ON and, using the "break" line through the top brick, invoke the TRIM command to trim off the brick faces extending past the "break" line. Invoke the PLINE command and trace over the parts of the "break" line between the brick faces. Set the Wall layer to OFF. The layout should look like Figure P9–7.

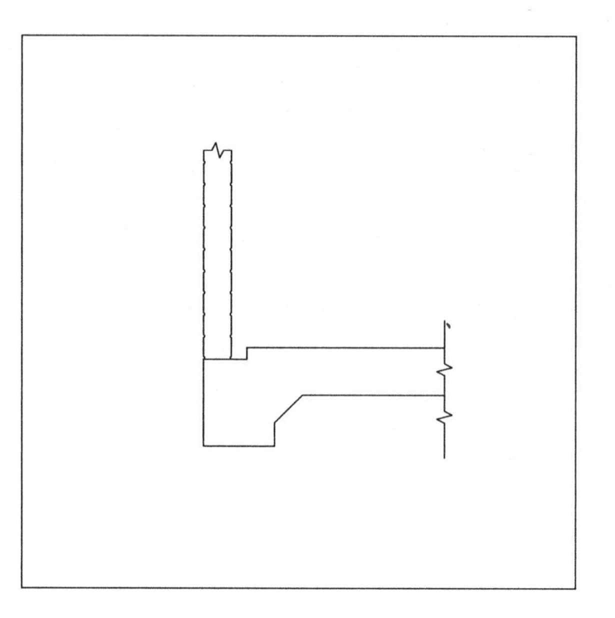

**Figure P9–7** *Layout of the brick wall with mortar joint*

**Step 11:** Set Bricksection as the current layer. Invoke the BHATCH command and, using the ANSI31 pattern at a scale of 6 and a hatch angle of zero (0), select a point in the boundary of brick faces and mortar arcs just created. Invoke the LINE command, and draw the necessary lines as shown in Figure P9–8 to complete the boundaries for the Sand and Earth hatch patterns.

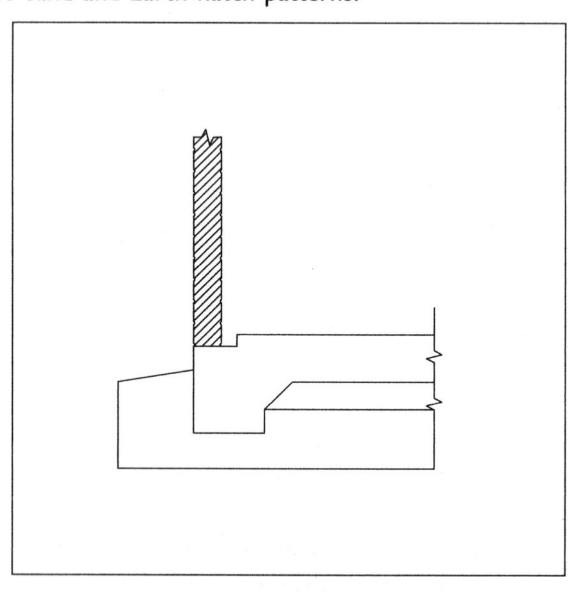

**Figure P9–8** *Hatch pattern for brick faces and mortar arcs*

**Step 12:** Set Concrete as the current layer. Invoke the BHATCH command and, using the AR-CONC pattern at a scale of 1/2 and a hatch angle of zero (0), select a point in the boundary representing the concrete slab.

**Step 13:** Set Sand as the current layer. Invoke the BHATCH command and, using the AR-SAND pattern at a scale of 1/2 and a hatch angle of zero (0), select a point in the boundary representing the sand beneath the concrete slab.

**Step 14:** Set Earth as the current layer. Invoke the BHATCH command and, using the EARTH pattern at a scale of 9 and a hatch angle of 45, select a point in the boundary representing the earth beneath the sand and concrete slab.

**Step 15:** Set Insulation as the current layer. Invoke the BHATCH command and, using the FLEX pattern at a scale of 3 and a hatch angle of 45, select a point in the boundary representing the frame wall between the exterior and interior sheathings. Set the Hatchboundary layer to OFF. The layout should look as shown in Figure P9–9.

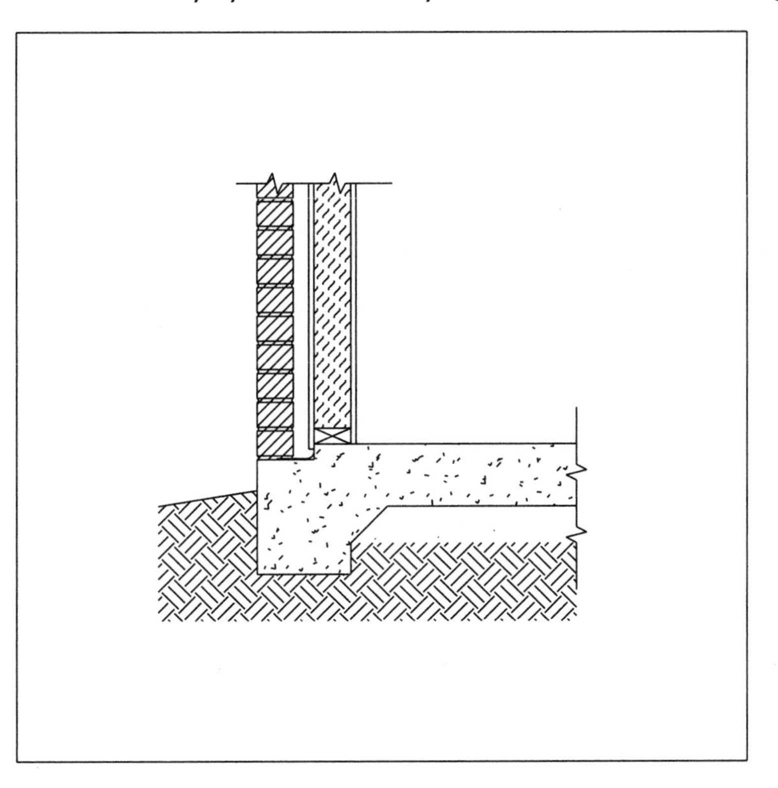

**Figure P9–9** *Layout of the exterior wall section*

**Step 16:** Set Text as the current layer. Invoke the LEADER command, and draw the text object with the leader lines as shown in Figure P9–10.

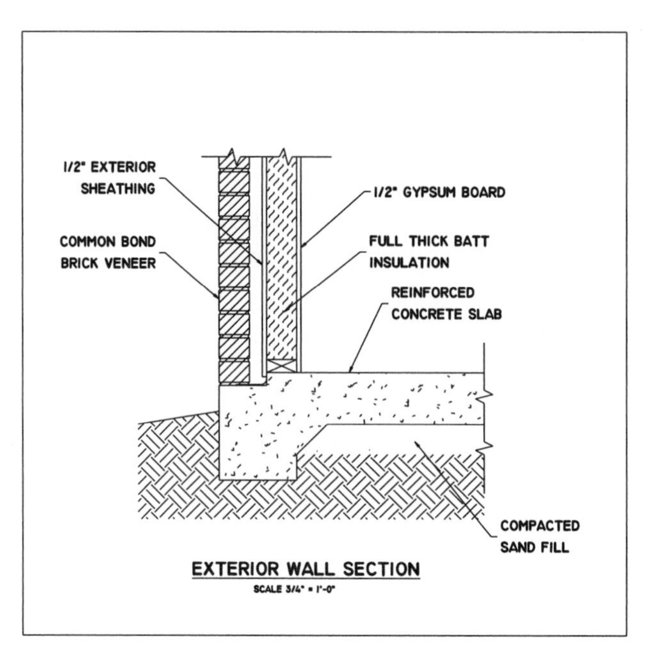


**Figure P9–10** *Layout of the exterior wall section with leader lines*

**Step 17:** Set Dimension as the current layer. Draw all the required dimensioning to complete the exterior wall section as shown in Figure P9–11.

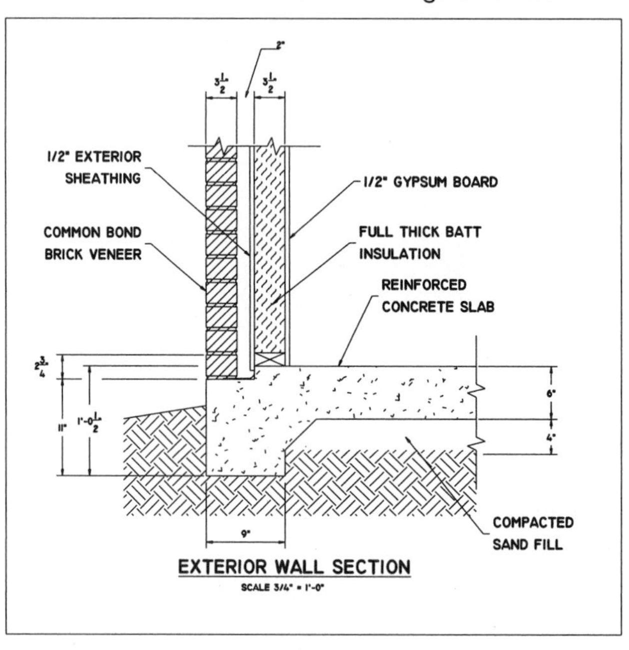


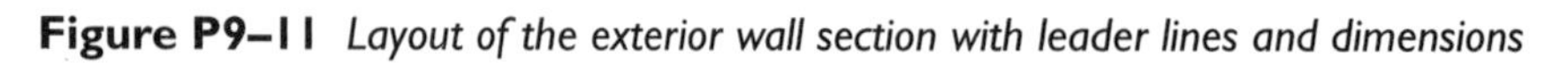

**Figure P9–11** *Layout of the exterior wall section with leader lines and dimensions*

## EXERCISE 9–1

Create the drawing shown in accordance with the settings given in the following table:

| Settings | Value | | | |
|---|---|---|---|---|
| 1. Units | Decimal | | | |
| 2. Limits | Lower left corner: 0,0 | | | |
| | Upper right corner: 12,9 | | | |
| 3. Grid | 0.5 | | | |
| 4. Snap | 0.25 | | | |
| 5. Layers | *NAME* | *COLOR* | *LINETYPE* | |
| | Construction | Cyan | Continuous | |
| | Object | White | Continuous | |
| | Center | Green | Center | |
| 6. Hatch Patterns | *NAME* | *LAYER* | *SCALE* | *ANGLE* |
| | ANSI31 | Hatch | 1.0 | 0 |
| | ANSI32 | Hatch | 1.0 | 90 |
| | ANSI33 | Hatch | 1.0 | 67.5 |

**Hints:** Use the PLINE command to draw half the object without the rounded corners. Grid lines are spaced at 0.5 in the figure.

Use the FILLET command to round the corners of the object.

Use the MIRROR command to complete the object.

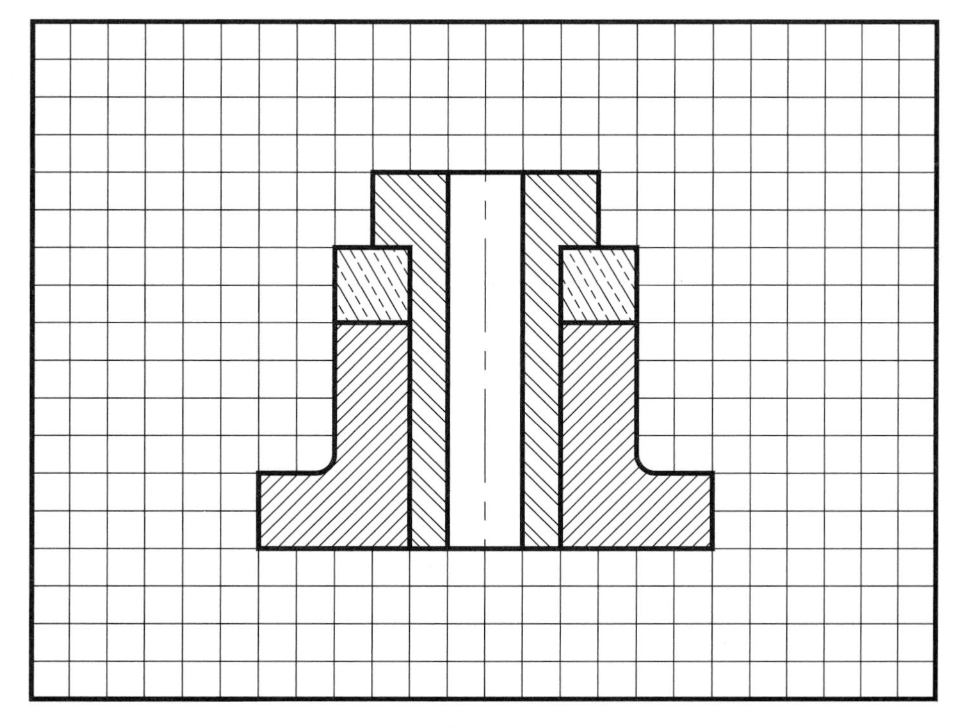

**Figure Ex 9–1** *Completed project drawing (with grid added for reference)*

# EXERCISE 9–2

Create the drawing shown in Figure Ex9–2, including dimensions, in accordance with the settings given in the following table:

| Settings | Value | | | |
|---|---|---|---|---|
| 1. Units | Architectural | | | |
| 2. Limits | Lower left corner: 0",0" | | | |
| | Upper right corner: 17",11" | | | |
| 3. Grid | 0.5" | | | |
| 4. Snap | 0.25" | | | |
| 5. Layers | *NAME* | *COLOR* | *LINETYPE* | |
| | Object | Green | Continuous | |
| | Center | Red | Center | |
| | Text | Blue | Continuous | |
| | Hatch | White | Continuous | |
| 6. Hatch Pattern | *NAME* | *LAYER* | *SCALE** | *ANGLE* |
| | ANSI31 | Hatch | 1.0 | 0 |
| | * Using a value of 1.0 adjusts the hatch pattern for a plotscale of 1=1 or Full scale plot | | | |

**Hint:** All three areas can be selected at once, thus creating a single hatch object.

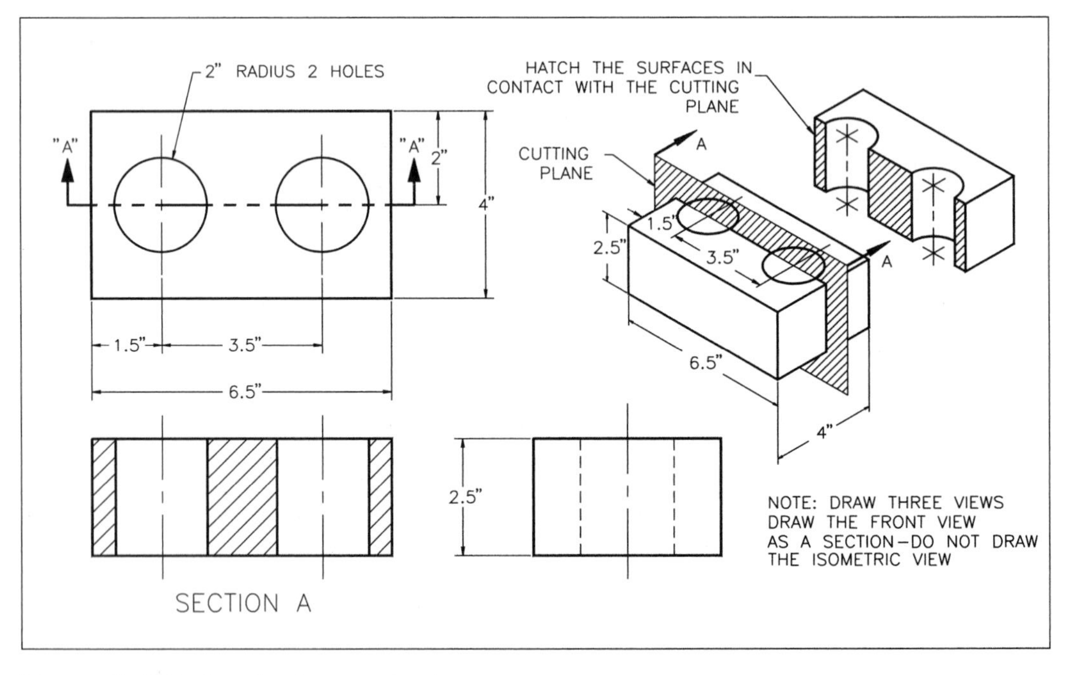

**Figure Ex9–2** *Completed project drawing*

## EXERCISE 9–3

Create the drawing shown in Figure Ex9–3, including dimensions, in accordance with the settings given in the following table:

| Settings | Value | | | |
|---|---|---|---|---|
| 1. Units | Decimal | | | |
| 2. Limits | Lower left corner: 0,0 | | | |
| | Upper right corner: 17,11 | | | |
| 3. Grid | 0.5 | | | |
| 4. Snap | 0.25 | | | |
| 5. Layers | *NAME* | *COLOR* | *LINETYPE* | |
| | Object | Green | Continuous | |
| | Center | Red | Center | |
| | Hidden | Magenta | Hidden | |
| | Text | Blue | Continuous | |
| | Hatch | White | Continuous | |
| 6. Hatch Pattern | *NAME* | *LAYER* | *SCALE** | *ANGLE* |
| | ANSI31 | Hatch | 1.0 | 0 |
| | * Using a value of 1.0 adjusts the hatch pattern for a plot scale of 1=1 or Full scale plot | | | |

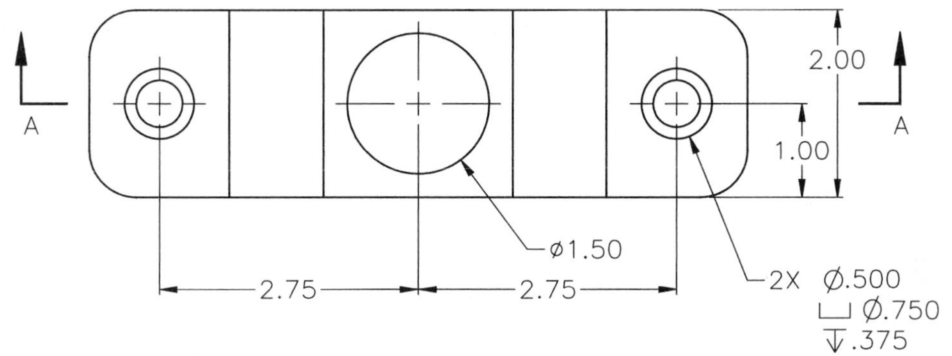


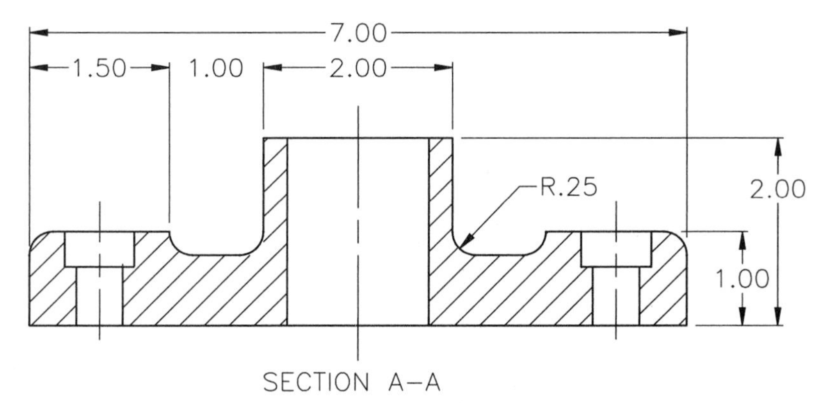


**Figure Ex9–3** *Completed drawing*

## EXERCISE 9–4

Create the drawing as shown in Figure Ex9–4, including dimensions, in accordance with the settings given in the following table:

| Settings | Value | | | |
|---|---|---|---|---|
| 1. Units | Decimal | | | |
| 2. Limits | Lower left corner: 0,0 | | | |
| | Upper right corner: 9,12 | | | |
| 3. Grid | 0.1 | | | |
| 4. Snap | 0.1 | | | |
| 5. Layers | *NAME* | *COLOR* | *LINETYPE* | |
| | Object | Green | Continuous | |
| | Center | Red | Center | |
| | Hidden | Magenta | Hidden | |
| | Text | Blue | Continuous | |
| | Hatch1 | White | Continuous | |
| | Hatch2 | White | Continuous | |
| | Knurl | White | Continuous | |
| 6. Hatch Patterns | *NAME* | *LAYER* | *SCALE** | *ANGLE* |
| | ANSI31 | Hatch1 | 1.0 | 0 |
| | ANSI31 | Hatch2 | 1.0 | 90 |
| | ANSI37 | Knurl | 1.0 | 15 |
| | * Using a value of 1.0 adjusts the hatch pattern for a plot scale of 1=1 or Full scale plot | | | |

**Hints:** The angle pattern areas can use the same pattern, ANSI31, but with the hatch angles 90 degrees from each other.

The knurled pattern needs added lines to close the boundary. These can be drawn on a separate layer. That layer can be set to OFF for viewing and plotting the drawing.

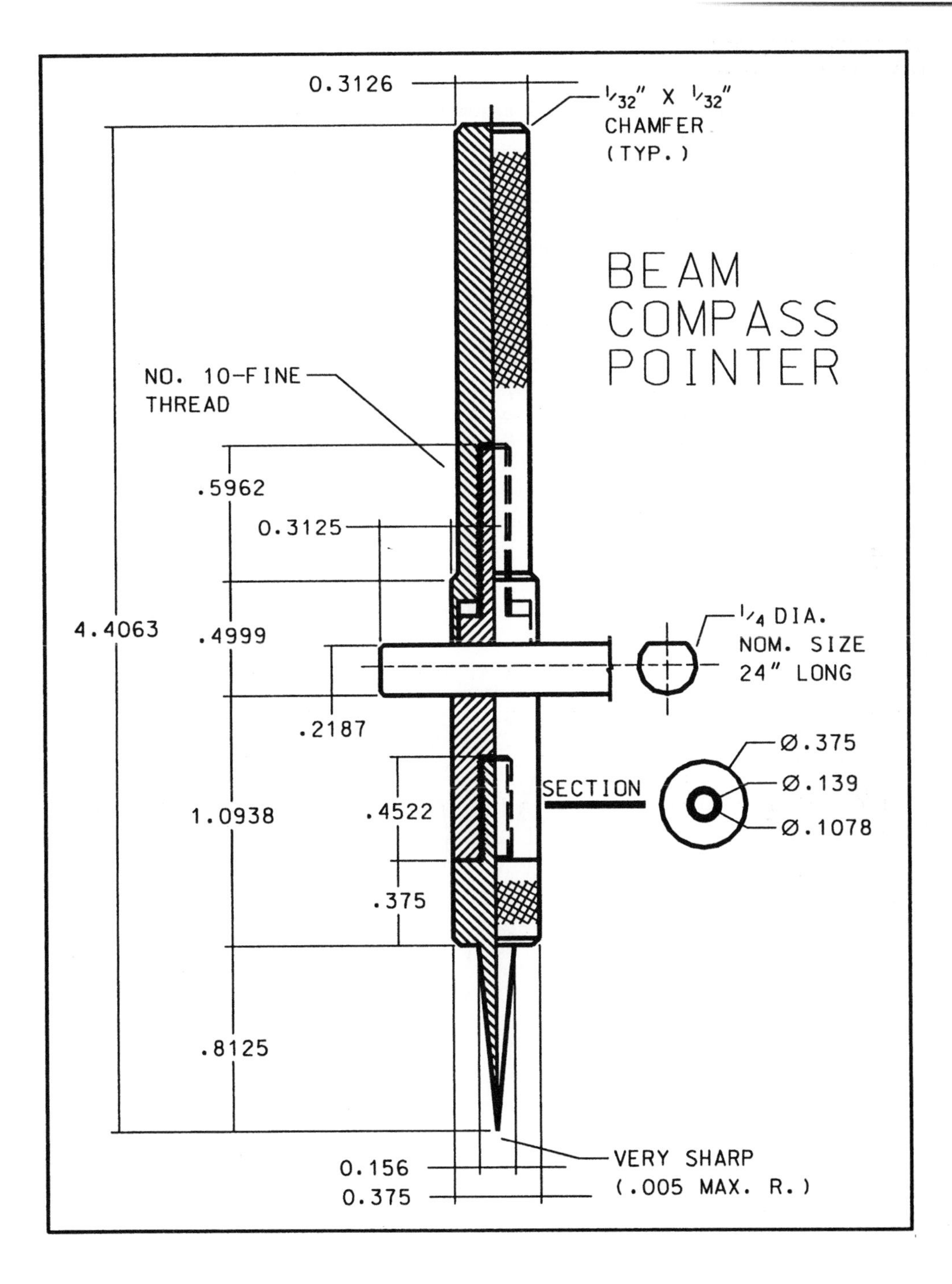


**Figure Ex9–4** *Completed drawing*

## EXERCISE 9–5

Create the drawing shown in Figure Ex9–5, including dimensions, in accordance with the settings given in the following table:

| Settings | Value | | | |
|---|---|---|---|---|
| 1. Units | Decimal | | | |
| 2. Limits | Lower left corner: 0,0 | | | |
| | Upper right corner: 17,11 | | | |
| 3. Grid | 0.5 | | | |
| 4. Snap | 0.25 | | | |
| 5. Layers | *NAME* | *COLOR* | *LINETYPE* | |
| | Object | Green | Continuous | |
| | Center | Red | Center | |
| | Hidden | Magenta | Hidden | |
| | Text | Blue | Continuous | |
| | Hatch1 | White | Continuous | |
| 6. Hatch Pattern | *NAME* | *LAYER* | *SCALE** | *ANGLE* |
| | ANSI31 | Hatch | 1.0 | 0 |
| | * Using a value of 1.0 adjusts the hatch pattern for a plot scale of 1=1 or Full scale plot | | | |

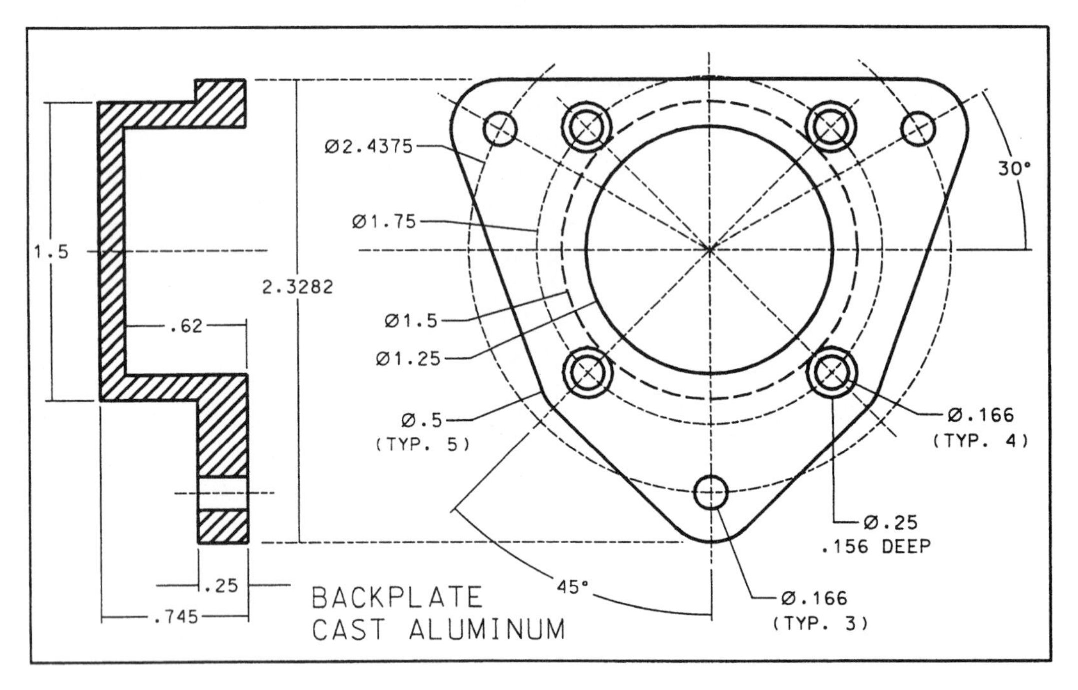


**Figure Ex9–5** *Completed drawing*

## EXERCISE 9–6

Create the drawing shown in Figure Ex9–6a, including dimensions, in accordance with the settings given in the following table:

| Settings | Value | | | |
|---|---|---|---|---|
| 1. Units | Architectural | | | |
| 2. Limits | Lower left corner: 0,0 | | | |
| | Upper right corner: 48',32' | | | |
| 3. Grid | 6" | | | |
| 4. Snap | 12" | | | |
| 5. Layers | *NAME* | *COLOR* | *LINETYPE* | |
| | Border | Cyan | Continuous | |
| | Object | White | Continuous | |
| | Text | Blue | Continuous | |
| | Brick | White | Continuous | |
| | Roof | White | Continuous | |
| 6. Hatch Patterns | *NAME* | *LAYER* | *SCALE** | *ANGLE* |
| | BRICK | Brick | 16" | 0 |
| | AR-RSHKE | Roof | 1/2" | 0 |
| | * Using a value of 16.0 adjusts the brick hatch pattern for a plotscale of 1=32 or 3/8"=1'-0" plot, yielding a 4" x 8" brick face | | | |

**Hint:** Using the dimensions in Figure Ex9–6b, draw the front elevation of the house. The main concern about being able to draw hatch patterns within the proper boundaries is to make sure the boundaries are closed. Overlapping at the intersections of the segments that make up the boundary is acceptable. But AutoCAD will not complete the hatch pattern if there is a gap between any of the boundary segments.

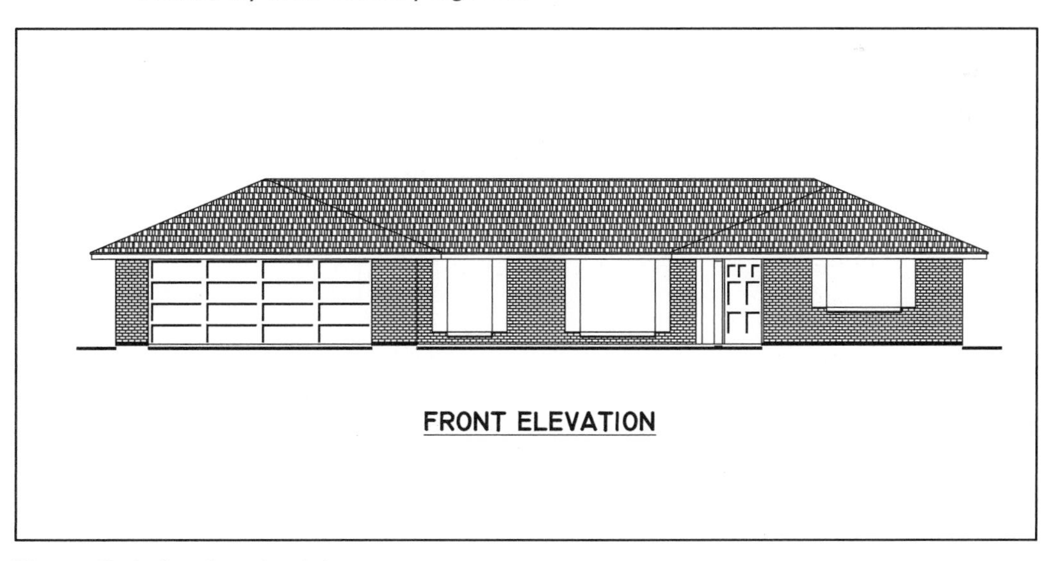


**Figure Ex9–6a** *Completed drawing*

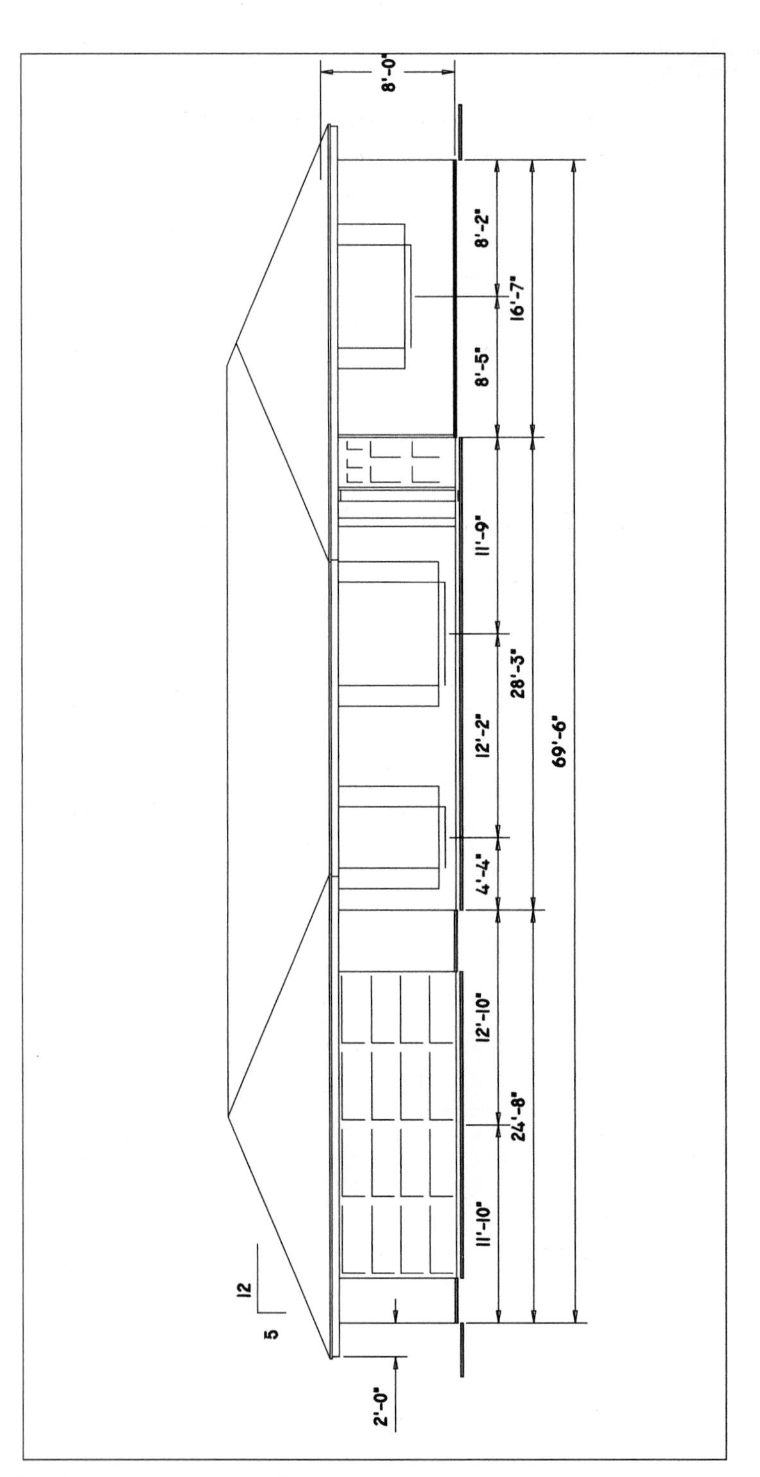


**Figure Ex9–6b** *Completed drawing with dimensions*

## EXERCISE 9–7

Open the masonry wall and footing section view drawing that was drawn in Exercise 2–19 and add the hatch patterns (see figure below) to the given specifications:

| Settings | Value | | | |
|---|---|---|---|---|
| 1. Hatch Pattern | NAME | LAYER | SCALE* | ANGLE |
| | AR-CONC | FtgHatch | 0'-0 1/2" | 0 |
| | ANS137 | CMUHatch | 0'-1" | 0 |
| | ANS137 | BbmHatch | 0'-1" | 0 |

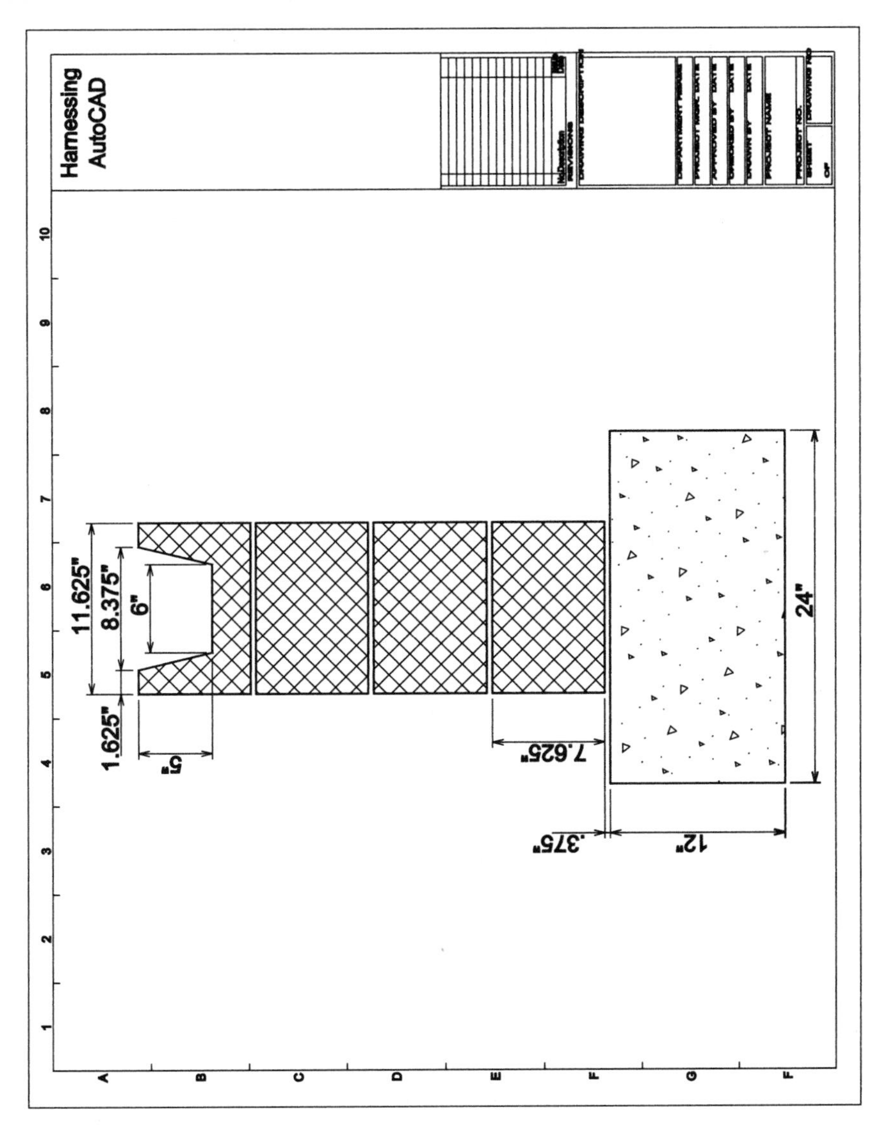


**Figure Ex9–7** *Masonry wall and footing section view*

## EXERCISE 9–8

Open the pre-fabricated walk-in refrigerator design that was drawn in Exercise 3–9, add the hatch pattern as shown in Figure Ex9–8.

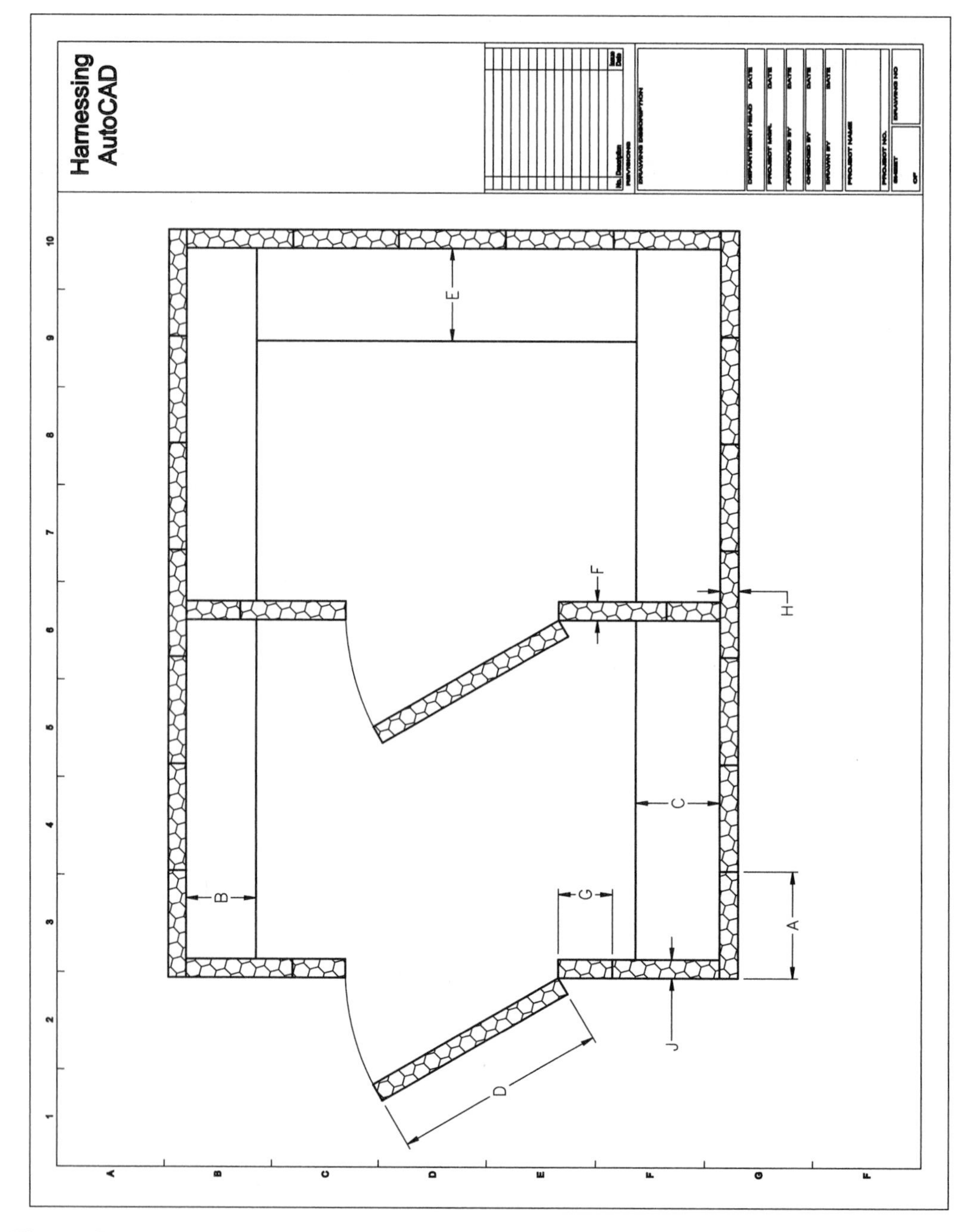

**Figure Ex9–8** *Pre-fabricated walk-in refrigerator*

## EXERCISE 9–9

Open the drawing of the recessed store front (plan and elevation views) that was drawn in Exercise 7–9 and add the hatch patterns (see Figure Ex9–9) to the given specifications:

| Settings | Value | | | |
|---|---|---|---|---|
| 1. Hatch Pattern | NAME | LAYER | SCALE* | ANGLE |
| | ANSI37 | Hatchplan | 0'-3" | 0 |
| | Brick | Hatchelev | 0'-3" | 0 |
| | ANSI34 | Hatchglass | 0'-4" | 70 |

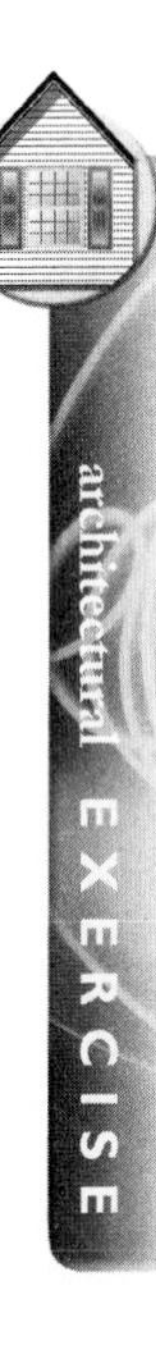

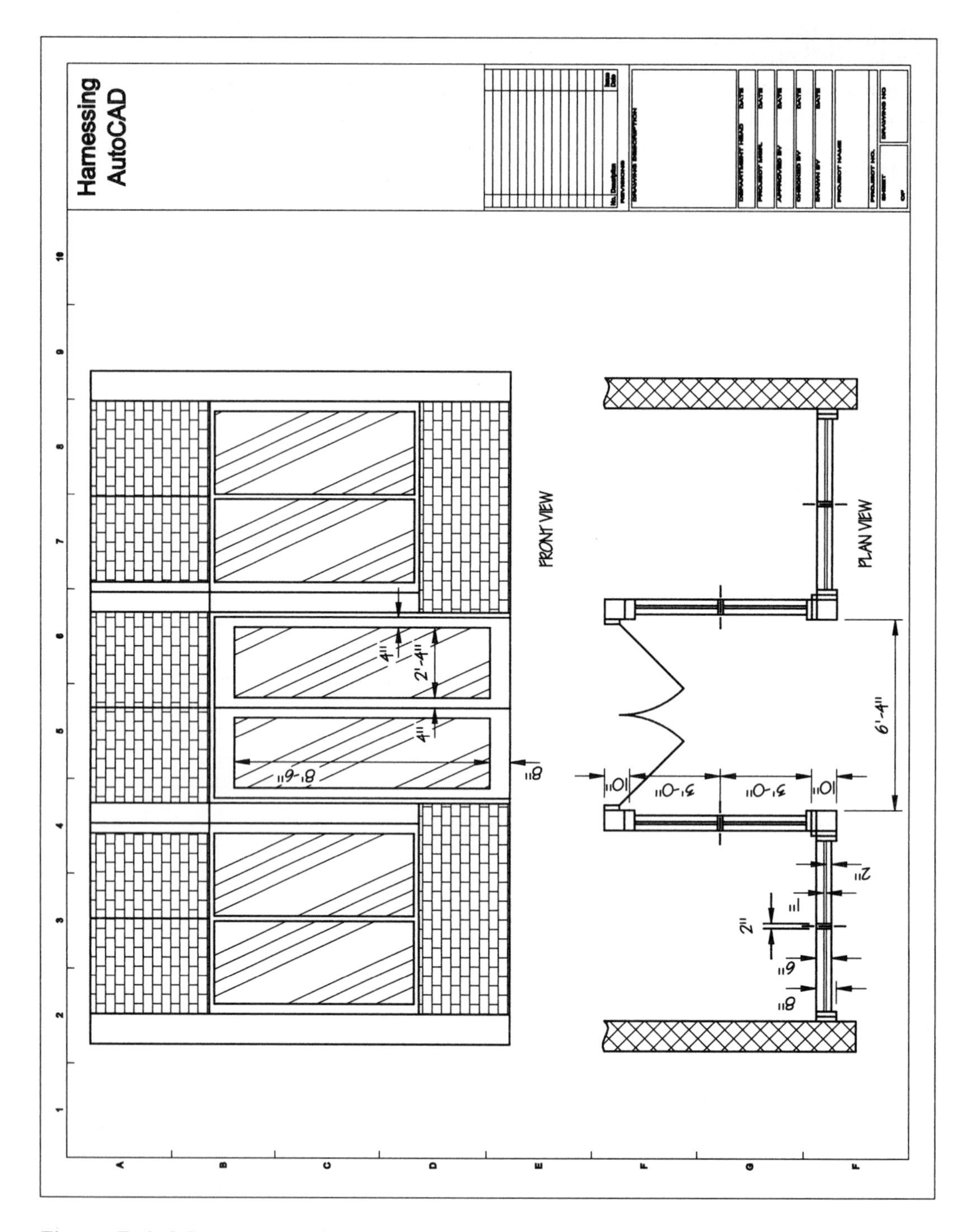


**Figure Ex9–9** *Recessed store front*

## EXERCISE 9–10

Create the drawing of a concrete slopped retaining wall (section view) shown in Figure Ex9–10, according to the settings given in the following table and then add the hatch patterns:

| Settings | Value | | |
|---|---|---|---|
| 1. Units | Decimal | | |
| 2. Limits | Lower left corner: 0,0 | | |
| | Upper right corner: 12,9 | | |
| 3. Grid | .5 | | |
| 4. Snap | .25 | | |
| 5. Layers | *NAME* | *COLOR* | *LINETYPE* |
| | Border | White | Continuous |
| | Concrete | White | Continuous |
| | Gravelbed | Magenta | Continuous |
| | Hatchearth | Yellow | Continuous |
| | Hatchgravel | Magenta | Continuous |
| | Hatchconc | White | Continuous |

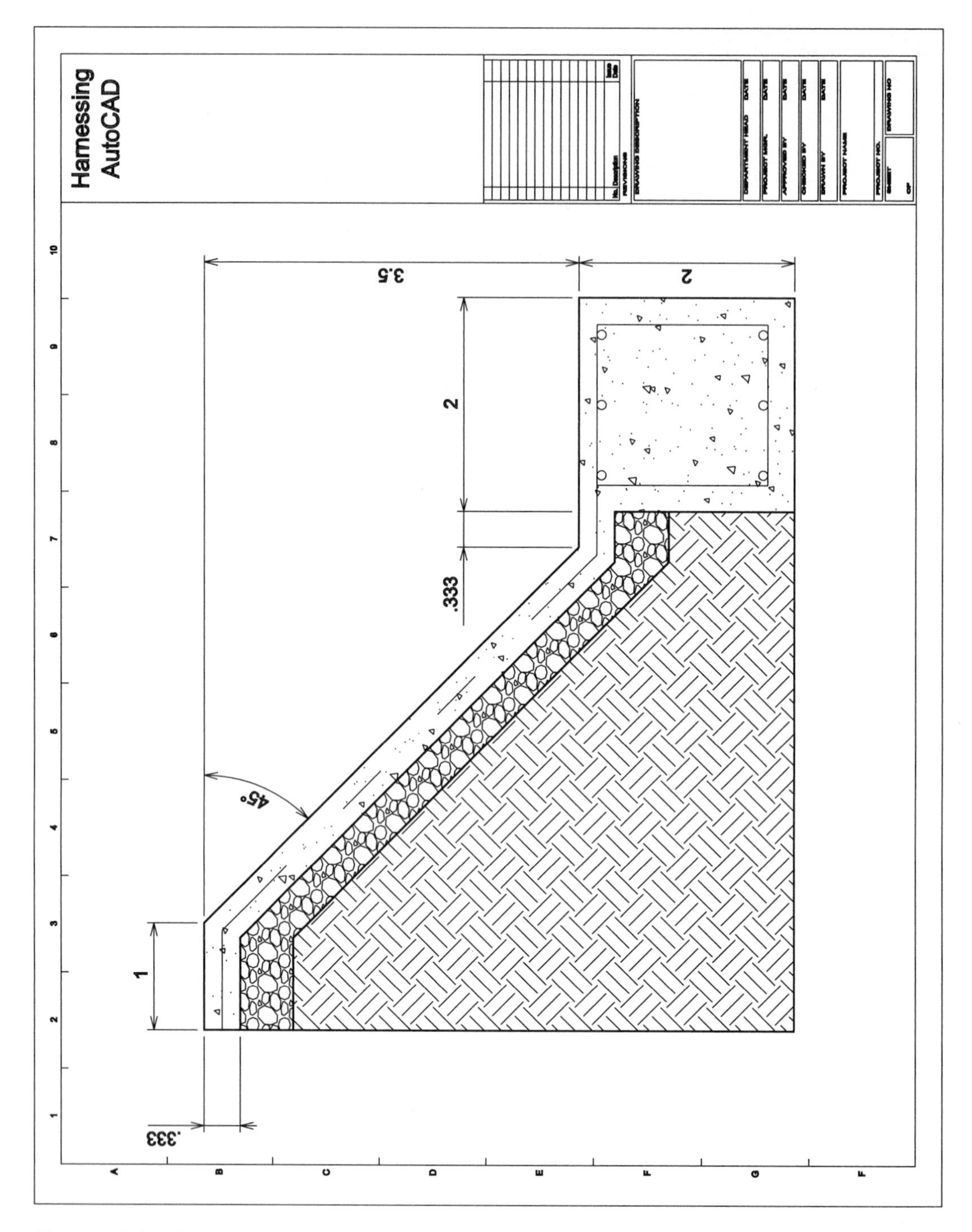


**Figure Ex9–10** *Slopped retaining wall*

## EXERCISE 9–11

Create the drawing of a section through a security retaining wall shown in Figure Ex9–11, according to the settings given in the following table, and then add the hatch patterns:

| Settings | Value | | |
|---|---|---|---|
| 1. Units | Architectural | | |
| 2. Limits | Lower left corner: 0,0 | | |
| | Upper right corner: 20', 15' | | |
| 3. Grid | 1" | | |
| 4. Snap | .5" | | |
| 5. Layers | *NAME* | *COLOR* | *LINETYPE* |
| | Border | Red | Continuous |
| | Center | Green | Center |
| | Wallfooting | Magenta | Continuous |
| | Reinforcing | White | Continuous |
| | Earth | Blue | Continuous |
| | Road | Yellow | Continuous |
| | Gravelbed | White | Continuous |
| | Dimensions | White | Continuous |
| | Hatchearth | Blue | Continuous |
| | Hatchroad | Yellow | Continuous |
| | HatchGravel | White | Continuous |

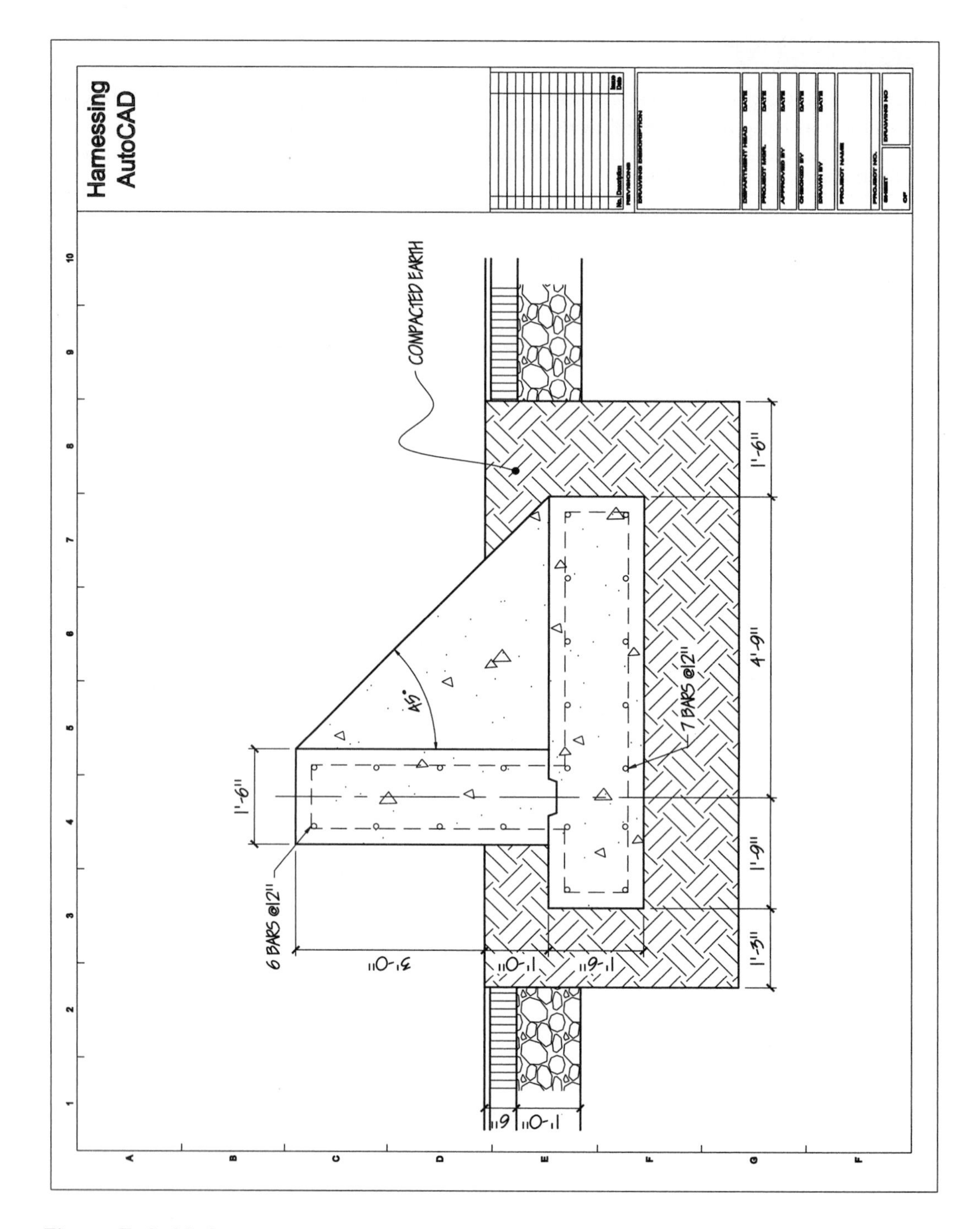


**Figure Ex9–11** *Security retaining wall*

## EXERCISE 9–12

Create the drawing of a porous paving and stone reservoir detail as shown in Exercise 9–12, add the hatch patterns as shown in Figure Ex9–12:

| Settings | Value | | |
|---|---|---|---|
| 1. Units | Architectural | | |
| 2. Limits | Lower left corner: 0,0 | | |
| | Upper right corner: 22', 17' | | |
| 3. Grid | 1" | | |
| 4. Snap | .5" | | |
| 5. Layers | *NAME* | *COLOR* | *LINETYPE* |
| | Border | White | Continuous |
| | Curbs | Green | Continuous |
| | Paving | Red | Continuous |
| | Reservoir | Blue | Continuous |
| | Dimensions | White | Continuous |
| | Notes | White | Continuous |
| | Settingbed | White | Continuous |
| | Dimensions | White | Continuous |
| | Hatchcurbs | Green | Continuous |
| | Hatchpaving | Red | Continuous |
| | Hatchreservoir | Blue | Continuous |
| | Hatchsettingbed | White | Continuous |
| | Hatchearth | White | Continuous |

civil EXERCISE

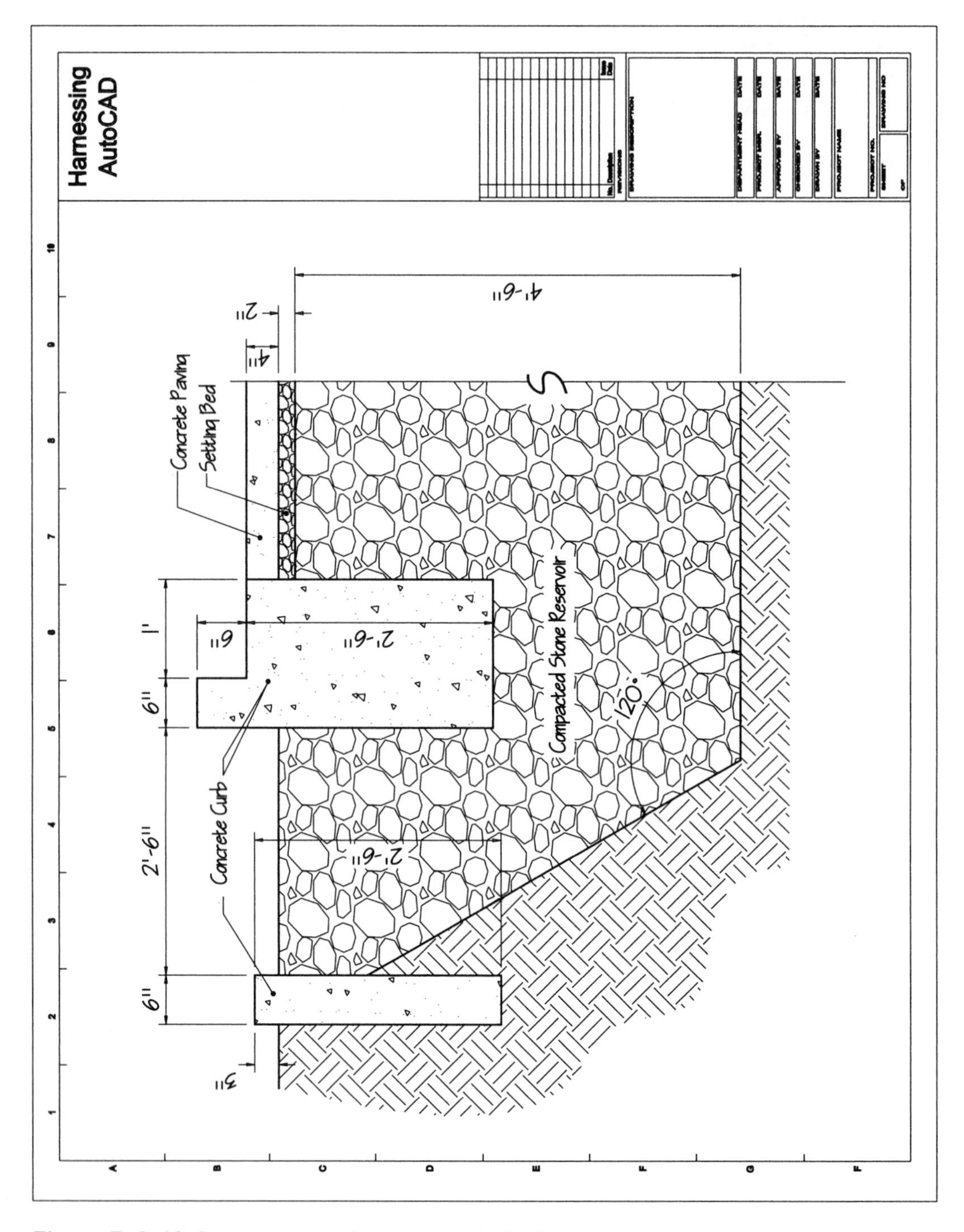


**Figure Ex9–12** *Porous paving and stone reservoir detail*

## EXERCISE 9–13

Open the drawing of the lighting switch circuit and wiring diagram that was drawn in Exercise 4–19, add the hatch patterns (see figure below) to the given specifications:

| Settings | Value | | | |
|---|---|---|---|---|
| 1. Hatch Pattern | <u>NAME</u> | <u>LAYER</u> | <u>SCALE*</u> | <u>ANGLE</u> |
| | ANSI37 | Hatch1 | 0'-2" | 0 |
| | ANSI31 | Hatch2 | 0'-2" | 0 |
| | ANSI31 | Hatch3 | 0'-2" | 90 |



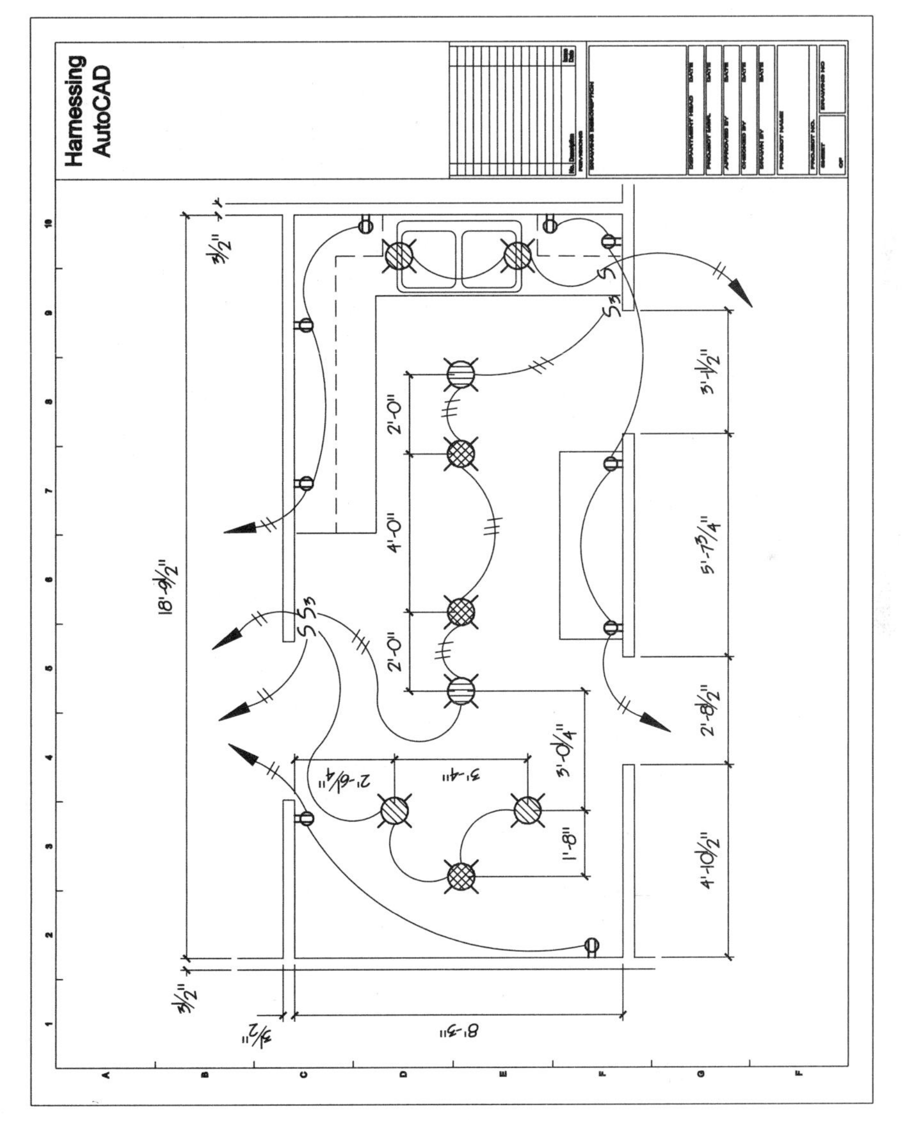


**Figure Ex9–13** *Lighting switch circuit and wiring diagram*

## EXERCISE 9–14

Create the schematic diagram of the partial electrical riser as shown in the figure and add the hatch patterns (see figure below) to the given specifications:

| Settings | Value | | | |
|---|---|---|---|---|
| 1. Hatch Pattern | *NAME* | *LAYER* | *SCALE** | *ANGLE* |
| | ANSI37 | Hatch1 | 0'-4" | 45 |
| | ANSI31 | Hatch2 | 0'-2 1/2" | 45 |
| | ANSI37 | Hatch3 | 0'-2 1/2" | 60 |



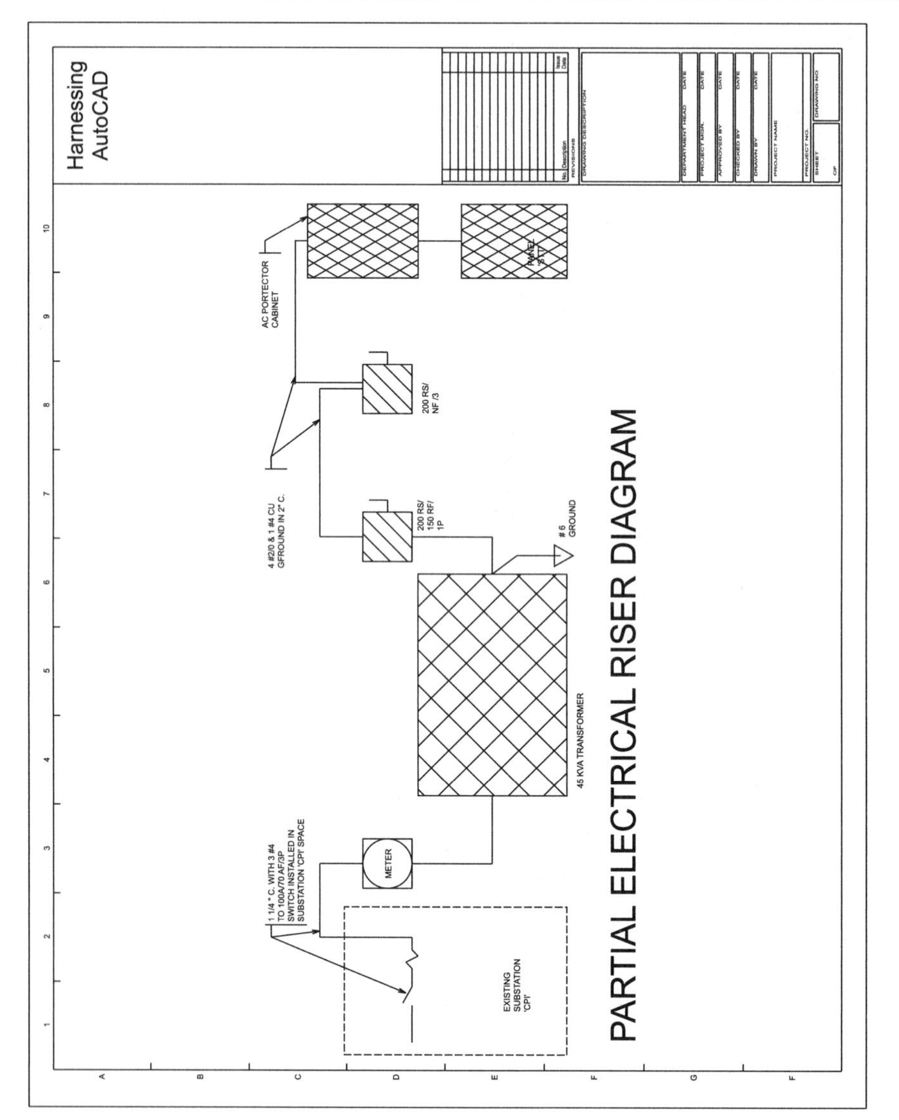


**Figure Ex9–14** *Schematic diagram of partial electrical riser*

## EXERCISE 9–15

Open the schematic diagram of the simple Ethernet peer to peer network that was drawn in Exercise 5–15. Add the hatch patterns (see figure below) to the given specifications:

| Settings | Value | | | |
|---|---|---|---|---|
| 1. Hatch Pattern | NAME<br>STARS | LAYER<br>Hatch1 | SCALE*<br>0'-2" | ANGLE<br>0 |

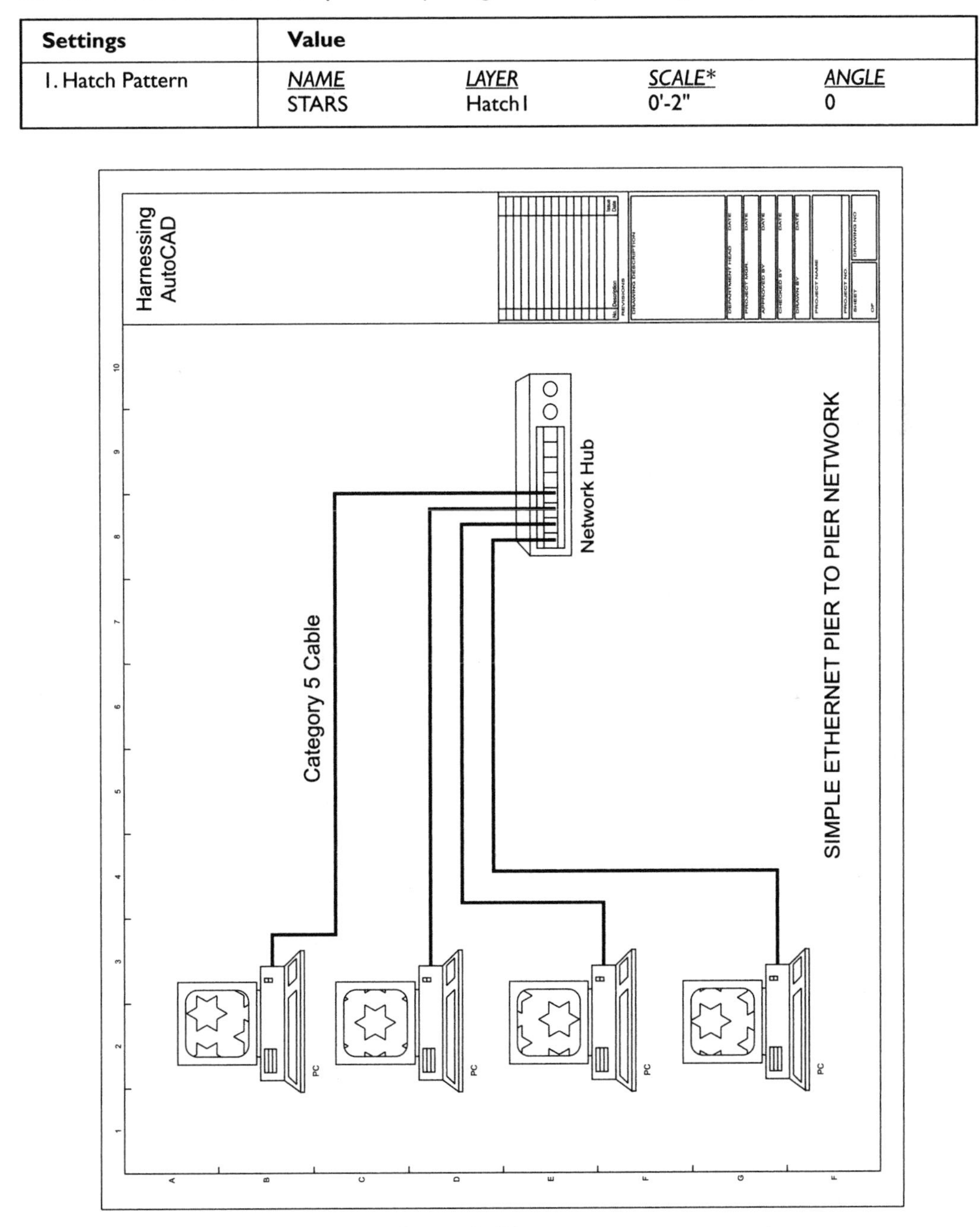


**Figure Ex9–15** *Schematic diagram of a simple Ethernet peer to peer network*

electrical EXERCISE

## EXERCISE 9–16

Create the high power electrical conduits encased in a concrete vault that was drawn in Ex6–15, according to the settings given in the following table. Then add the patterns:

| Settings | Value | | | |
|---|---|---|---|---|
| 1. Units | Architectural | | | |
| 2. Limits | Lower left corner: 0,0 | | | |
| | Upper right corner: 30', 24' | | | |
| 3. Grid | .25" | | | |
| 4. Snap | .125" | | | |
| 5. Layers | *NAME* | *COLOR* | *LINETYPE* | |
| | Border | White | Continuous | |
| | Vaultwall | Yellow | Continuous | |
| | Conduit | Yellow | Continuous | |
| | Panels | Green | Continuous | |
| | Mainswitch | Red | Continuous | |
| | Ladder | White | Continuous | |
| | Accesspanel | White | Hidden | |
| | Notes | White | Continuous | |
| | Dimensions | White | Continuous | |
| | HatchVaultwall | Yellow | Continuous | |
| | HatchConduit | Yellow | Continuous | |
| | HatchMainswitch | Red | Continuous | |
| | HatchPanels | White | Continuous | |
| 6. Hatch Pattern | *NAME* | *LAYER* | *SCALE** | *ANGLE* |
| | AR-CONC | HatchVaultwall | 0'-2" | 0 |
| | AR-CONC | HatchConduit | 0'-2" | 0 |
| | ANSI31 | HatchMainswitch | 0'-2 ½" | 90 |
| | ANSI31 | HatchPanels | 0'-1" | 90 |

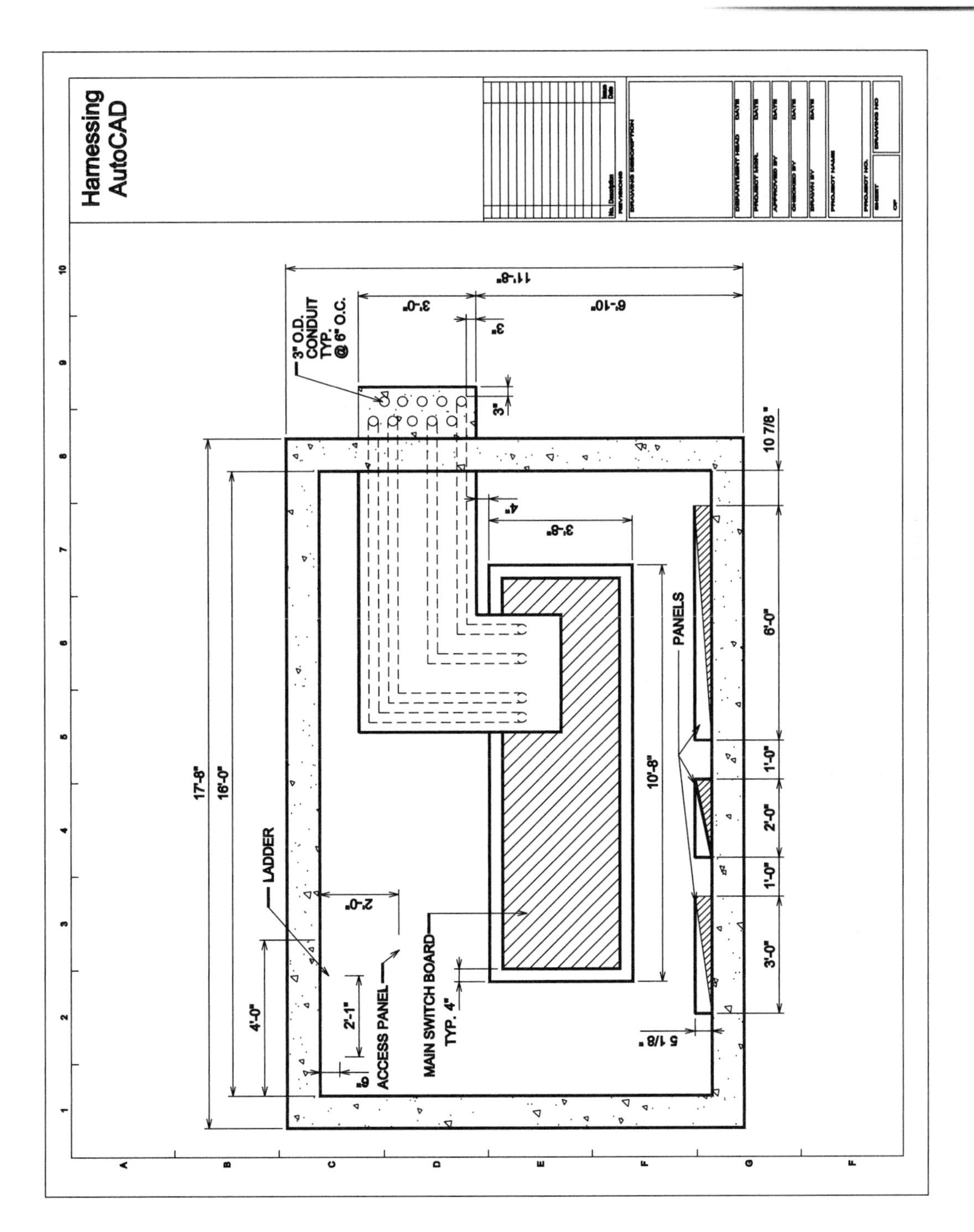


**Figure Ex9–16** *High power electrical conduits encased in a concrete vault*

# Block References and Attributes



## PROJECT EXERCISE

This project exercise provides point-by-point instructions for creating the objects shown in Figure P10–1. In this exercise you will apply the skills acquired in Chapters 1 through 10.

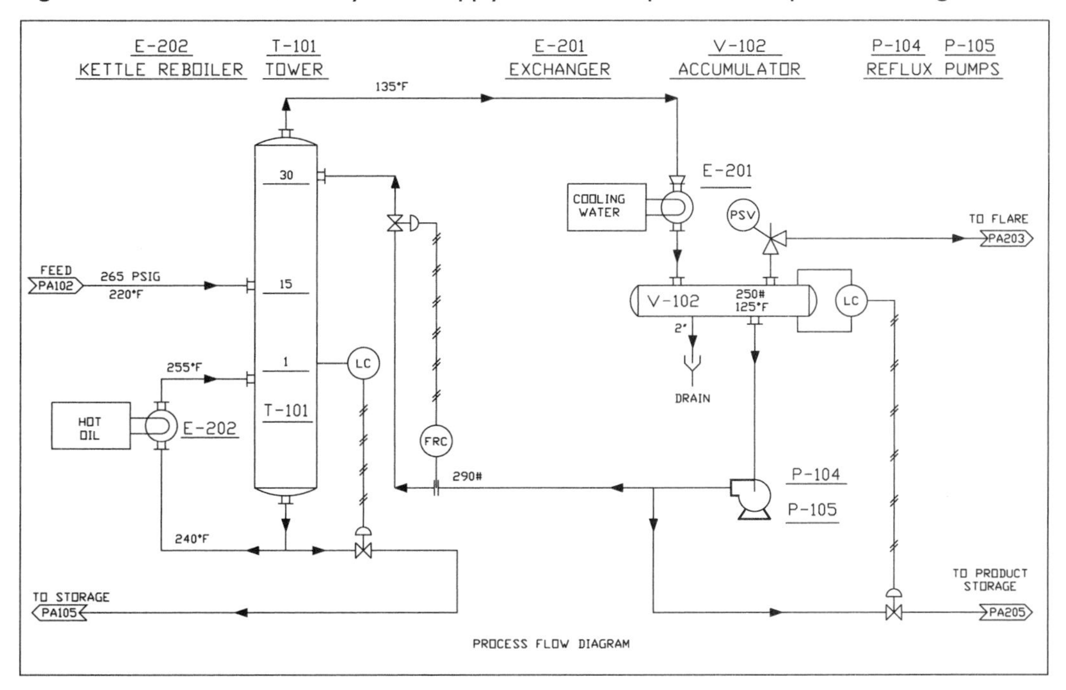


**Figure P10–1** *Completed project drawing*

In this project you will learn to do the following:

- Set up the drawing, including Limits, and Layers.
- Use the LINE, PLINE, CIRCLE, ARC, and SOLID commands to create objects.

- Use the BLOCK command to save objects for later insertion.
- Use the ATTDEF command to create an attribute and include it in the block definition. This enables you to attach a text string to the block reference when it is inserted.
- Use the INSERT command to insert the block references created earlier.

**SET UP THE DRAWING AND DRAW A BORDER**

**Step 1:** Start the AutoCAD program.

**Step 2:** To create a new drawing, invoke the NEW command from the Standard toolbar or choose New from the File menu. AutoCAD displays the **Create New Drawing** dialog box. Choose the **Start From Scratch** option, and choose **OK** to create a new drawing.

**Step 3:** Invoke the LIMITS command, and set the limits to the lower left and upper right coordinates as shown in the SETTINGS/VALUES table. Set the grid and snap values as shown also.

| SETTINGS | VALUE | | |
|---|---|---|---|
| UNITS | Decimal | | |
| LIMITS | Lower left corner: 0,0 | | |
| | Upper right corner: 18,12 | | |
| GRID | 0.5 | | |
| SNAP | 0.25 | | |
| LAYERS | *NAME* | *COLOR* | *LINETYPE* |
| | Blocks | Red | Continuous |
| | Border | Cyan | Continuous |
| | Equipment | Green | Continuous |
| | Instrument | Magenta | Continuous |
| | Pipeline | Blue | Continuous |
| | Text | White | Continuous |

**Step 4:** Invoke the LAYER command from the Object Properties toolbar, or choose Layer from the Format menu. AutoCAD displays the Layer Properties Manager dialog box. Create six layers, rename them as shown in the table, and assign the appropriate color and linetype.

**Step 5:** Set Border as the current layer and set the Ortho, Grid, and Snap options all to ON. Invoke the ZOOM ALL command to display the entire limits on screen.

**Step 6:** Invoke the RECTANGLE command from the Draw toolbar to draw the border (17" x 11"), as shown in Figure P10–2.

Command: **rectangle**
Specify first corner point or [Chamfer/Elevation/Fillet/Thickness/Width]: **.5,.75**
Specify other corner point or [Dimensions]: **@17,11**

**Figure P10–2** *Border for piping flowsheet*

**Step 7:** Set Blocks as the current layer.

**Step 8:** Invoke the ZOOM command and use the Window option to zoom in on a small area of the display, as shown in Figure P10–3.

Command: **zoom**
Specify corner of window, enter a scale factor (nX or nXP), or
[All/Center/Dynamic/Extents/Previous/Scale/Window] <real time>: **window**
Specify first corner: **3,3**
Specify opposite corner: **6,6**

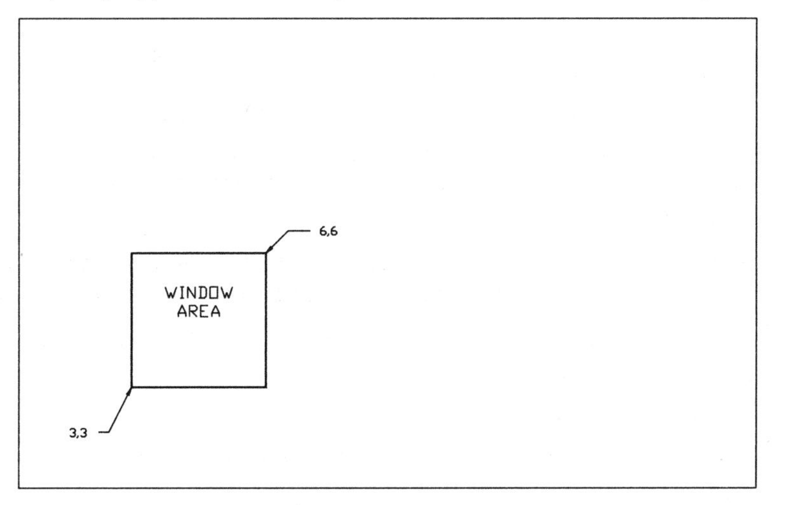


**Figure P10–3** *Zoom in on a portion of the display*

**Step 9:** Open the Drafting Settings dialog box from the Tools menu, and set the grid to 0.125 and the snap to 0.0625, and set the grid and snap to ON.

**Step 10:** Draw the gate valve as shown in Figure P10–4.

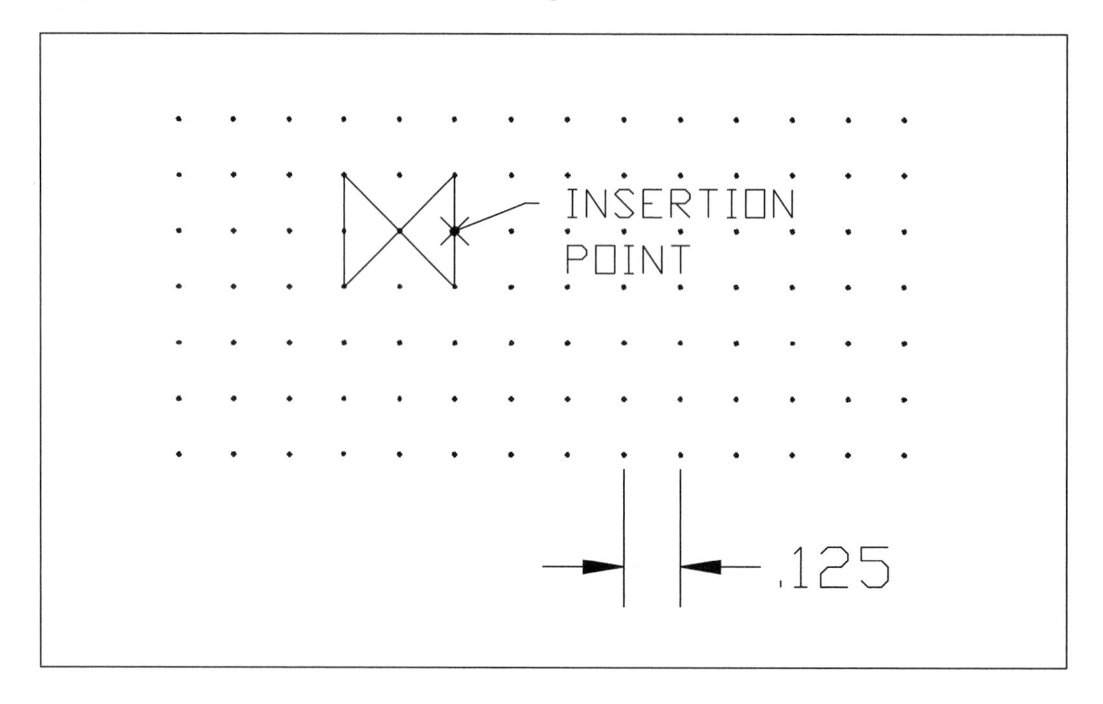


**Figure P10–4** *The gate valve*

**Step 11:** Invoke the BLOCK command from the Draw toolbar, as shown in Figure P10–5.

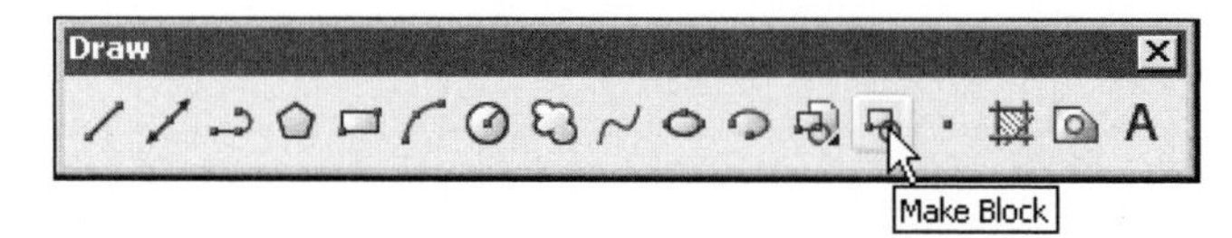


**Figure P10–5** *Invoking the* BLOCK *command from the Draw toolbar*

AutoCAD displays the Block Definition dialog box shown in Figure P10–6.

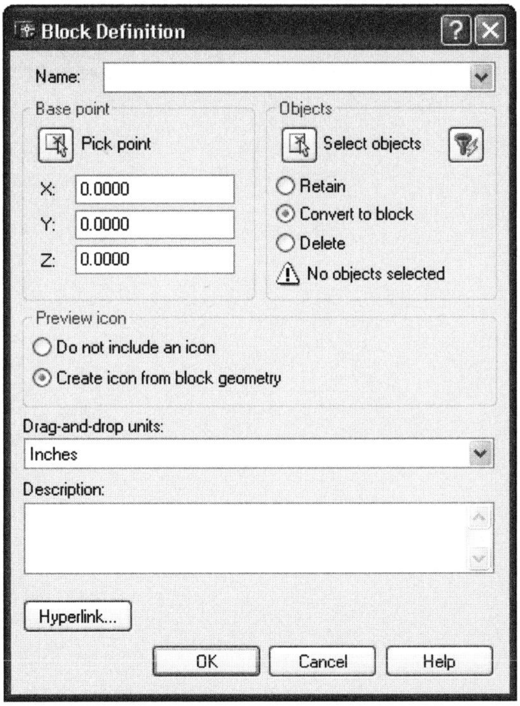

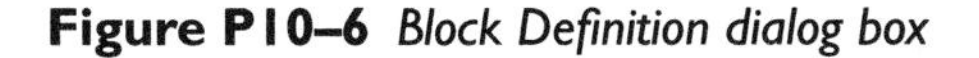


**Figure P10–6** *Block Definition dialog box*

Enter **GATE** in the **Block name:** text box. Select the **Pick Point** button. AutoCAD prompts:

Specify insertion base point: *(specify the insertion point as shown in Figure P10–4)*

Once you specify the insertion point, AutoCAD redisplays the Block Definition dialog box. Choose the **Select Objects** button. AutoCAD prompts:

Select objects: *(select all the objects that comprise the gate valve)*

Once you select the objects, AutoCAD redisplays the Block Definition dialog box. Choose the **Delete** radio button in the Objects section of the dialog box. Choose the **OK** button to create the block called GATE and close the Block Definition dialog box.

**Step 12:** Draw the remaining symbols required for the piping flow sheet on the appropriate layers, as shown in the Figure P10–7.

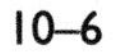

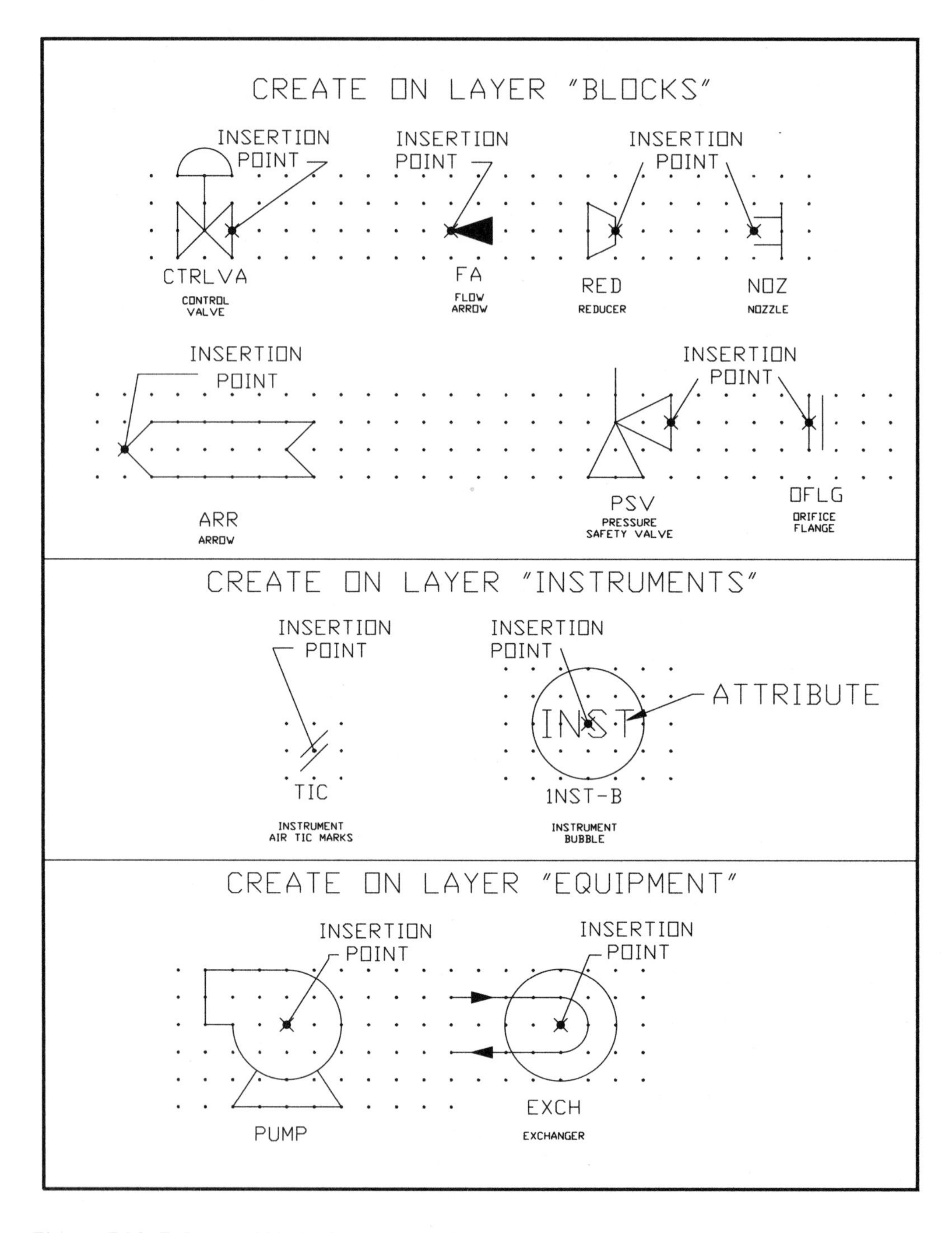


**Figure P10–7** *Required blocks for the piping flowsheet drawing*

Create individual blocks for all the symbols by providing appropriate insertion points and block names (as shown in uppercase letters in Figure P10–7). If necessary, refer to Step 11 for the Step-by-Step procedure to create a block.

Before you create a block for the instrument bubble, define an attribute.

Invoke the ATTDEF command. AutoCAD displays the Attribute Definition dialog box. Set appropriate attributes and the required prompts, as shown in Figure P10–8. Choose the **Pick Point** button to place the attribute in the center of the instrument bubble, as shown in Figure P10–7.

Once you have defined the attribute, create the block for the instrument bubble. Make sure to include the attribute tag as part of the block when selecting objects.

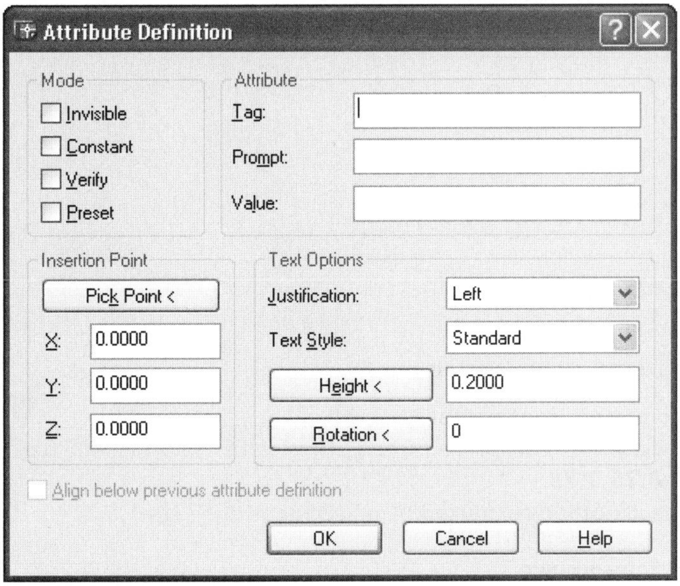

**Figure P10–8** *Attribute Definition dialog box*

**Step 13:** Invoke the ZOOM ALL command to display the entire drawing.

**Step 14:** Open the Drafting Settings dialog box from the Tools menu and set the Grid to 0.25, the Snap to 0.125, and set the Grid and Snap to ON.

**Step 15:** Set Equipment as the current layer.

**Step 16:** Draw the vertical and horizontal vessels and the boxes for the exchangers as shown in Figure P10–9. Following are the prompt sequences to draw the vessels and boxes for the exchangers.

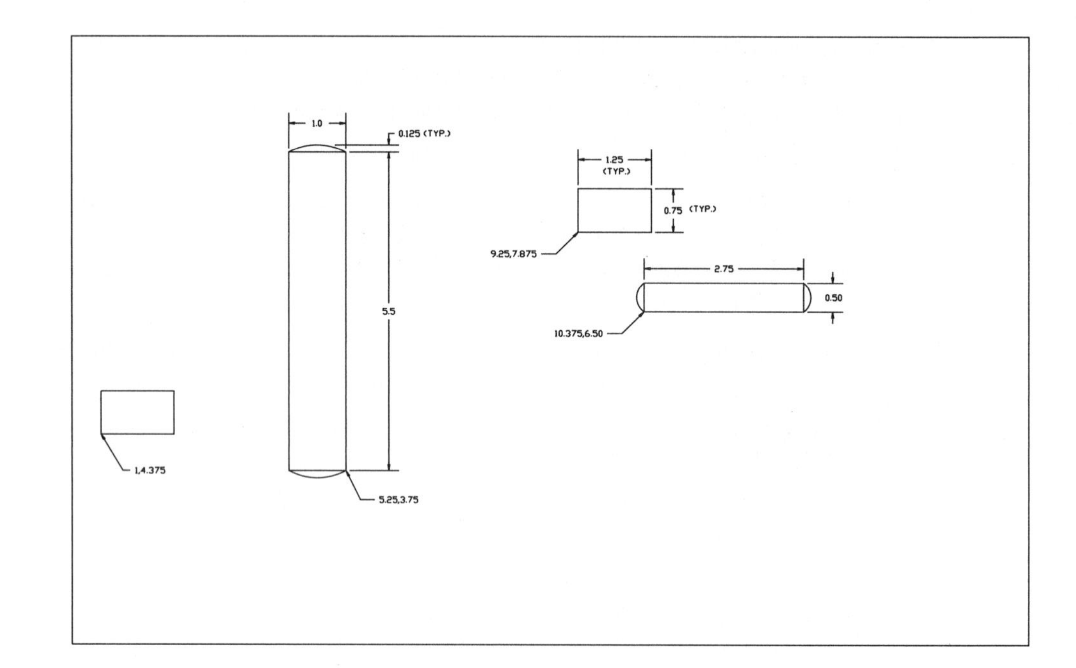


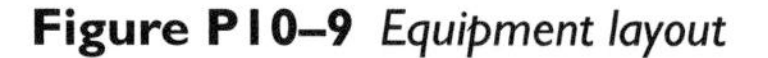
**Figure P10–9** *Equipment layout*

To draw the vertical vessel:

Command: **rectangle**
Specify first corner point or [Chamfer/Elevation/Fillet/Thickness/Width]:
**5.25,3.75**
Specify other corner point or [Dimensions]: **@–1,5.5**

Command: **arc**
Specify start point of arc or [CEnter]: **4.25,3.75**
Specify second point of arc or [CEnter/ENd]: **4.75,3.625**
Specify end point of arc: **5.25,3.75**

Command: **arc**
Specify start point of arc or [CEnter]: **5.25,9.25**
Specify second point of arc or [CEnter/ENd]: **4.75,9.375**
Specify end point of arc: **4.25,9.25**

To draw the horizontal vessel:

Command: **rectangle**
Specify first corner point or [Chamfer/Elevation/Fillet/Thickness/Width]:
**10.375,6.5**
Specify other corner point or [Dimensions]: **@2.75,0.50**

Command: **arc**
Specify start point of arc or [CEnter]: **10.375,7**
Specify second point of arc or [CEnter/ENd]: **10.25,6.75**
Specify end point of arc: **10.375,6.5**

Command: **arc**
Specify start point of arc or [CEnter]: **13.125,6.5**
Specify second point of arc or [CEnter/ENd]: **13.25,6.75**
Specify end point of arc: **13.125,7**

To draw the boxes for the exchangers:

Command: **rectangle**
Specify first corner point or [Chamfer/Elevation/Fillet/Thickness/Width]: **9.25,7.875**
Specify other corner point or [Dimensions]: **@1.25,0.75**

Command: **rectangle**
Specify first corner point or [Chamfer/Elevation/Fillet/Thickness/Width]: **1,4.375**
Specify other corner point or [Dimensions]: **@1.25,0.75**

**Step 17:** Insert the exchangers and pump as shown in Figure P10–10 by invoking the INSERT BLOCK command from the Draw toolbar (see Figure P10–11).

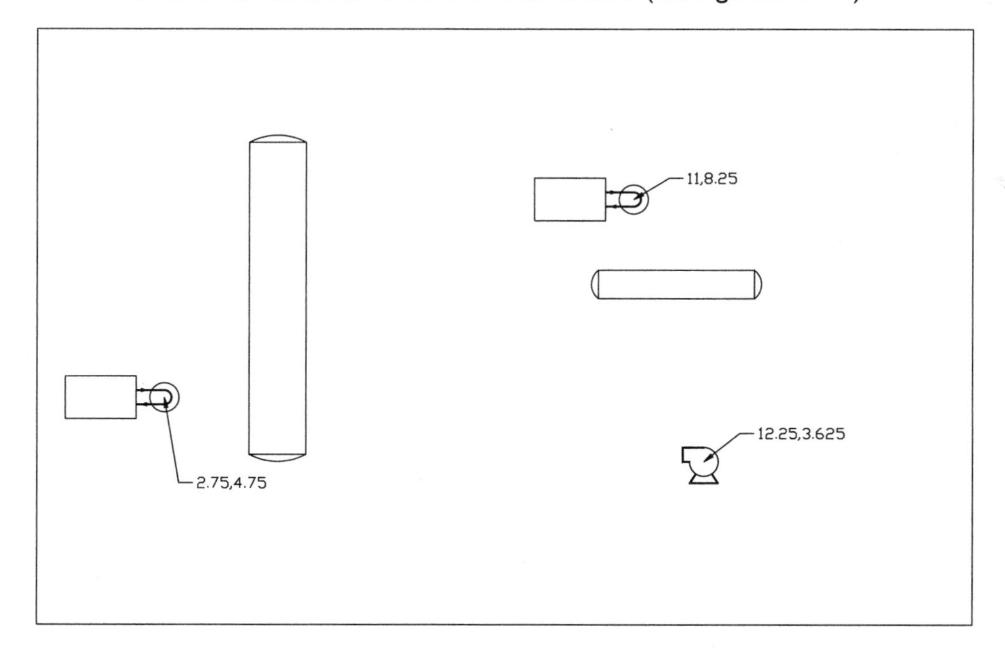


**Figure P10–10** *Layout with exchangers and pump*

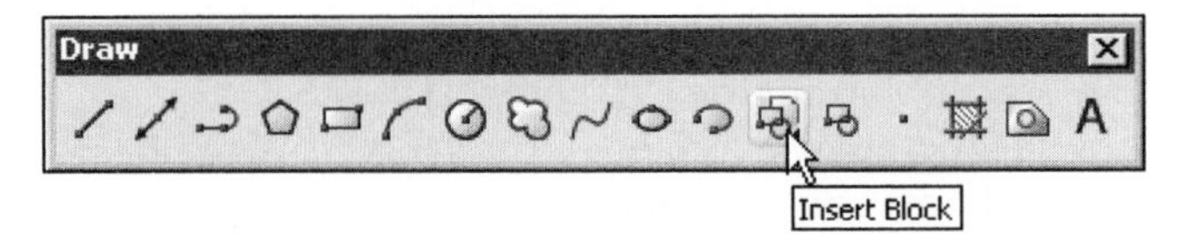


**Figure P10–11** *Invoking the* INSERT BLOCK *command from the Draw toolbar*

AutoCAD displays the Insert dialog box shown in Figure P10–12.

Insert
Name:
Browse...
Path:
Insertion point
Specify On-screen
X: 0.0000
Y: 0.0000
Z: 0.0000
Scale
Specify On-screen
X: 1.0000
Y: 1.0000
Z: 1.0000
Uniform Scale
Rotation
Specify On-screen
Angle: 0
Explode
OK
Cancel
Help

**Figure P10–12** *Insert dialog box*

Select EXCH block name from the **Name:** list box. Set the **Specify On-screen** check box to OFF in the **Insertion point** section of the dialog box. In the Insertion Point section of the dialog box, enter **2.75** in the **X:** text box and **4.75** in the **Y:** text box. In the Scale section of the dialog box, enter **1.00** in the **X:** text box and **1.00** in the **Y:** text box. In the Rotation section of the dialog box, enter **0** in the **Angle:** text box. Choose the **OK** button to insert the block reference and close the dialog box.

Similarly, insert the EXCH block reference again at the insertion point **11,8.25**, with scale set to **1.00** and rotation to **0** degrees. Insert the PUMP block reference at the insertion point **12.25,3.625**, with scale set to **1.00** and rotation set to **0** degrees.

**Step 18:** Insert the NOZ block reference at the appropriate locations, as shown in Figure P10–13.

Draw a horizontal line (representing a tray) in the vessel just below the highest side, and then copy it three times, as shown in Figure P10–13.

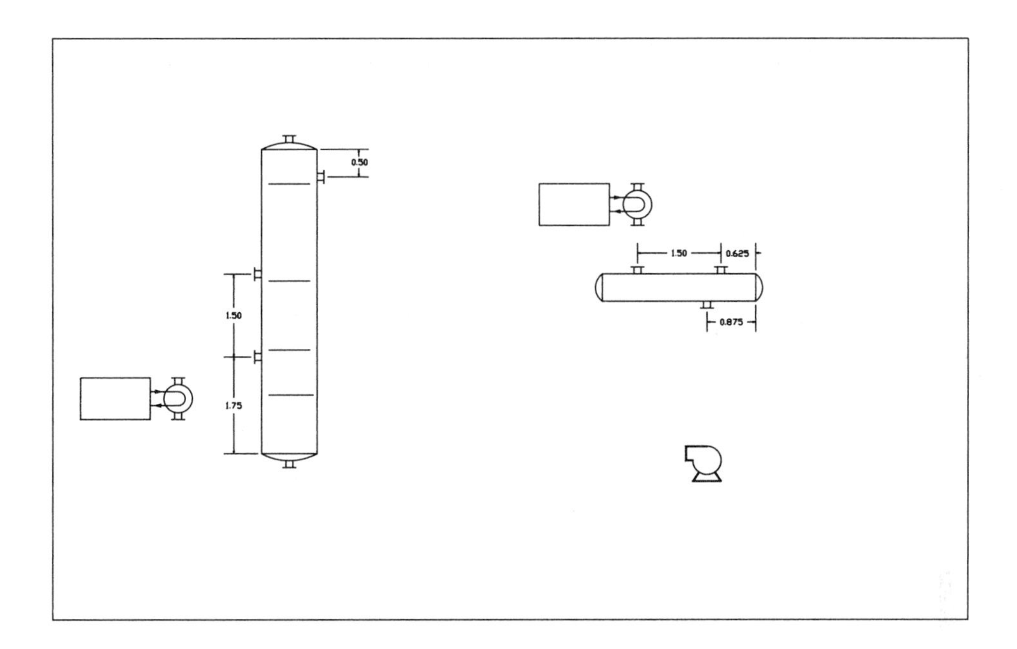

**Figure P10–13** *Layout with nozzles*

**Step 19:** Set Pipeline as the current layer.

**Step 20:** Lay out the pipelines as shown in Figure P10–14 by invoking the LINE command.

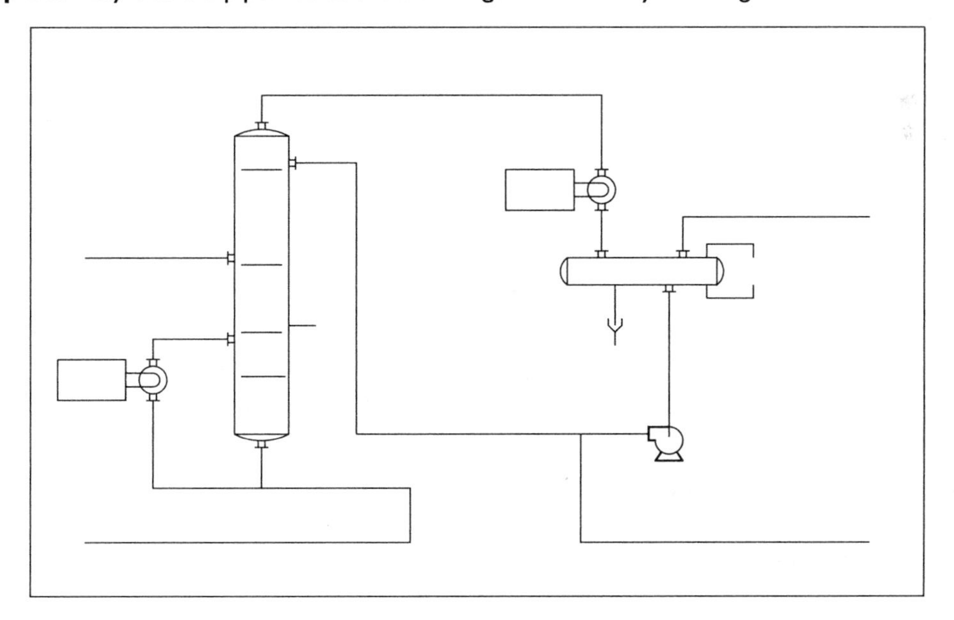

**Figure P10–14** *Layout with pipelines*

**Step 21:** Insert the flowsheet symbols at the appropriate places by invoking the INSERT command, as shown in Figure P10–15. Make sure to provide appropriate attribute values while inserting the instrument bubble block reference. After inserting the block references, use the BREAK command to break out the pipeline passing through the valve symbols.

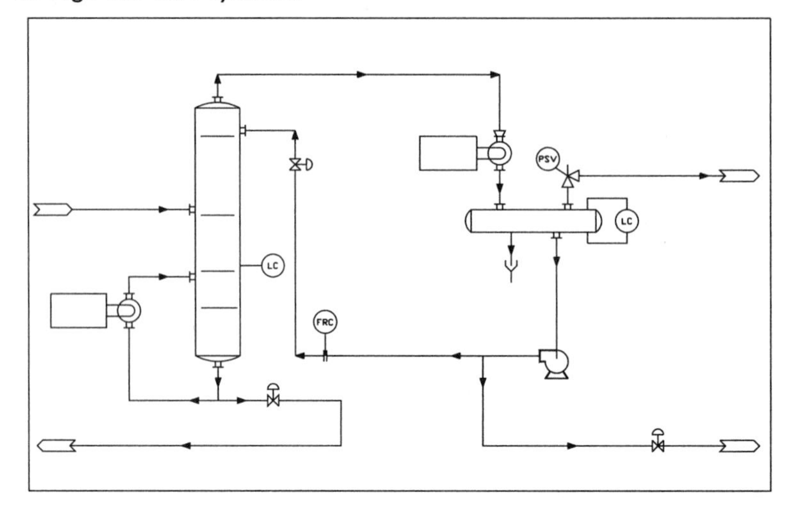


**Figure P10–15** *Layout with flowsheet symbols*

**Step 22:** Set Instruments as the current layer.

**Step 23:** Invoke the PLINE command to draw LINE 1, LINE 2, and LINE 3 instrument air lines as shown in Figure P10–16.

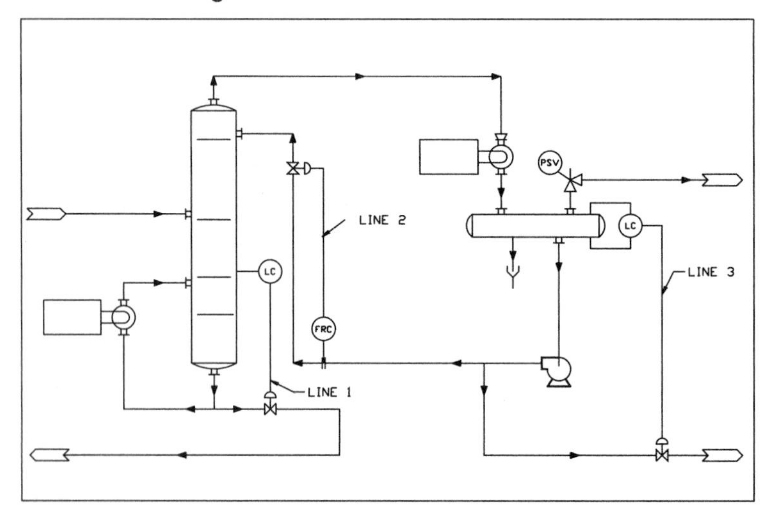


**Figure P10–16** *Layout with instrument air lines*

**Step 24:** Invoke the DIVIDE command to insert the block reference tic on the instrument air lines as shown in Figure P10–17. AutoCAD prompts:

Command: **divide**
Select object to divide: *(select LINE 1)*
Enter the number of segments or [Block]: **b**
Enter name of block to insert: **tic**
Align block with object? [Yes/No] <Y>: **n**
Enter the number of segments: **5**

Command: **divide**
Select object to divide: *(select LINE 2)*
Enter the number of segments or [Block]: **b**
Enter name of block to insert: **tic**
Align block with object? [Yes/No] <Y>: **n**
Enter the number of segments: **7**

Command: **divide**
Select object to divide: *(select LINE 3)*
Enter the number of segments or [Block]: **b**
Enter name of block to insert: **tic**
Align block with object? [Yes/No] <Y>: **n**
Enter the number of segments: **8**

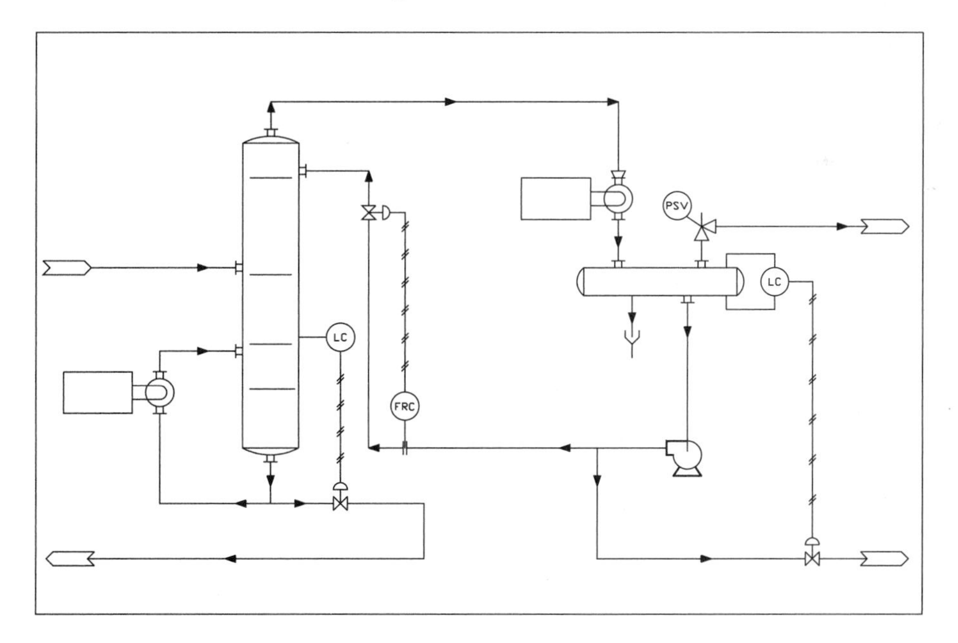


**Figure P10–17** *Layout with tic marks on the instrument air lines*

**Step 25:** Set Text as the current layer.

**Step 26:** Invoke the TEXT command to add the text as shown in Figure P10–18. The larger text height is set to 0.18 and the smaller text height is set to 0.125.

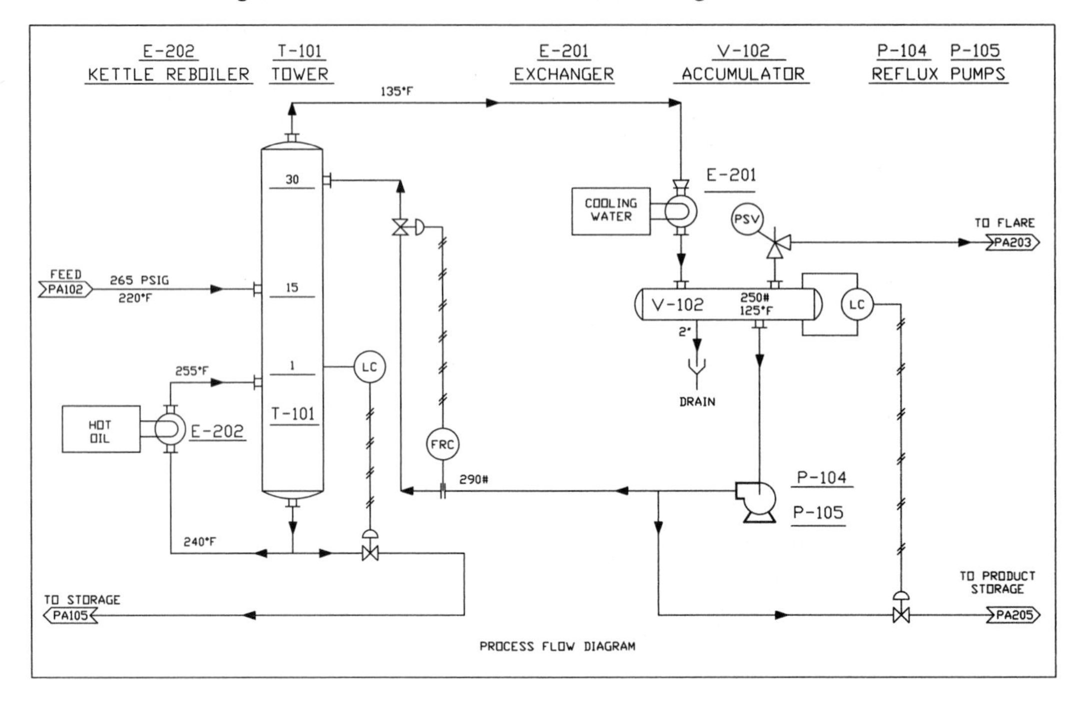


**Figure P10–18** *Layout with text*

**Step 27:** Save the drawing as CH10-PROJ and exit the AutoCAD program.

Congratulations! You have just successfully applied several AutoCAD concepts in creating a simple schematic drawing.

## EXERCISE 10–1

Create a kitchen cabinet layout, as shown in Figure Ex10–1, according to the settings given in the following table. Create all the required blocks by referring to the architectural symbols in Figure Ex10–2a.

| Settings | Value |
|---|---|
| 1. Units | Architectural |
| 2. Limits | Lower left corner: 0,0<br>Upper right corner: 20', 15' |

| Settings | Value | | |
|---|---|---|---|
| 3. Layers | NAME | COLOR | LINETYPE |
| | Border | Red | Continous |
| | Counter | Green | Hidden |
| | Sink | Blue | Hidden |
| | Cabinets | Yellow | Continuous |
| | Notes | White | Continuous |
| | Dimensions | Yellow | Continuous |
| | Doorswings | White | Phantom |
| | Notes | White | Continuous |

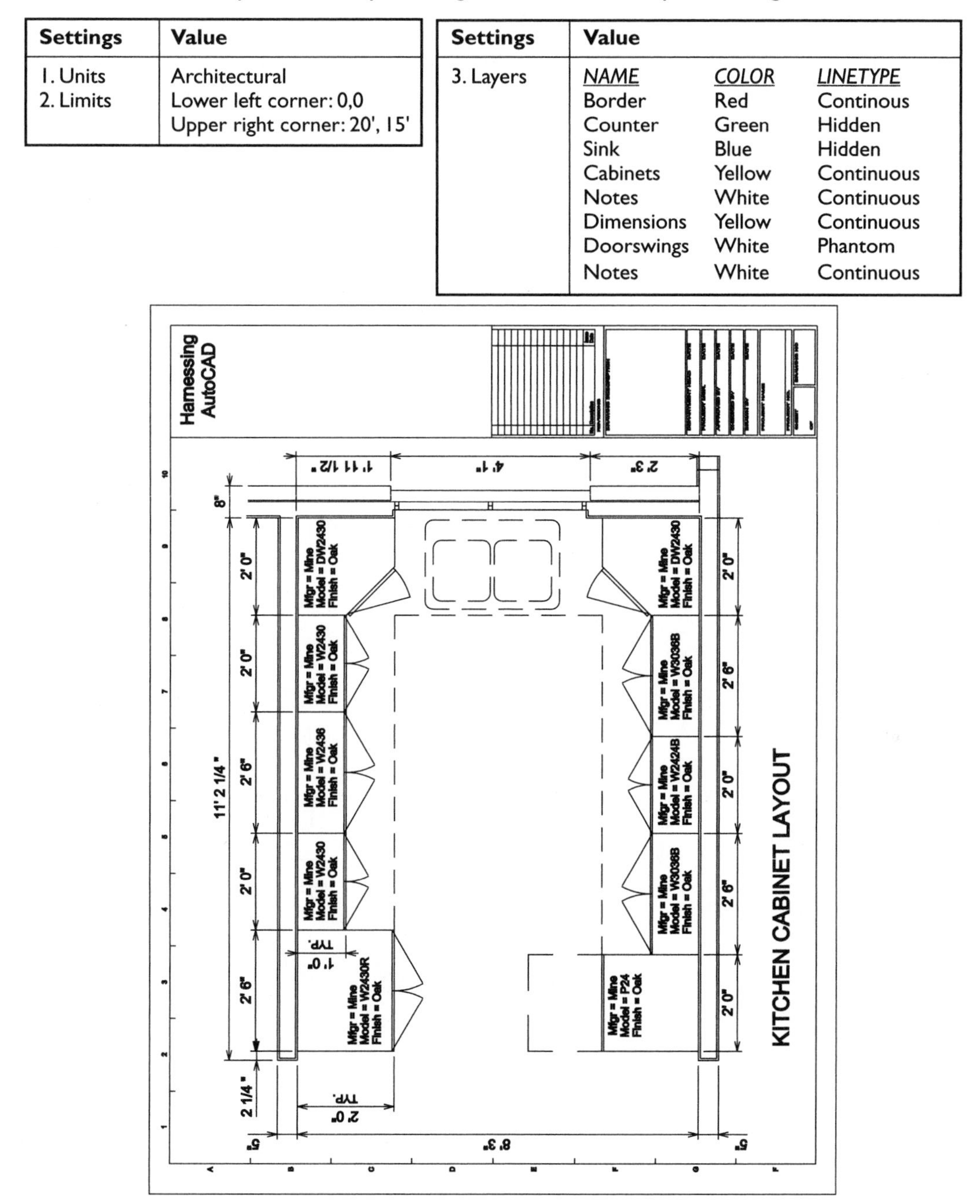


**Figure Ex10–1** *Kitchen cabinet layout*

## EXERCISE 10–2

Create the partial ground floor plan drawing shown in Figure Ex10–2 according to the settings given in the following table. Create all the required blocks by referring to the architectural symbols in Figure Ex10–2a.

| Settings | Value |
|---|---|
| 1. Units | Architectural |
| 2. Limits | Lower left corner: 0',0'<br>Upper right corner: 100', 75' |

| Settings | Value | | |
|---|---|---|---|
| 3. Layers | *NAME* | *COLOR* | *LINETYPE* |
| | Border | White | Continuous |
| | Centerline | Blue | Center |
| | Columns | Green | Continuous |
| | Walls | Yellow | Continuous |
| | Door | White | Continuous |
| | Furniture | Red | Continuous |
| | Notes | White | Continuous |
| | Dimensions | White | Continuous |

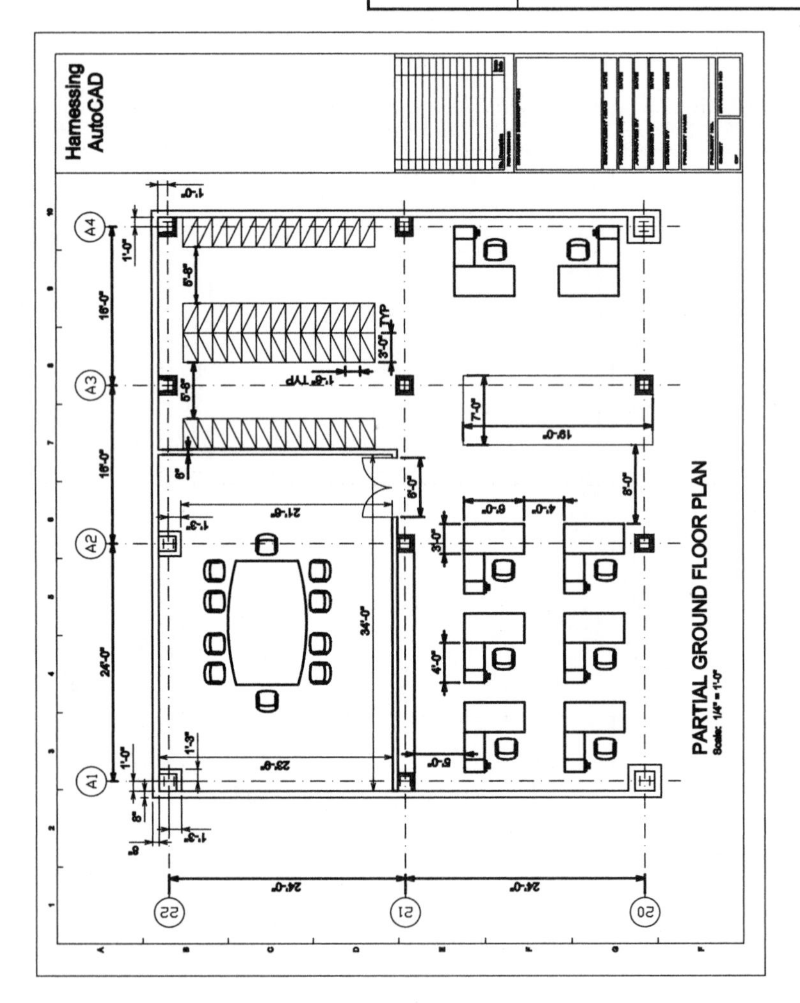

**Figure Ex10–2** *Partial ground floor plan*

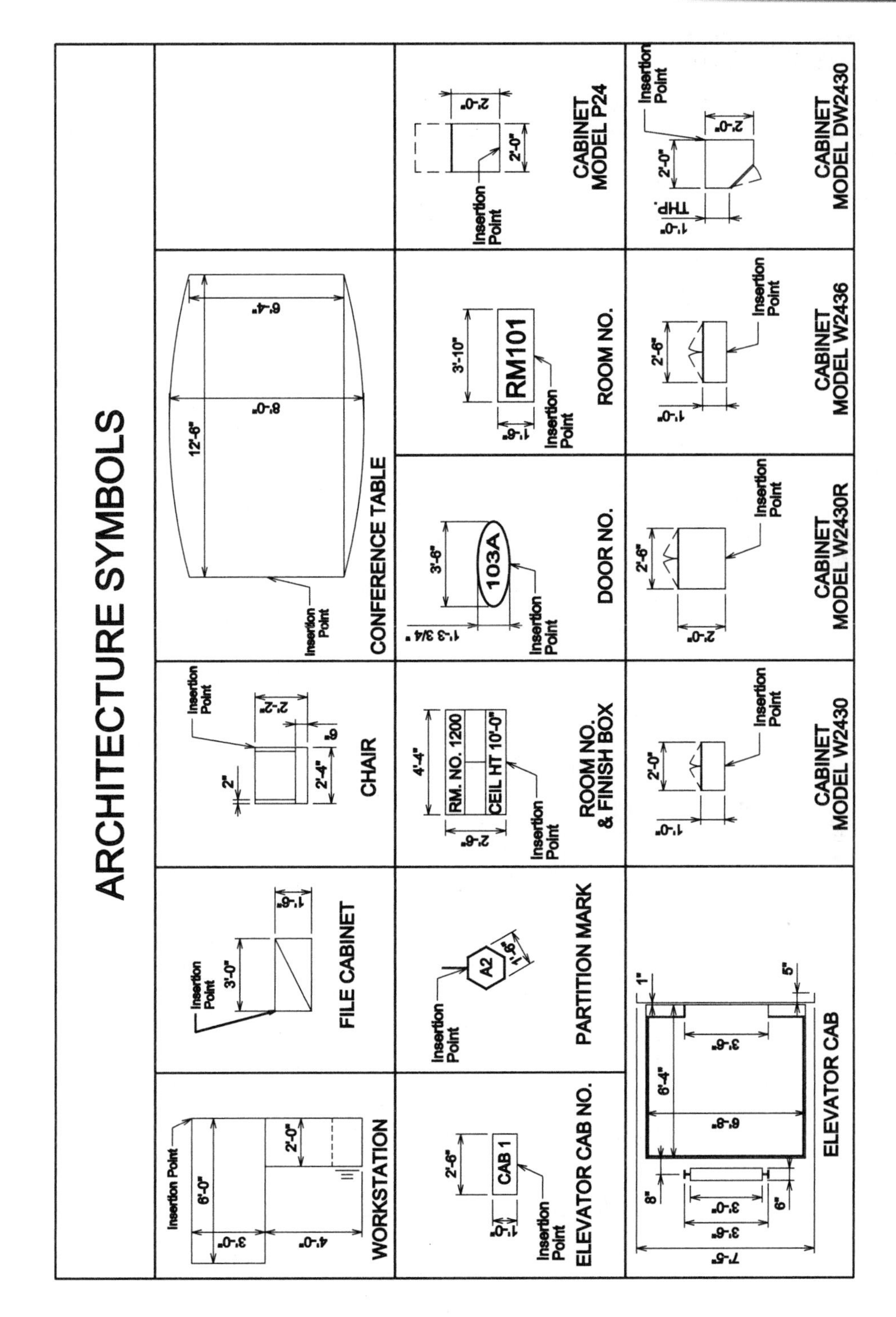


**Figure Ex10–2A** *Architectural Symbols*

## EXERCISE 10–3

Create the elevator lobby drawing shown in Figure Ex10–3 according to the settings given in the following table. Make blocks of all items except the shaft walls and shear walls. Create all the required blocks by referring to the architectural symbols in Figure Ex10–2a.

| Settings | Value |
|---|---|
| 1. Units | Architectural |
| 2. Limits | Lower left corner: 0,0<br>Upper right corner: 50', 38' |

| Settings | Value | | |
|---|---|---|---|
| 3. Layers | *NAME* | *COLOR* | *LINETYPE* |
| | Border | White | Continuous |
| | Shear wall | Blue | Continuous |
| | Elevators | White | Continuous |
| | Partition | Blue | Continuous |
| | Frames | Green | Continuous |
| | Notes | Green | Hidden |
| | Beams | Yellow | Continuous |
| | Centerline | White | Phantom |
| | Dimensions | White | Continuous |

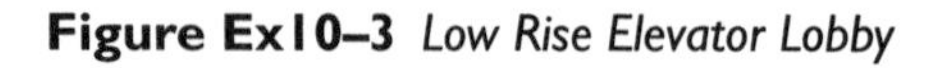

**Figure Ex10–3** *Low Rise Elevator Lobby*

## EXERCISE 10–4

Create a layout of a building drawing shown in Figure Ex10–4 according to the settings given in the following table. Create all the required blocks by referring to the architectural symbols in Figure Ex10–2a.

| Settings | Value | | |
|---|---|---|---|
| 1. Units | Architectural | | |
| 2. Limits | Lower left corner: 0,0 | | |
| | Upper right corner: 83',64' | | |
| 3. Layers | *NAME* | *COLOR* | *LINETYPE* |
| | Border | Red | Continous |
| | Holes | Green | Continuous |
| | Center | Blue | Phantom |
| | Ext wall | Blue | Continuous |
| | Ext door | White | Continuous |
| | Partition | Yellow | Continuous |
| | Int doors | White | Continuous |
| | Notes | White | Continuous |

**Hint:** For the exterior wall and interior partitions sizes (see the tables below). Use the multi-line command to create the wall and partitions.

Exterior Wall Dimensions

| Brick | Insulation | Air Space | Concrete Masonry Unit | Total Wall Thickness |
|---|---|---|---|---|
| 3 5/8" | 2" | 1" | 7 5/8" | 14 ¼" |

Interior Wall Dimensions

| Metal Stud Width | Gypsum Board Thickness | Total Wall Thickness |
|---|---|---|
| 3 5/8" | 5/8" | 4 7/8" |

architectural EXERCISE

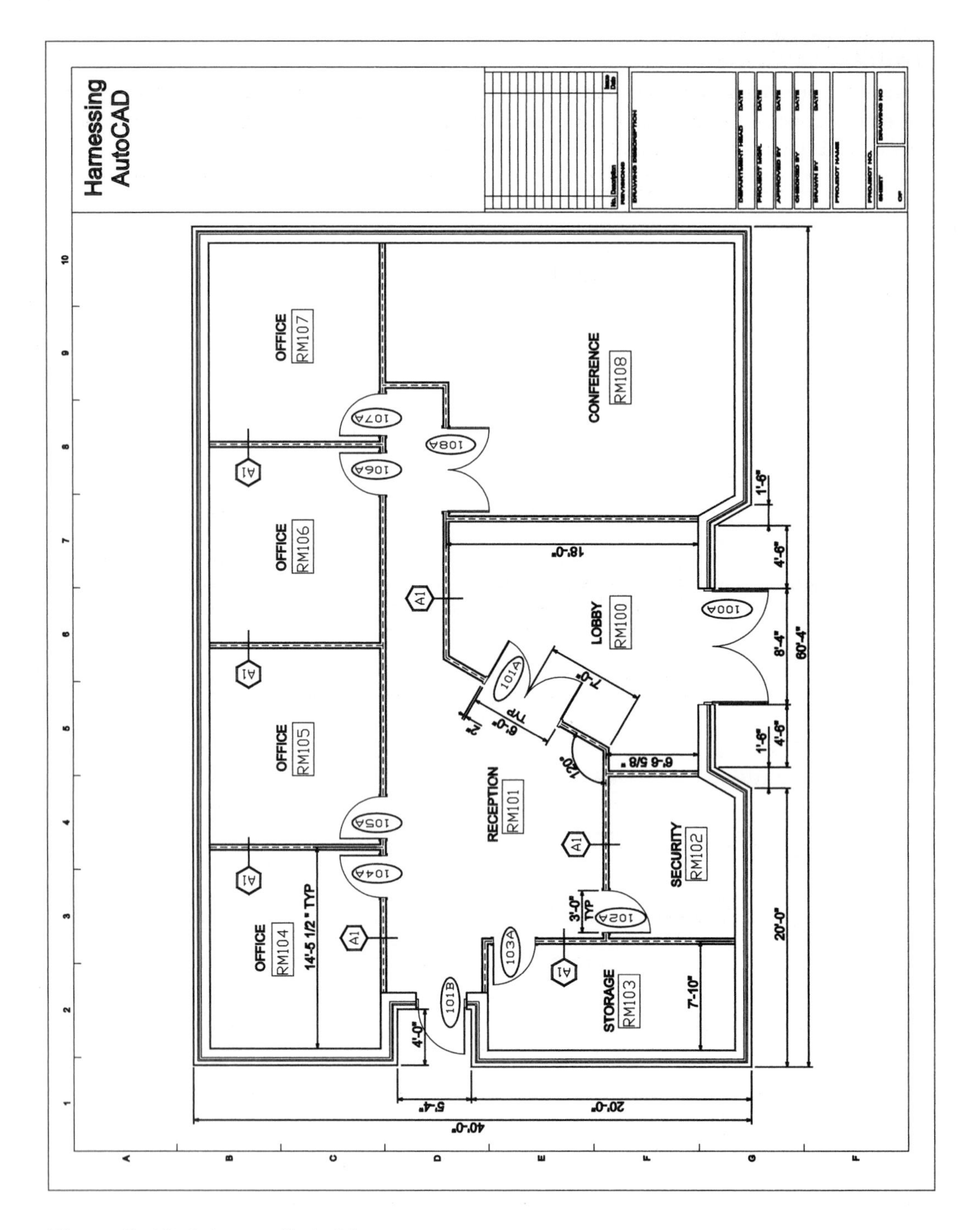


**Figure Ex10–4** *Layout of a building*

## EXERCISE 10–5

Create the drawing shown in Figure Ex10–5a according to the settings given in the following table. Create all the required blocks by referring to the electrical symbols in Figure Ex10–5b. (Set the grid to 0.125 and the snap to 0.0625 to draw the symbols.)

| Settings | Value | | |
|---|---|---|---|
| 1. Units | Decimal | | |
| 2. Limits | Lower left corner: 0,0 | | |
| | Upper right corner: 12,9 | | |
| 3. Grid | 0.5 | | |
| 4. Snap | 0.125 | | |
| 5. Layers | *NAME* | *COLOR* | *LINETYPE* |
| | Blocks | Red | Continuous |
| | Border | Cyan | Continuous |
| | Lines | Magenta | Continuous |
| | Text | White | Continuous |
| | Construction | Red | Continuous |

**Hints:** Note that the insertion points are located on the symbol where they join the objects to which they are connected.

Draw the circuits with continuous lines on the Construction layer. Insert the block references by using the Osnap mode Nearest. Then set the Construction layer to OFF and draw the lines between the symbols, connecting to their endpoints.

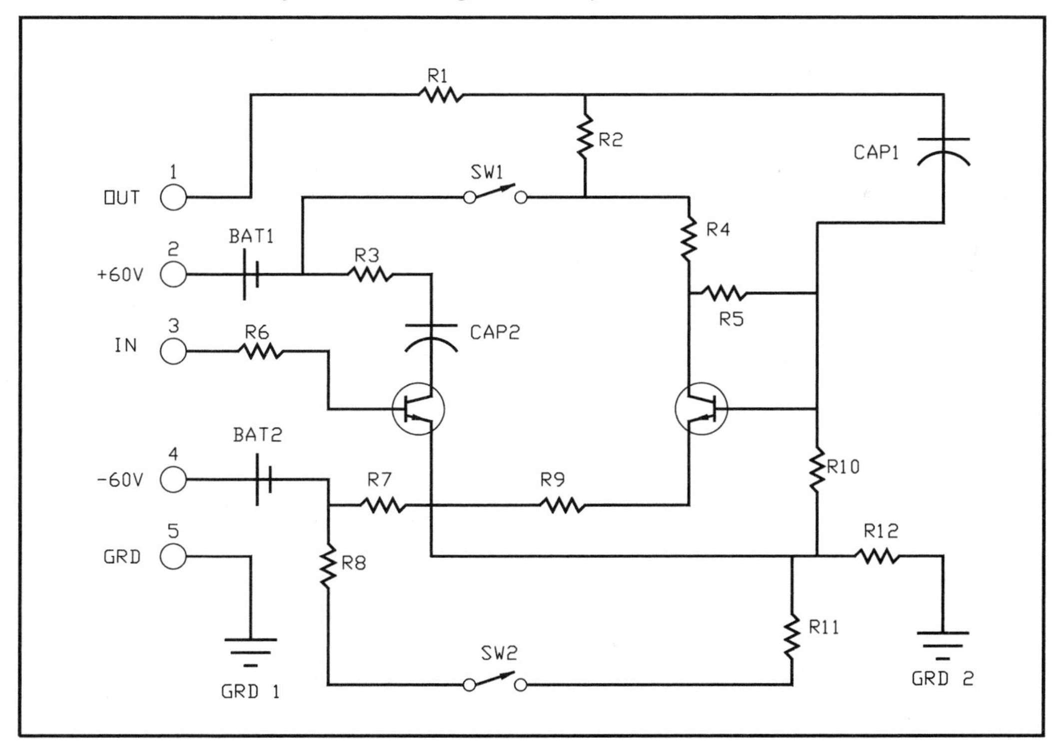


**Figure Ex10–5a** *Electrical Schmatic*

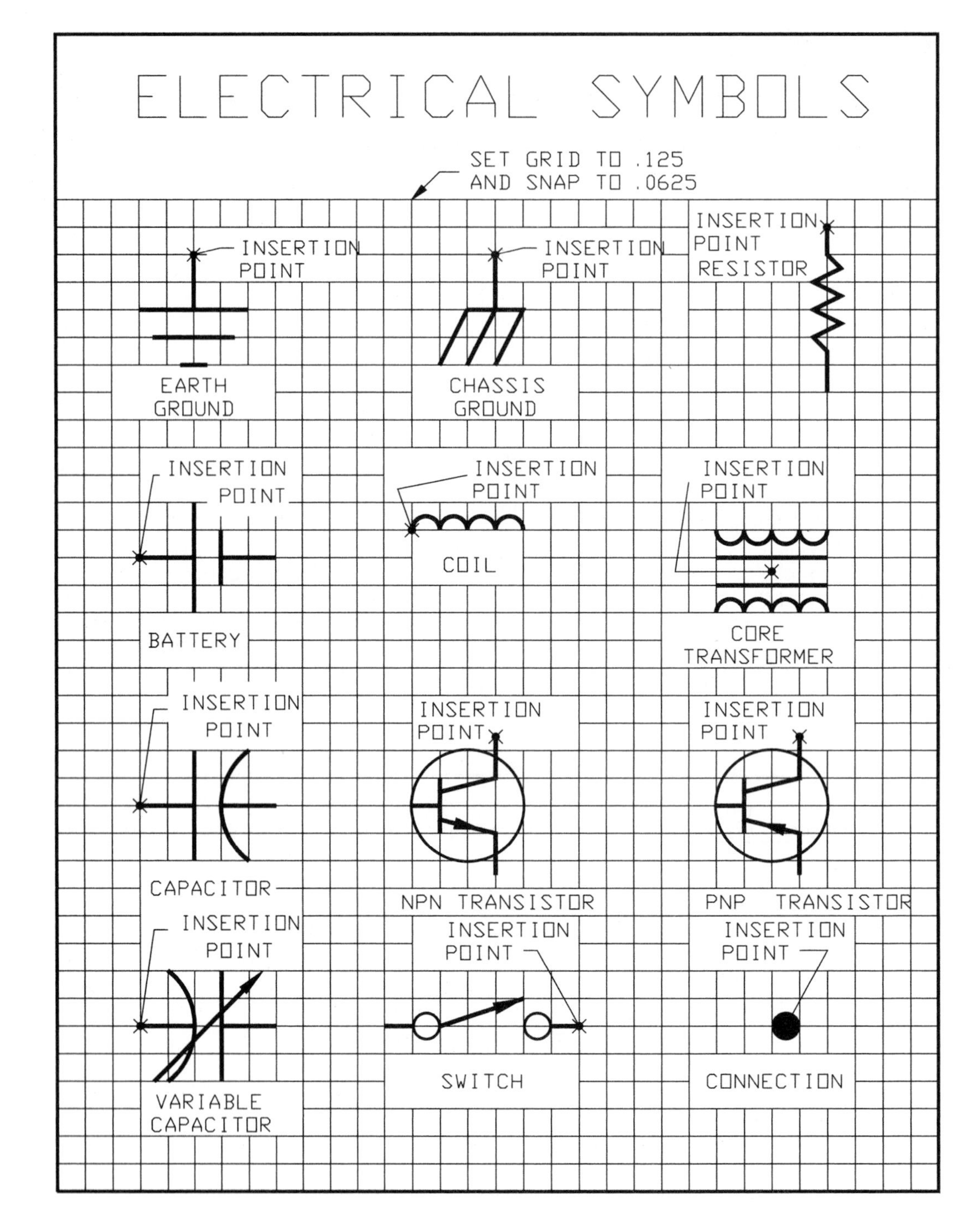


**Figure Ex10–5b** *Electrical symbols*

## EXERCISE 10–6

Create ground-fault protection diagram as shown in the figure according to the settings given in the following table:

| Settings | Value | | |
|---|---|---|---|
| 1. Units | Decimal | | |
| 2. Limits | Lower left corner: 0,0 | | |
| | Upper right corner: 17,11 | | |
| 3. Layers | *NAME* | *COLOR* | *LINETYPE* |
| | Border | White | Continuous |
| | Relays | Green | Continuous |
| | Tripcoil | Blue | Continuous |
| | Switchs | White | Continuous |
| | Sensor | White | Continuous |
| | Notes | White | Continuous |
| | WiresA | White | Continuous |
| | WiresB | White | Hidden |
| | StoredMech | White | Continuous |

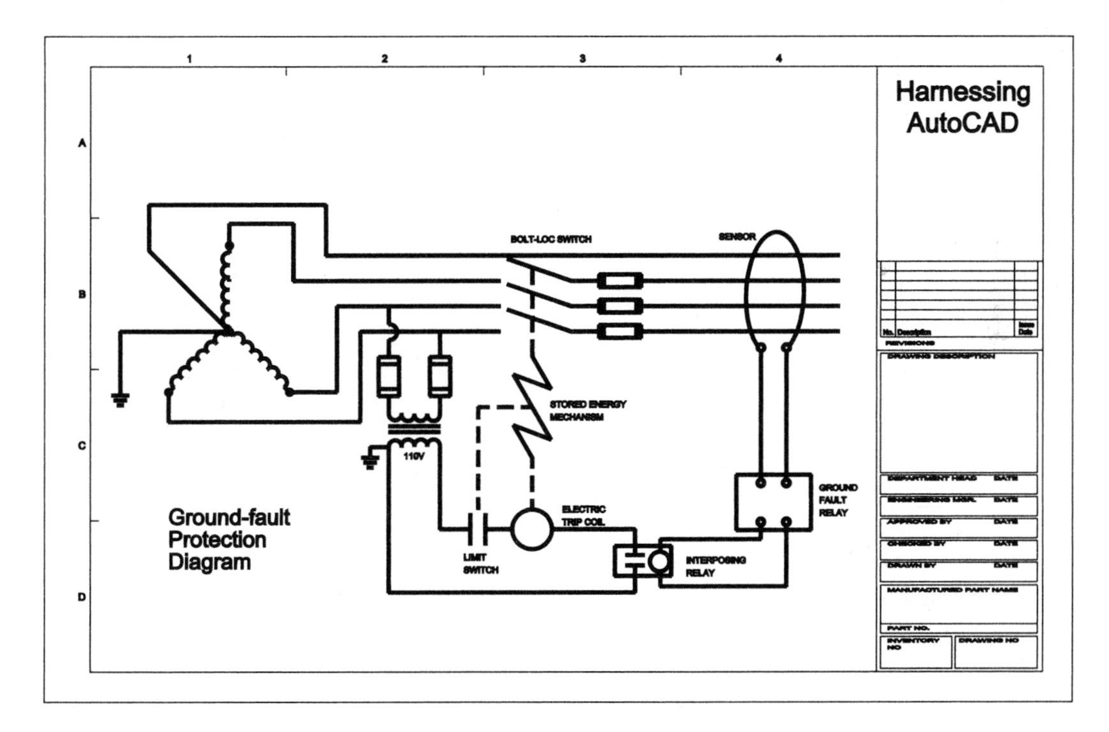


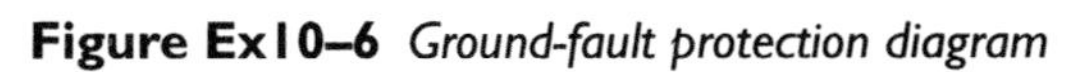
**Figure Ex10–6** *Ground-fault protection diagram*

## EXERCISE 10–7

Create the drawing shown in Figure Ex10–7 according to the settings given in the following table:

| Settings | Value | | |
|---|---|---|---|
| 1. Units | Decimal | | |
| 2. Limits | Lower left corner: 0,0<br>Upper right corner: 17,11 | | |
| 3. Layers | *NAME* | *COLOR* | *LINETYPE* |
| | Border | White | Continuous |
| | Breaker | Green | Center |
| | Receptacles | Blue | Continuous |
| | Wires | White | Continuous |
| | Notes | White | Hidden |

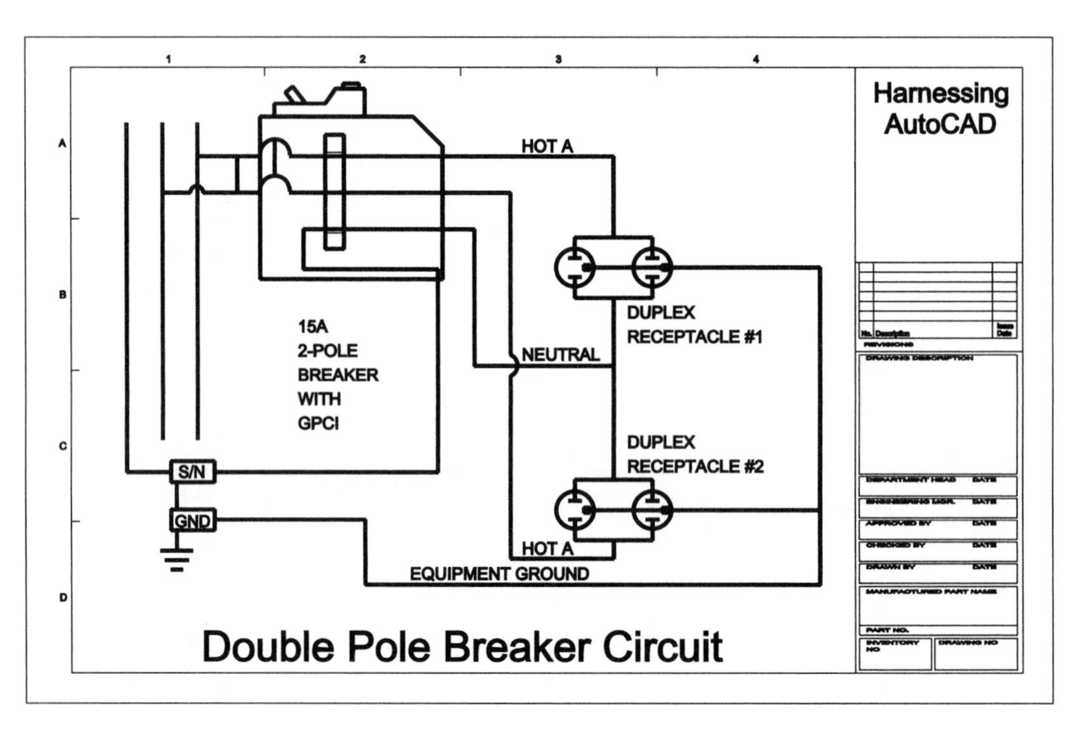


**Figure Ex10–7** *Double Pole Breaker Circuit*

CHAPTER 11

# External References

## PROJECT EXERCISE

This project exercise provides point-by-point instructions for creating two drawings and then using them as external references in the third drawing, shown in Figure P11–1.

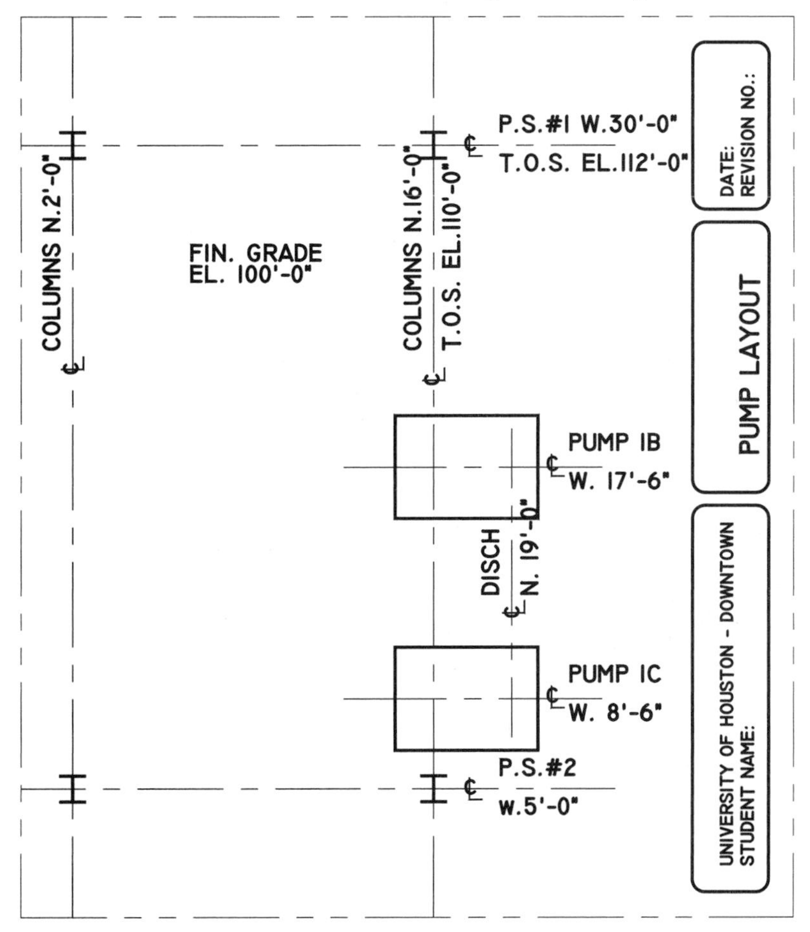


**Figure P11–1** *Completed project drawing*

**Step 1:** Start the AutoCAD program.

**Step 2:** Create a new drawing with the parameters given in the following table.

| SETTINGS | VALUE | | |
|---|---|---|---|
| UNITS | Architectural | | |
| LIMITS | Lower left corner: 0,0 | | |
| | Upper right corner: 28'-0",35'-0" | | |
| GRID | 1' | | |
| SNAP | 6" | | |
| LAYERS | *NAME* | *COLOR* | *LINETYPE* |
| | Centerline | Green | Center |
| | Object | White | Continuous |
| | Text | Blue | Continuous |
| | Border | Red | Phantom |

Set Centerline as the current layer and set up LTSCALE factor to 32.

**Step 3:** Invoke the LINE command from the Draw toolbar and draw the centerlines as shown in Figure P11–2.

Command: **line**
Specify first point: **2',0**
Specify next point or [Undo]: **@35'<90**
Specify next point or [Undo]: (ENTER)
Command: **line**
Specify first point: **0,5'**
Specify next point or [Undo]: **@23'<0**
Specify next point or [Undo]: (ENTER)

Invoke the OFFSET command from the Modify toolbar, and offset the vertical line 14' to the right and the horizontal line 25' upward.

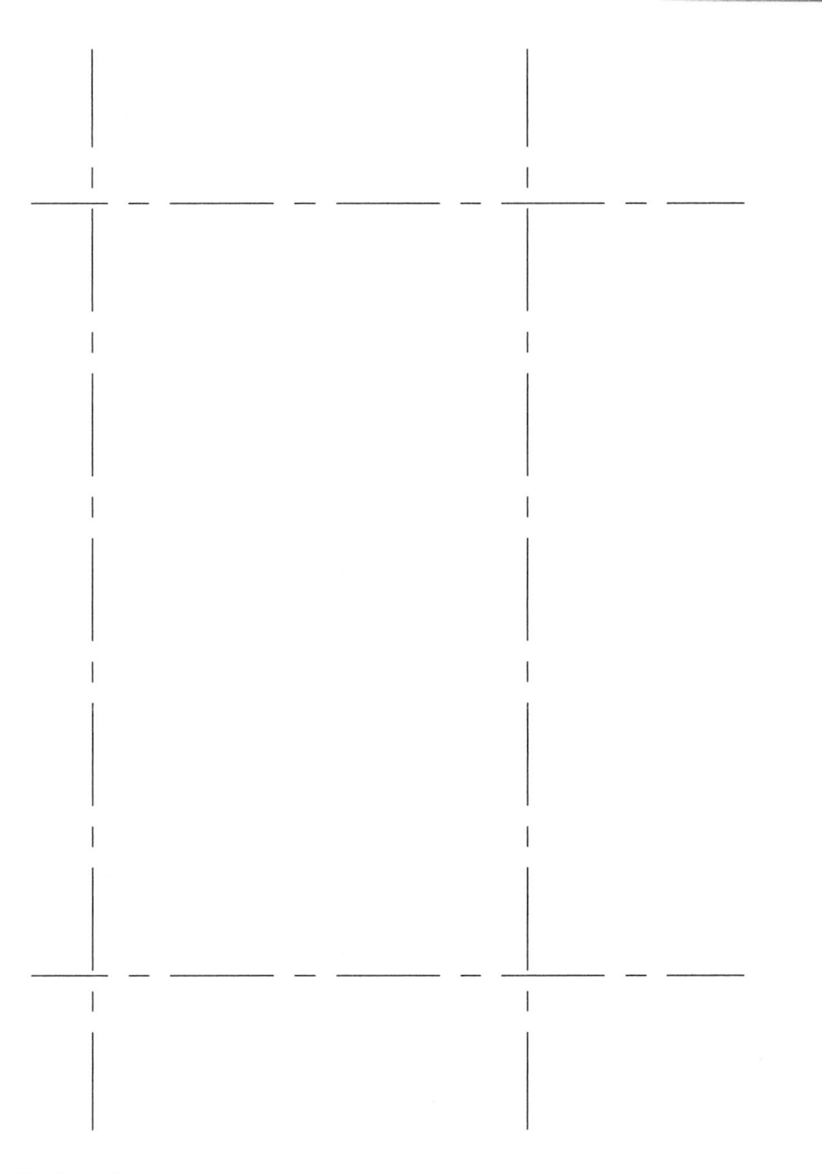

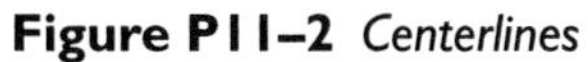

**Figure P11–2** *Centerlines*

**Step 4:** Set Object as the current layer, and use the PLINE command with the width set to 1" to draw the steel columns at the points shown in Figure P11–3. The steel columns are 12" × 12".

**Note:** Draw one column and then copy it to the other locations. Make sure the grid and snap are set to ON.

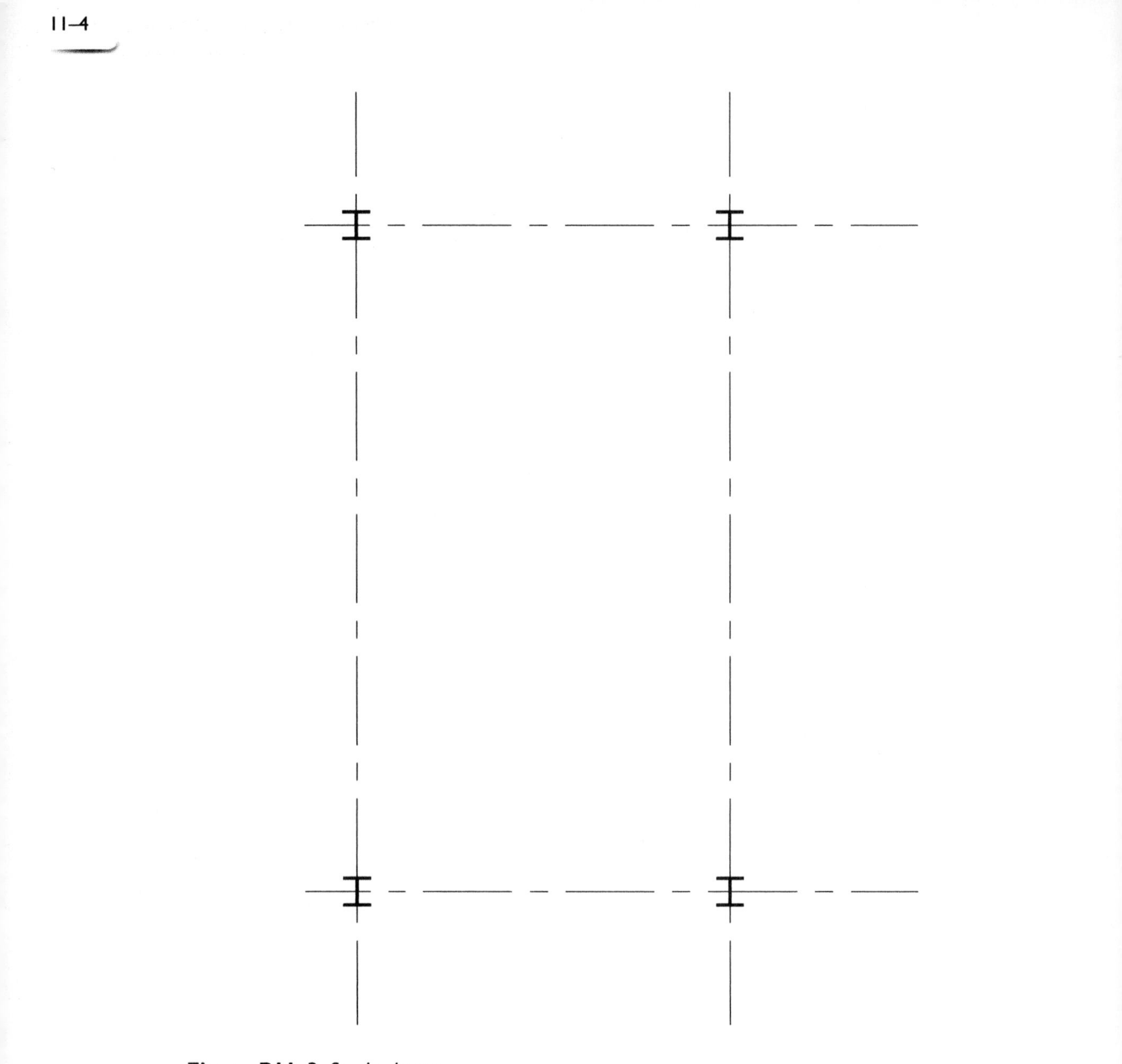

**Figure P11–3** *Steel columns*

**Step 5:** Set Text as the current layer, and invoke the TEXT command to draw in the coordinate callouts as shown in Figure P11–4. Set the text height to 8"; this will cause the text to be plotted out 0.25" high when plotted at 3/8" = 1'-0".

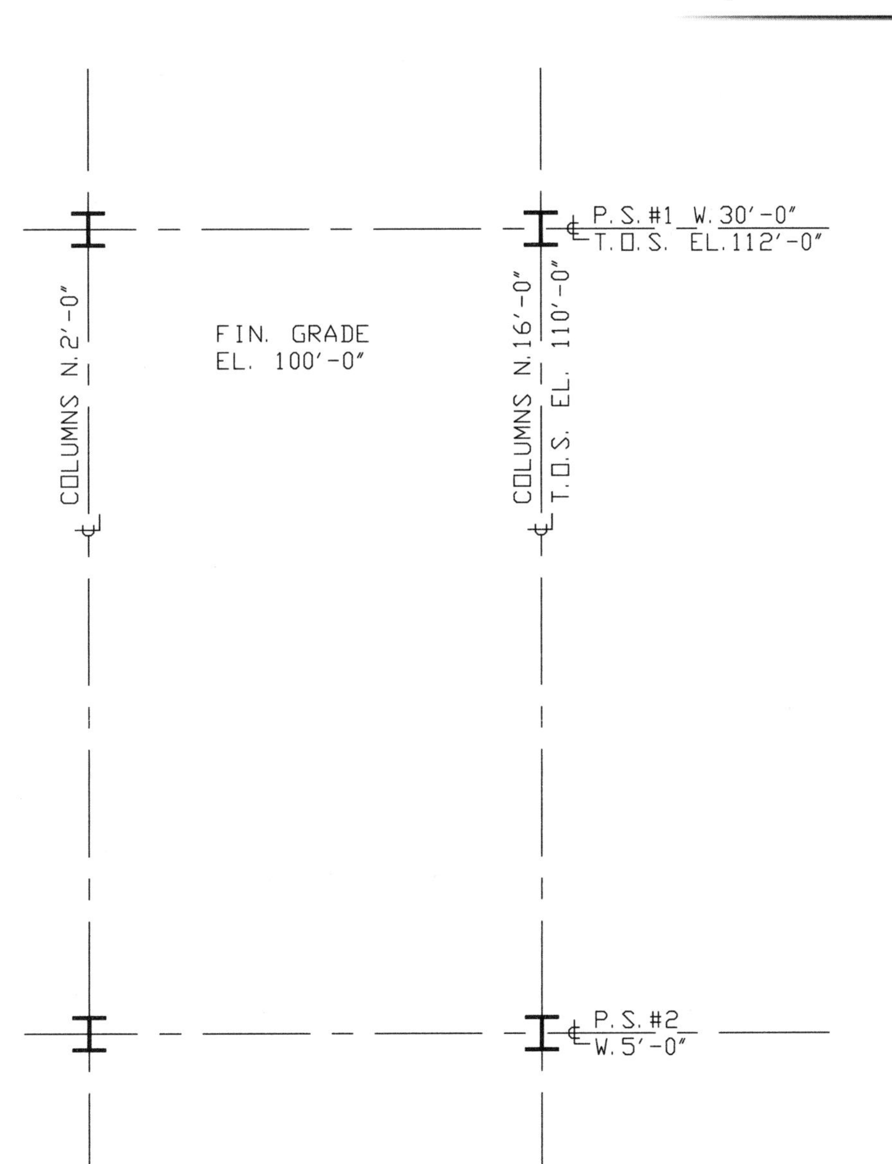


**Figure P11–4** *Add text to the drawing*

**Step 6:** Save the drawing with the name "Piperack" and close the drawing.

**Step 7:** Create another new drawing with the parameters given earlier in the settings table. Set Centerline as the current layer.

**Step 8:** Invoke the LINE command from the Draw toolbar and draw the centerlines as shown in Figure P11–5.

Command: **line**
Specify first point: **12'6,8'6**
Specify next point or [Undo]: **@10'<0**
Specify next point or [Undo]: (ENTER)
Command: **line**
Specify first point: **19',7'**
Specify next point or [Undo]: **@12'<90**
Specify next point or [Undo]: (ENTER)

Invoke the OFFSET command from the Modify toolbar and offset the horizontal centerline 9' upward.

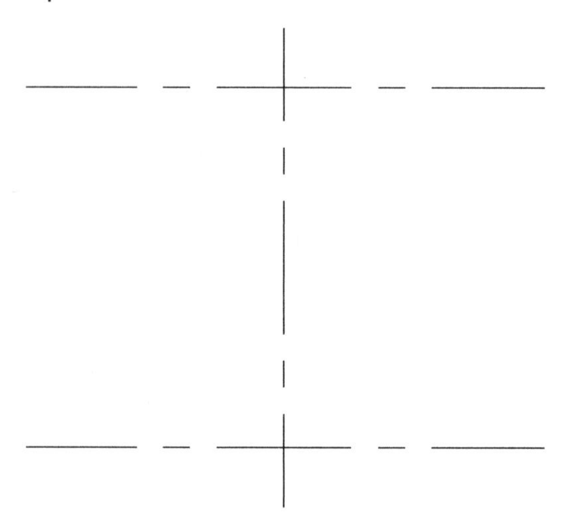

**Figure P11–5** *Pump centerlines*

**Step 9:** Set Object as the current layer and use the LINE command to draw the pump foundation as shown in Figure P11–6.

Command: **line**
Specify first point: **14'6,6'6**
Specify next point or [Undo]: **@5'6<0**
Specify next point or [Undo]: **@4'<90**
Specify next point or [Close/Undo]: **@5'6<180**
Specify next point or [Close/Undo]: **c**

Command: **line**
Specify first point: **14'6,15'6**
Specify next point or [Undo]: **@5'6<0**
Specify next point or [Undo]: **@4'<90**
Specify next point or [Close/Undo]: **@5'6<180**
Specify next point or [Close/Undo]: **c**

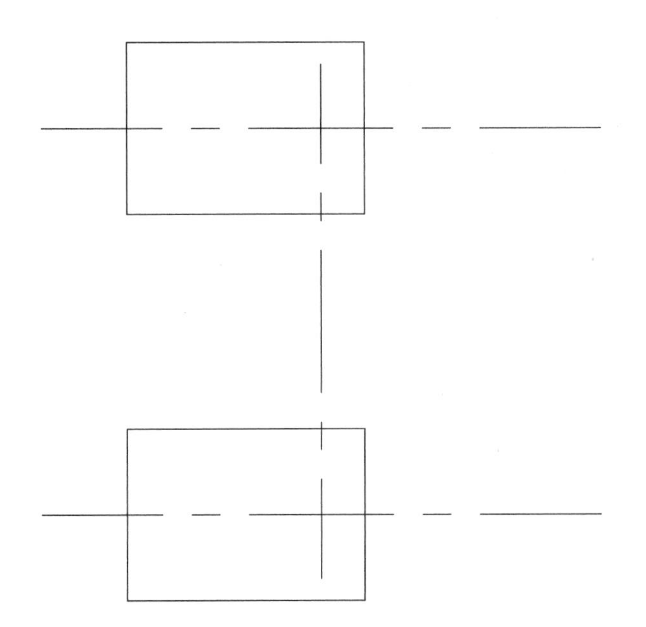

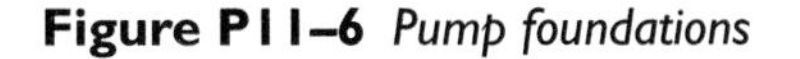

**Figure P11–6** *Pump foundations*

**Step 10:** Set Text as the current layer and invoke the TEXT command from the Draw toolbar to draw in the coordinate callouts as shown in Figure P11–7. Set the text height to 8". This will cause the text to be plotted out 0.25" high when plotted at 3/8=1'-0".

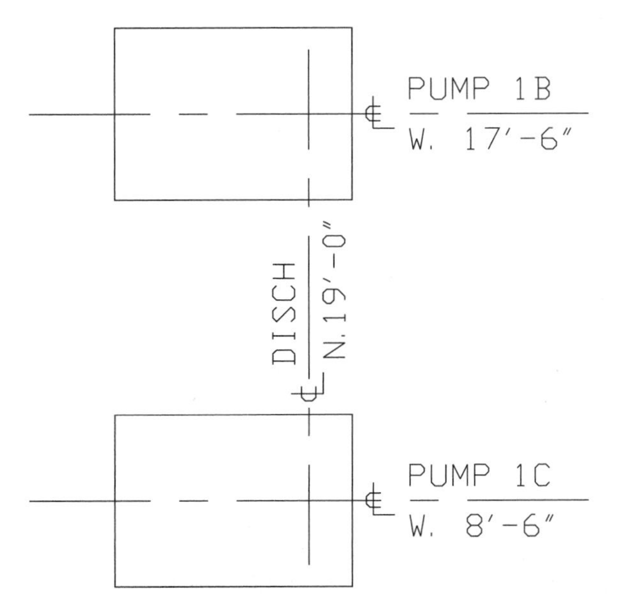


**Figure P11–7** *Add text to the drawing*

**Step 11:** Save the drawing with the name "Pump" and close the drawing.

**Step 12:** Create another new drawing with the parameters given earlier in the settings table. Set Border as the current layer.

**Step 13:** Invoke the LINE command from the Draw toolbar and draw the border line as shown in Figure P11–8.

Command: **line**
Specify first point: **0,0**
Specify next point or [Undo]: **@28'<0**
Specify next point or [Undo]: **@35'<90**
Specify next porin or [Close/Undo]: **@28'<180**

Specify next point or [Close/Undo]: **c**

**Figure P11–8** *Border drawing*

**Step 14:** Attach the piperack drawing as a reference file to the current drawing with an insertion point of 0,0, a scale factor of 1.0, and a rotation angle of 0 degrees.

After attaching the piperack drawing, your drawing will appear as shown in Figure 11–9.

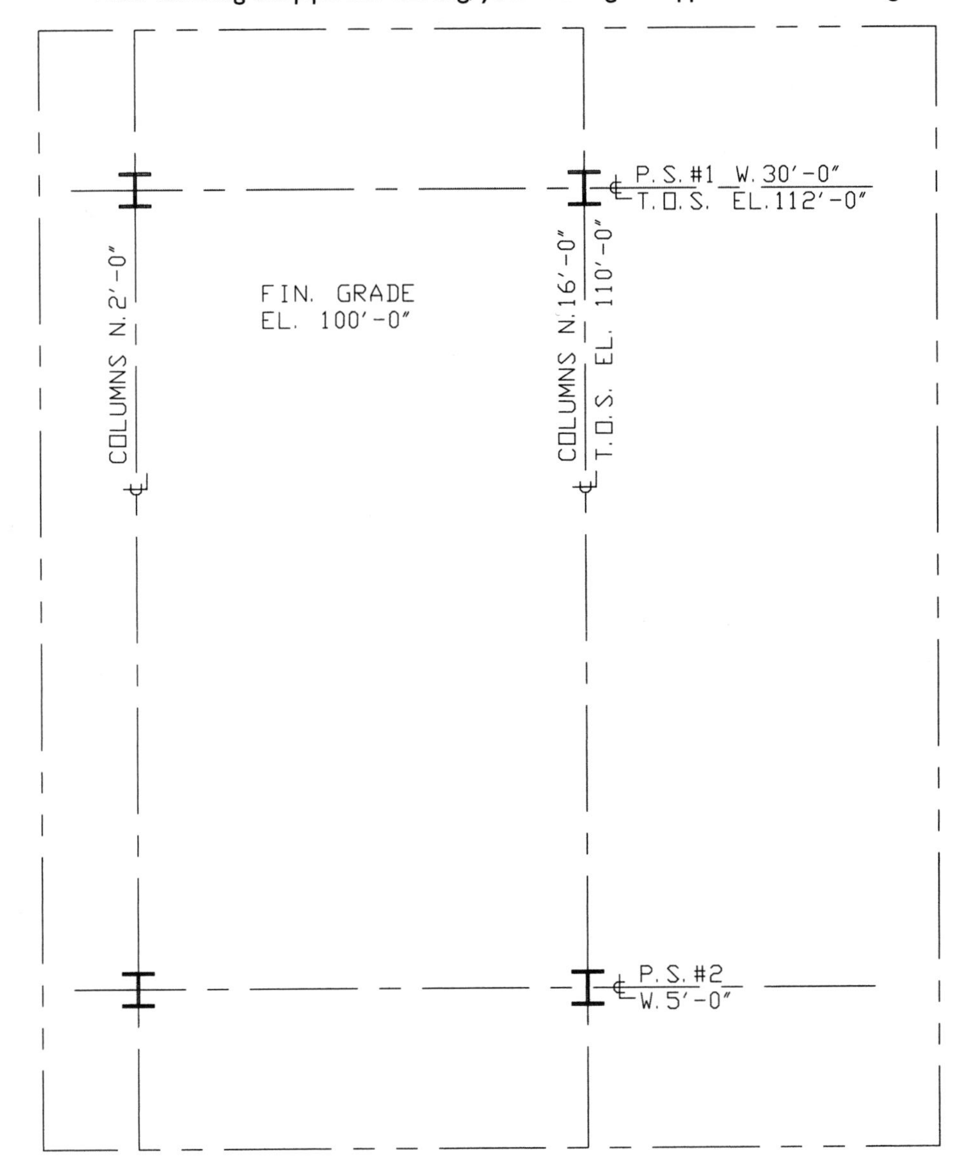


**Figure P11–9** *Attach the pipe rack drawing*

**Step 15:** Attach the pump drawing as a reference file to the current drawing with an insertion point of 0,0, a scale factor of 1.0, and a rotation angle of 0 degrees.

After you attach the pump drawing, your drawing will appear as shown in Figure P11–10.

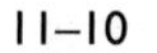

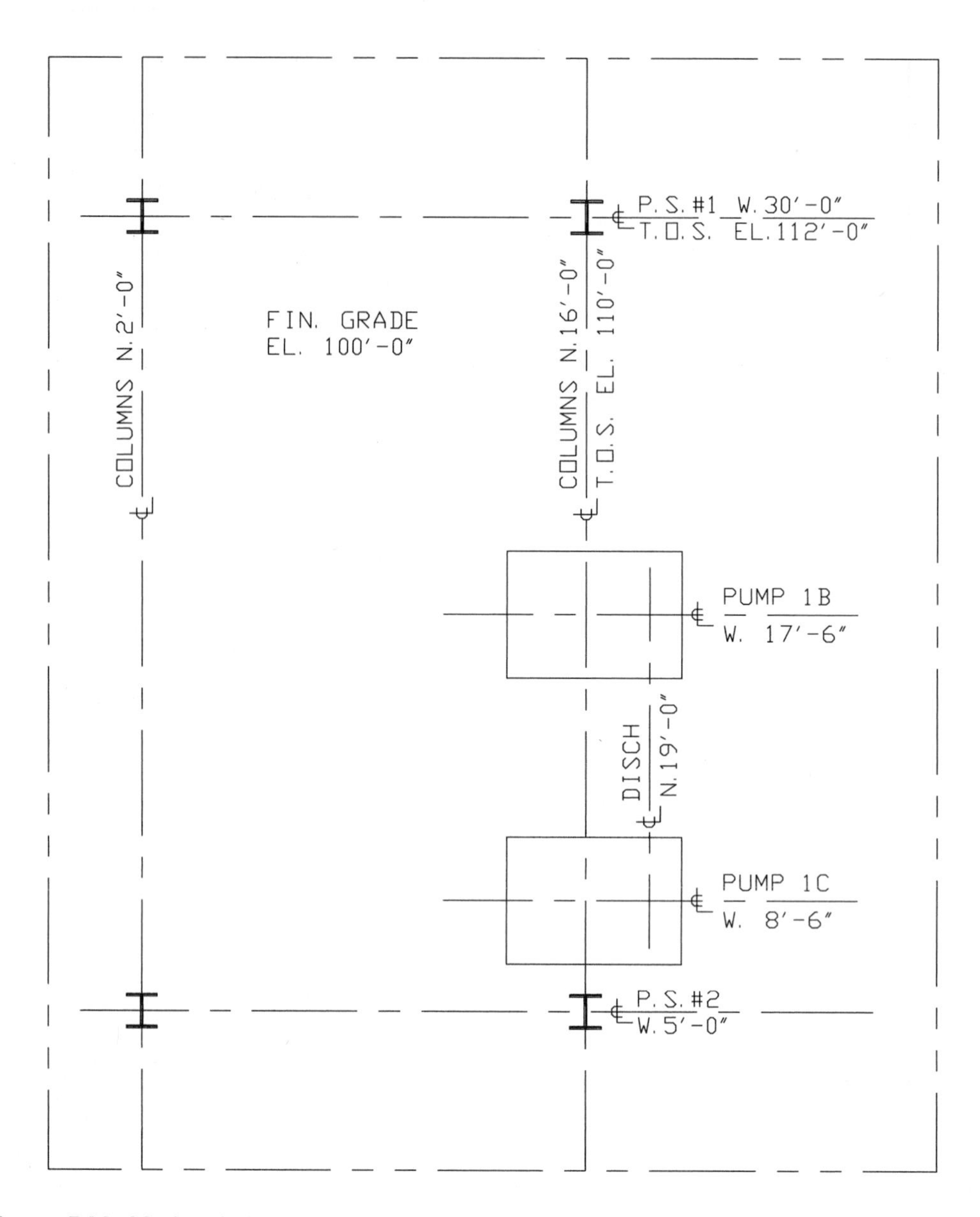


**Figure P11–10** *Attach the pump drawing*

**Step 16:** Complete the title block as shown in Figure P11–11, with your name and the date filled in. Save the drawing as PROJCH11, and close the session.

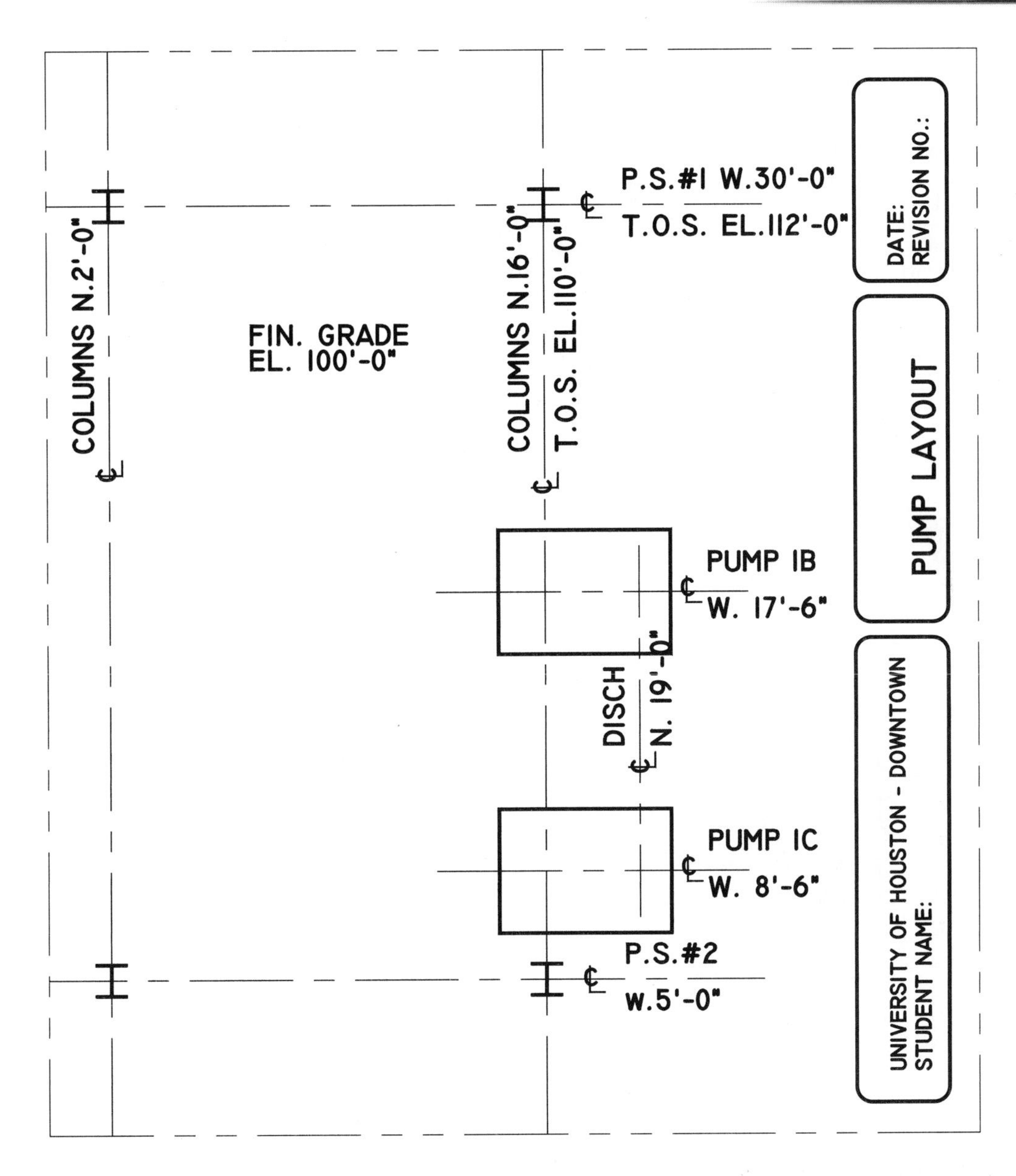


**Figure P11–11** *Completed drawing*

## EXERCISE 11–1

Create a new drawing according to the parameters given in the table, and draw the vessel shown in Figure Ex11–1a. Save the drawing with the name EX11–1A.

| Settings | Value | | |
|---|---|---|---|
| 1. Units | Architectural | | |
| 2. Limits | Lower left corner: 0,0 | | |
| | Upper right corner: 28'-0",35'-0" | | |
| 3. Grid | 1' | | |
| 4. Snap | 6" | | |
| 5. Layers | *NAME* | *COLOR* | *LINETYPE* |
| | Centerline | Green | Center |
| | Object | White | Continuous |
| | Text | Blue | Continuous |

**Hint:** This exercise uses a drawing that already has external references attached to it, and those external references must also be accessible for this exercise.

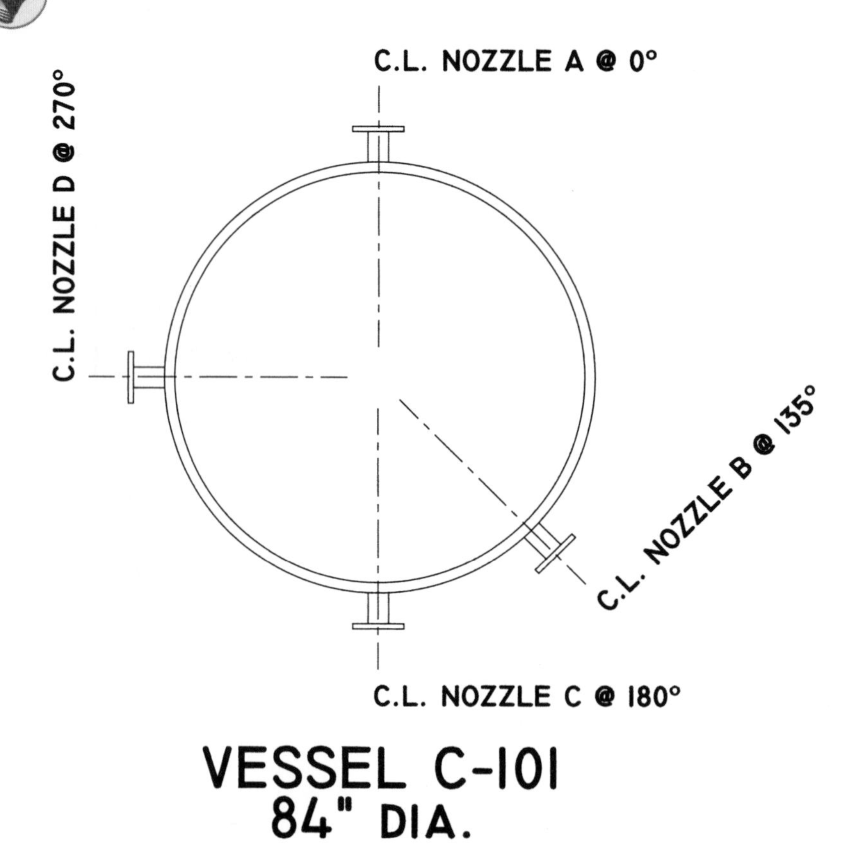


**Figure Ex11–1a** *Vessel layout*

Create another new drawing according to the parameters given in the table, and draw an octagonal foundation as shown in Figure Ex11–1b. Do *not* add the dimensions, which are provided here only for reference.

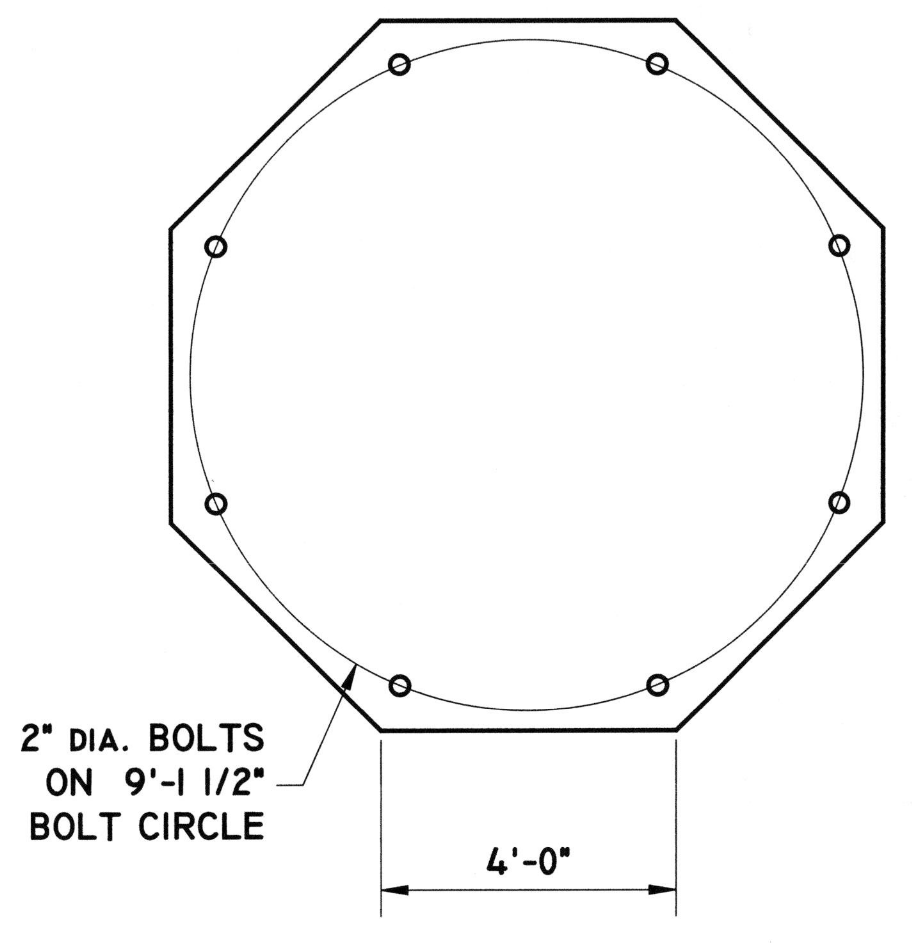


**Figure Ex11–1b** *Octagonal foundation*

Attach the file Ex11–1A as a reference file, with an insertion point of 0,0, a scale factor of 1.0, and a rotation angle of 0 degrees. The drawing should appear as shown in Figure Ex11–5c. Save the drawing as EX11–5C.

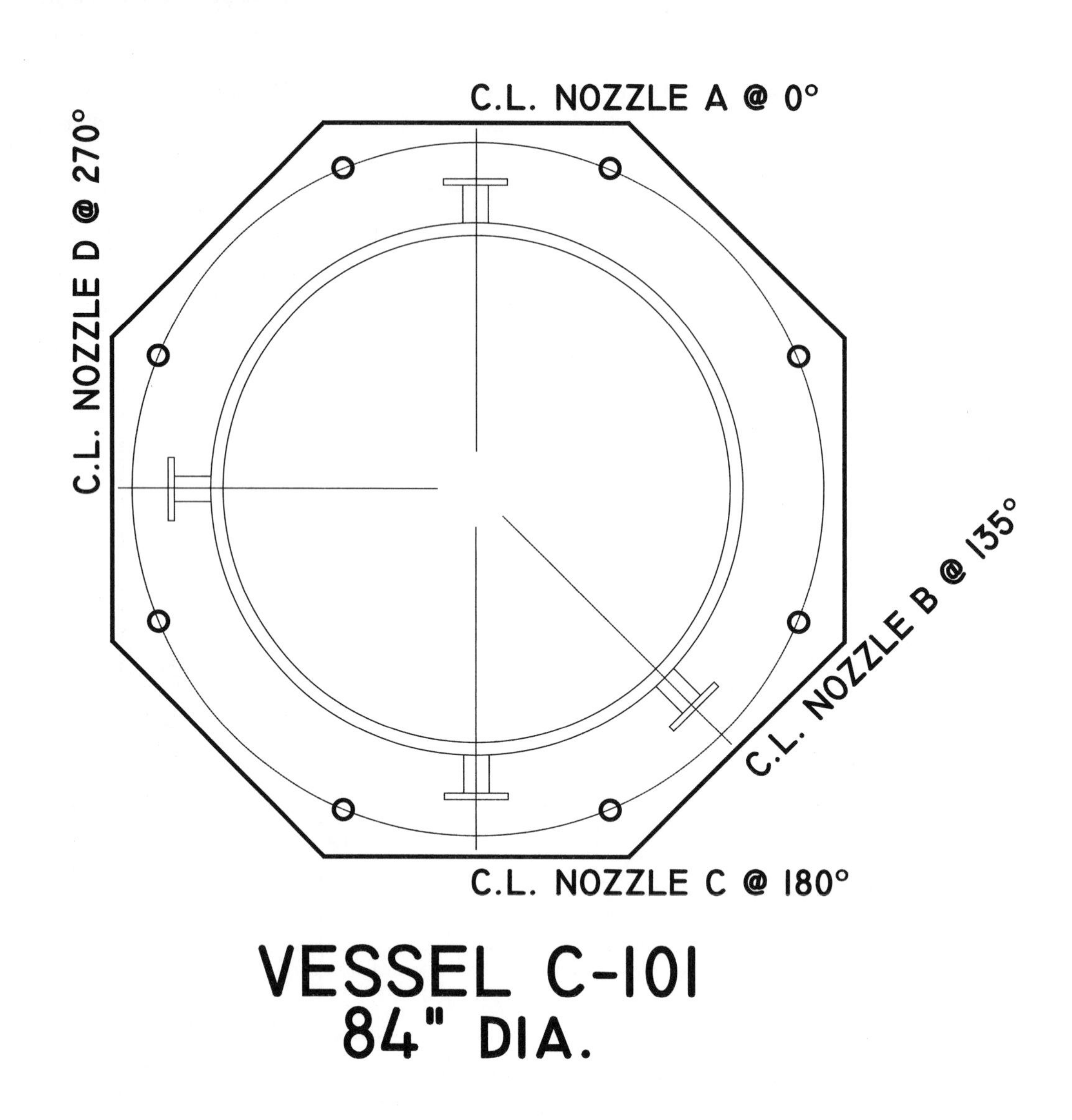


**Figure Ex11–1c** *Vessel in an octagonal foundation*

Open the drawing of the vessel and pump plan, PROJCH11.DWG, the project exercise completed earlier in this chapter. Attach the file Ex11–1C as a reference file, with an insertion point of 7'-0",8'-0", a scale factor of 1.0, and a rotation angle of 0 degrees. The drawing should appear as shown in Figure Ex11–1d. Save the drawing as EX11–1D.

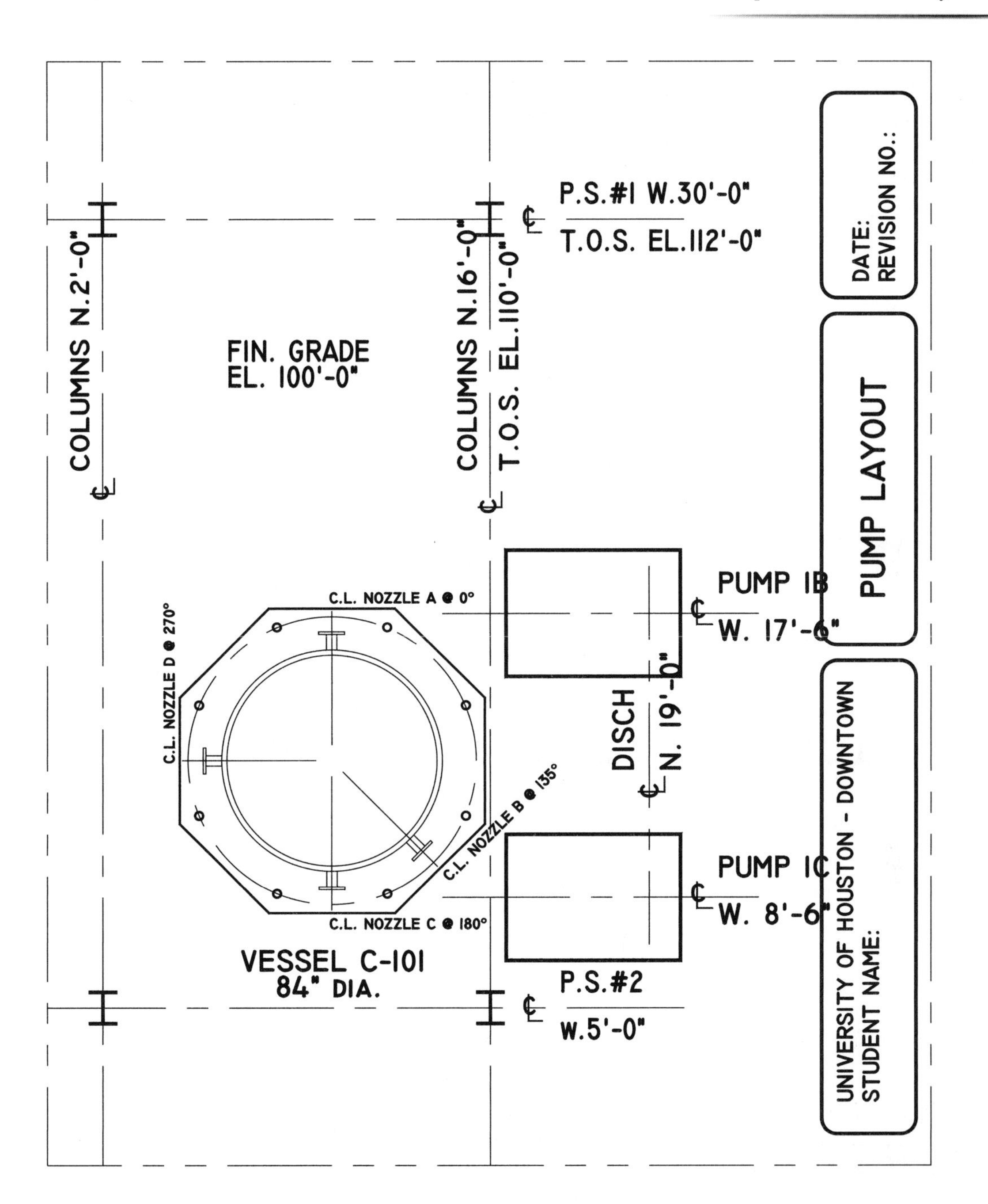


**Figure Ex11–1d** *Layout of the vessel and pump with a vessel in an octagonal foundation*

## EXERCISE 11–2

Create a new drawing and draw check valve bonnet with nuts as shown in figure EX11-2a to the settings given below. Save the drawing as EX11–2A.

| Settings | Value | | |
|---|---|---|---|
| 1. Units | Architectural | | |
| 2. Limits | Lower left corner: 0,0,<br>Upper right corner: 22",15" | | |
| 3. Grid | .0625" | | |
| 4. Snap | .03125" | | |
| 5. Layers | *NAME* | *COLOR* | *LINETYPE* |
| | Center | White | Center |
| | Bonnet | Green | Continuous |
| | Bolt holes | Green | Hidden |
| | Nuts | Yellow | Continuous |
| | Dimensions | White | Continuous |

**Hint:** Begin the drawing as point XY=9.75, 6.50 as shown in Figure

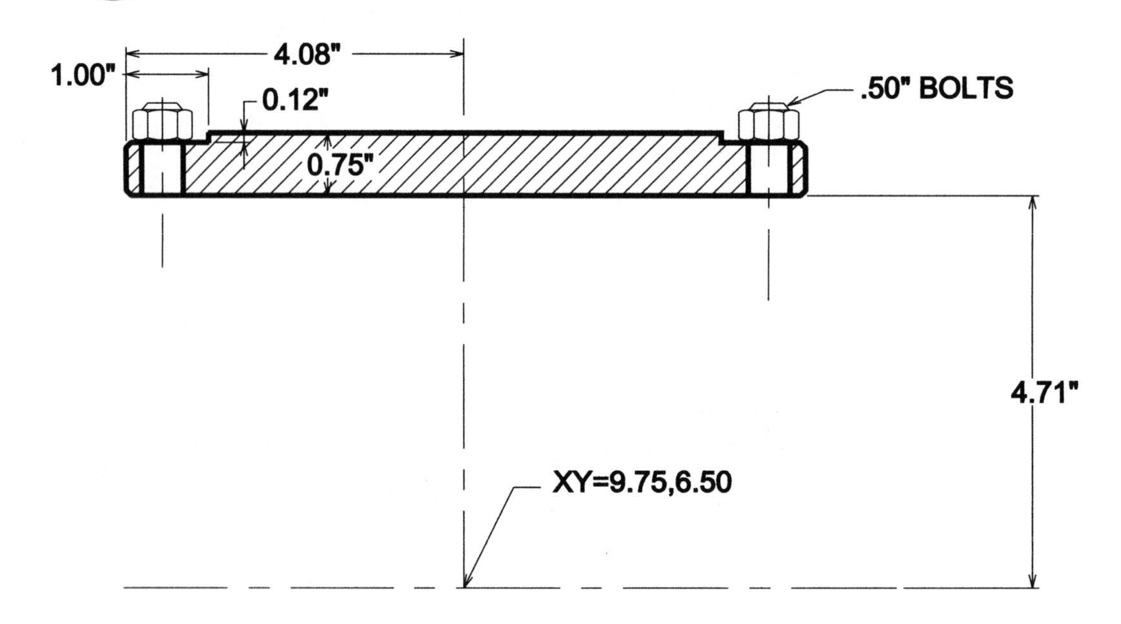


**Figure Ex11–2a** *Check Valve Bonnet*

Create another new drawing and draw the check valve body with nuts as shown in figure EX11–2b to the settings given below. Save the drawing as EX11–2B.

| Settings | Value | | |
|---|---|---|---|
| 1. Units | Architectural | | |
| 2. Limits | Lower left corner: 0,0,<br>Upper right corner: 22",15" | | |
| 3. Grid | .0625" | | |
| 4. Snap | .03125" | | |
| 5. Layers | NAME | COLOR | LINETYPE |
| | Center | White | Center |
| | Body | Blue | Continuous |
| | Bolt holes | Blue | Hidden |
| | Nuts | Yellow | Continuous |
| | Dimensions | White | Continuous |
| | Text | White | Continuous |

**Hint:** Begin the drawing as point XY=9.75, 6.50 as shown in Figure.

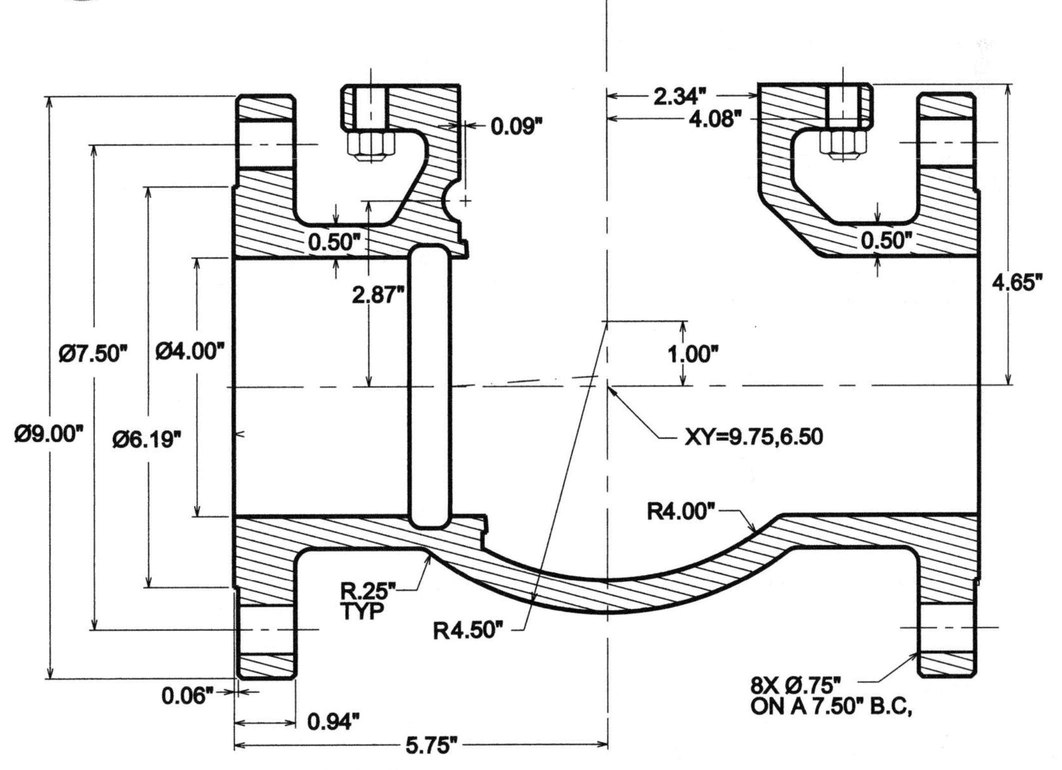


**Figure Ex11–2B** *Check Valve Body*

Create another new drawing and draw check valve clapper as shown in Figure EX11–2C to the settings given below. Save the drawing EX11–3C.

| Settings | Value | | |
|---|---|---|---|
| 1. Units | Architectural | | |
| 2. Limits | Lower left corner: 0,0,<br>Upper right corner: 22",15" | | |
| 3. Grid | .0625" | | |
| 4. Snap | .03125" | | |
| 5. Layers | *NAME* | *COLOR* | *LINETYPE* |
| | Center | White | Center |
| | Clapper | Red | Continuous |
| | Dimensions | White | Continuous |

**Hint:** Begin the drawing as point XY=9.75, 6.50 as shown in Figure.

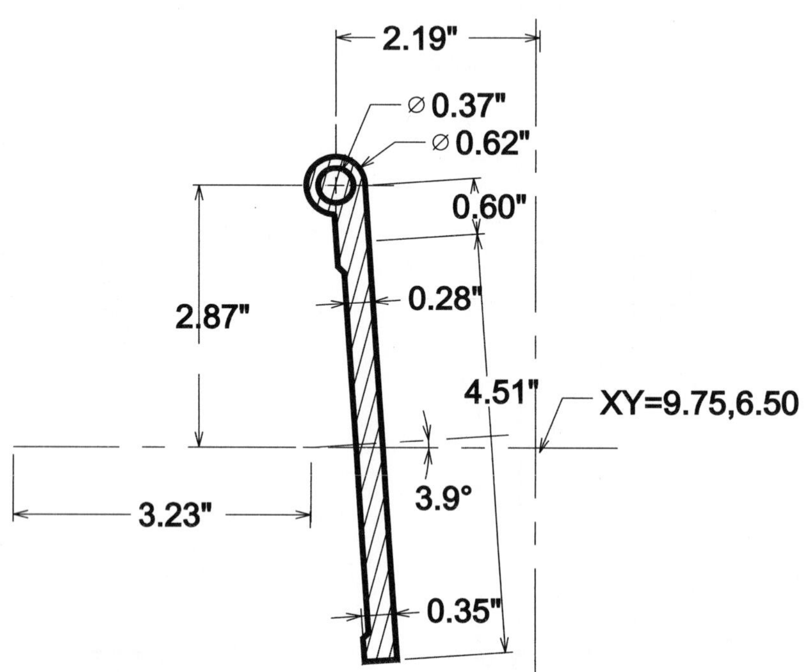


**Figure Ex11–2C** *Check valve clapper*

Create another new drawing and draw the title block as shown Figure EX11–2D to the settings given below. Save the drawing CHECK VALVE.

| Settings | Value | | |
|---|---|---|---|
| 1. Units | Architectural | | |
| 2. Limits | Lower left corner: 0,0,<br>Upper right corner: 29",22" | | |
| 3. Grid | .0625" | | |
| 4. Snap | .03125" | | |
| 5. Layers | *NAME* | *COLOR* | *LINETYPE* |
| | Title block | White | Continuous |

Attach the file EX11–2A as a reference file, with an insertion point of XY=9.75, 6.50 with a scale factor of 1.0 and a rotation angle of 0 degrees. Then attach the file EX11–2B as a reference file, with an insertion point of XY=9.75, 6.50 to a scale factor of 1.0 and a rotation angle of 0 degrees. And then attach the file EX11–2C as reference file to complete the assembly with an interstion point of XY=9.75,6.50 to a scale factor of 1.0 and rotation angle of 0 degrees. Completed assembly should appear as shown in EX11–2D and save the drawing as EX11–2D.

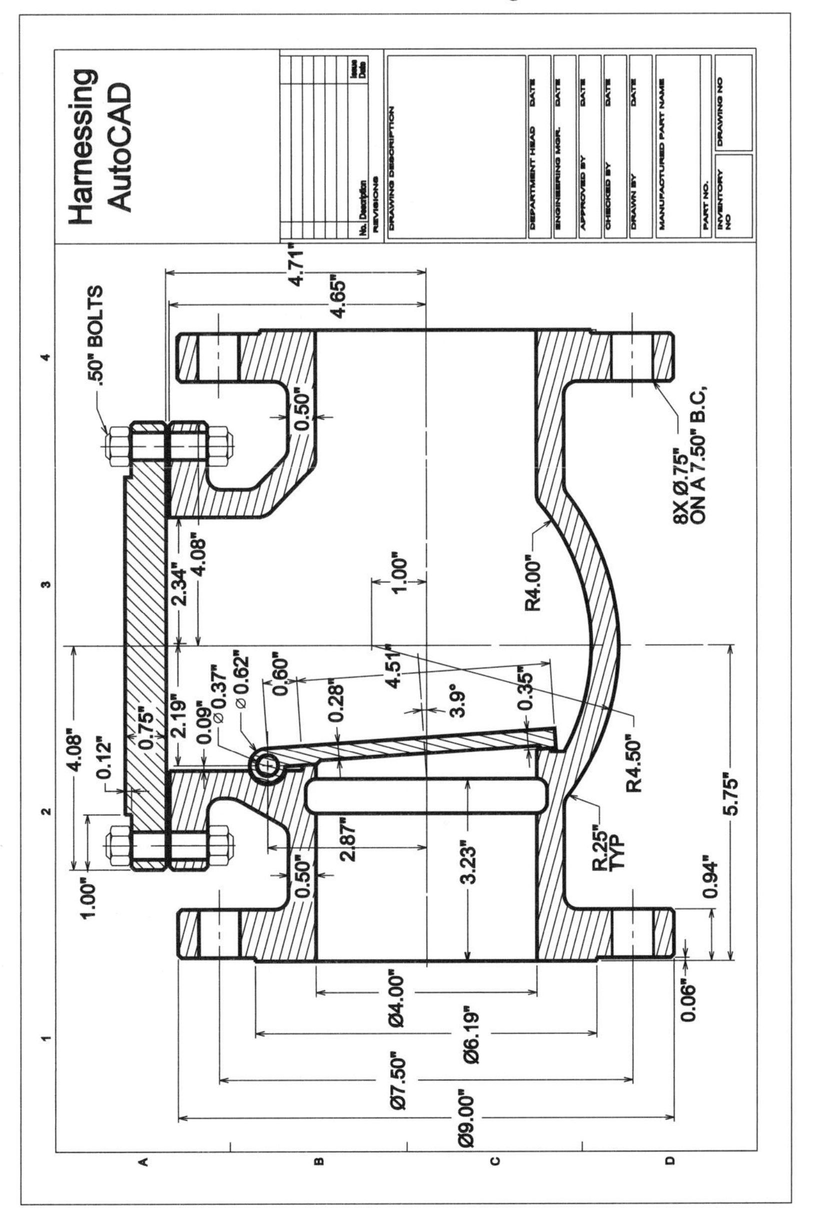


**Figure Ex11–2D** *Completed assembly*

## EXERCISE 11–3

Create a new drawing according to the parameters given in the table, and draw the column layout as shown in Figure Ex11–3a. Add the necessary dimensions and save the drawing as EX11–3A.

| Settings | Value | | |
|---|---|---|---|
| 1. Units | Architectural | | |
| 2. Limits | Lower left corner: –10'-0",–10'-0"<br>Upper right corner: 50'-0",35'-0" | | |
| 3. Grid | 12" | | |
| 4. Snap | 6" | | |
| 5. Layers | *NAME* | *COLOR* | *LINETYPE* |
| | Border | Red | Continuous |
| | Beams | Cyan | Continuous |
| | Columns | Green | Continuous |
| | Text | Blue | Continuous |
| | Dimension | White | Continuous |

**Hint:** The columns on the corners are 8" x 8", and the remaining columns are 10" x 10". They can be drawn as polylines with a width of 0.5.

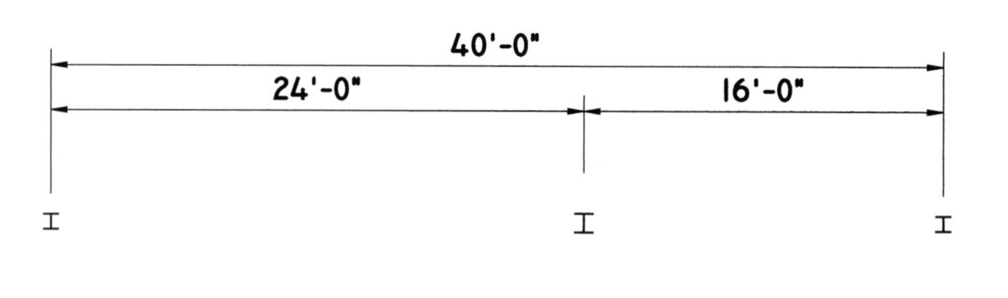


**Figure Ex11–3a** *Layout of the columns*

Create another new drawing according to the parameters given in the table.Attach the file EX11–1a as a reference file. Draw the walls and doors as shown in Figure Ex11–3b. Add the necessary dimensions, text, and save the drawing as EX11–3B.

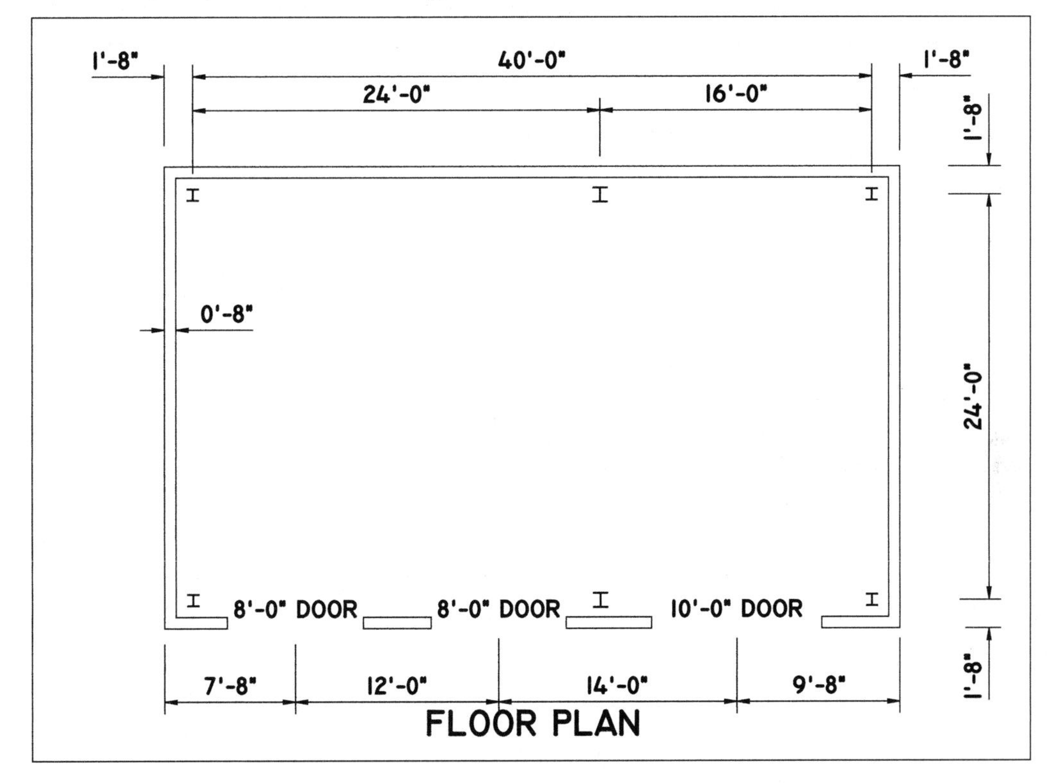


**Figure Ex11–3b** *Layout of the walls, doors, and columns*

Open the drawing EX11–3A and make the necessary changes for column spacing as shown in Figure Ex11–3c. Save the drawing.

Open the drawing EX11–3B; the changes in the column layout show up automatically. The walls and doors will have to be altered to conform to the revised external reference column layout. Make the necessary changes, including eliminating one of the doors, as shown in Figure Ex11–3d. The wall along the column line on the left has to be relocated, and the dimension of the remaining door in the left bay changed from 8'-0" to 10'-0". Save the drawing as Ex11–3D.

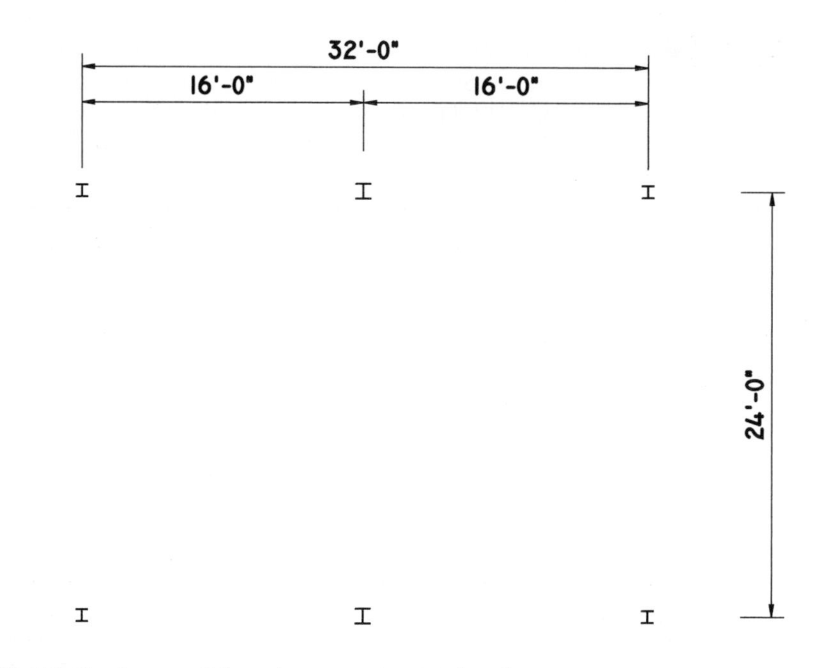


**Figure Ex11–3c** *Layout of the column spacing (updated)*

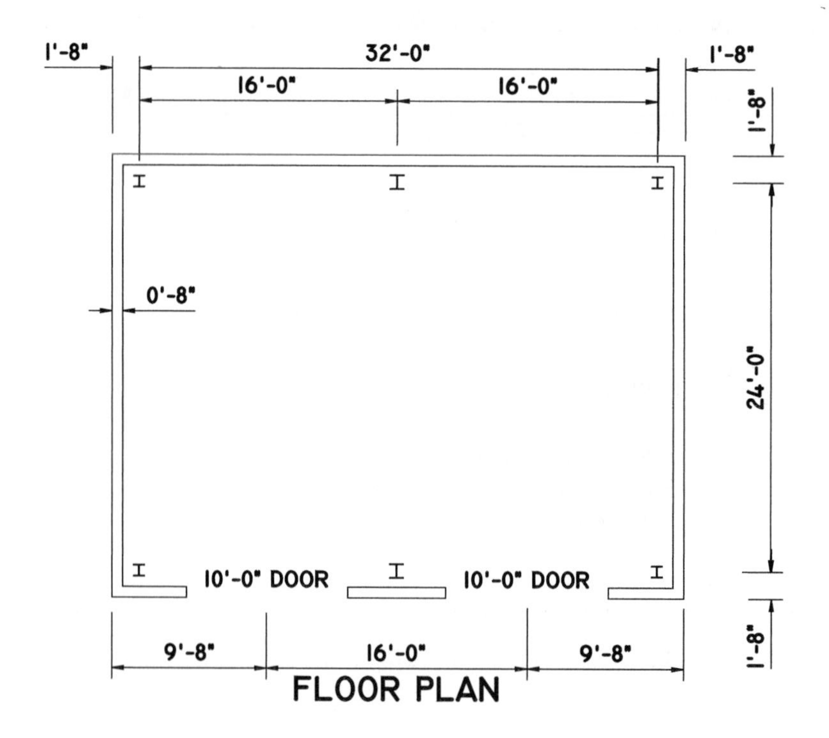


**Figure Ex11–3d** *Layout of the columns (final revision)*

## EXERCISE 11–4

Open the drawing of the laundry and storage floor plan, *PROJCH6.DWG (Project Exercise from Chapter 6)*. Set the Construction, Hidden, and Text layers to OFF. The drawing should appear as shown in Figure Ex11–4a. Save the drawing with the name EX11–4A.

| Settings | Value | | |
|---|---|---|---|
| 1. Units | Architectural | | |
| 2. Limits | Lower left corner: –5'-0",–6'-0" | | |
| | Upper right corner: 24'-4",15'-4" | | |
| 3. Grid | 6" | | |
| 4. Snap | 2" | | |
| 5. Layers | NAME | COLOR | LINETYPE |
| | Border | Red | Continuous |
| | Columns | Green | Continuous |
| | Text | Blue | Continuous |
| | Dimension | White | Continuous |
| | Electrical | Magenta | Continuous |
| | Lighting | Red | Continuous |

**Hint:** The electrical outlet symbols (110V and 220V) can each be drawn one time and then copied to the other locations. The curved line from the switch to the light fixture can be drawn on the same layer as the switch and the fixture. You can then use the Properties button on the standard toolbar and change the Linetype to hidden.

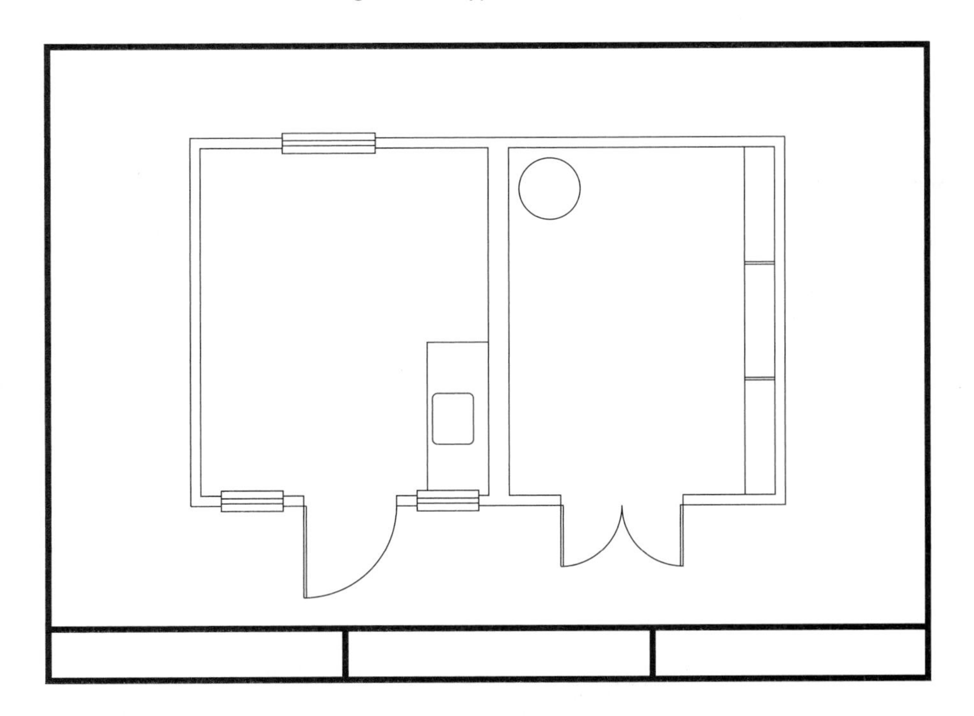

**Figure Ex11–4a** *Laundry and storage floor plan with the Construction, Dim, Hidden, and Text layers set to OFF*

Create a new drawing according to the parameters given in the table. Attach the file EX11–4A as a reference file. Turn on the layer EX11–4a hidden to show squares on the left side of the wall. Draw the lighting and electrical symbols as shown in Figure Ex11–4b. Save the drawing as EX11–4B.

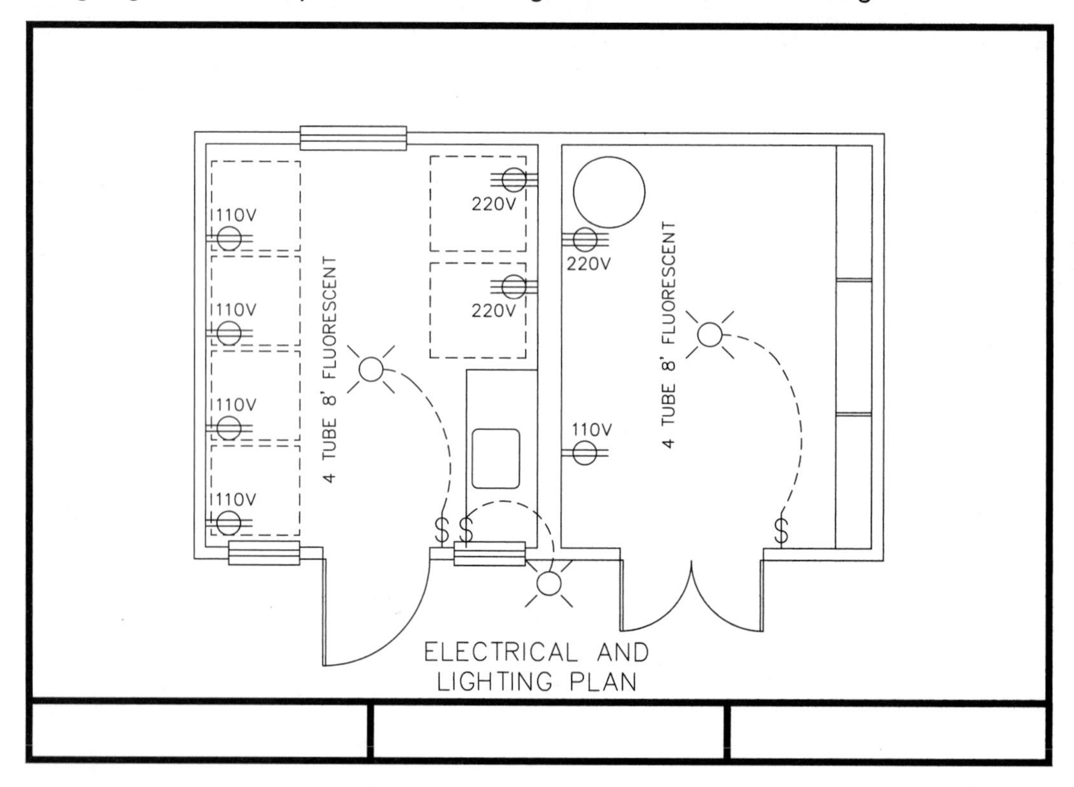


**Figure Ex11–4b** *Layout of the walls, doors, and columns*

## EXERCISE 11–5

Open the drawing *EX5-8.dwg* that was completed in Chapter 5. Erase Detail A, Detail B, and the details of the stair riser. Also erase the interior footings (the 24 double rectangles inside the perimeter grade beams), the dimensions, and two interior lines of the five perimeter lines. Change the linetype of the perimeter grade beam lines to hidden linetype. The drawing should appear as shown in Figure Ex11–5a. Save the drawing with the name EX11–5.

| Settings | Value |
|---|---|
| 1. Units | Architectural |
| 2. Limits | Lower left corner: 0'-0",0'-0"<br>Upper right corner: 60'-0",45'-0" |
| 3. Grid | 12" |
| 4. Snap | 6" |

| Settings | Value | | |
|---|---|---|---|
| 5. Layers | NAME | COLOR | LINETYPE |
| | Construction | Cyan | Continuous |
| | Border | Red | Continuous |
| | Object | Green | Continuous |
| | Text | Blue | Continuous |
| | Dim | White | Continuous |

**Hint:** The foundation layout can be used from the external reference as the basis for the walls and stair for the roofing plan. Once the roofing plan is completed, the layers in the external reference can be turned off. If any later changes are made in the foundation layout, they will automatically show up in the roofing plan, which can then be revised to suit the changes.

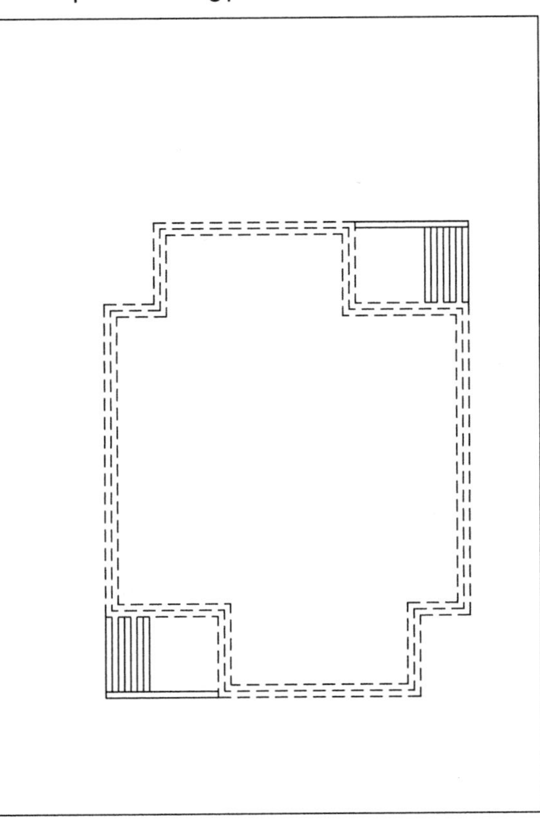

**Figure Ex11–5a** *Foundation plan*

Create a new drawing according to the parameters given in the table. Attach the file EX11–5a as a reference file, with an insertion point of 0,0, a scale factor of 1.0, and a rotation angle of 0 degrees. Draw the roof as shown in Figure Ex11–5b. Save the drawing as EX11–5B.

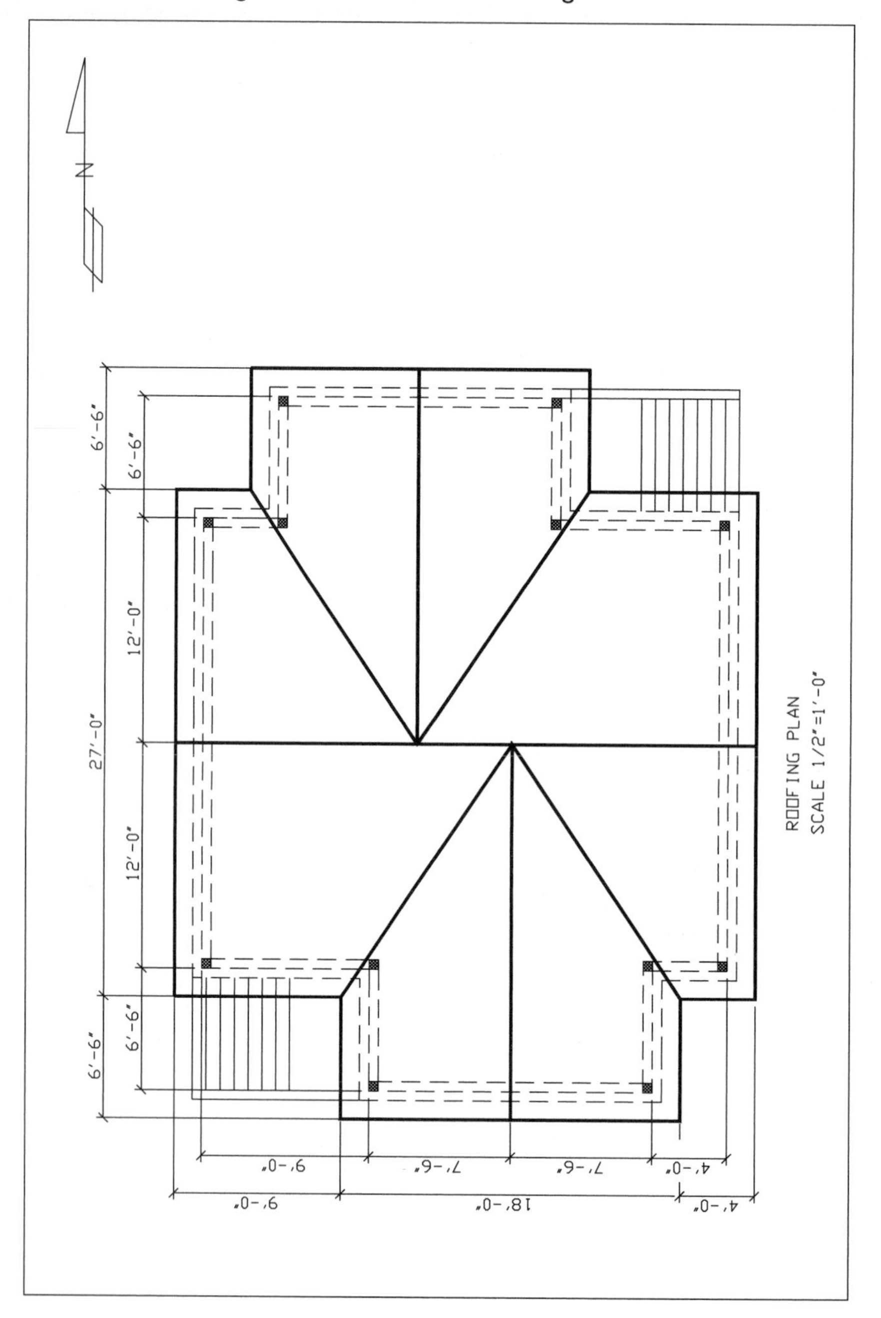


**Figure Ex11–5b** *Roof layout*

## EXERCISE 11–6

Create a new drawing according to the parameters given in the table, and draw the floor plan as shown in Figure Ex11–6a. Do *not* add the dimensions, which are provided here only for reference. Save the drawing as EX11–6A.

| Settings | Value | | |
|---|---|---|---|
| 1. Units | Architectural | | |
| 2. Limits | Lower left corner: –2'-0",–2'-0"<br>Upper right corner: 70'-0",50'-0" | | |
| 3. Grid | 6" | | |
| 4. Snap | 2" | | |
| 5. Layers | *NAME* | *COLOR* | *LINETYPE* |
| | Walls | Red | Continuous |
| | Ductwork | White | Continuous |
| | Text | Blue | Continuous |

**Hint:** The ductwork can be drawn with the MLINE command.

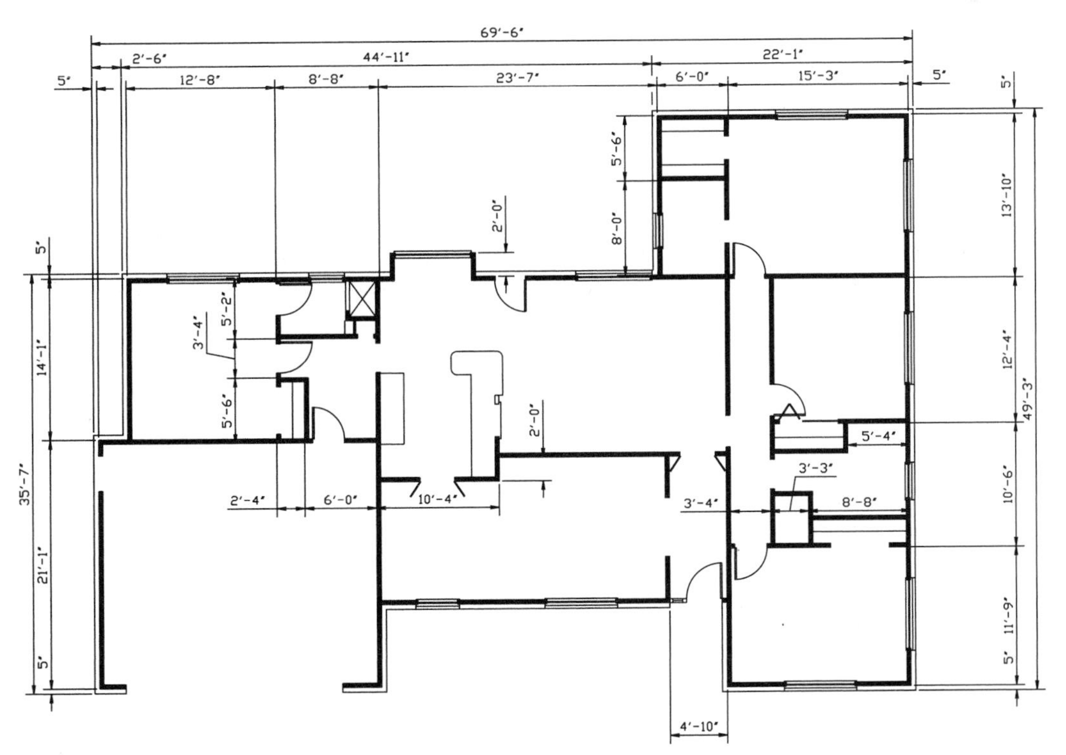


**Figure Ex11–6a** *Floor plan*

Create a new drawing according to the parameters given in the table. Attach the file EX11–6A as a reference file. Draw the A/C duct layout as shown in Figure Ex11–6b. Save the drawing as EX11–6B.

**Figure Ex11–6b** *Floor plan with the A/C duct layout*

## EXERCISE 11–7

Create a new drawing according to the settings below and draw the site plan as shown in EX11–7a. Save the drawing as EX11–7A.

| Settings | Value |
|---|---|
| 1. Units | Architectural |
| 2. Limits | Lower left corner: 0,0<br>Upper right corner: 500', 400' |

| Settings | Value | | |
|---|---|---|---|
| 3. Layers | *NAME* | *COLOR* | *LINETYPE* |
| | Road | White | Continuous |
| | Road Center | White | Center |
| | Curbs | Orange | Continuous |
| | Proplines | White | Phantom |
| | Monuments | Yellow | Continuous |
| | Contours | Green | Dashdot2 |
| | Trees | Green | Continuous |
| | Wooded area | Green | Continuous |
| | Turning Symbols | White | Continuous |
| | Notes | White | Continuous |
| | Dimensions | White | Continuous |
| | North Arrow | White | Continuous |

**Hint:** Begin the drawing at point XY= 31'-6", 264'-0" as shown in Figure.

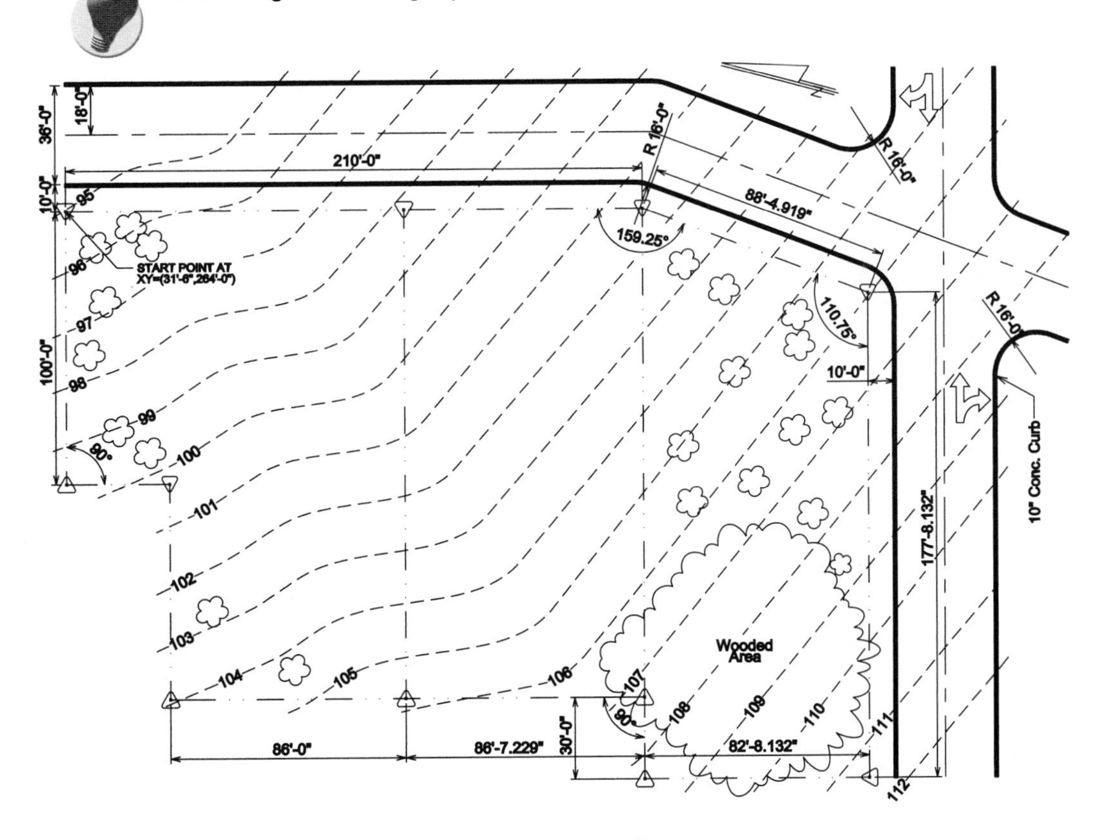


**Figure Ex11–7a** *Existing roads, property boundaries and contours*

Create a new drawing according to the settings below and draw the sub surface utilities layout as shown in the Figure EX11–7b. Save the drawing as EX11–7B.

| Settings | Value | | |
|---|---|---|---|
| 1. Units | Architectural | | |
| 2. Limits | Lower left corner: 0,0 | | |
| | Upper right corner: 500' , 400' | | |
| 3. Layers | *NAME* | *COLOR* | *LINETYPE* |
| | RoadCenter | White | Center |
| | Power Poles | White | Continuous |
| | Water main | Blue | Continuous |
| | Firehydrant | Blue | Continuous |
| | Storm sewer pipe | Red | Continuous |
| | Storm sewer manhole | Red | Continuous |
| | Catc basins | Yellow | Continuous |
| | Gasmain | Orange | Continuous |
| | Gas manhole | Orange | Continuous |
| | Cable & Tele | Purple | Phantom2 |

**Hint:** Begin the drawing as point XY= 31'-6", 264'-0" as shown in Figure.

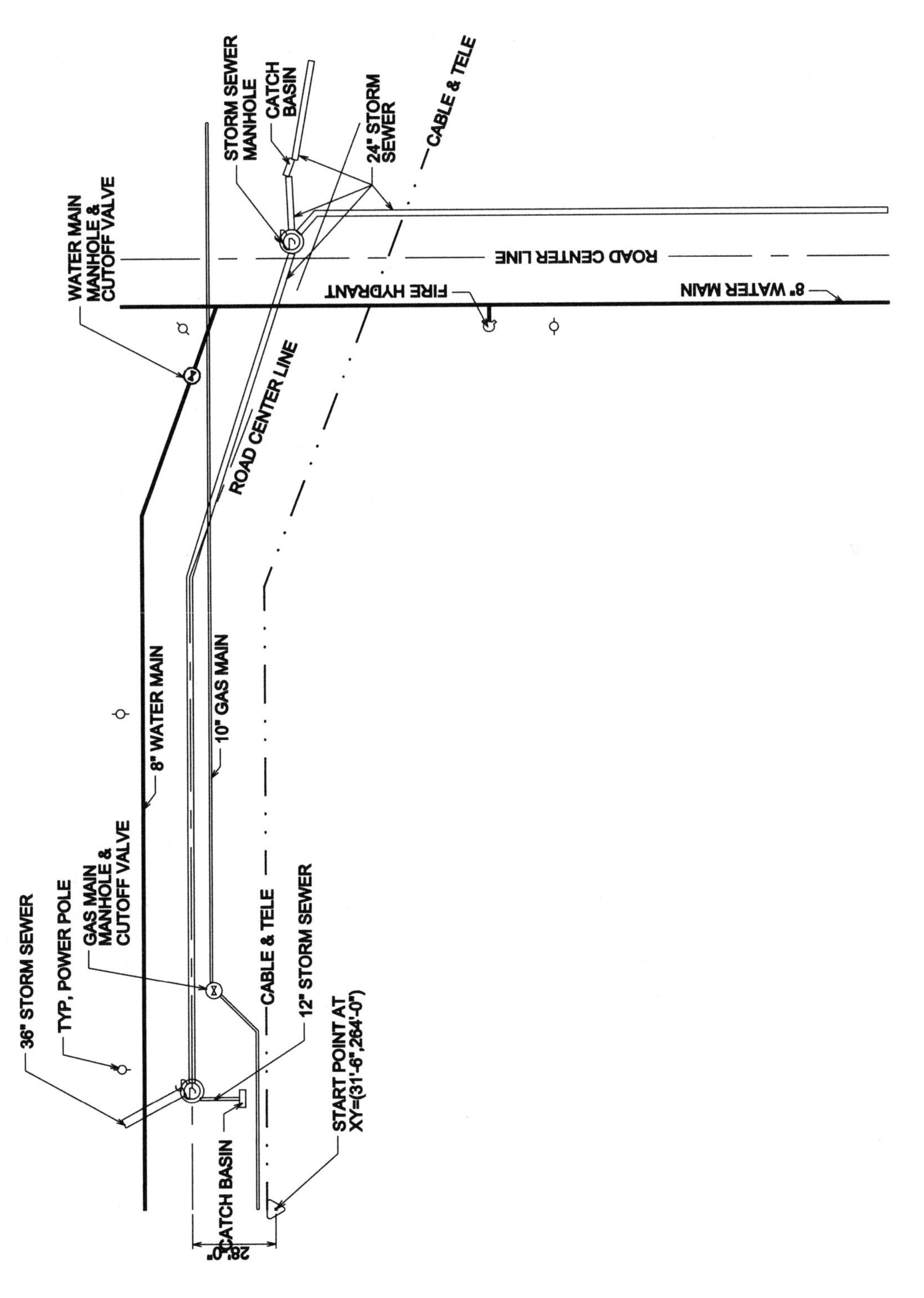


**Figure Ex11–7B** *Sub-surface utilities*

Create a new drawing according to the settings given below and draw the building footprint, entrance driveway and walk patterns for this site plan as shown in EX11–7C. Save the drawing as EX11–7C.

| Settings | Value | | |
|---|---|---|---|
| 1. Units | Architectural | | |
| 2. Limits | Lower left corner: 0,0 | | |
| | Upper right corner: 500', 400' | | |
| 3. Layers | *NAME* | *COLOR* | *LINETYPE* |
| | Road Center | White | Center |
| | Side walk | White | Continuous |
| | Entrance drive | White | Continuous |
| | Entrance drive center | White | Center |
| | Bldg footprint | Blue | Continuous |
| | Patter walks | Yellow | Continuous |

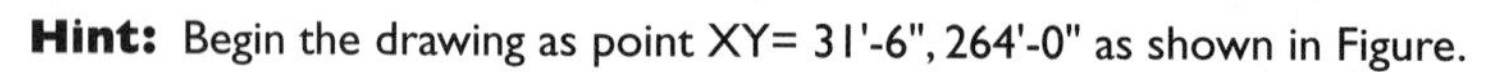

**Hint:** Begin the drawing as point XY= 31'-6", 264'-0" as shown in Figure.

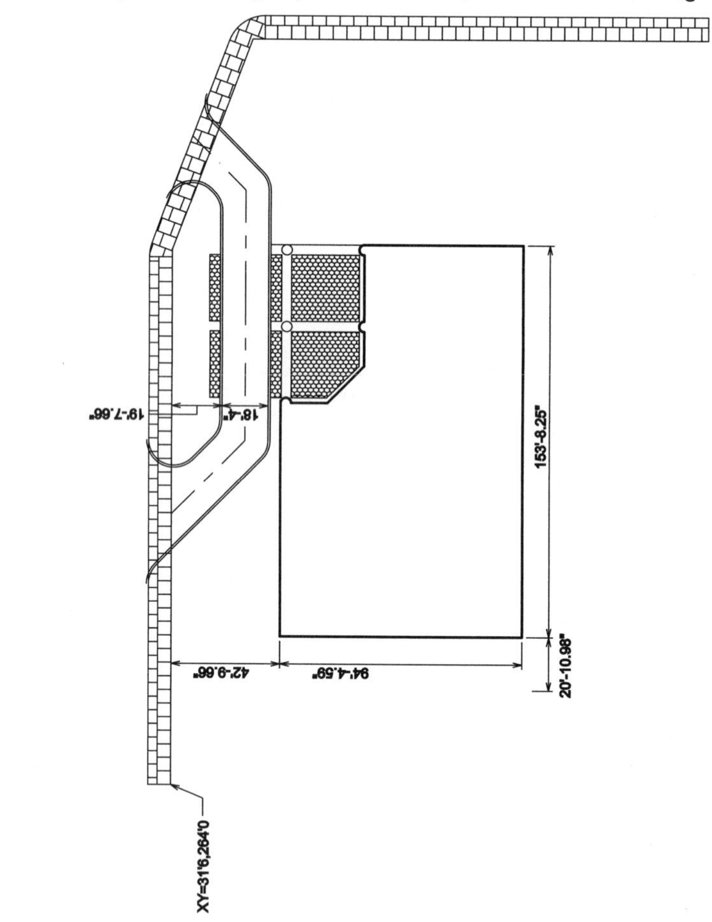


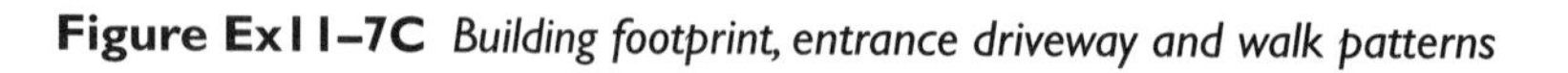

**Figure Ex11–7C** *Building footprint, entrance driveway and walk patterns*

Create another new drawing and draw the title block as shown Figure EX11–7D to the settings given below. Save the drawing as EX11–7D.

| **Settings** | **Value** | | |
|---|---|---|---|
| 1. Units | Architectural | | |
| 2. Limits | Lower left corner: 0,0,<br>Upper right corner: 540',400' | | |
| 3. Grid | 10' | | |
| 4. Snap | 5' | | |
| 5. Layers | *NAME* | *COLOR* | *LINETYPE* |
| | Title block | White | Continuous |

Attach the file EX11–7A as a reference file, with an insertion point of XY= 31'-6", 264'-0" with a scale factor of 1.0 and a rotation angle of 0 degrees. Then attach the file EX11–7B as a reference file, with an insertion point of XY= 31'-6", 264'-0" to a scale factor of 1.0 and a rotation angle of 0 degrees. And then attach the file EX11–7C as reference file to complete the assembly with an interstion point of XY= 31'-6", 264'-0" to a scale factor of 1.0 and rotation angle of 0 degrees. Completed plot plan should appear as shown in EX11–7D and save the drawing as EX11–7D.

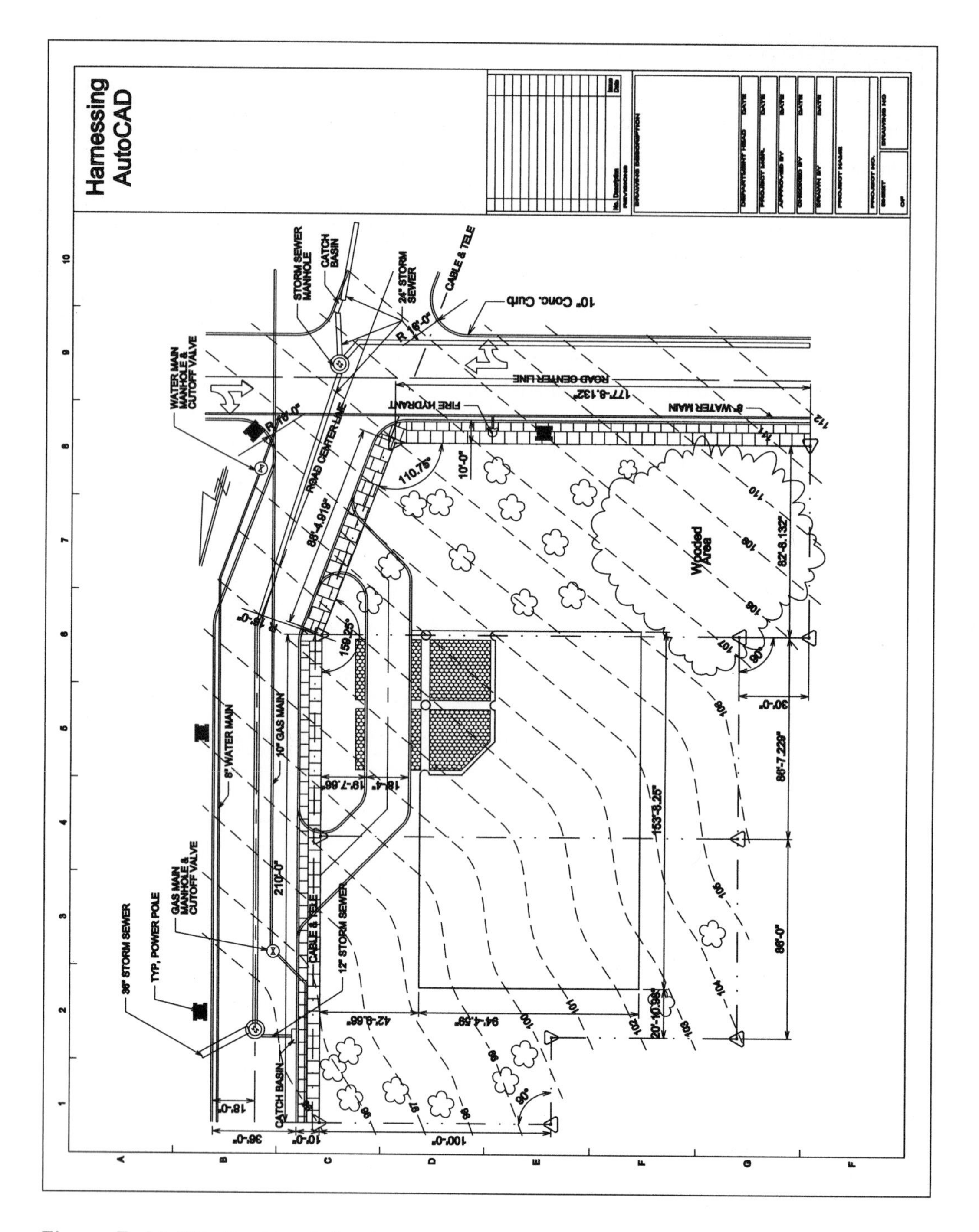


**Figure Ex11–7D** *Completed plot plan*

## EXERCISE 11–8

Create a new drawing according to the settings given below and draw the structure, entrance and elevator lobby background as shown in EX11–8A. Save the drawing as EX11–8A.

| Settings | Value |
|---|---|
| 1. Units | Architectural |
| 2. Limits | Lower left corner: 0,0<br>Upper right corner: 120', 90' |

| Settings | Value | | |
|---|---|---|---|
| 3. Layers | NAME | COLOR | LINETYPE |
| | Elevators | White | Hidden |
| | Partitions | white | Hidden |
| | Columns | White | Hidden |
| | Centerline | White | Phantom |
| | Dimensions | White | Continuous |

**Hint:** Begin the drawing at point XY= 19'-0", 12'-9" (intersection of A and 5) as shown in Figure.

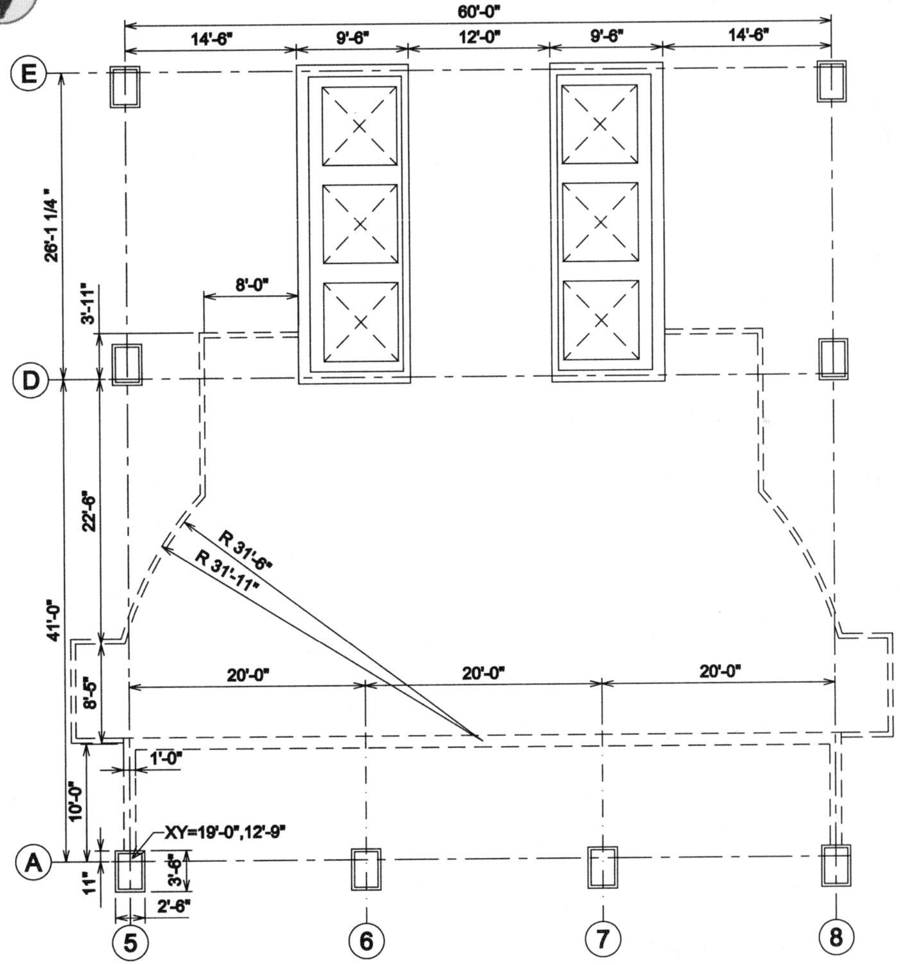

**Figure Ex11–8A** *Structure, entrance and elevator lobby background*

Create another new drawing to the given settings and draw the entrance and elevator lobby ceiling as shown in Figure 11–8B. Save the drawing as EX11–8B.

| Settings | Value | | |
|---|---|---|---|
| 1. Units | Architectural | | |
| 2. Limits | Lower left corner: 0,0 | | |
| | Upper right corner: 120', 90' | | |
| 3. Layers | *NAME* | *COLOR* | *LINETYPE* |
| | Ceiling plan | Blue | Continuous |
| | Centerline | White | Center |
| | Dimensions | White | Continuous |
| | Text | White | Continuous |

**Hint:** Begin the drawing at point XY= 19'-0", 12'-9" (intersection of A and 5) as shown in Figure.

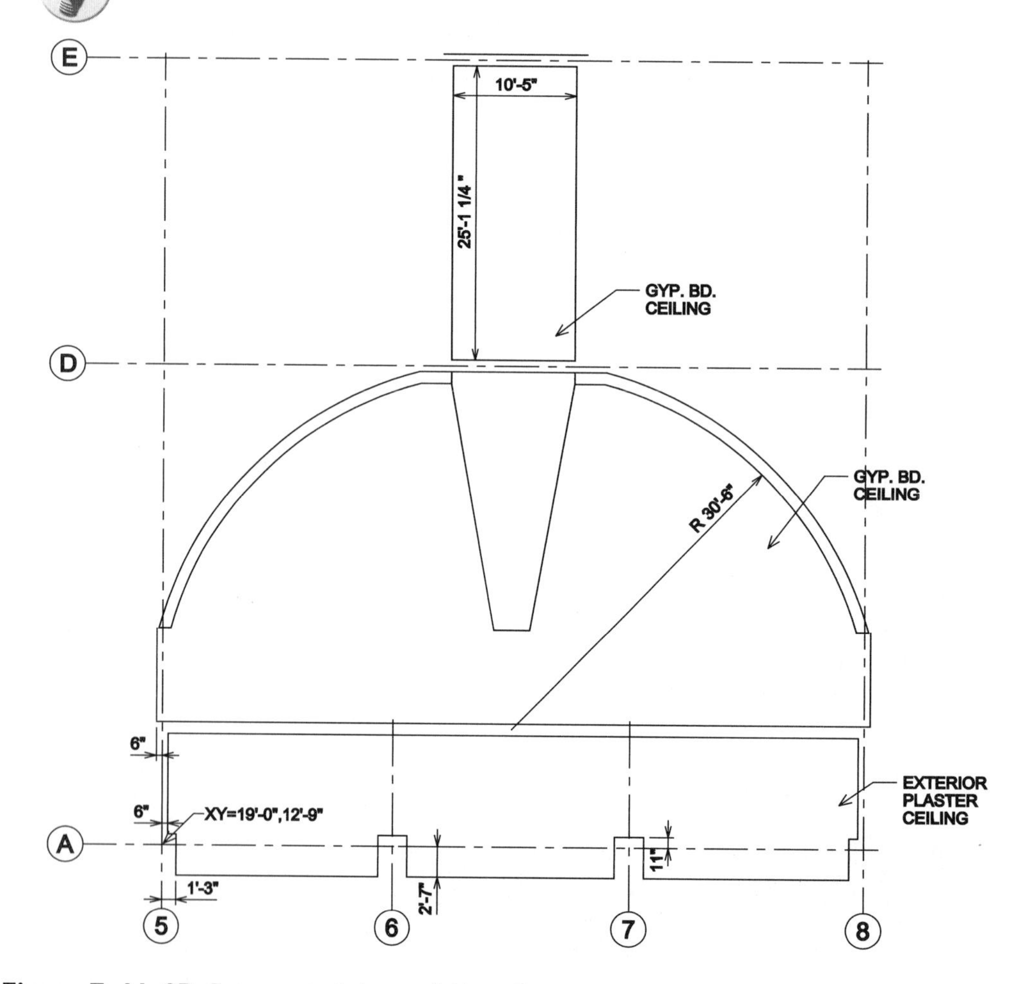


**Figure Ex11–8B** *Entrance and elevator lobby ceiling*

Create a new drawing to the settings given below. Attach the file EX11–8A as a reference file, with an insertion point of XY= 19'-0", 12'-9", a scale factor of 1.0 and a rotation angle of 0 degrees. Attach the file EX11–8B as a reference file, with an insertion point of XY= 19'-0", 12'-9", a scale factor of 1.0 and a rotation angle of 0 degrees. Add entrance and elevator lobby lighting as shown in Figure 11–8C. See Figure 11–8D for completed drawing. Save the drawing as EX11–8D.

| **Settings** | **Value** | | |
|---|---|---|---|
| 1. Units | Architectural | | |
| 2. Limits | Lower left corner: 0,0 | | |
| | Upper right corner: 120', 90' | | |
| 3. Layers | *NAME* | *COLOR* | *LINETYPE* |
| | Light-N | Green | Continuous |
| | Light-L | Orange | Continuous |
| | Flourscent-F | Yellow | Continuous |
| | Wiring | White | Continuous |
| | Wire circuits | White | Continuous |
| | Text | White | Continuous |

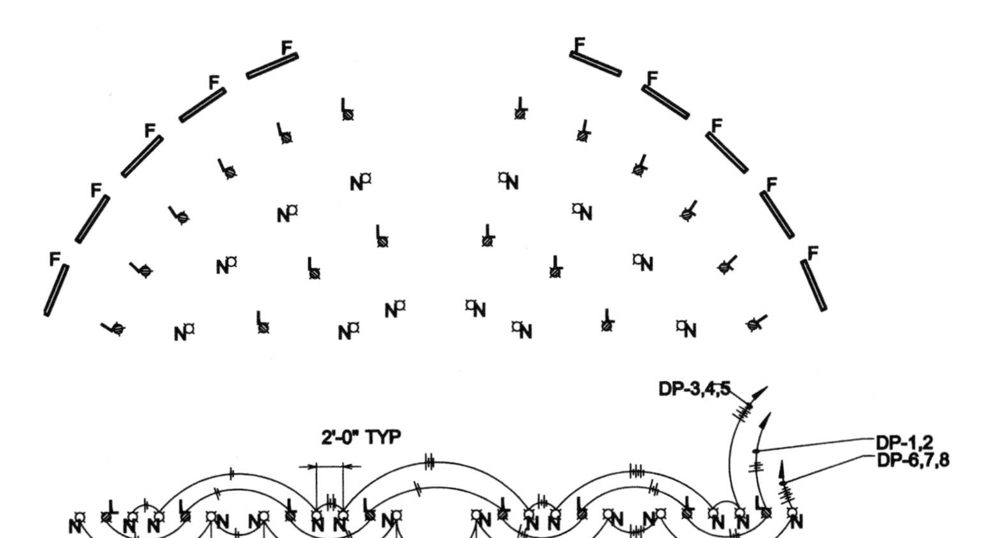


**Figure Ex11–8C** *Entrance and elevator lobby lighting*

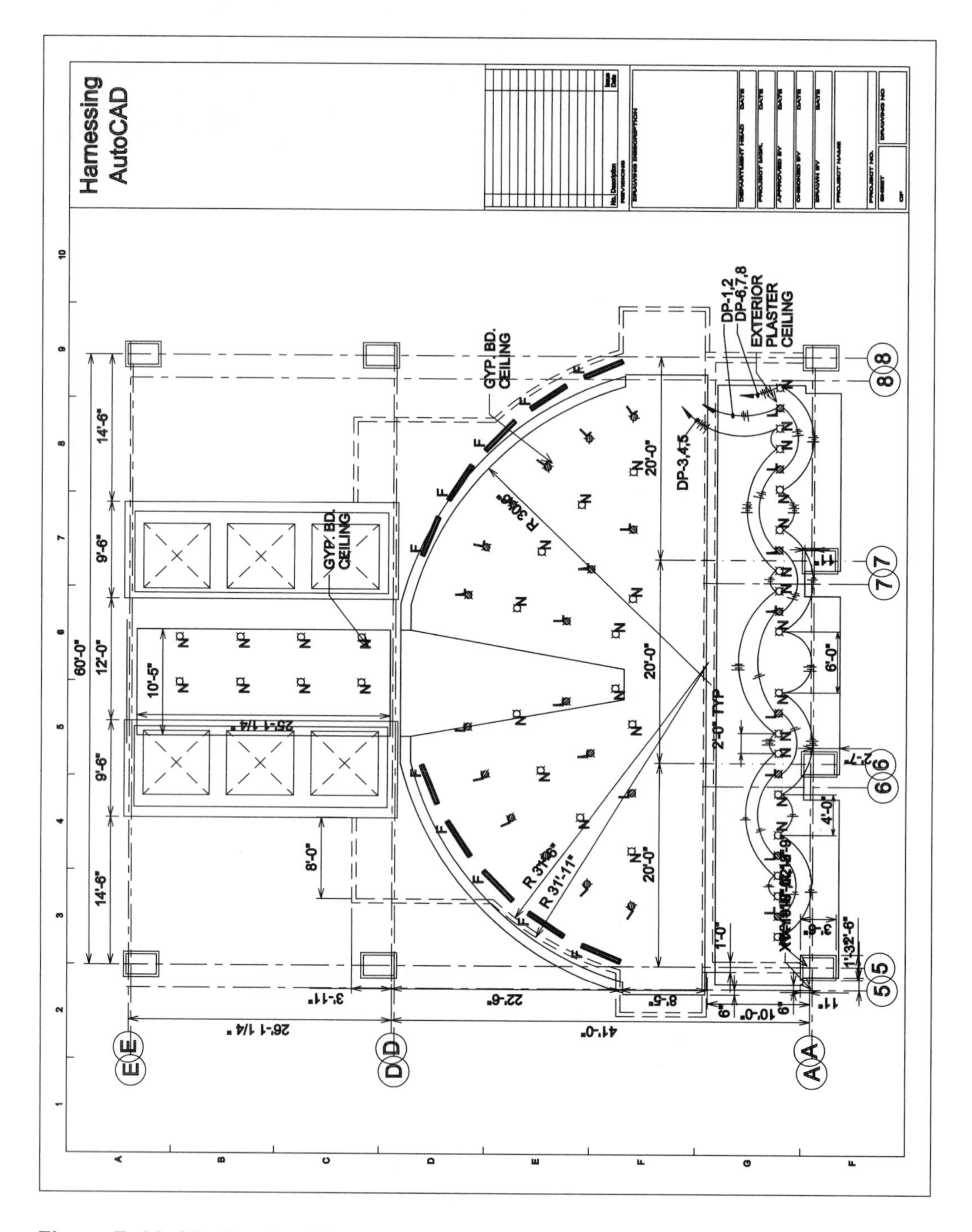


**Figure Ex11–8D** *Completed drawing*

CHAPTER 15

# AutoCAD 3D

project EXERCISE

## PROJECT EXERCISE

This project creates the bracket shown in Figure P15–1. The bracket is drawn entirely by means of AutoCAD solid-modeling features. Follow the steps, and you will be able to build the model by using various commands available in AutoCAD solid modeling.

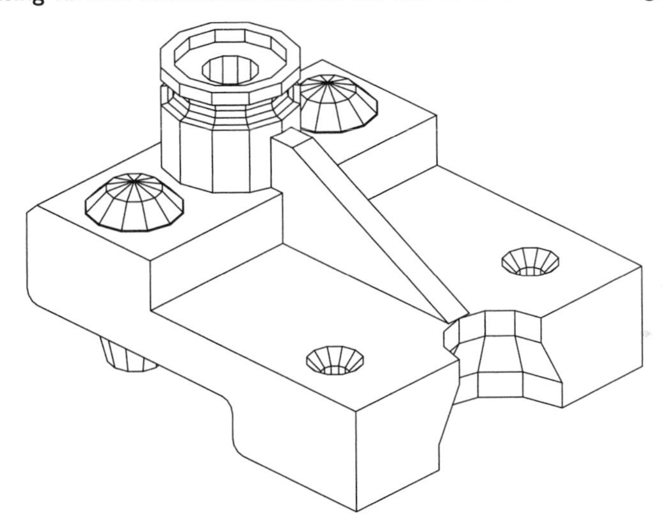

**Figure P15–1** *Creating a bracket using solid modeling*

**Step 1:** Begin a new drawing using the Quick Setup Wizard. Set Units to Decimal and Area to 22 by 17.

**Step 2:** Create the following layers with appropriate colors and linetypes:

| Layer Name | Color | Linetype |
|---|---|---|
| Object | Red | Continuous |
| Border | Green | Continuous |
| Dim | Blue | Continuous |
| Viewports | Cyan | Continuous |

**Step 3:** Invoke the VPORTS command to create four equal viewports. AutoCAD displays the Viewports dialog box, as shown in Figure P15–2.

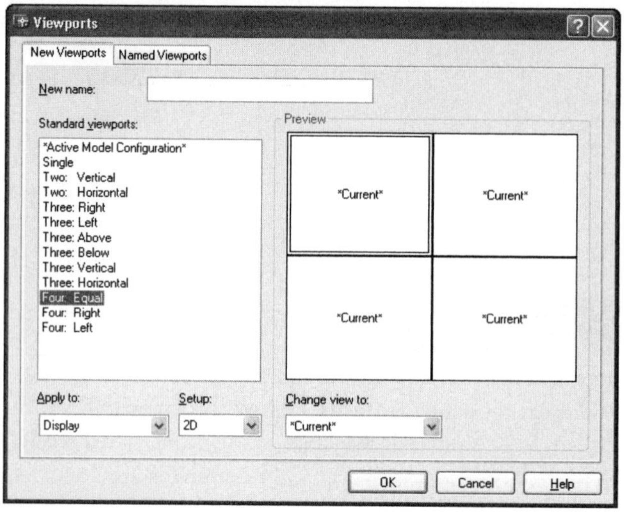

**Figure P15–2** *Viewports dialog box*

Select Four: Equal from the **Standard viewports:** list box and *3D* from the **Setup** option menu, as shown in Figure P15–2. Choose the **OK** button to create the viewports and close the Viewports dialog box. AutoCAD creates four viewports with orthographic views and sets the appropriate UCS orientation, as shown in Figure P15–3.

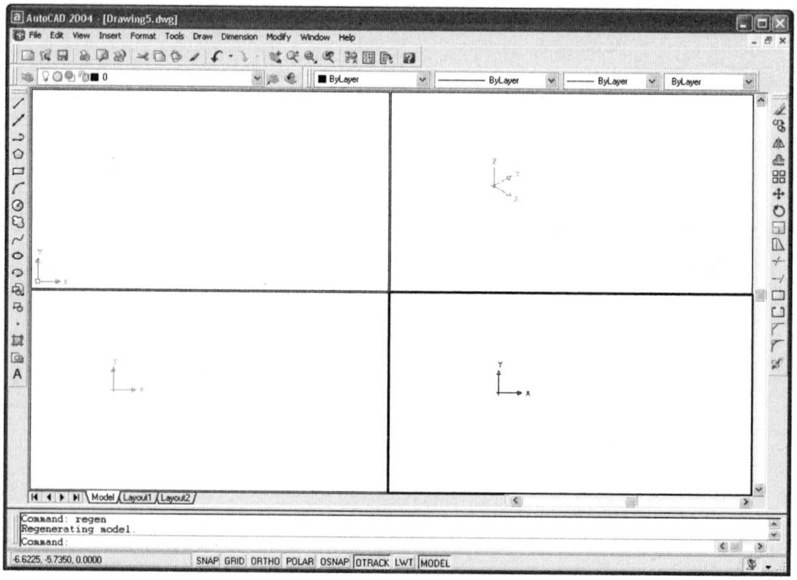

**Figure P15–3** *Four viewports and appropriate UCS orientation*

**Step 4:** Make the upper right viewport current. Invoke the **Display UCS Dialog** from the UCS toolbar, as shown in Figure P15–4.

**Figure P15–4** *Invoke the Display UCS Dialog from the UCS toolbar*

AutoCAD displays the UCS dialog box, as shown in Figure P15–5.

UCS
Named UCSs | Orthographic UCSs | Settings
UCS Icon settings
☑ On
☑ Display at UCS origin point
☐ Apply to all active viewports
UCS settings
☑ Save UCS with viewport
☐ Update view to Plan when UCS is changed
OK | Cancel | Help

**Figure P15–5** *UCS dialog box*

Select the Settings tab and set the **Save UCS with viewport** toggle button to off, as shown in Figure P15–5. Choose the **OK** button to close the UCS dialog box.

**Step 5:** Set Object as the current layer.

Begin the layout of the drawing by drawing four boxes using the BOX command:

Command: **box**
Specify corner of box or [CEnter] <0,0,0>: **0,0,-2**
Specify corner or [Cube/Length]: **l**
Length: **8**
Width: **7**
Height: **1**

Command: **box**
Specify corner of box or [CEnter] <0,0,0>: **0,0,-1**
Specify corner or [Cube/Length]: **l**
Length: **3**
Width: **7**
Height: **1**

Command: **box**
Specify corner of box or [CEnter] <0,0,0>: **5,0,-3**
Specify corner or [Cube/Length]: **l**
Length: **3**
Width: **7**
Height: **1**

Command: **box**
Specify corner of box or [CEnter] <0,0,0>: **2.5,3.25,-1**
Specify corner or [Cube/Length]: **l**
Length: **.75**
Width: **.5**
Height: **2**

The preceding box constructions form the basic shape of the bracket, as shown in Figure P15–6.

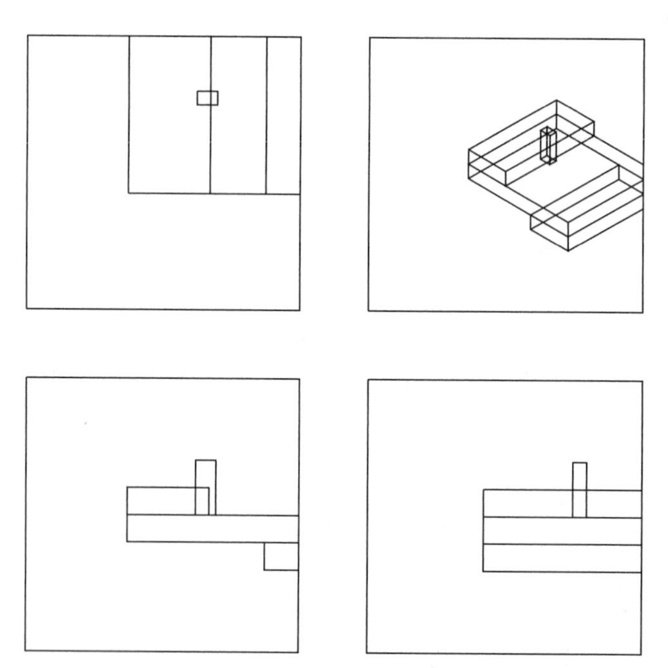

**Figure P15–6** *Creating the basic shape of the bracket*

**Step 6:** Invoke the CYLINDER command to create a cylinder:

Command: **cylinder**
Specify center point for base of cylinder or [Elliptical] <0,0,0>: **1.5,3.5**
Specify radius for base of cylinder or [Diameter]: **1.25**
Specify height of cylinder or [Center of other end]: **2**

**Step 7:** Invoke the WEDGE command to create a wedge as shown in Figure P15–7:

Command: **wedge**
Specify first corner of wedge or [CEnter] <0,0,0>: **3.25,3.25,-1**
Specify corner or [Cube/Length]: **l**
Length: **3.75**
Width: **.5**
Height: **2**

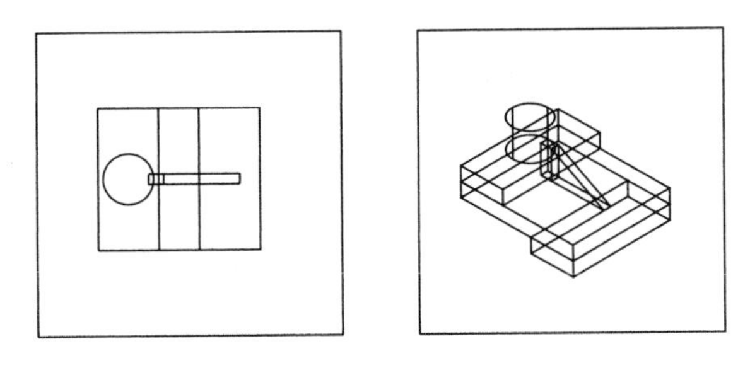

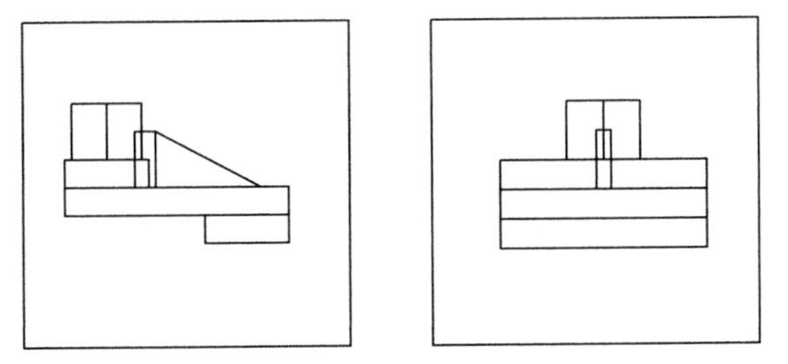

**Figure P15–7** *Creating the basic shape with cylinder and wedge of the bracket*

**Step 8:** Invoke the UCS command to define UCS as follows:

Command: **ucs**
Enter an option New/Move/orthoGraphic/Prev/Restore/Save/Del/Apply/?/
World]
<World>: **n**
Specify origin of new UCS or [ZAxis/3point/OBject/Face/View/X/Y/Z]
<0,0,0>: **3**
Specify new origin point <0,0,0>: *(select point 1 by using the object snap ENDpoint, as shown in Figure P15–8)*

Specify point on positive portion of X-axis: *(select point 2 by using the object snap ENDpoint, as shown in Figure P15–8)*
Specify point on positive-Y portion of the UCS XY plane: *(select point 3 by using the object snap ENDpoint, as shown in Figure P15–8)*

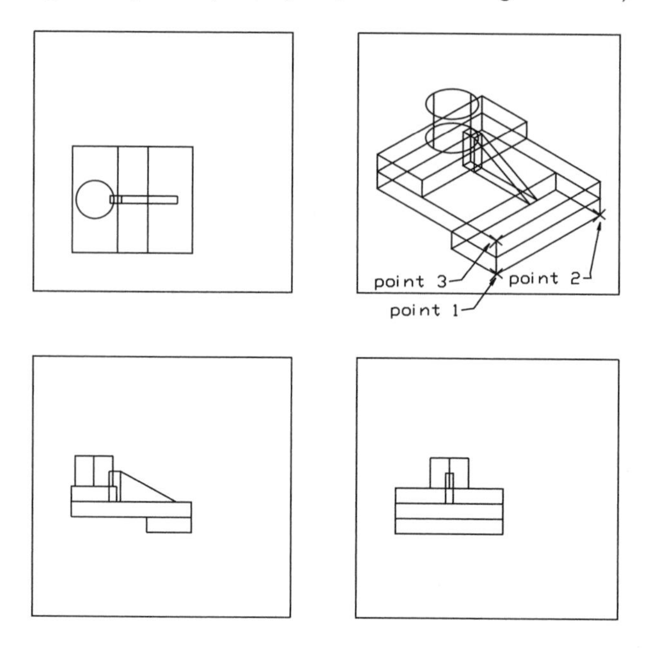


**Figure P15–8** *Defining a UCS by 3 points*

**Step 9:** Invoke the PLINE command to draw a polyline to the given coordinates, as shown in Figure P15–9:

Command: **pline**
Specify start point: **3.5,2**
Current line-width is 0.0000
Specify next point or [Arc/Close/Halfwidth/Length/Undo/Width]: **@1<180**
Specify next point or [Arc/Close/Halfwidth/Length/Undo/Width]: **@0.5<270**
Specify next point or [Arc/Close/Halfwidth/Length/Undo/Width]: **@-0.5,-1**
Specify next point or [Arc/Close/Halfwidth/Length/Undo/Width]: **@0.5<-90**
Specify next point or [Arc/Close/Halfwidth/Length/Undo/Width]: **@1.5<0**
Specify next point or [Arc/Close/Halfwidth/Length/Undo/Width]: **c** (ENTER)

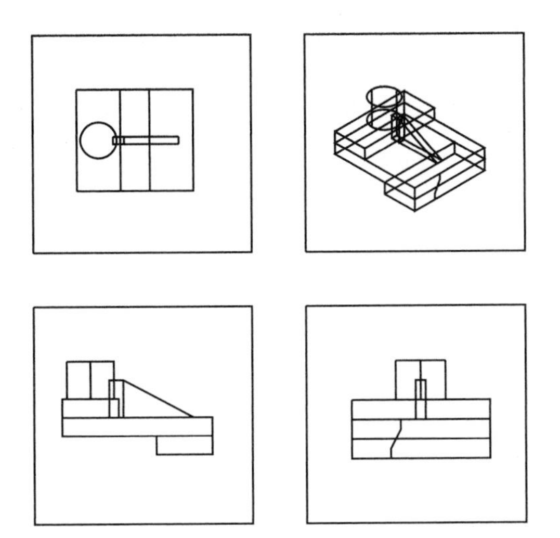

**Figure P15–9** *Drawing a polyline to specified coordinates*

**Step 10:** Invoke the REVOLVE command to revolve the polyline just created into a solid, as shown in Figure P15–10:

Command: **revolve**
Current wire frame density: ISOLINES=4
Select objects: **l**
Select objects: (ENTER)
Specify start point for axis of revolution or define axis by [Object/X (axis)/Y (axis)]: **3.5,2**
Specify endpoint of axis: **@2<270**
Specify angle of revolution <360>: **180**

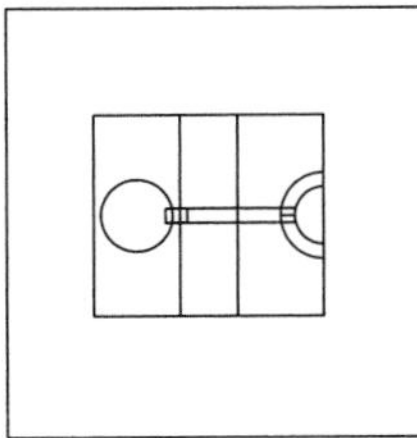

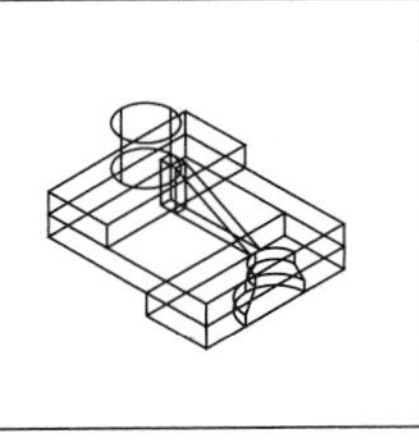

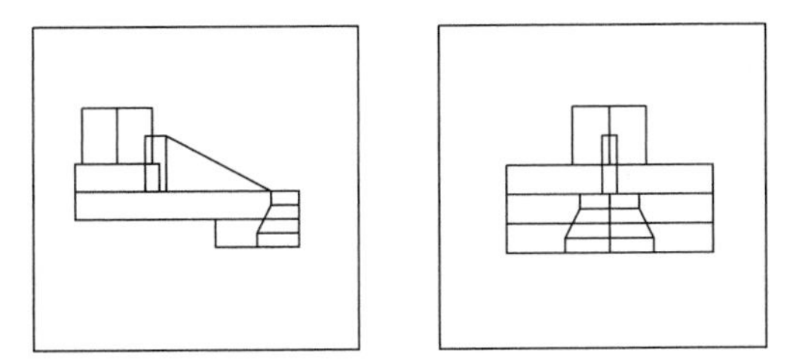

**Figure P15–10** *Revolving a polyline into a solid*

**Step 11:** Set the UCS to world coordinates systems and invoke the SPHERE command to create two spheres.

Command: **ucs**
Enter an option New/Move/orthoGraphic/Prev/Restore/Save/Del/Apply/?/ World]
<World>: (ENTER)
Command: **sphere**
Specify center of sphere <0,0,0>: **1.5,1.125,-0.5**
Specify radius of sphere or [Diameter]: **1**

Copy the sphere to a displacement of 0,4.75, as shown in Figure P15–11.

Command: **copy**
Select objects: **l**
Select objects: (ENTER)
Specify base point or displacement, or [Multiple]: **0,0**
Specify second point of displacement or <use first point as displacement>: **0,4.75**

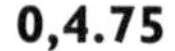

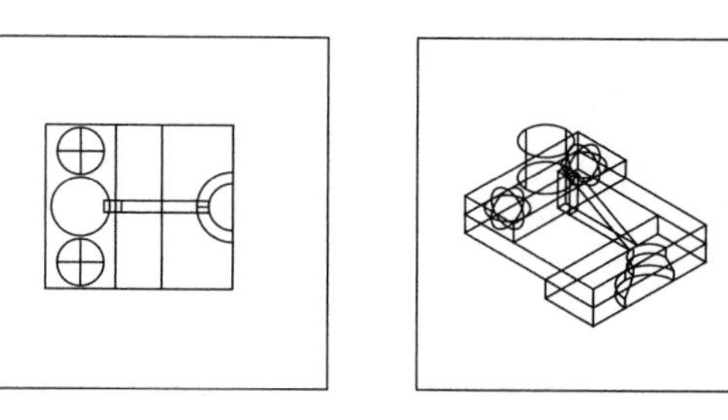

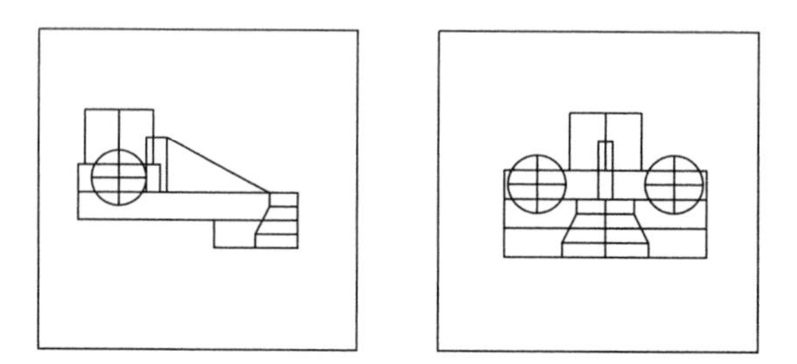

**Figure P15–11** *Copying a sphere to a specified displacement*

**Step 12:** Invoke the CONE command to draw two cones, as shown in Figure P15–12.

Command: **cone**
Specify center point for base of cone or [Elliptical] <0,0,0>: **1.5, 1.125,-2**
Specify radius for base of cone or [Diameter]: **0.75**
Specify height of cone or [Apex]: **-3**

Copy the cone to a displacement of 0,4.75.

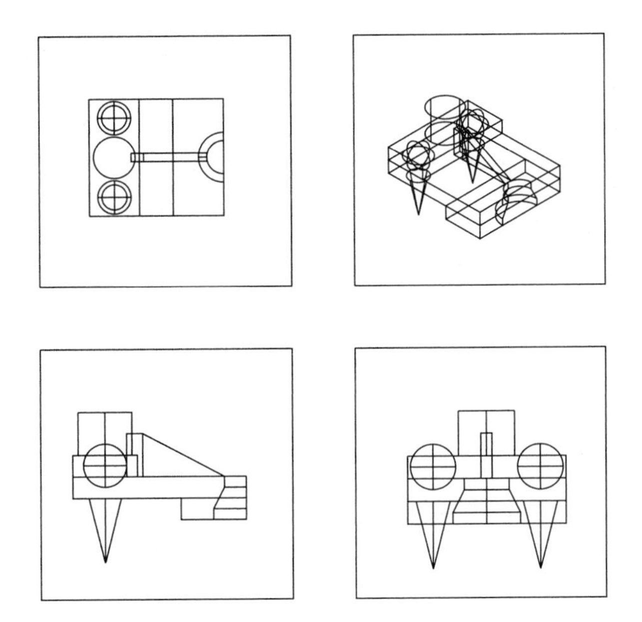

**Figure P15–12** *Drawing two cones using the* CONE *command*

**Step 13:** Starting at 0,0,–5, create a box that is 3 x 7 x 2, as shown in Figure P15–13.

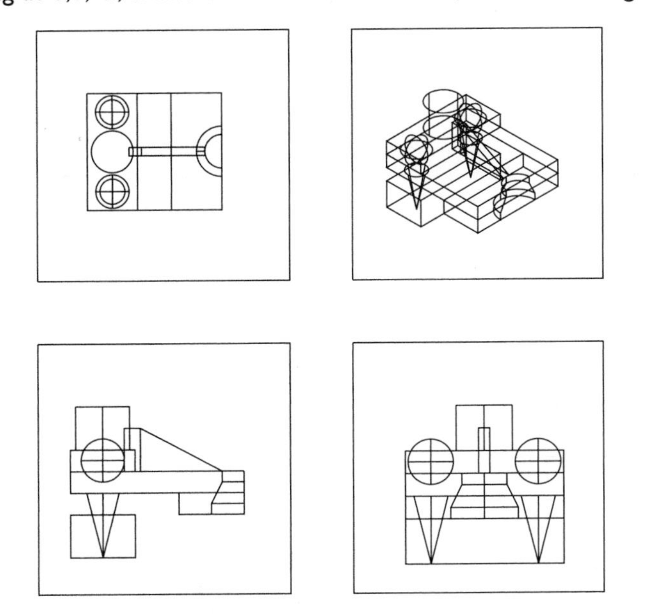

**Figure P15–13** *Creating a box with starting point 0,0,–5 and dimensions of 3 x 7 x 2*

**Step 14:** Create a 0.5-radius cylinder, centered at 1.5,3.5,–2, to a height of 4, as shown in Figure P15–14.

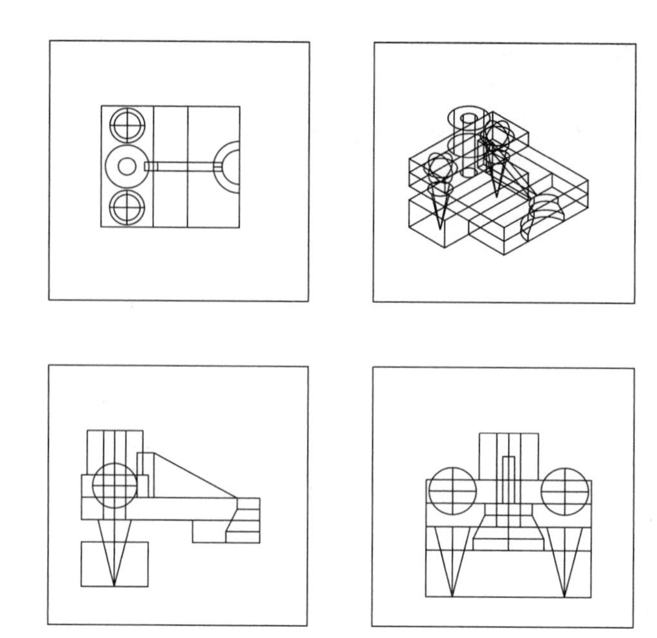

**Figure P15–14** *Creating a 0.5 cylinder centered at 1.5,3.5,–2 and with a height of 4*

**Step 15:** Create a cylinder with radius 1, centered at 1.5,3.5,1.75 to a height of 0.25, as shown in Figure P15–15.

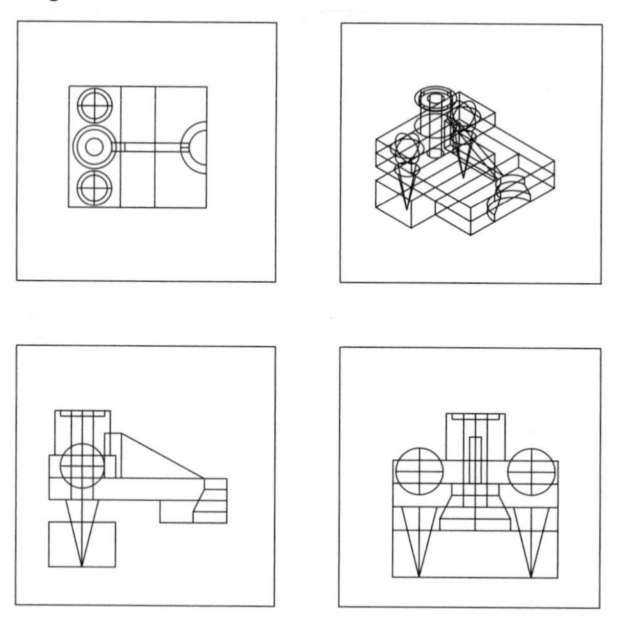

**Figure P15–15** *Creating a 0.5 cylinder centered at 1.5,3.5,1.75, with a height of 0.25 and a radius of 1*

**Step 16:** Create a cylinder with radius 0.25, centered at 6.5,1.0,–3 to a height of 2, as shown in Figure P15–16.

Copy the cylinder to a displacement of 0,4.

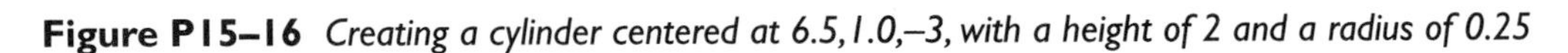

**Figure P15–16** *Creating a cylinder centered at 6.5,1.0,–3, with a height of 2 and a radius of 0.25*

**Step 17:** Invoke the TORUS command to create a torus, as shown in Figure P15–17:

Command: **torus**
Specify center of torus <0,0,0>: **1.5,3.5,1.5**
Specify radius of torus or [Diameter]: **1.25**
Specify radius of tube or [Diameter]: **0.25**

**Step 18:** Select the connected boxes (except the box that was drawn in Step 13), the wedge, the large cylinder, the spheres, and the cones for use with the UNION command.

Command: **union**
Select objects: *(select the boxes, wedge, large cylinder, spheres, and cones, and press* ENTER*)*

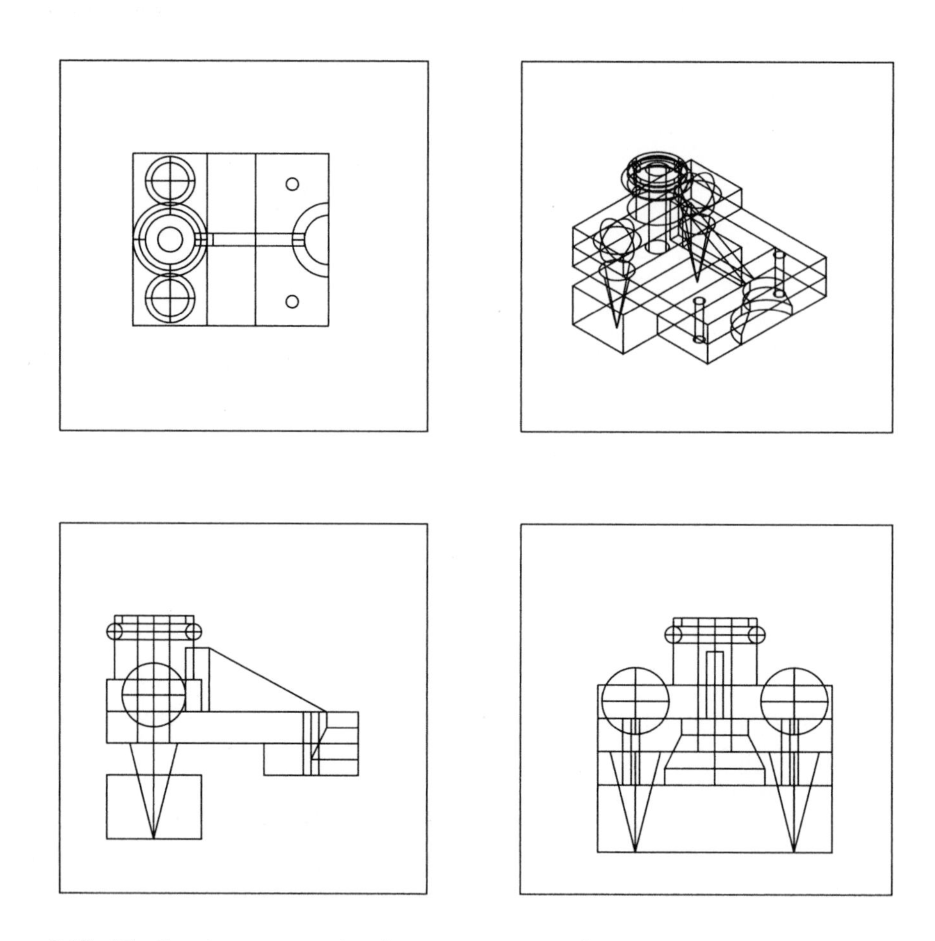

**Figure P15–17** *Creating a torus using the* TORUS *command*

**Step 19:** Select the resulting solid in response to the first SUBTRACT prompt, and then selecting the remaining primitives to be subtracted from it.

Command: **subtract**
Select solids and regions to subtract from ...
Select objects: *(select the solid resulting from STEP 18)*
Select objects: (ENTER)
Select solids and regions to subtract ...
Select objects: *(select the remaining primitives)*
Select objects: (ENTER)

The drawing should look like Figure P15–18.

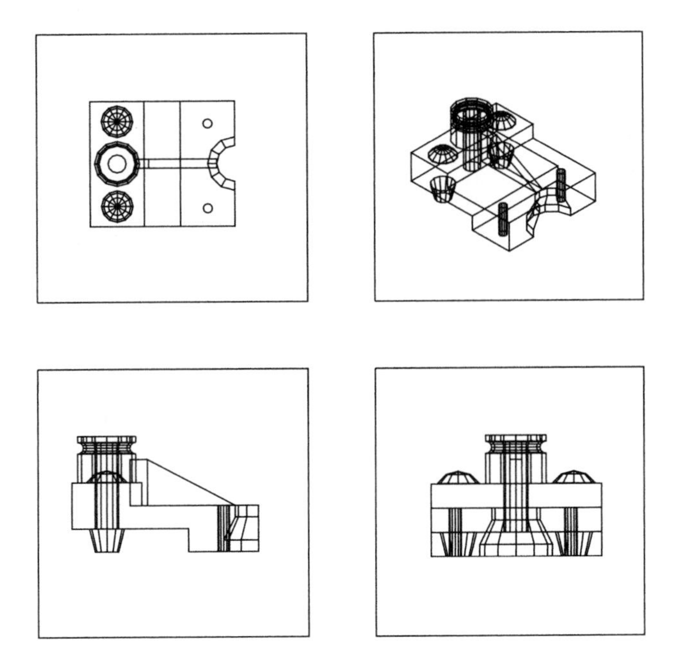

**Figure P15–18** *Subtracting the primitives from the newly created solid using the* SUBTRACT *command*

**Step 20:** Select the faces, as shown in Figure P15–19, for the chamfer and fillet. Use the CHAMFER and FILLET commands with 0.25 as the chamfer values and the radii on the respective selected objects. The end result should look like Figure P15–20.

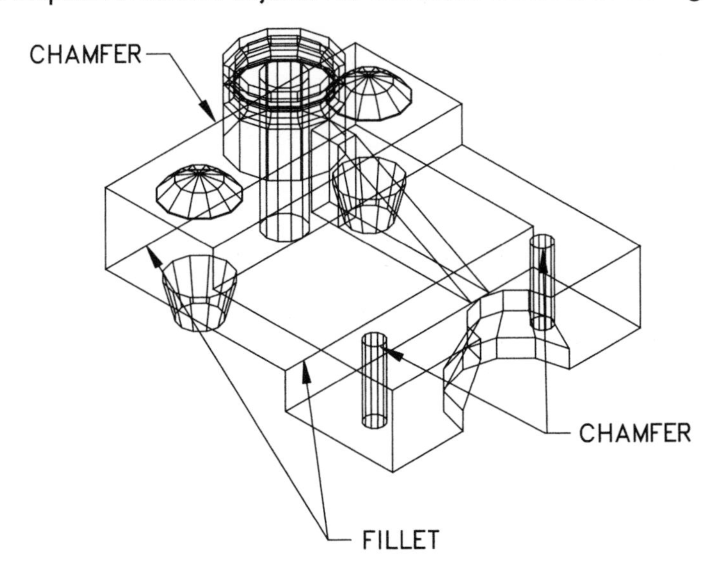

**Figure P15–19** *Using the* CHAMFER *and* FILLET *commands to chamfer and fillet the faces*

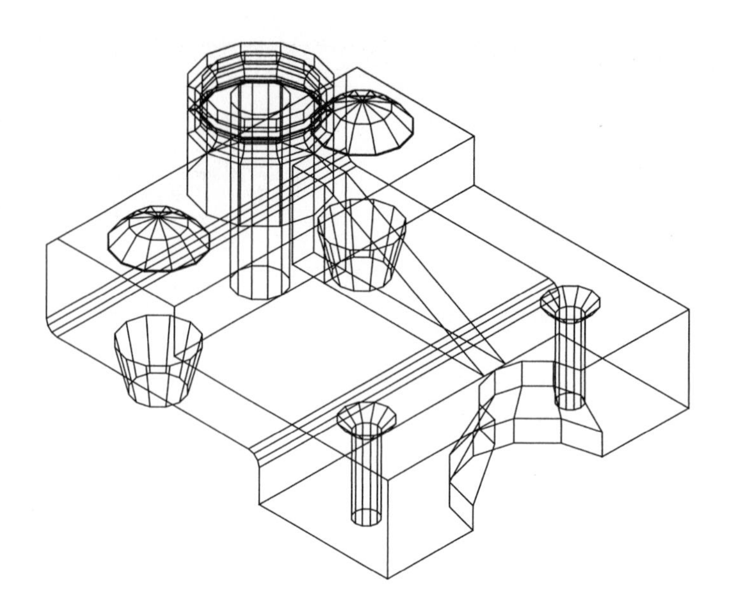

**Figure P15–20** *The solid after chamfering and filleting the faces*

**Step 21:** Make the upper right viewport active. Invoke the HIDE command, and the result is as shown in Figure P15–21.

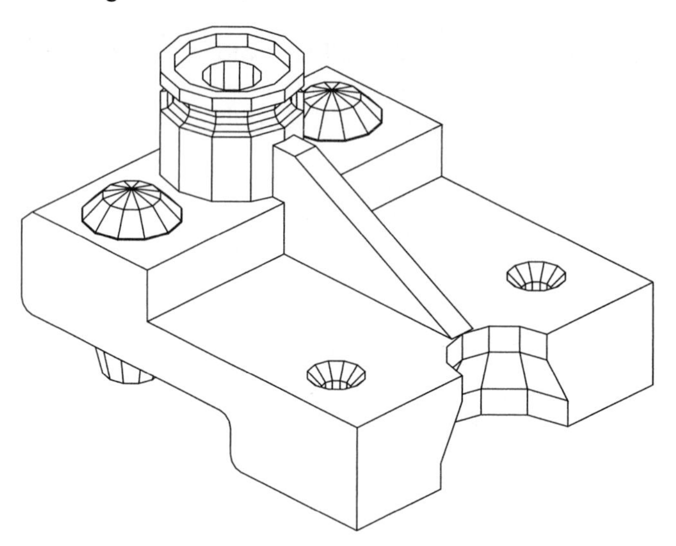

**Figure P15–21** *Completed solid after using the* HIDE *command*

**Step 22:** Set Viewports as the current layer. Invoke the LAYOUTWIZARD command to create a layout to plot model. AutoCAD displays the Create Layout - Begin dialog box, as shown in Figure P15–22.

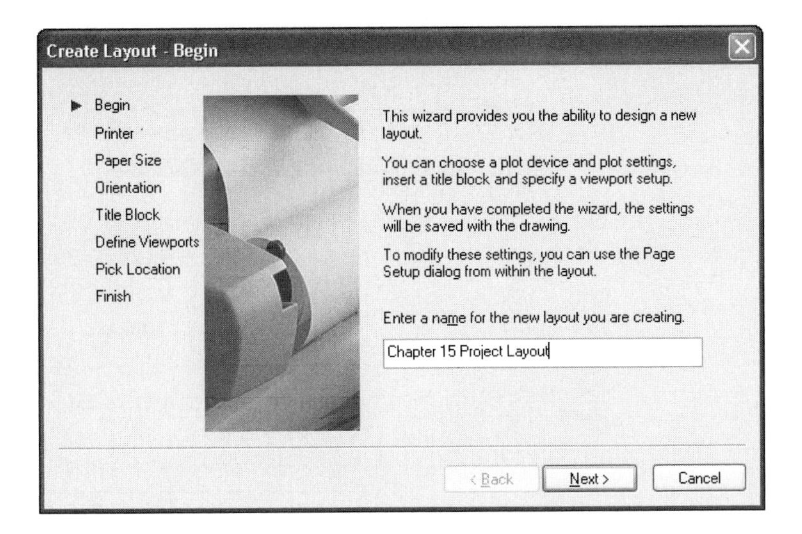


**Figure P15–22** *Create Layout - Begin dialog box*

Type **Chapter15 Project Layout** in the **Enter a name for the new layout you are creating** edit field, as shown in Figure P15–22. Choose the **Next >** button, and AutoCAD displays the Create Layout - Printer dialog box, as shown in Figure P15–23.

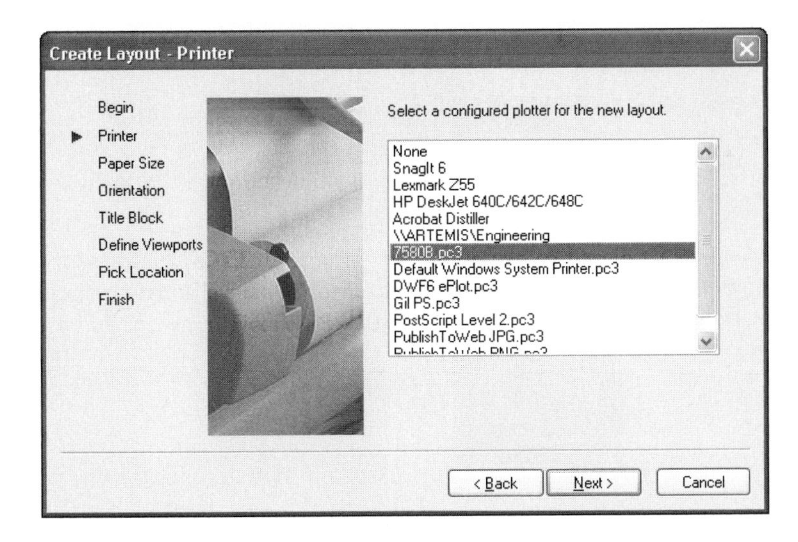


**Figure P15–23** *Create Layout - Printer dialog box*

Select a printer from the **Select a configured for a new layout** list box. In Figure 15–24, hp7580b.pc3 is selected. Choose the **Next >** button, and AutoCAD displays the Create Layout – Paper Size dialog box, as shown in Figure P15–24.

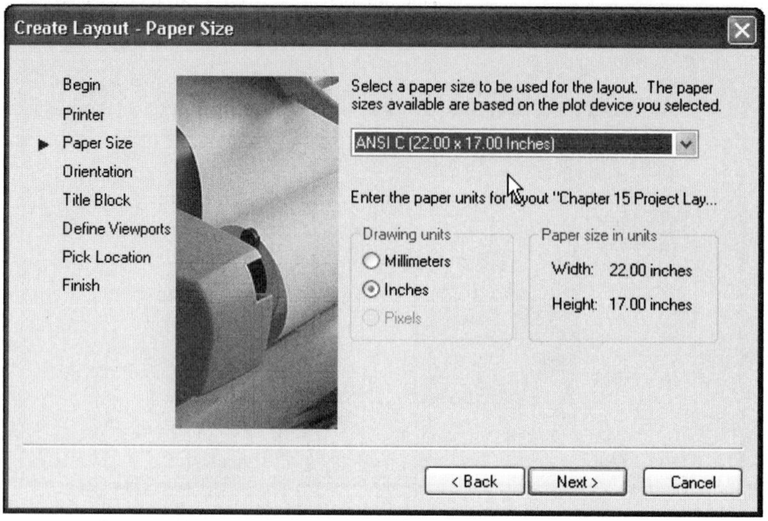

Create Layout - Paper Size

Begin
Printer
▶ Paper Size
Orientation
Title Block
Define Viewports
Pick Location
Finish

Select a paper size to be used for the layout. The paper sizes available are based on the plot device you selected.

ANSI C (22.00 x 17.00 Inches)

Enter the paper units for layout "Chapter 15 Project Lay...

Drawing units
Millimeters
Inches
Pixels

Paper size in units
Width: 22.00 inches
Height: 17.00 inches

< Back | Next > | Cancel

Figure P15–24 *Create Layout – Paper Size dialog box*

Select a paper size to be used in plotting. In Figure 15–25, ANSI C (22.00 x 17.00 Inches) is selected. Choose the **Next** > button, and AutoCAD displays the Create Layout – Orientation dialog box, as shown in Figure P15–25.

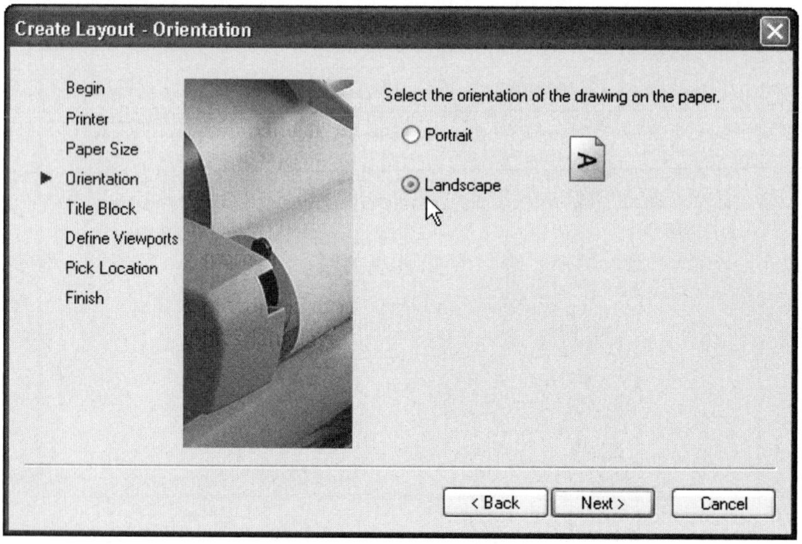

Create Layout - Orientation

Begin
Printer
Paper Size
▶ Orientation
Title Block
Define Viewports
Pick Location
Finish

Select the orientation of the drawing on the paper.

Portrait
Landscape

< Back | Next > | Cancel

Figure P15–25 *Create Layout – Orientation dialog box*

Select the **Landscape** radio button, as shown in Figure P15–25. Choose the **Next** > button, and AutoCAD displays the Create Layout – Title Block dialog box, as shown in Figure P15–26.

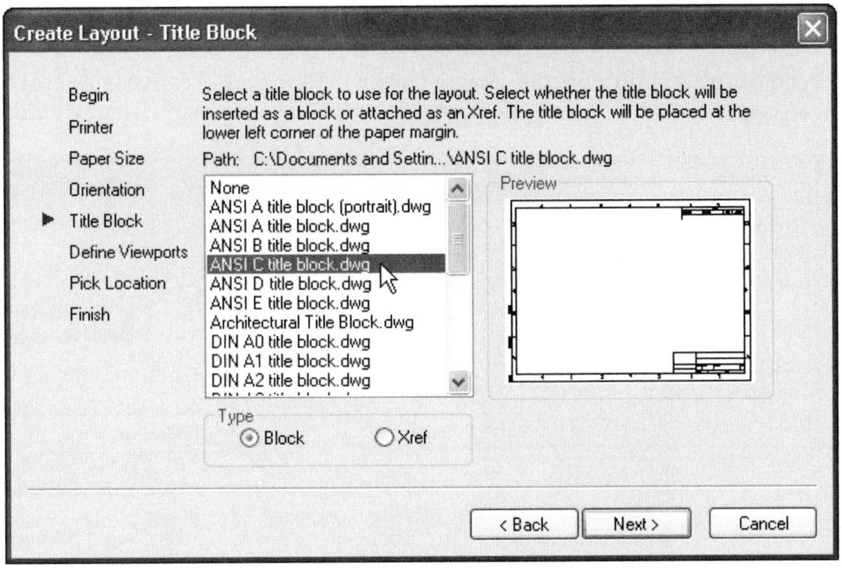

Create Layout - Title Block

Begin
Printer
Paper Size
Orientation
▶ Title Block
Define Viewports
Pick Location
Finish

Select a title block to use for the layout. Select whether the title block will be inserted as a block or attached as an Xref. The title block will be placed at the lower left corner of the paper margin.

Path: C:\Documents and Settin...\ANSI C title block.dwg

None
ANSI A title block (portrait).dwg
ANSI A title block.dwg
ANSI B title block.dwg
ANSI C title block.dwg
ANSI D title block.dwg
ANSI E title block.dwg
Architectural Title Block.dwg
DIN A0 title block.dwg
DIN A1 title block.dwg
DIN A2 title block.dwg

Preview

Type
Block
Xref

< Back
Next >
Cancel

**Figure P15–26** *Create Layout – Title Block dialog box*

Select the appropriate title block for the selected paper size. In Figure P15–26, **ANSI C title block.dwg** is selected. Choose the **Next** > button, and AutoCAD displays the Create Layout – Define Viewports dialog box, as shown in Figure P15–27.

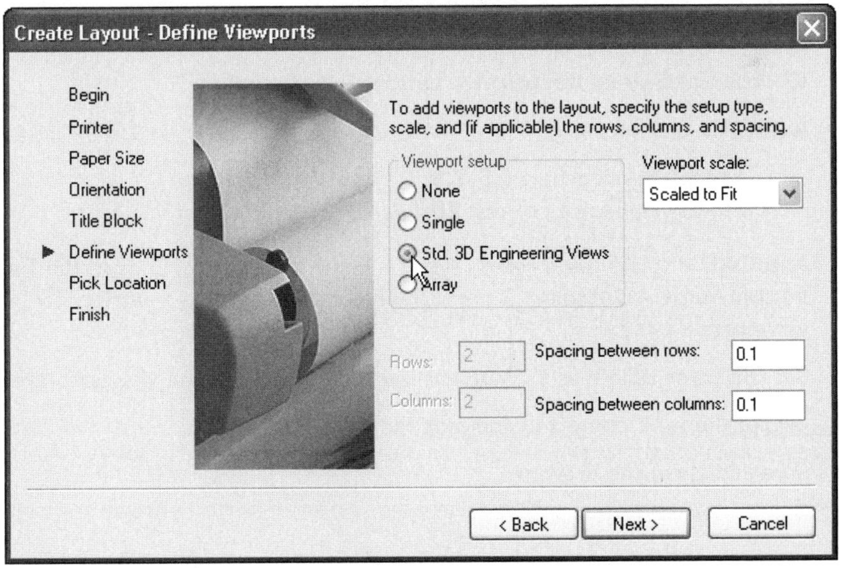

Create Layout - Define Viewports

Begin
Printer
Paper Size
Orientation
Title Block
▶ Define Viewports
Pick Location
Finish

To add viewports to the layout, specify the setup type, scale, and (if applicable) the rows, columns, and spacing.

Viewport setup
None
Single
Std. 3D Engineering Views
Array

Viewport scale:
Scaled to Fit

Rows: 2
Spacing between rows: 0.1
Columns: 2
Spacing between columns: 0.1

< Back
Next >
Cancel

**Figure P15–27** *Create Layout – Define Viewports dialog box*

Select the **Std. 3D Engineering Views** radio button in the Viewport setup section of the dialog box, as shown in Figure P15–27. Choose the **Next** > button, and AutoCAD displays the Create Layout – Pick Location dialog box, as shown in Figure P15–28.

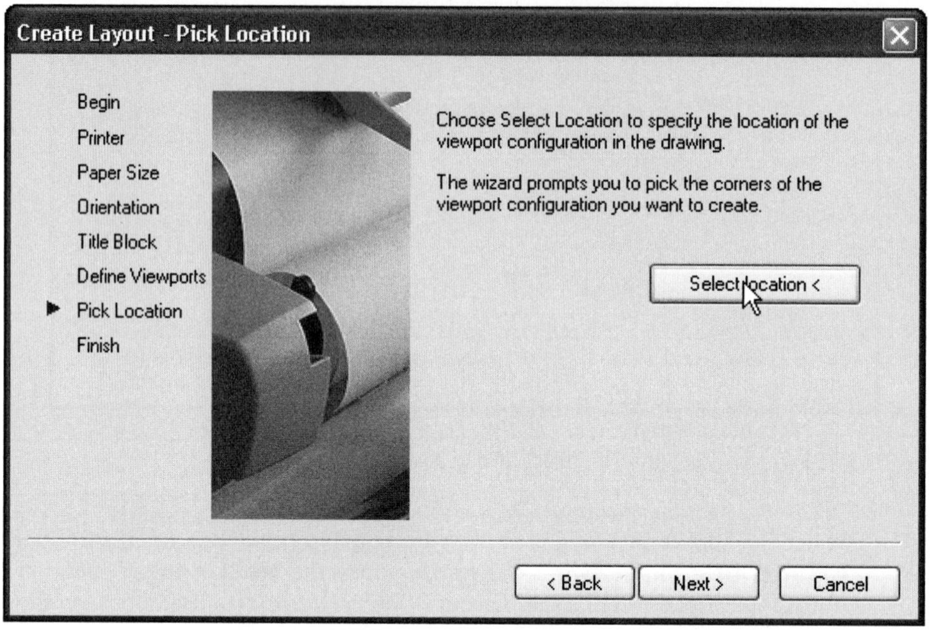

**Figure P15–28** *Create Layout – Pick Location dialog box*

Choose the **Select Location <** button.

AutoCAD prompts:

Specify first corner: **1.5,2.5**
Specify opposite corner: **20,14**

AutoCAD displays the Create Layout – Finish dialog box. Choose the **Finish** button. AutoCAD displays 3 orthographic views and an isometric view in four viewports.

Set the Layer object as the current layout and turn off the viewport layer.

Invoke the PLOT command and plot the drawing.

**Step 22:** Save and close the drawing.

## EXERCISE 15–1

Layout the objects shown in *3D* form for Exercises 15–1 to 15–5. Create the drawings to the given dimensions. Display the drawing with VPOINT in four different views. Select the HIDE command for one of the views.

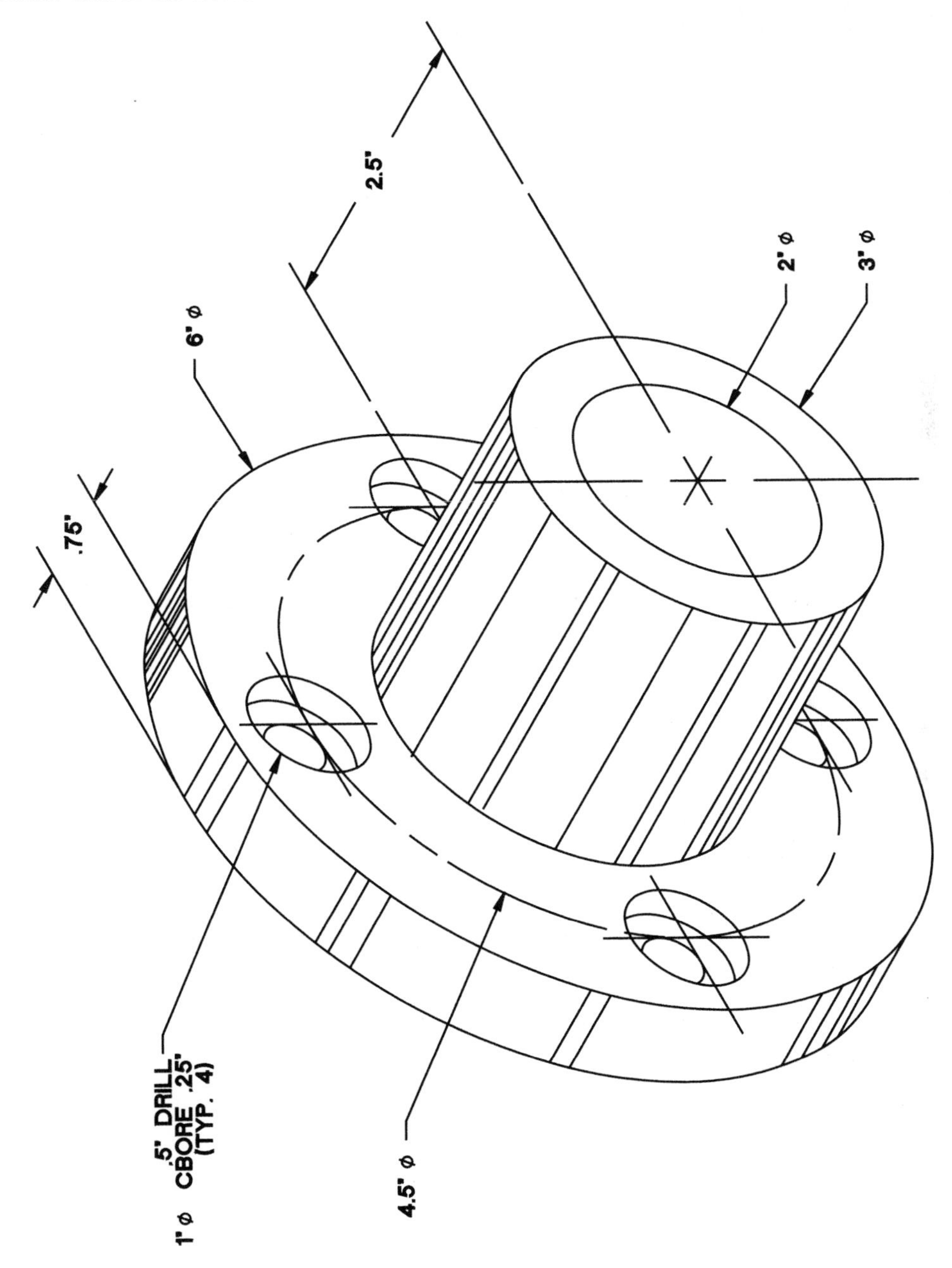

**Figure Ex15–1**

## EXERCISE 15–2

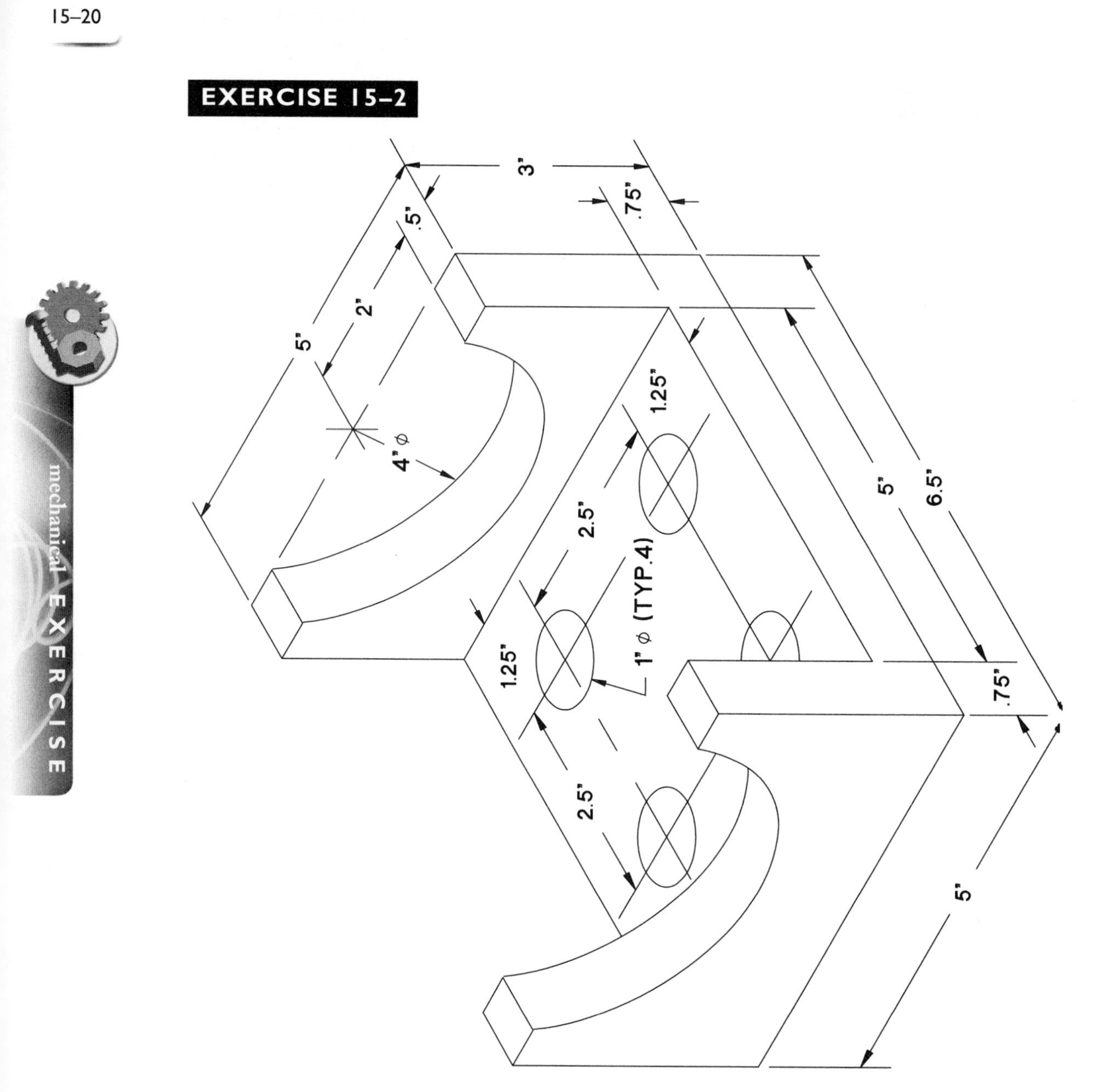

**Figure Ex15–2**

## EXERCISE 15–3

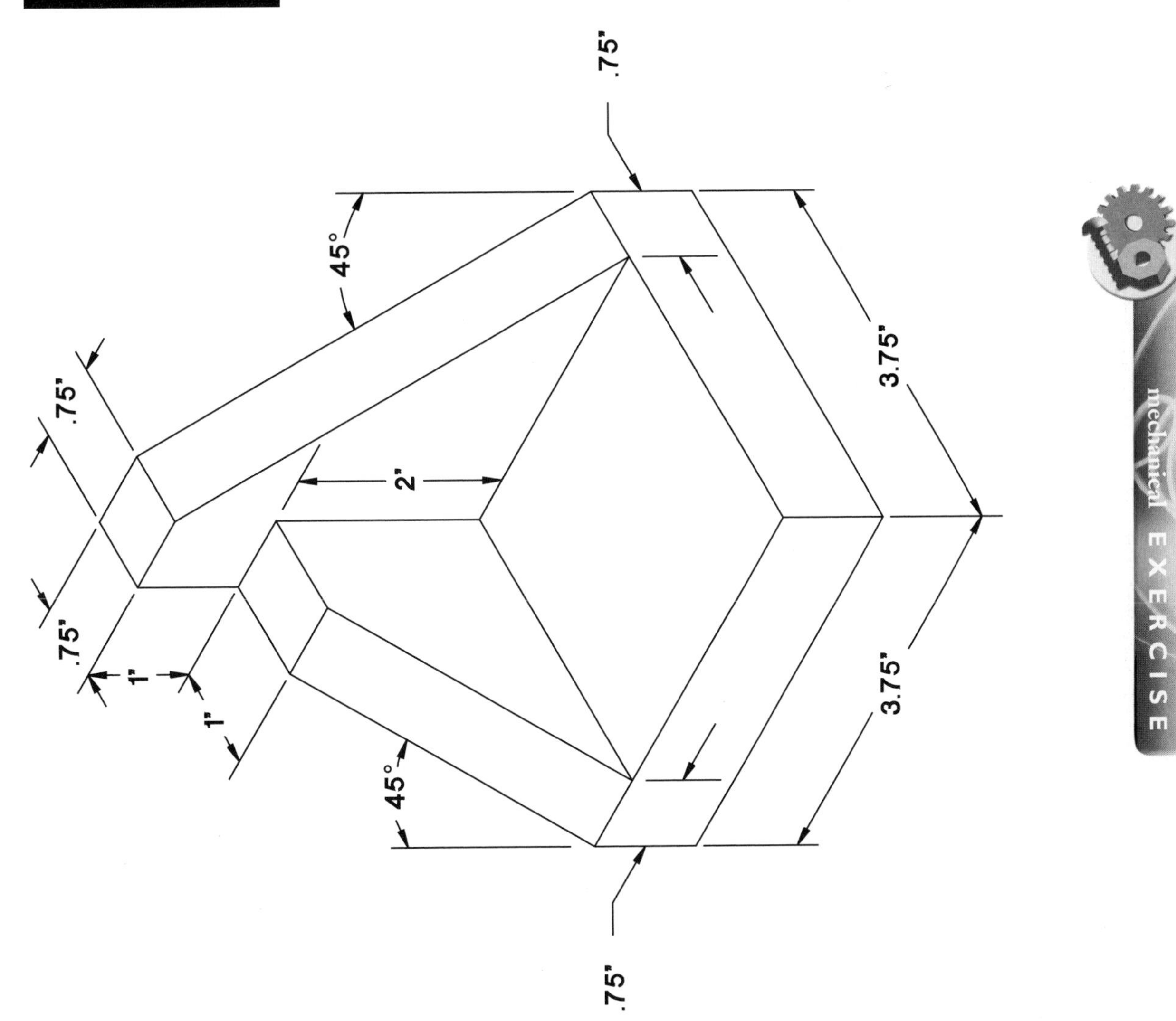


**Figure Ex15–3**

## EXERCISE 15–4

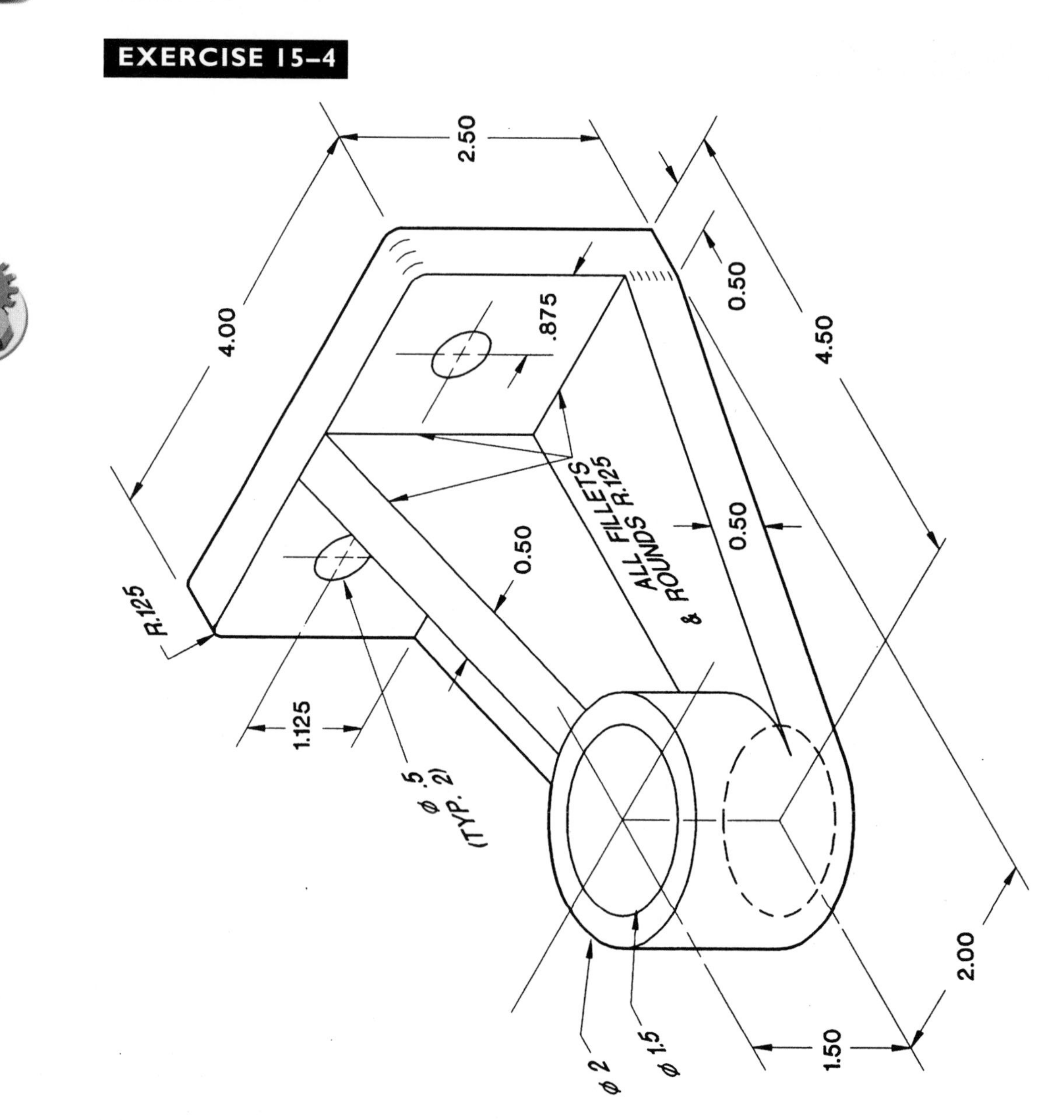


**Figure Ex15–4**

# EXERCISE 15–5

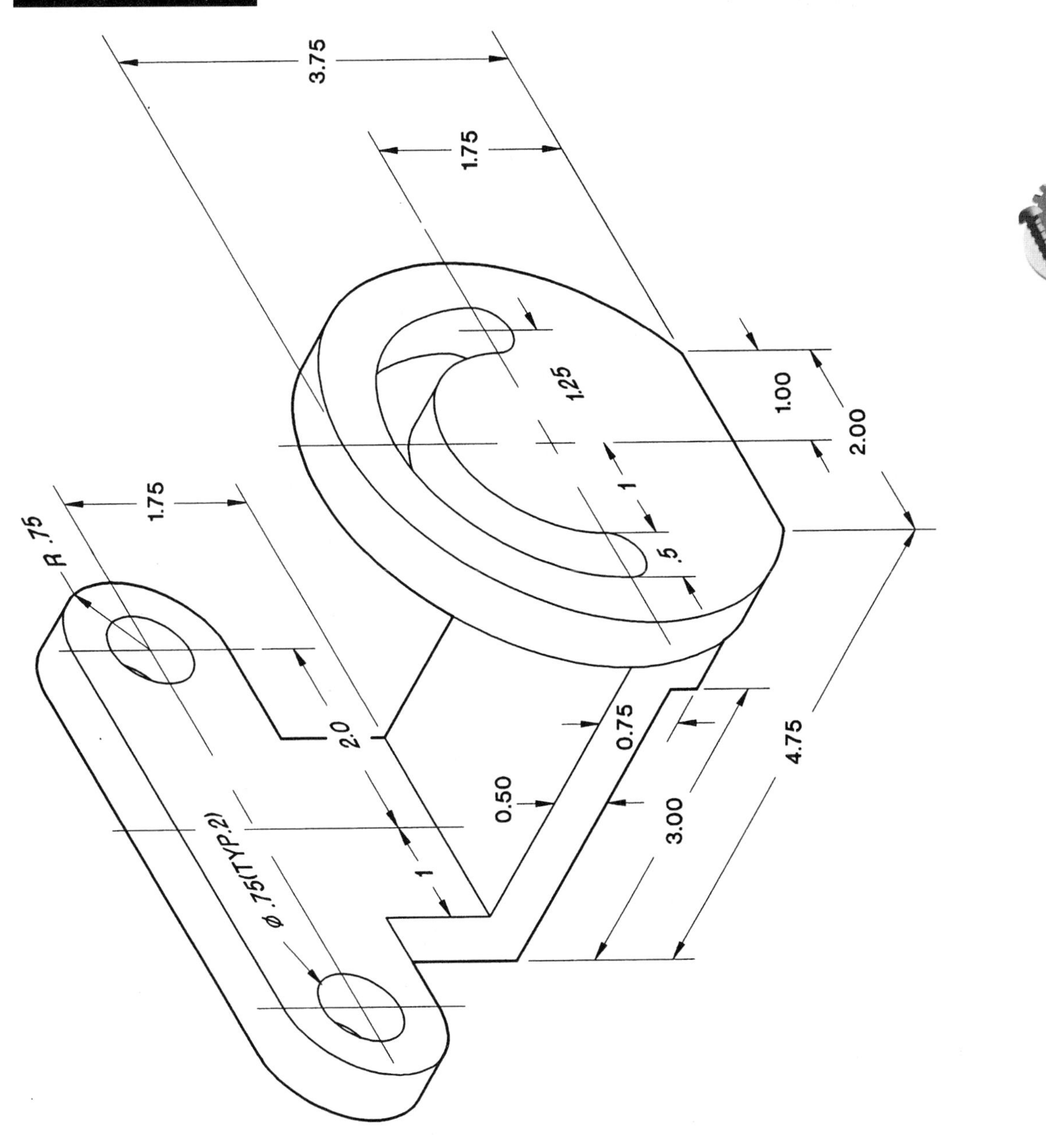

**Figure Ex15–5**

## EXERCISE 15–6

Layout the 4" 150# Slip-on flange drawing shown in Figure 15–6 to the given dimensions.

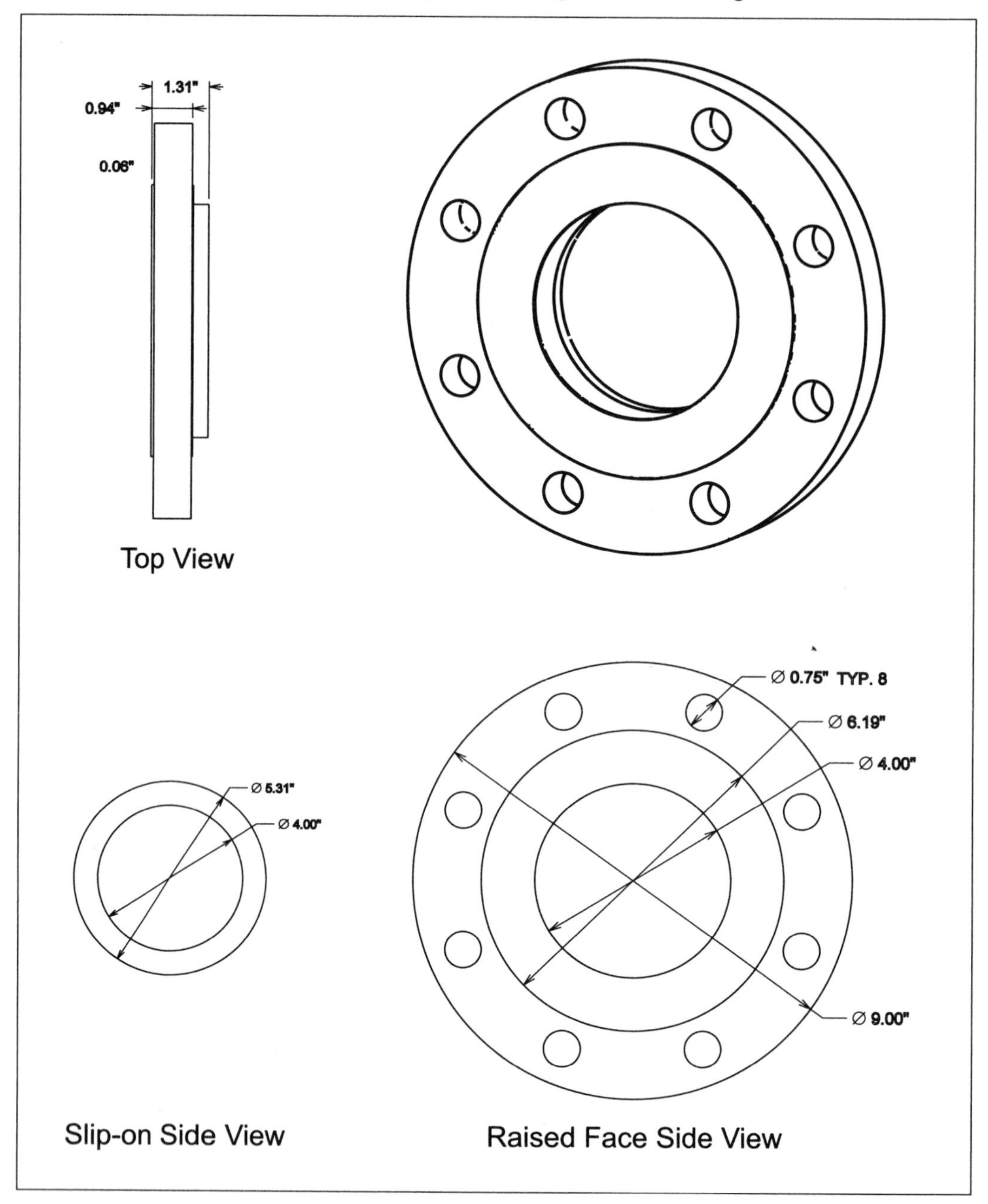


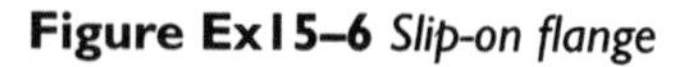

**Figure Ex15–6** *Slip-on flange*

## EXERCISE 15–7

Layout the Clevis drawing shown in Figure 15–7 to the given dimensions.

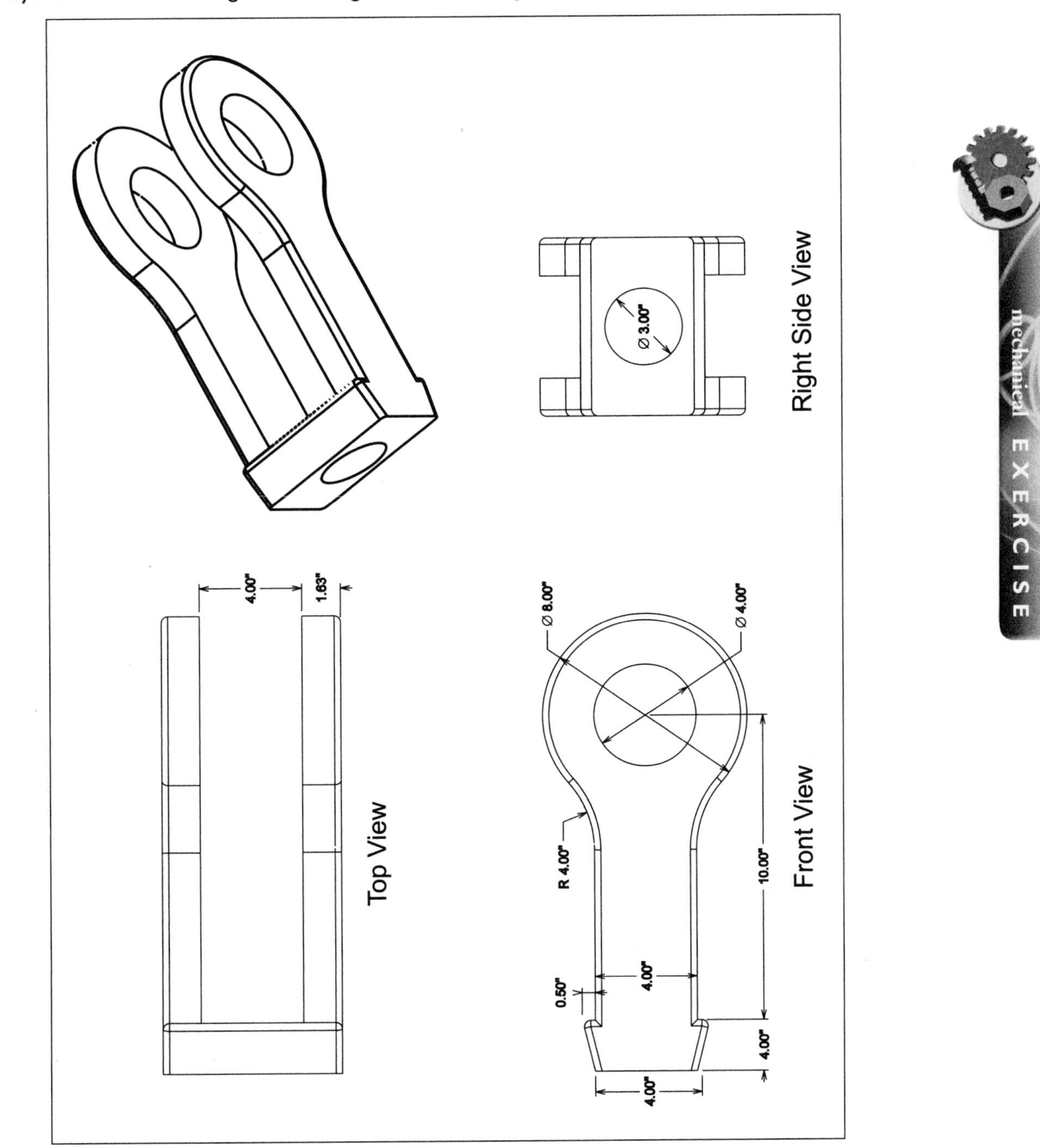


**Figure Ex15–7** *Clevis*

## EXERCISE 15–8

Layout the drawing shown in Figure 15–8 to the given dimensions.

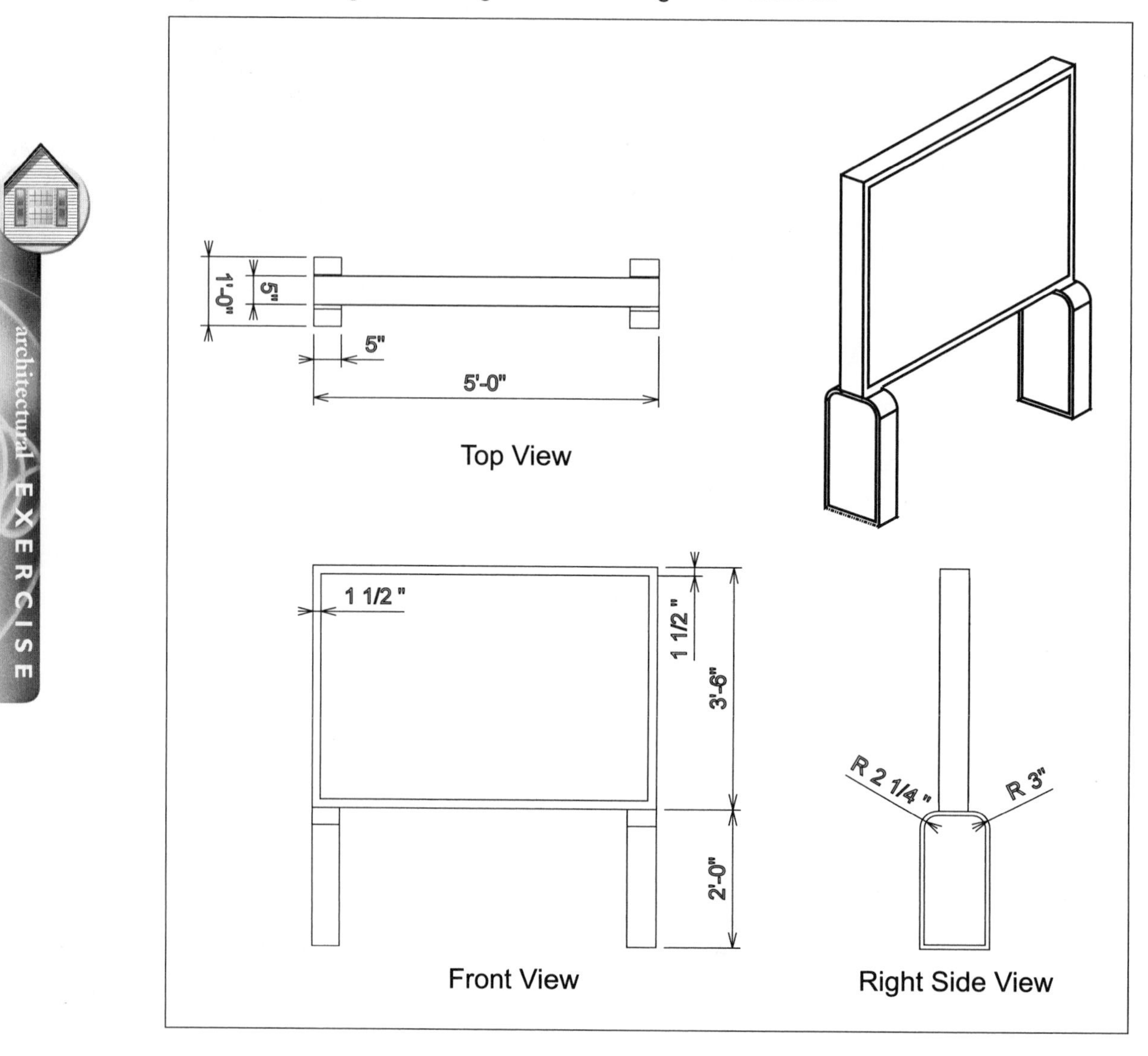

**Figure Ex15–8**

## EXERCISE 15–9

Layout the drawing shown in Figure 15–9 to the given dimensions.

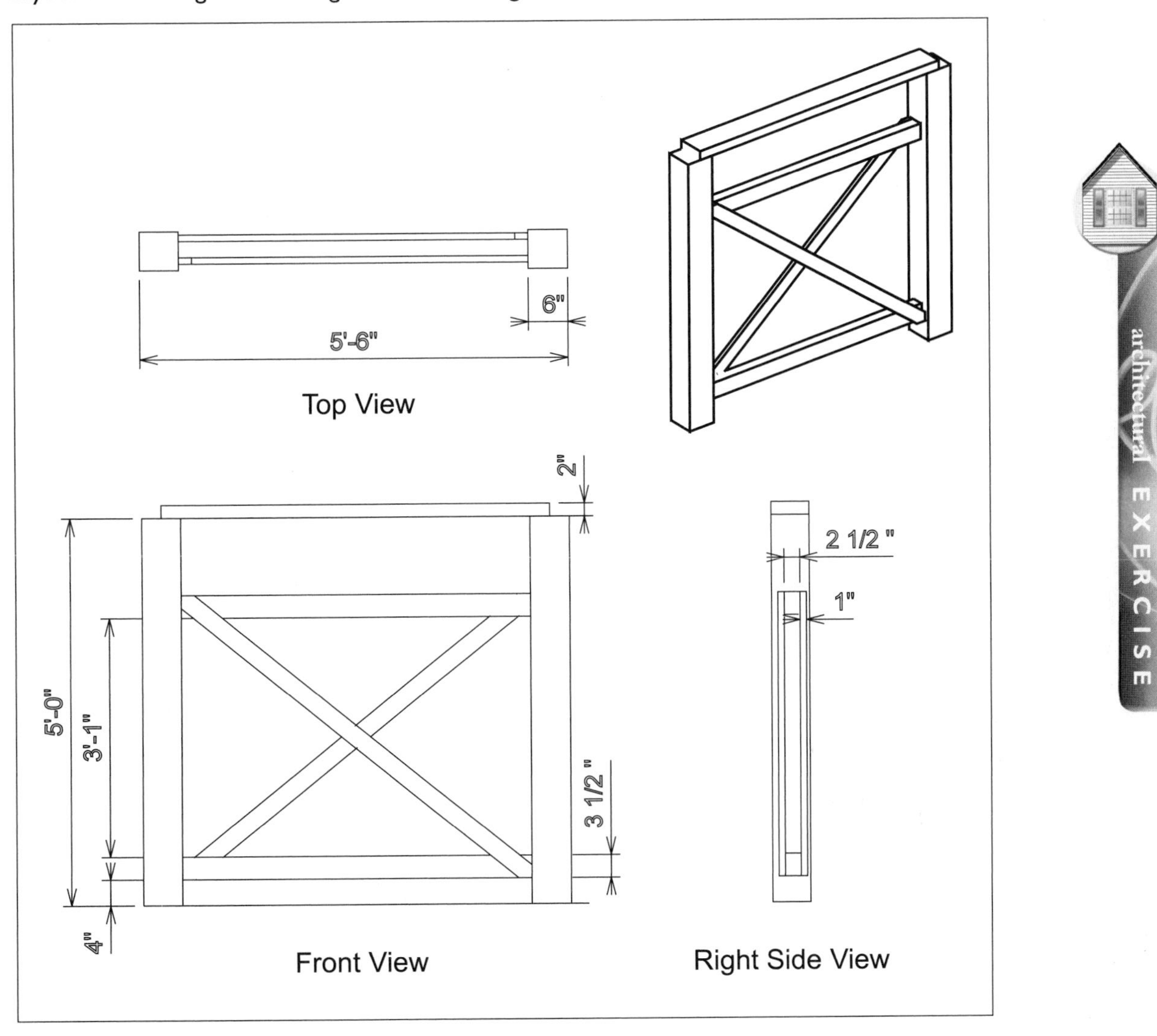

**Figure Ex15–9**

## EXERCISE 15–10

Layout the drawing shown in Figure 15–10 to the given dimensions.

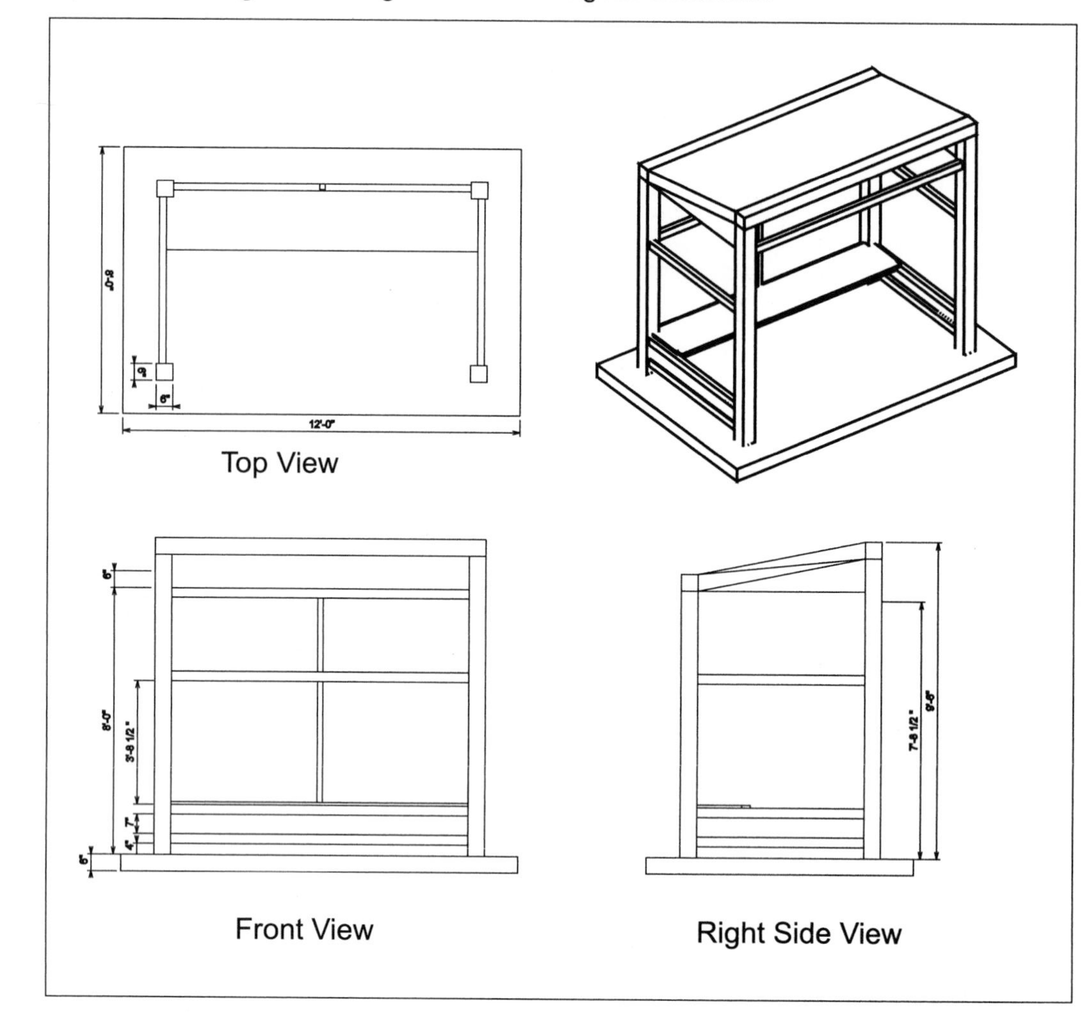


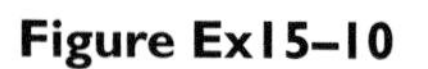

**Figure Ex15–10**

## EXERCISE 15–11

Layout the drawing shown in Figure 15–11 to the given dimensions.

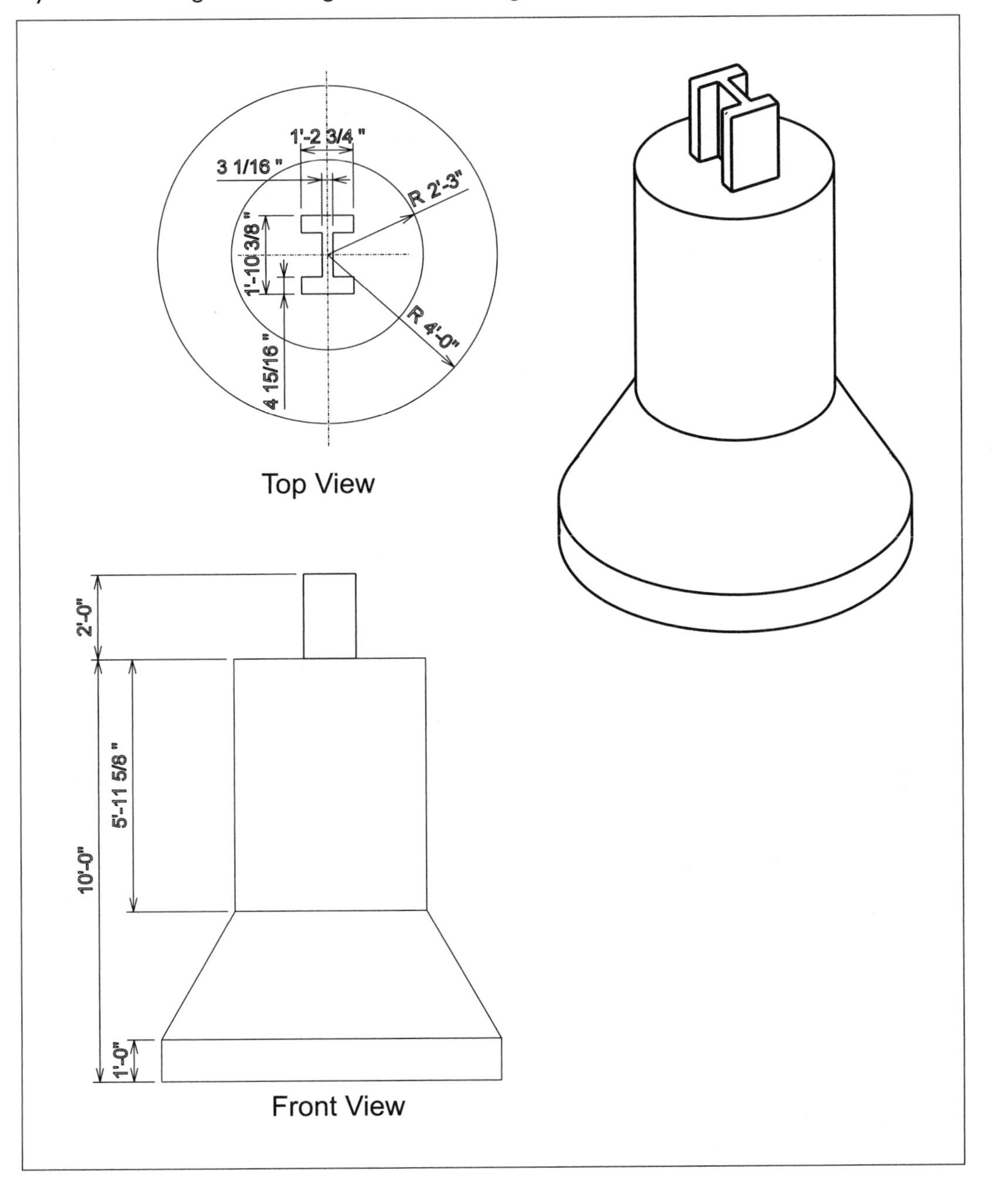

**Figure Ex15–11**

## EXERCISE 15–12

Layout the drawing shown in Figure 15–12 to the given dimensions.

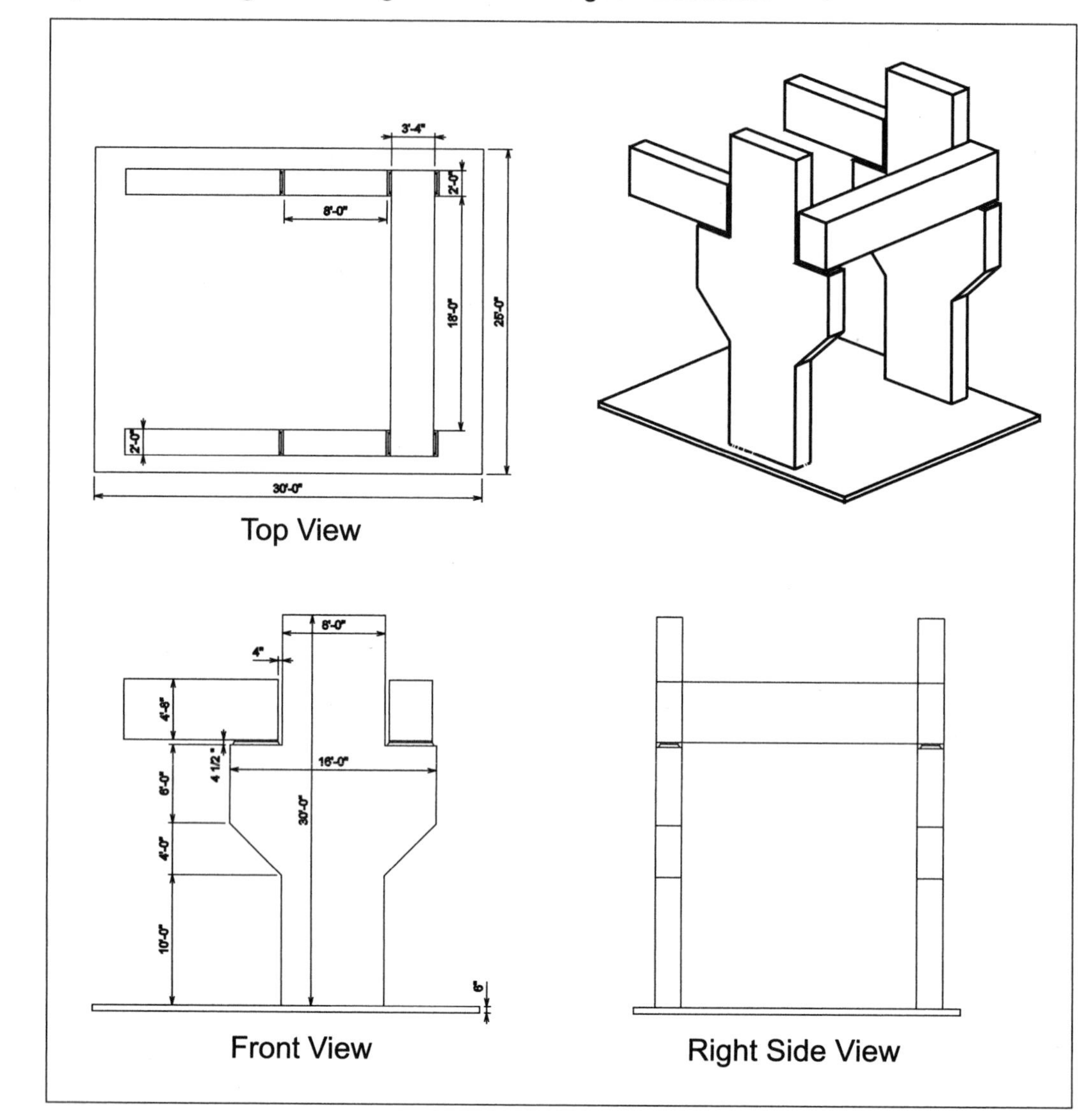

**Figure Ex15–12**

civil EXERCISE

autodesk Press

# *Harnessing* AutoCAD® 2004 *Exercise Manual*

Thomas A. Stellman • G. V. Krishnan

Student-priced Autodesk® software available at *www.cadcampus.com*! See site for eligibility details.

**The companion Exercise Manual to *Harnessing AutoCAD® 2004* provides more than 200 discipline-specific exercises and projects for learning how to use today's leading desktop design and drawing software. Updated to AutoCAD 2004, the manual features problems in complete project format that help readers practice concepts and commands modeled on AutoCAD use in the architectural, mechanical, electrical, and civil fields. Plenty of easy-to-follow instructions and must-know techniques in the core book set readers up for success when working through the exercises and projects of varying difficulty that are featured in this workbook. These exercises are also included in PDF format on the CD-ROM in the back of *Harnessing AutoCAD 2004*.**

**Harness the true functionality of AutoCAD® 2004!**

***Also Available from Autodesk Press:***

***Harnessing AutoCAD® 2004,*** Stellman and Krishnan
Order #1-4018-5079-0

***Introducing AutoCAD® 2004,*** Stellman and Krishnan
Order #1-4018-5059-6

***About the Authors:***

**Thomas A. Stellman** conducts AutoLISP seminars, has published in CADENCE magazine, and is currently Project Coordinator for a state-of-the-art engineering firm. He is the author of Delmar's Practical AutoLISP and has been teaching for more than eight years.

**G.V. Krishnan** is the Director of the Applied Business & Technology Center at the University of Houston (Downtown). He has been teaching basic through ...urses for more than 10 years at ...gest authorized Autodesk

***Bookshelf Categories:***

Architecture, AutoCAD, CADD, Drafting, Engineering

ISBN 1-4018-5080-4
90000
9 781401 850807

Visit ***www.delmarlearning.com***
or ***www.autodeskpress.com*** for your lifelong learning solutions

For more learning solutions by Thomson:
***www.thomson.com/learning***

THOMSON
DELMAR LEARNING